The Natural Human Diet? The Optimal Human Diet?

The known and the unknown, the possible and the probable – an analysis of the evidence.

First e-edition Nov 10, 2010, this 1st Edition 2011 with all responses from right of reply offers and peer invitations to comment as at May 2011 and on to this revision Aug 2012.

David J Vance, Multifactor Health and Education Initiative:

BSc (med physiol, biochem, chem) (Hons: nutrition), James Cook U
BA (Psychology), James Cook University
Master of Medical Science (Epidemiology), University of Queensland
Master of Public Health (Nutrition), University of Queensland
Master of Medical Statistics, University of Newcastle,
PGrad Diploma of Public Health & Tropical Medicine, James Cook U
Grad Dipl of Health Economics & Policy, Monash University
Naturopathic Certificate, North Coll Nat (KS Jaffery, 'Nature Cure')
Dipl Clin Nutrition, Internat Acad Nutr (R Buist, 'Orthomolecular Nutr')
Level 2 Sports Coach, (ACHPER) TAFE

Self-published by: David Joseph Vance, **Multifactor Health and Education Initiative** (MFHEI)

Self-publisher: David Joseph Vance, **Multifactor Health and Education Initiative** (MFHEI)

ISBN 978-0-9871469-0-8

Dedication:

To those sufferers, of any species, unlucky enough to bear the burden of more suffering than the average, for example cage-farmed animals, among the worst so affected, and testimony to our collective mental deficiencies that the practice is allowed to continue.

Acknowledgements:

To my mother for her excellent early education of me.

To my father for some genes, and cross-cultural experiences, and for his share of my food and lodging in early life.

To Steven L Hardie, BPsych, my thanks for gifts from his personal library, and for condescending to again eat durian with me notwithstanding his years of regression to a scavenger diet, pending possible rehabilitation on further thought, I hope.

To the University of Queensland Herston Medical Library Staff, for their competence.

To the better of the scientists whose work I have examined in my work here, for their shoulders to stand on.

To Kenneth S Jaffery, 'nature curist' (his more appalling misconceptions notwithstanding, he seems to have been correct about some things), for his shoulders to stand on.

To professor(!!...) _oger _ughes and those gormless minions of his that colluded to bring about my exclusion from the master of nutrition and dietetics course at _riffith U, not relishing the prospect of having to compete with me nor of having me point out the manifold defects in their own work, for making me look so good in comparison – none of you will ever produce a work of science to match this; your paltry intellects will be honed in the fields of networking to secure each other's job and paycheck, clabbering together like rancid scum.

To the chaotic and causally predetermined universe for my greater share than average of good luck, the disparity being unfair nevertheless; a disparity that I will try to reduce nonetheless the predetermination, because that will make me feel like my life has 'meaning', because if I do not try, then certainly my life will have little if any 'meaning', and accordingly I will ultimately feel less satisfied than if I do try.

Contents in Brief (PLEASE!! Make your best use of the Summary(ies)):

Chapters, Contents in Detail: p6

Introduction: arguments, opinions, beliefs, habits, bias, rationalization, socialization, group membership, nutritional understanding, psychological understanding, logical, probabilistic analysis of the evidence. p14

Notes: intended audience, method of reviewing (quoting " " and commenting [ ]), layout, omission of part of a sentence (...) or of sentences (......), reviewed authors' references, my references. p15

Summary: a 47-page summary of the most essential points and conclusions of this book. p17

Chapter 1. Evolutionary Origins, Directions, Milestones and Timeframes p65

Chapter 2. Our closest animal relatives: apes (chimpanzees, bonobos, gorillas, orangutans, and gibbons) and monkeys p212

Chapter 3 Diet-related biochemistry and mutations – indicators of evolutionary adaptations to and for diet p263

Chapter 4 Food and food group nutrient supply sufficiencies and deficiencies according to Recommended Daily Allowances (RDAs) – indicators of evolutionary adaptations to and for diet food group proportions p289

Chapter 5. Essential amino acid complementarities among food groups p343

Chapter 6. The difference between 'natural' and 'optimal' diets: natural statistical distributions of optimality – perfection is unusual; the average is far from perfect p350

Chapter 7. Hunter-gatherers and/or gatherer-hunters p351

Chapter 8. The disappearance of large animals in recent prehistory and the Overkill Hypothesis p375

Chapter 9. Examples of human performance: Dave Scott, vegan Hawaii Ironman Triathlon (3.8 km swim, 180 km cycle, 42 km run) 6 times world champion; Carl Lewis, vegan 100 m & 200 m sprint and longjump world champion p383

Chapter 10. An example of human performance after 20 years of experimental exclusively raw vegan diet with absolutely no supplemental vitamin B12: the author, in the top 15% of the 2007 Ironman Korea Triathlon (Swim 3.8 km, bicycle 180 km, run 42 km), a lifetime best at age 45 carrying hamstring and spinal disc injuries, and nervous system apparently still intact. The details p383

Chapter 11. Breast Milk: nearly always the right concentration in nearly all climatic conditions p388

Chapter 12. Sense of taste: natural taste; habit-formed taste; proper hunger; sensible comparisons (raw, cooked, garnished) on taste amongst foods p388

Chapter 13. Dental health, optimization of digestion, and food group eating order p393

Chapter 14. Ecology, ethics and animal welfare/ rights p398

Chapter 15. The psychology of diet optimization p400

Chapter 16. Natural and synthetic food contaminants; organic food; parasites p402

Chapter 17. Balanced meal and balanced diet; vegetarian diet and health; caloric restriction and longevity p402

Chapter 18. Individual health vs. population health - deleterious mutations and the effect of medical treatments and nutritional supplements on their retention and prevalence rate in the gene pool; Individuality within species, populations p404

Chapter 19. Eating for weight (surplus body fat) loss, blood sugar and fat levels, fat metabolism and body fat mobilization for fuel p406

Chapter 20 Homocysteine, vitamin B12 and folate and associated foods, atherosclerosis p408

Chapter 21 Afterthoughts: The China Study; Ornish's and Esselstyn's heart disease reversal diets; Inflammation Index and other potential refinements of diet p409

Right of reply offers to Professors/ Doctors whose work has been critiqued in my work here, and invitations to comment, and all such replies and comments p410

Chapters, Contents in Detail (PLEASE! make your best use of the Summaries):

Introduction: p14
Notes: p15
Summary: a 47-page summary of this book p17

Chapter 1. Evolutionary Origins, Directions, Milestones and Timeframes p65

1.1 Summary of Chapter 1 p66

1.2 Human and ape genetic and evolutionary relationship; ape-like human origin p88

1.3 Earlier insectivore origin, persistence of insectivory, vitamin B12 p89

1.4 Colour vision, sweetness preference and fruit p90

1.5 Archaeological remains biases p90

1.6 Archaeological interpretation and attribution biases due to lack of appropriate quantification and epidemiological consideration p90

1.7 Archeological site discovery biases p95

1.8 Archaeological burial soil contamination of remains (diagenesis), Sr/Ca; Ba/Ca; $^{13}C/^{12}C$ p96

1.9 Some archaeological sites, their evidence, and interpretations p98

1.10 Feet, bipedalism, arboreality p134

1.11 Teeth, the supra-pharyngeal vocal tract and language p139

1.12 The gastrointestinal/ digestive tract p159

1.13 Tooth isotopic ^{13}C from C_4 plant origins p160

1.14 Cooking p164

1.15 Energetic models p176

1.16 The brain and intelligence p179

1.17 Cracking and grinding tools, digging tools p183

1.18 Defense needs and weapons p184

1.19 Scavenging, power scavenging, and hunting p184

1.20 Proposed implications of studies of early hominin diets: Cordain; Eaton; Ungar; in Ungar (2007) Eaton, Shostak and Konner (1988) (includes sodium:potassium ratios (p192) and acid/base contributions (p191) of foods) p185

References for chapter 1 p207

Chapter 2. Our closest animal relatives: apes (chimpanzees, bonobos, gorillas, orangutans, and gibbons) and monkeys p212

2.1 Summary of chapter 2 p214

2.2 Different types of primates and their diets. Harding (1981) p218

2.3 Gombe chimpanzee detailed food list; dubious calculation of meat consumption (13 % of calories) without any detailing of denominator group, season (i.e. of preferred food plenty or scarcity), or time trend across the 13 year period covering the four periods sampled; human influence; cooperative hunting; only adult males hunt; strong lack of carrion-eating; power-scavenging of fresh kills; transportation of food and objects; 20-fold increase of severe attacks on other chimpanzees at predatory events; travel preferred across more open ground on trails; terrestrial travel with arboreal feeding; stone thrown as weapon. Teleki (1981) p221

2.4 Omnivory not a very informative term regards position in that continuum; digestibility of food types. Chivers, Hladik (1984) p226

2.5 Legumes as important diet mainstays in many primate species in varied climates; faulty statement of broadness of omnivory due to non-differentiation of apes from other larger primates; failure to properly address frugivore diet seeds, nuts, legumes and grains; brain size and sociality; failure to address need for large brain in memorizing seasonal timing and spatial location of fruits, leaves, seeds, nuts, legumes and grains, and in tool-using processing of some of these; failure to note that nuts, oily seeds, legumes and grains are *more* energy-dense than flesh. Ripley (1984) p227

2.6 Lowland gorillas' and chimpanzees' diet of preference is mostly fruit; in seasons of fruit scarcity gorillas eat more non-fruit plant parts whereas chimpanzees eat more meat; highland gorillas eat 80% non-fruit plant parts, highland chimpanzees eat 40% non-fruit foods. Yamagiwa, Maruhashi, Yumoto, Mwanza (1996) p229

2.7 Borneo orangutans' diet complex, and mostly fruit other than in some months; no meat-eating recorded. Galdikas (1988) p229

2.8 Chimpanzee meat-eating mostly a fallback in seasons of scarcity of preferred plant foods in more seasonal habitats; gorilla and orangutan meat-eating approximately nil; initial disgust and eventual habituation of captured apes to meat as a food; recalcitrance and illogicality of paleoanthropologists in failing to properly extrapolate these facts into theories of ancestral hominin diet. Kortlandt (1984) p230

2.9 Exaggeration and sensationalization of Gombe chimpanzee carnivory. Stanford (1999) p233

2.10 Better detailing of the Gombe chimpanzee situation and less than 5% of diet calories as meat. Stanford (1998) p236

2.11 Failures: to quantify Kibale Forest chimpanzee carnivory and to note its seasonality (My quantification suggests less than 0.5 – 1 gram monkey per chimp per day), to detect and quantify insectivory, and to mention nut-cracking and grain-eating. Wrangham, Conklin, Chapman, Hunt (1991) p238

2.12 Gabon chimpanzees, diet fruits 68%, leaves 28%, animals 4%, great hunger in fruit shortage season (growth stops), uneven distribution of vertebrate meat. Hladik (1981) p239

2.13 Gabon chimpanzees and gorillas: similar teeth, guts and gut transit times; no nut-cracking with tools here at Lope Reserve; both prefer fruit-dominated diet with leaves, stems, piths, seeds and insects; gut entodiniomorph ciliates more present in gorillas; gorillas always eat some foliage; in seasons of fruit scarcity gorillas fall back on leaves, stems and piths, whereas chimpanzees split up to continue seeking fruit. Tutin, Fernandez, Rogers, Williamson, McGrew (1991) p240

2.14 Gombe chimpanzees: infanticidal cannibalism due to poor mothering; eating of human babies. Montgomery (2009) p241

2.15 Kenya Pumphouse Gang baboons, predator (lions, leopards) reduction by humans, baboon and ungulate population growth, long-distance hunting by baboons catalyzed by ordinarily dangerous behavior of unorthodox individual male evolves into cooperative hunting which disappears with immigration and emigration of particular males and increased ungulate vigilance. Strum (1981) p243

2.16 Monkeys and apes; tooth shape and enamel thickness; digestive systems; energy-density of foods; chimpanzee hammer, anvil, and mortar and pestle use; capuchin monkey hammer, anvil, mortar and pestle, and digging tool use; seasonality and climate shift, fallback foods versus preferred foods. Lambert (2007) p248

2.17 Tooth structure and function have variable relationships; bilophodont molars associated with high-fiber diet; high, pointed cusps for puncture-crushing hard and brittle foods, e.g. bone and cartilage of vertebrates. Chivers, Andrews, Preuschoft, Bilsborough, Wood (1984) p251

2.18 Primate elongated canine teeth, dominance hierarchies, defense, prey killing and eating. Mace (1992) p252

2.19 Bearded saki monkey canine teeth used to crack hard-shelled fruits and nuts. Van Roosmalen (1984) p252

2.20 Gelada baboons: grass seeds most of diet; enlarged premolar and molar teeth for grinding; high manual dexterity; erect sitting position; fatty deposits on buttocks; buttock-shuffling feeding locomotion. Foley (1992) p253

2.21 Tooth enamel ^{13}C study - extinct baboons ate more grass than modern South African savanna baboons; faulty statement of broad-spectrum omnivory due to non-differentiation of apes from other larger primates. Codron, Luyt, Lee-Thorp, Sponheimer, de Ruiter, Codron (2005) p253

2.22 Pleistocene African savanna baboons were nearly as large as modern gorillas; became extinct with the appearance of man. Coe (1984) p254

(chapter 2 contents continues below)

2.23 Protein content (dry weight) of legumes and leaves is high, of cereals, oily nuts and seeds and (underground) storage organs is substantial but generally lower, of fruit flesh is 20% that of leaves; diet of primates normally ensures an adequate supply of water; B vitamins; vitamin C; gut flora of all mammals capable of some cellulose digestion; plant defenses, tannins, lignins, proteinase inhibitors, anti-microbials, toxins. Waterman (1984) p254

2.24 Gibbons defend territory, baboon groups avoid each other in extensive areas of territorial overlap, langurs have intergroup dominance relationships Harvey (1992); Apes savagely defend territory, other than bonobos; hunting versus combat violence; population stress on food supplies; infanticide; generalisability of the psychology of killing p258

2.25 Chimpanzees dominance, leverage, cooperation, friendship rules and outrage from their breakage. De Waal (1996) p259

2.26 Corralled chimpanzees: reciprocity with food sharing for grooming, and sharing for sharing. De Waal (1989) p259

References for chapter 2 p260

Chapter 3 Diet-related biochemistry and mutations – indicators of evolutionary adaptations to and for diet p263

3.1 Summary of chapter 3 p264

3.2 Introduction: evolutionary loss of function vs. development of a new function p267

3.3 Vitamin C biosynthesis evolutionarily allowed to deteriorate to nil function p268

3.4 Vitamin B12 uptake, transport and storage, very well conserved; saturation of intestinal B12 uptake at 1.5-2.0 micrograms/meal in the context of the body's ability to store 3000-5000 micrograms; low breast milk B12 and low newborn liver B12 p268

3.5 Human dietary protein requirements related to food protein contents; human dietary protein requirements very low (20g purified amino acids/ 40g food protein / day/ person; food type protein/energy ratios; human breast milk protein/energy ratio lower than that of all other food types (less than 1/10 that of meat), and much lower (only a fraction) than milk of other mammals; human water requirements related to food water and food protein contents considered together p271

3.6 Protein metabolism for energy – limited to 50% of calories before overt toxicity; the metabolism of protein itself uses more energy than that of CHO and fats. Nitrogenous waste byproducts are diuretic p278

3.7 Starch digestion and evolutionary triplication of salivary amylase; starchy tubers and cereal grains p279

3.8 Supposed human inefficiency in biosynthesis of taurine, as in cats (obligate carnivores); vegan breast milk 30 times the taurine level of cow's milk-based infant formula indicates otherwise p280

3.9 Supposed human inefficiency in elongating and desaturating 18 carbon fatty acids to their product 20 and 22 carbon fatty acids, docosohexaenoic acid (DHA) and arachidonic acid (AA), similarly to obligate carnivores; comparisons of vegan breast milk with captive and wild apes and monkeys (similar levels found) indicates otherwise p281

3.10 Preformed Vitamin A from animal food is far in excess of requirements in western diets; ability to synthesize vitamin A from pre-vitamin beta-carotene from plant food sources retained through evolution p282

3.11 Lactose intolerance / adult lactase absence in most of the world's population; unsuitability of milk outside of infancy p282

3.12 Inborn errors of fructose metabolism are very low in frequency (1/20,000 or less); very common fructose malabsorption seems mostly due to unnatural modern context of rapid ingestion of some high fructose:glucose ratio foods in the absence of other meal-balancing protein and fat foods, and to be otherwise beneficial p283

3.13 Celiac disease 1% prevalence is only very weak evidence, or none at all, for humans not being evolutionarily suited to eating cereal grains p286

References for chapter 3 p287

Chapter 4 Food and food group nutrient supply sufficiencies and deficiencies according to Recommended Daily Allowances (RDAs) – indicators of evolutionary adaptations to and for diet food group proportions p289

4.1 Introduction p289

4.2 Summary of chapter 4 p292

4.3 Calcium (Ca), Phosphorus (P) & Magnesium (Mg), and Fiber Content, and Weight, per 2500 kcal of Foods, and their sufficiencies or deficiencies related to the Recommended Daily Allowances (RDAs) / Adequate Intakes (AIs) p306

4.4 Vitamin B12 (VitB12), Folate, Iron (Fe), and Zinc (Zn) Content, and Weight, per 2500 kcal of Foods, and their sufficiencies or deficiencies related to the Recommended Daily Allowances (RDAs) / Adequate Intakes (AIs) p315

4.5 Vitamin C, Vitamin A, Vitamin E, and Selenium (Se) Content, and Weight, per 2500 kcal of Foods, and their sufficiencies or deficiencies related to the Recommended Daily Allowances (RDAs) / Adequate Intakes (AIs) p325

4.6 Thiamin (VitB1), Riboflavin (VitB2), Niacin (VitB3) & Pyridox(al/ine/amine) (VitB6) Content, and Weight, per 2500 kcal of Foods, and their sufficiencies or deficiencies related to the Recommended Daily Allowances (RDAs) / Adequate Intakes (AIs) p335

References for chapter 4 p341

Chapter 5. Essential amino acid complementarities among food groups p343

5.1 Introduction p343

5.2 Summary of chapter 5 p344

5.3 Table 15: The lowest and highest essential amino acids of foods relative to the proportion required by the human body p345

5.4 Interpretation of the data p348

References for chapter 5 p349

Chapter 6. The difference between 'natural' and 'optimal' diets: natural statistical distributions of optimality – perfection is unusual; the average is far from perfect p350

Chapter 7. Hunter-gatherers and/or gatherer-hunters p351

7.1 Summary of chapter 7 p352

7.2 Presentism: over-extrapolation of evidence towards conformity with our own lives and lives of our times (Walker 2007) p354

7.3 Survey of 104 gatherer-hunter groups; dubious dietary data; meat nutritionally secondary; plant food nutritionally primary; extreme habitat exceptions; seasonal exceptions; insects; infanticide; cannibalism; group size; division of labour (Hayden 1981); hunting as an escape from drudgery by males; hunting as a proxy display of martial power, and meat as its symbol; maintenance of territory; meat sharing in bonding; cuckoldry, prostitution, and pimping; far too little mention of defense needs p355

7.4 Californian and Australian gatherer-hunters; diet mostly vegetarian; egalitarian meat sharing (Gould 1981) p361

7.5 Kalahari G/Wi gatherer-hunters; diet mostly plant foods; seasons; water from melons; seasonal plant food plenty allows male freedom from gathering; nitrogen metabolism endproducts diuretic; power scavenging; hide acquisition and use; hunting ages and activities; male prestige of hunting; high symbolic value of meat; substantially/ sufficiently egalitarian sharing of meat (Silberbauer 1981) p362

7.6 Mbuti pygmies, Zaire; interaction with agricultural civilization; diet seasonality; hunting; meat-sharing protocol includes all group members; prestige of elephant killing (Harako 1981); big-game hunting as a proxy display of martial power; no mention of infanticide, warfare, cannibalism, or disease p365

7.7 Hadza of Africa; seasonal food types; tubers, wild vs. cultivated (Schoeninger 2007) p368

7.8 Australian Aborigines; diet and season; tubers; native millet, a summer staple (O'Dea 1991); fixation on animal hunting and failure to address well-documented plant food use as proportion of diet; failure to address fallback foods; hunting as a proxy display of martial power; territory acquisition and defense p369

7.9 Hadza of Africa; large prey as common goods with extensive sharing and continuity of supply (Hawkes, O'Connell, Blurton Jones 1991); failure to consider big-game hunting as a proxy display of martial power with value within the group as a mating and power display and as an outlet for competitive and aggressive urges, and between groups in the acquisition and defense of territory p371

7.10 Amazonian Araya, Parakana, Arawete (neighboring), and Mayoruna natives; different carbohydrate-rich staples and tooth decay; arbitrary prey type selection and avoidance and cultural distancing; fat- and protein-rich palm larvae (Milton 1991); high dietary proportion of animal food (including human-like monkeys) associated with high degree of inter-tribal warfare and killing; tenability of arbitrary prey type selection indicates energy- not protein-limitation of available foods p372

References for chapter 7 p374

Chapter 8. The disappearance of large animals in recent prehistory and the Overkill Hypothesis p375

8.1 Summary of chapter 8 p375

8.2 Overkill Hypothesis; debate; disappearance of large animals; timing; climate change; human migration and population increase; faulty ethnographic analogy; archaeological bias (Webster 1981) big-game hunting as a proxy display of martial power; territory acquisition and defense p376

References for chapter 8 p383

Chapter 9. Examples of human performance: Dave Scott, vegan Hawaii Ironman Triathlon (3.8 km swim, 180 km cycle, 42 km run) 6 times world champion; Carl Lewis, vegan 100 m & 200 m sprint and longjump world champion p383

Chapter 10. An example of human performance after 20 years of experimental exclusively raw vegan diet with absolutely no supplemental vitamin B12: the author, in the top 15% of the 2007 Ironman Korea Triathlon (Swim 3.8 km, bicycle 180 km, run 42 km), a lifetime best at age 45 carrying hamstring and intervertebral disc injuries, and nervous system apparently still intact. The details p383

Chapter 11. Breast Milk: nearly always the right concentration in nearly all climatic conditions p388

Chapter 12. Sense of taste: natural taste; habit-formed taste; proper hunger; sensible comparisons (raw, cooked, garnished) on taste amongst foods p388

Chapter 13. Dental health, optimization of digestion, and food group eating order p393

13.1 The causes of dental disease p393

13.2 Individual dental health management requirements p394

13.3 Nutrition/ diet in dental health management: food type eating sequence p394

13.4 Physical exercise in dental health management p396

13.5 Pharmacy in dental health management p397

Chapter 14. Ecology, ethics and animal welfare/ rights p398

Chapter 15. The psychology of diet optimization p400

Chapter 16. Natural and synthetic food contaminants; organic food; parasites p402

Chapter 17. Balanced meal and balanced diet; vegetarian diet and health; caloric restriction and longevity p403

Chapter 18. Individual health vs. population health - deleterious mutations and the effect of medical treatments and nutritional supplements on their retention and prevalence rate in the gene pool; Individuality within species, populations p404

Chapter 19. Eating for weight (surplus body fat) loss, blood sugar and fat levels, fat metabolism and body fat mobilization for fuel p406

Chapter 20 Homocysteine, vitamin B12 and folate and associated foods, atherosclerosis p408

Chapter 21 Afterthoughts: The China Study; Ornish's and Esselstyn's heart disease reversal diets; Inflammation Index and other potential refinements of diet p409

Right of reply offers to Professors/Doctors whose work has been critiqued in my work here, and invitations to comment, and all such replies and comments p410

Introduction:

There has been much argument about what is the best diet for humans. Most of the opinions expressed on this matter incorporate some degree of belief, as against a logical analysis of evidence.

Belief, especially on dietary matters, is most often very much derived from the personal preferences (habits) that the opinion-giver has grown up with or evolved to have, rather than being derived from a logical, probabilistic analysis of the relevant evidence.
The growing up with, or the evolving to, the opinions usually incorporates a large component of enculturation and psychosocial programming, to the detriment of a logical, probabilistic analysis of the relevant evidence.

Many of those espousing an omnivorous diet with a high content of animal foods are unable to see their own psychological rationalization of their lifelong and cherished eating habits and their involvement thereby with various types of mistreatment of animals. Many of those espousing a vegetarian diet are similarly unable to see their own psychological rationalization of their need to feel different from or similar to some group of people, their need for individuality or group membership, their need for moral sufficiency or superiority, and/or their aesthetically based aversion to killing animals (however, note that the aesthetic basis could have both genetic and psychological, social, psychosocial or cultural components)….

Eating is a very society-centered and social activity for the large majority of us. Deviation from the dietary norm prevailing in our own particular situation in our own particular society is likely to be accompanied by some amount of discomfort of some sort, even if it fulfils some need for individualism. The net discomfort may increase or decrease over time, dependent on various factors, and our own psychological state will always be one of them.

Two people may make the same dietary change, one for good reasons, the other for bad reasons, one in the right time and place, the other in the wrong time and place, one with an appropriate understanding of the nutritional facts, the other without, one with an appropriate understanding of their own and others' psychology, the other without.

Yet as members of the one species, with little exception arising from evolution, we will be more similar regards what would be the best diet for us as individuals, and less similar with regard to our psychosocial situations as relevant to dietary choice, in my opinion.

I suggest that the three magic words of science are, after all, not "I love …….", but "I don't know".

This is a quote from the summary of the 1991 Royal Society of London symposium which gathered together scientists investigating different aspects of archaeology as it relates to human diet: "The meeting was useful for the way in which it drew together workers from diverse disciplines to address the analysis of the feeding process in primates. … With hindsight it became apparent that schemes of this kind, which attempt to integrate diverse kinds of evidence, require considerable familiarity with the rules of logic and logical procedures. Some of us felt that previous training had scarcely prepared us for such activities. … many participants were aware that they might not be making the best use of the evidence that was already available… ."

So, hoping that my background of training, study, and personal experience has sufficiently prepared me for the attempt, I turn to what I hope will be a logical,

probabilistic analysis of the relevant evidence, with the minimum clouding from any personal biases of my own.
Hopefully many years of studying various sciences and knowledge disciplines (one of which, of particular relevance here, is psychology) will sufficiently override my personal preferences and biases, for the purposes of helping others to improve their understanding and selection of what they eat.

The title of this book is a little similar in its last part to that of Peter S. Ungar's (Editor) (2007) book "Evolution of the human diet. The known, the unknown, and the unknowable" - which book draws together work by 30 or so specialists in various related fields, in 20-odd chapters. When I first saw Ungar and Co's book, as available for online purchase, the last part of its title, being such a very sensible thing to have there, and it's implication of scholarship of a level well above much of what I had read on subjects related to the question of what the optimum diet for humans might be, initially caused me to think my own work here might have been preempted into redundancy by it – thankfully, from the lower-level 'egocentric' position, and regrettably, from the higher-level world-centric position, I found, on reviewing it, that it had not. And so I am pleased to be still able to usefully offer my work here for consideration.

Notes:

This book is written for a lay audience of average mental ability, as well as for professionals and academics, and so some things are explained or repeated in different places to a greater extent than would be the case if it were written for a purely academic audience.
Extensive use is made of summaries:
Often chapter section titles constitute a sort of summary.
Each major chapter has a summary at its front.
There is a summary of the whole book at its front, including shortened versions of the summaries of the major chapters.
This is to make the crucial findings of the book available with as little effort as possible to even those who have not the time to read the whole book.
Reading the whole book will give a fuller understanding of the topic. If one has not the time to do that, the next best is to read all the major chapter summaries and the smaller chapters. If one has not the time to do that, then at least the 47-page summary of the whole book should be read.
Chapters 13, 17, 19, 12 & 15, in lieu of an abstract, are general recommendations for diet.

The review goes through reviewed works mostly in their own order, from start to finish.

I do not adhere to a standard paragraph structure, most often starting a section of the reviewed author(s)' words, with quotation marks ("Abc) in one line, and my comment(s) on them, even if I interrupt the reviewed author(s)' sentence to do so, immediately below on a new line, enclosed in square brackets ([Abc … xyz.]) – this is intended to provide a balance between presenting exactly what the reviewed author(s) wrote (rather than what I might have thought they meant, or might like them to mean); providing my comment immediately next to the relevant portion of the reviewed author(s)' words; and providing the opportunity to read the reviewed author(s)' words in a connected manner by more readily bypassing my comments.
If I make no comment, I accept what the author has said as plausible enough for the present purposes – and I almost always reproduce the author's text verbatim, rather than the usual process of rearranging it so as to be more, or more resembling of, my own work – I consider my own contribution here to be significant enough as to spare me the tedium of doing that, and to spare the reader the added mental strain of attempting to assess the

probability that the reviewed author(s)' meaning was as I present it, and, as my comments are often destructively critical, it is important that what I am criticizing is represented here exactly. Note that in nearly all scientific writing, as here, "reportedly" is underused.

Also, within a paragraph I will often put a new point on a new line, when there is no great essential continuity of sequence between the subject matter of sentences in a block of text – my work here is foremost a work of science, and science is not always best written in the way usually prescribed for prose and narrative things, in my opinion, notwithstanding common academic requirements – a little space between facts that are important to remember on their own (although the meaning of all the facts is ultimately in the context of them all being considered together) is a good thing in scientific writing, in my opinion. The effect is halfway between a 'dot-point' presentation and the usual paragraph format.

In the writing of this review, I write in the first person (i.e. "It seems to me that …" rather than "It seems that …", as I think that this is the most accurate representation of the reality – the writer (some person, myself) is expressing their thoughts and opinions on some matter, rather than the writing being generated by an infallible scientific analyzing machine, much as though that is what I aspire to be as close as possible to…

My comments and other writings here are always enclosed in square brackets [… …].

Usually 3 dots (…) means that a portion of a reviewed author(s)' sentence has not been reproduced by me, other than where it indicates a 'pregnant pause' (implying that at least some of the (, and substantial,) significance of my words is understood or understandable by the reader without my needing to write it all out); and 6 dots (……) means that at least 1 sentence of a reviewed author(s)' has not been reproduced by me.

Subject matters in a reviewed author(s)' text, reproduced by me, that are more separated in subject matter have a one line space in between them, whereas those that are more connected will have no line spaces.

I do not provide the full reproduction of the reviewed author(s)' own references in their work, just the Author's name and the date – partly to save myself scarce time and effort, and partly because I greatly dislike the citing and referencing of works that have not been properly read (much less understood…) (often in vast numbers) by writers – to make part of this perfectly clear, in the large majority of cases, I have not read the works that the reviewed author(s) cite in their references, and so they will not appear fully detailed anywhere among my references, or in my work here.
When I have read only a scientific journal abstract, I indicate that by placing "(abstract only read)" next to the reference.

A 'hominid' is a not fully human animal resembling a modern human more than an ape, all of which are related relatively closely to us, but only some of which are our direct ancestors.
A 'hominoid' is an ape, not a 'hominid'.
A 'hominin' includes both modern humans and hominids, but not hominoids.

The authors of some of the work critically reviewed here, including but not limited to those most destructively so reviewed, will be provided with a copy of this work in early November 2010, and any response they provide at any time thereafter, including an indication of intent to respond at some stated or unstated time in the future, will be added at the end of the book, verbatim as provided, no matter how negative, within 2 weeks of my receiving it.
For further details see the full offer at the present end of the book on p410 below.

47-Page Summary of the main points and conclusions of this book

Chapters 13, 17, 19, 12 & 15, in lieu of an abstract, are general recommendations for diet.

Chapter 1. Evolutionary Origins, Directions, Milestones and Timeframes (briefest summary)

[1.2 Genetically we remain very similar to the anthropoid apes that are our closest relatives: chimpanzees, bonobos, gorillas, orangutans and gibbons - DNA studies show approximately 97 % exact similarity in DNA between humans and apes.
It seems reasonable to think that the diet of our distant ancestors in the times when our line diverged from the apes (6 million years ago) was more like that of present-day apes than like that of present-day humans.]

[1.3 All present-day apes and monkeys, in fact all or very nearly all primates, continue to have a substantial, regular, daily deliberate or incidental (in plant foods) ingestion of insects and their larvae (or pupae) and eggs.
This has a particular relevance for vitamin B12, which is generally taken to be available to humans only (other than from supplemental vitamins) from foods of animal origin.
It is not ethical to subject an infant to the risk of permanent brain damage through a purely vegan diet without supplementary vitamin B12.

Like present-day apes and monkeys, even our most distant hominin ancestors very probably used naturally shaped stones and sticks as tools – sticks for digging up root vegetables, natural hammer and anvil stones for cracking nuts, and naturally shaped mortar and pestle and grindstones for grinding seeds, grains and root vegetables. Soaking was probably also used.

That the canine teeth of all hominins are/were no longer elongated as they are/were in the apes and monkeys suggests that humans are the least naturally carnivorous among humans and apes (chimpanzees eating a small amount of meat as a fallback food in seasons of fruit scarcity, and gorillas, orangutans and bonobos eating almost no meat).

That human premolar teeth are molarized (more like the molars), that human molars are flatter, and that the jawbone has been moved to the rear, compared to our ape relatives, suggests an adaptation to more chewing of small seeds like cereal grains, and of other things that require more crushing than shearing, perhaps tubers.

That human/hominin tooth enamel is thicker than that of the thickest-enameled apes, the orangutans, and much thicker than that of the chimpanzees and gorillas, is concordant with evolutionary suitedness to crushing hard things, as well as to grinding things that grow close to or in the ground and therefore have more abrasive dirt particles on them, such as cereal grains, seeds of small herbaceous plants, and tubers.

Human digestive tract proportions, and physiological parameters such as stomach digestive juice pH and emptying time, are far closer to those of the apes than they are to those of more naturally carnivorous animals like the cats and dogs. The slight reduction in relative size of the large intestine and slight increase in relative size of the small intestine are more probably related to increased consumption of macronutrient (protein, fat and carbohydrate) –dense plant foods than of meat (as the nuts, seeds, legumes and grains are much more calorically dense than lean or even regular modern farmed meat).

Processing of foodstuffs by hammer and anvil, mortar and pestle, and grinding stone, and soaking, are more likely to have facilitated the meeting of nutritional requirements than

the use of fire (the making on demand of which probably came only very late in human evolution).
It may be that the human intelligence was a further evolution in some ways from the relatively high mental functioning continued in evolution into the orangutans – they remember not only where the different edible-fruited trees in their range are, but also what their fruiting season is, and they use relatively advanced logic to decide on efficient routes between different trees in a day's, or in a series of days', foraging for fruit and the green leaves and insects eaten also in their natural diet.

Substantial evolutionary changes (other than *removal* of DNA-*degenerative* changes) of long-lived species like humans (of the sort related to evolution of a suitability to at least general dietary food group proportions, and probably also to other, food- or nutrient-specific proportions or amounts) probably do not take place over periods of time so short as a thousand or ten thousand years, as the natural mutation rate of our DNA is not fast enough.

The many proponents of human dietary suitability who refer to archeological evidence (in itself flawed by vegetable matter being *far* more decomposable than bones…) from the last ten thousand or so years (bones of animals eaten by humans) in proposing that humans have evolved into an optimally substantially meat eating animal do not properly relate meat as a proportion of diet (They grossly overestimate it, on no good scientific assessment of the evidence) to the 6 million years of our evolution since our common ancestors with the chimpanzees.

Nor do they properly address the cycles of climatic variation within that 6 million years – ice ages and the periods between them, and the warmer (interglacial) and colder (glacial) periods within an ice age.

Nor do they take into account the varying periods of time, in most (and particularly in some major) human populations, where not much (and sometimes very little) animal-food was eaten in the last 7,000 years since the advent of agriculture as a major source of human food, wherein DNA-degenerative changes (i.e., like loss of vitamin C synthesis) that increased meat eating may have allowed to accumulate (for example in vitamin B12 absorption, transport and storage) would have been once again removed from the gene pool, potentially in a much shorter time than it took for their accumulation, back down to a substantially lower level.]

[1.4 Various authors have suggested that the evolutionary regaining of red colour vision, lost in the evolution of mammals, by a few species of primates, was related to a dietary shift to greater consumption of fruits and young leaves with a red colour.]

[1.5 The archaeological record is highly biased; some materials (such as stone and bone) preserve very well and are routinely found during excavation, while others (organic material such as plant material (i.e., food remains and digging sticks) and non-bone body parts) either rot entirely away or leave only such subtle traces as are missed.
Much dietary theorizing, in particular that done without properly considering not only what biased remains there are (as well as how they could or are likely to be biased), but also the other relevant facts known from fields of knowledge such as nutrition, physiology, biochemistry, chemistry, genetics, epidemiology, primatology, and anthropology, produces, accordingly, very faulty theories of what the diet(s) of our hominin ancestors were.
In particular, theories grossly overestimating the proportion of animal food in particular the meat of hunted animals, and underestimating the proportion of plant foods such as fruits, leaves, legumes, nuts, oily seeds, cereal grains and tubers and corms.]

[1.6 Archaeological interpretation and attribution biases due to lack of appropriate quantification and epidemiological consideration are very common, perhaps even usual.

Almost invariably, in studies of animal bones left in concentrated deposits, with stone tool cutmarks on some but not all bones, by our hominin ancestors, there is no attempt made to perform what calculations are able to be performed to estimate what the amount of meat thus gained per hominin per unit of time might be, and so to estimate what proportion of the diet that meat might have been. Almost invariably, large deposits of bones such as those at the Solutre site (see section 1.6 below), are taken to indicate a most ferocious level of carnivory – but when the period of time, thousands or tens of thousands of years, and the number of humans involved (some smallish group), are taken into consideration such as to be used in calculating an estimate of what the amount of meat thus gained per hominin per unit of time might be, such calculation indicates that the meat gained through such serial slaughtering was very likely to have been only a *very* small proportion of the diet.

Regarding the supposed larger size of preagricultural humans compared to agricultural humans, attributed for example to the diets, commonly in what literature I have read, the theorists fail to consider the possibility that in general, organized warfare between specialized societies led to preferential selection of larger and stronger men for the armies, who would then frequently kill off large proportions of each other, such that this could quite easily reduce the population gene pool of genes for largeness, strength and robustness, which is exactly the opposite of what would happen in combat between random individuals fighting each other, or in small groups where all males took part in fighting – where the smaller, weaker and less robust are more likely to die, increasing the population gene pool of genes for largeness, strength and robustness.
Therefore claims that agriculture lead to a reduction in body size due to the supposed unsuitability of agricultural diets for humans are likely to be mostly false, and in any case are made on the basis of a very defective consideration of the evidence.

Also, there is the effect of increased population density in agricultural societies (itself more suggestive of benefit than detriment, one might have thought…) increasing the rate of infectious diseases, such as those known to result in pathological bone porosities often mistakenly attributed to agricultural diet.]

[1.7 Archeological site discovery biases include that very often it is the initial finding of collections of animal bones and manufactured stone tools, not the finding of hominid bone remains, that is more likely to be the trigger for archaeologists to excavate a site, biasing findings towards carnivory.

Also, dry arid environments (other than silty river beds) favor fossilization, so that the sites found and studied may represent the fringes (by population or season) of the hominid environments rather than the main hominid environments.
And, as modern gatherer/hunters, and baboons and chimpanzees, exploit a number of habitats during a yearly (i.e., seasonal) round, early hominins may likewise have seasonally lived in various habitats, where they ate different diets.
So, very possibly the bias here again leads to *finding* evidence of carnivory rather than of plant eating.]

[1.8 Archaeological burial involves long periods of time, and soil contamination of fossil remains (diagenesis) has been found to be significant/substantial for all chemical constituents commonly assayed (i.e., Sr/Ca; Ba/Ca; $^{13}C/^{12}C$).]

[1.9 There have been enough archaeological sites examined, and the interpretation of the evidence they provide is reasonable enough for the purpose (if not for the quantification of the meat eating as a proportion of diet), to be able to state as a matter of virtual fact that our hominin ancestors did eat some meat at some times.
And, that at some evolutionarily very recent times (for example within the last 40,000 years), probably under climatic and population stress, in a substantial number of areas, meat did constitute a substantial proportion of their diet.
However, the theorizing on the meaning of the findings at the various sites is plagued with defects of the sort I have mentioned above, as well as the failure to consider other relevant evidence from other areas of knowledge as I have done here in this book.

A few more enlightened commentators have noted the strong tendency for theorists to interpret the data (incomplete as I note that the limited subset of it available that they have chosen to include in their analyses and theorizings is…) from a position strongly biased by their own cultural and personal dietetic positions.]

[1.10 Humans, and therefore all of our evolutionary ancestors since the common ancestor of humans and chimpanzees, still climb trees *quite* well, just not *as* well as the apes do - certainly modern humans and all of our hominin ancestors climbed trees well enough to be harvesting and eating fruit from them as a daily staple.

We humans are better evolved to travel longer distances between trees because we can cover open ground better than the apes, and our use of weapons to defend ourselves (strangely overlooked, as against for killing things, as a possibility, by many archaeological and anthropological theorists) offers us a defense while on the ground that the other apes don't have, and that the gorillas don't need due to their larger size. It may well be that the evolution of the use of defensive weapons enabled the evolution of the human feet.

Almost certainly our early hominin ancestors would have been able to carry clubs, babies, some (i.e. missile and hammer) useful stones, and some food, maybe in bags made of hide. Modern wild chimpanzees are reported to use all of these except the last.

Consideration of nut gathering and cracking with hammer and anvil stones, grain gathering and grinding with useable natural stones, and root vegetable finding, digging up, transport, and pounding/grinding, as well as fruit tree location, access and climbing, is necessary for sensible theorization about the evolution of bipedalism in hominins.

Consideration of highly proficient weapon use, which requires fast, precise, stable movements, and the driving of force from the ground through the body to the weapon, as appropriate, is essential in the question of the evolutionary drives for bipedalism.]

[1.11 Regards the final changes in the human mouth and dental structures, it needs to be borne in mind that in more modern humans, in a process only finalized sometime in the last 100,000 years, maybe as recently as 40,000 years ago, the roof of the mouth and the lower jaw have been reduced compared with non-human primates and archaic hominins. The reduced chewing ability is offset by the increased phonetic range of the supralaryngeal airway, such that the modern human vocal tract can seal off the air pathway that leads through the mouth from the nose, so yielding non-nasal sounds, resulting in clearer, more readily identified sounds.

The comparative anatomy of living primates and of hominid fossils suggests that the evolution of the human supralaryngeal vocal tract probably started in early (about 1.6 million years ago) African populations of *Homo erectus* and was not completed until the appearance of fully modern humans.]

[1.12 It seems that modern humans are less well adapted for digesting cellulose than the apes – it should be borne in mind, however, that the last third of the modern human large intestine favors for its energy supply the short chain fatty acids that result from the breakdown of cellulose, in contradistinction to all other human body tissues, which favor glucose, fatty acids or ketones for their energy supply.

Claims that the evolutionary changes leading to the slight reduction in the large intestine and slight increase in the small intestine (that plausibly took place since the human line diverged from the chimpanzee line) were based on a large intake of meat in our ancestral diet are generally quite defective in that they fail to consider that the oily nuts and seeds are actually clearly *more* energy-dense (600 kcal / 100 g) than meat (190 kcal / 100 g, lean; 230 kcal / 100 g, regular) is (as meat is approximately 70% water), undried cereal grains roughly as energy-dense dense as meat (dried cereal grains are 380 kcal / 100 g), and starchy tubers energy-dense enough to supply human energy needs without recourse to much cellulose digestion. And that these foods are fairly likely to have been used by our ancestors for the very large majority of the 6 million years since our line diverged from that of the chimpanzees. And that the taste of raw nuts, seeds and tubers (as well as fruits) was probably preferred to the taste of raw meat by our evolutionary ancestors.]

[1.13 The carbon isotopes in bone and tooth enamel reliably provide dietary information in modern animals, but only tooth enamel levels have proved reliable with fossils, as chemical exchange with the surrounding soil alters bone levels substantially and variably.

It is essential to understand that there are many diets that can produce the same stable carbon isotope ratio ($^{13}C/^{12}C$), and that it is impossible, for instance, to use the ratio from fossil teeth to distinguish between, for example, a purely vegetarian human ancestor that ate 50% C_3 fruits and 50% C_4 grass (including seeds and storage roots/ corms) and a highly predaceous human ancestor that ate nothing but C_3- and C_4-derived mammal flesh in roughly equal proportions.]

[1.14 Statements such as: “It is also known that a diet of raw plant food creates substantial energetic problems for humans under even the best conditions.” are the most arrant nonsense.
For example, near the end of 20-plus years of raw vegan diet unsupplemented by any vitamin B12, at the age of 45 years, I finished the 2007 Ironman Korea Triathlon (a 3.8 km swim followed by a 180 km bicycle ride followed by a 42km marathon run, the same distance as at the famous Hawaii Ironman Triathlon) at the 15th percentile, that is, with 15% of the other participants in front of me and 85% of the participants behind me.
And my higher brain functions seem to have survived intact, as a reader might judge from my work here.

In much the same vein of ignorance are statements such as:
“First, although raw plant food is evidently a poor diet for humans, a sufficient inclusion of raw meat, in theory, can create an energetically adequate diet.”
The basic absurdity of this sort of statement is obvious from a nutritional content analysis at a childish level of mental difficulty: raw meat is about 70% water; raw nuts and oily seeds have very little water, the large majority of the weight being taken up by about 20% protein on average, and the another 70-odd% being fully fat or carbohydrate – so obviously raw nuts (600 kcal / 100 g) or oily seeds (600 kcal / 100 g), and also cereal

grains (380 kcal / 100 g) for that matter, are more energy dense than raw meat (190 kcal / 100 g, lean; 230 kcal / 100 g, regular).

Furthermore note that high-protein diets are reported to increase the cost of digestion by about 30% compared to high-fat diets.

Statements to the effect that cooking results in a large improvement in the diet are generally not well considered, regarding i.e.: palatability vs. quality; rendering fiber more digestible and dental caries; the speed at which the carbohydrate in a food is able to be absorbed into the bloodstream ('glycemic index'); eating quickly and digestion, psychological satisfaction and social aspects; 'Maillard products' and reduced lysine absorption; oxidative damage of unsaturated fat and atherosclerosis; and increase of food allergenicity by roasting.
On the other hand there is the beneficial inactivation of some toxic legume lectins, but this needs to be considered in the context of raw legumes having been reliably reported to be one of the very main foods of a wide range of primates, allowing their ecological development as a group.
Cooking probably does increase the proportion of starch in a food able to be absorbed in particularly the *small* intestine, but this is not at all necessarily a good thing. Chewing food properly has much the same effect and is probably better for health.]

[1.15 Modern humans require more energy per unit of total body weight than chimpanzees to sustain their proportionally larger brain, but on the other hand they require less energy to travel from place to place per kilometer but more energy absolutely due to the much greater distances travelled.
However, if chimpanzees are spending more time climbing trees then this is likely to significantly affect the comparison, and often this aspect is not properly incorporated in analyses and theorizings. Likewise, the weight of ingested food to be carried, and the weight of gathered food to be carried, and the effort involved in travelling to and carrying drinking water if sufficient water is not taken in as part of foods like fruit and vegetables. Most energetic models do not come close to properly considering these factors.

Even among chimpanzees, the most predatory of the great apes, only about 5% of their calories are derived from animal foods, including insects.

The basic flaws in theorizing about the addition of meat to the evolutionary ancestral diet includes the failure to understand that: raw meat is about 70% water; raw nuts and oily seeds have very little water, the large majority of the weight being taken up by about 20% protein on average, and another 70-odd% being fully fat or carbohydrate – so obviously raw nuts (600 kcal / 100 g) (or oily seeds (600 kcal / 100 g), and also cereal grains (380 kcal / 100 g) for that matter…) are more energy dense than raw meat (190 kcal / 100 g, lean; 230 kcal / 100 g, regular) – getting repetitive, isn't it?…
Furthermore, the proportions of energy/protein in raw nuts are much closer to that optimal/required by apes or humans than is that of meat.

A certain amount of water is needed every day – the natural diet of apes and monkeys supplies them with sufficient water – it seems likely that food (fruits and vegetables) is a very good place for humans to be getting most of their required water from, on considerations of prolongation of water absorption and therefore retention time, decreased vulnerability to predation at drinking water sites, and decreased exposure to infectious disease.

And, without detailing of just *which* "nutrients" are better provided by animal foods (which is not at all the same thing as a nutrient being more present in a food, because there are other very important factors to consider, including other nutrients, and other foods in combination in the same meals and in the diet), isn't that part of their statements here just another motherhood saying about the diet they are habituated to? … the only nutrient of any real concern here on average is vitamin B12, and we need *very* little (2 to 3 *micro*grams, not *mille*grams, per day (RDI), closer to 1 microgr. more likely), of that…]

[1.16 & 1.17 Claims that consumption of more animal foods was necessary for the evolution of the larger human brain, on the basis that the composition of all mammalian brain tissue is similarly rich in two fatty acids, docosohexaenoic acid, DHA, and arachidonic acid, AA, and that these two fatty acids are richer in animal foods than in plant foods, might best be considered along with the fact that the elephant has the biggest brain on the planet, twice the size that of the blue whale and four times the size that of humans - and elephants are not generally known to eat meat, though they would ingest some insect matter incidentally in the plant matter they eat. And, that the apes are all relatively big-brained, yet eat generally very little meat, though again some insect matter, deliberately as well as incidentally.
Furthermore, frugivorous bat species' brain size, relative to body size, is typically almost twice as large as in aerial insectivore bat species, with no overlap between the two groups.
Nectarivorous bats have relatively large brains and generally overlap with the frugivores, whereas the remaining bats (carnivores, piscivores, sanguivores and foliage gleaners) typically have brain sizes intermediate between frugivores and insectivores.

Claims that in their raw form, the starch in tubers is not absorbed in the small intestine and is passed through the body are very incorrect - if one actually chews the stuff properly first, instead of bolting it unchewed like a dog, then the large majority of it will in fact be digested and absorbed – an easy enough experiment to try by oneself (a radical suggestion, apparently, for some armchair theorists and parrotic vocalizers in nutrition and dietetics, but there we go…), even without the simplest of blood sugar assay equipment…

Note that the ratio of protein/energy in human breast milk is very low, only a fraction of what it is for all of the several other mammals I've found data for, indicating that we remain after evolution a naturally physiologically and biochemically low-protein diet consuming animal.
Accordingly, most often protein will not be the limiting macronutrient factor in the diet, including that of the breast-feeding mother - energy will most often be the limiting factor. So it seems quite possible that the use of starchy root vegetables, added to the diet of the common ancestor of the chimpanzees and humans (which was probably a generally very ape-like diet), and increased nut, seed and grain consumption, which were all very likely to have been increased as a proportion of the diet of our hominin ancestors, taken together, could meet any increased energy requirement. And bear in mind that even in the absence of any increase in energy requirement, an increase in the macronutrient-density of the diet could result in the slightly increased small intestine and slightly decreased large intestine of humans in comparison with chimpanzees – bear in mind that for chimpanzees sweet juicy fruits are the preferred foods and nuts requiring cracking more often fallback foods for times (i.e. seasons) of scarcity of sweet juicy fruits – it is possible that our hominin ancestors ate more nuts, seeds and grains because that is what was more available, not necessarily because they needed a more energy-dense diet.

Alternatively, increased returns for food foraging and processing effort may be an important pre-requisite for evolving a larger brain, and a larger brain may be necessary to organize human food foraging and processing behavior, particularly in view of a diet of 'imbedded foods' (foods that are neither visually obvious nor available without expending considerable time and energy in processing them, being extracted from a secondary source) probably requiring more complex cognitive skills.

It seems obviously correct to note that tool use increases the efficiency of extraction of such foods from their coverings, and widens the range of resources available, and as such could enable an increase of net energy such as would contribute to evolution of a larger brain.
However, not enough consideration has been given to the possibility of the use of spears as a defensive weapon for ground foraging, and the use of (accurately thrown) spears and stones to defend tree fruit from competitors such as monkeys, apes and birds, and the cognitive requirements of this as a strategy requiring an evolutionary step up, but a plausibly achievable step up, in planning for the near future.
Furthermore, that the full exploitation of nuts ripe on the ground, that only some chimpanzee groups have learned to use anvil and hammer rocks to start doing, would be facilitated by spears, as hominids, with their feet differentiated from the apes and monkeys, while still able to take shelter in trees as unencumbered adults with sufficient speed to escape predation, would not be as fast at doing so as apes and monkeys, particularly if encumbered with infants and very young children, particularly as nut-cracking is likely to be a noisy process, potentially attracting predators, and our earlier hominin ancestors very probably relied on quiet movement and activity for survival.

Statements and theories to the effect that meat represents a high quality resource which could enable an increase of net energy such as would contribute to evolution of a larger brain should be viewed in the context that a food's being denser in energy and/or protein, is not the same thing as its being of higher quality, because there are other very important factors to consider, including other nutrients, and other foods in combination in the same meals and in the diet (something that a large number of more simplistic vocalizers and writers seem utterly unaware of in making statements that some food is a good source of some nutrient… *good* source and *rich* source are two very different things…).
Furthermore, to repeat from the summary of section 1.14 above: raw meat is about 70% water; raw nuts and oily seeds have very little water, the large majority of the weight being taken up by about 20% protein on average, and another 70-odd% being fully fat or carbohydrate – so obviously raw nuts (or oily seeds, and also cereal grains for that matter…) are more energy dense than raw meat – from any standard nutrition tables: obviously raw nuts (600 kcal / 100 g) (or oily seeds (600 kcal / 100 g), and also cereal grains (380 kcal / 100 g) for that matter…) are more energy dense than raw meat (190 kcal / 100 g, lean; 230 kcal / 100 g, regular) – getting really repetitive now, isn't it?…
Furthermore, the proportions of energy/protein in raw nuts are much closer to that optimal/required by apes or humans than is that of meat.

The evidence for claims of (even seasonally-varying) heavy consumption of meat by Homo erectus is very poor - nearly always this claim is pure assumption.

Also, taste preferences have a bearing on diet (surprise, surprise…), but are little addressed by many theorists in this field(s) of enquiry here.
If a more energy-dense diet was required for the evolution of the larger human brain, then an increase in intake of nuts would very certainly meet that criterion, as would an increase of smaller such things generally termed seeds, likewise grains (These are in order of decreasing ease of extraction, large nuts being very readily extracted with the correct use of hammer and anvil stones, seeds less so (though not all may require extraction prior to eating), and grains maybe less so again (though not all may require extraction prior to eating at all stages i.e. sprouting or soaked or ground up.)).
If the taste of these was preferred to the taste of raw meat, then meat may have been only a fallback food in times of scarcity of the preferred foods.

Aside from the requirements for a larger brain in more complex food gathering and processing, the other main factor is, very probably, increasing social complexity and its cognitive demands, including the following (and please see the full section 1.16 below for the authorship citation) in ordering of ascending difficulty, with apes able to do, and doing, more of the complex things than monkeys:
discovery behaviours (curiosity and exploration of physical objects, and social play for gathering information about social relationships);
innovation or tactical flexibility as responses to novel associations of events (spatio-temporal fluctuations in food availability, or in social interactions the abstraction of "conditional social rules" as a basis for social actions, in relation to specific actions by other con-sociates whose past social-interactive patterns can be stored in short-term and long-term memory);
cognitive representations of the intentionality and attentional states of others, perhaps involving attributions of several levels of intentions ("X thinks Y knows that X wants Y to do such-and-such", or in more complex tripartite interactions "X thinks Y knows that Z wants Y to do such-and-such", etc);
and (perhaps only in humans) *self-reflective monitoring* as a means of abstracting an explanatory model of human action.

Non-human primate grouping patterns and social systems are determined by three major factors: dietary characteristics, predation risk, and social competition.

Perhaps an increase in body-size, coupled with larger group sizes than were optimal simply for foraging due to an increased predation risk from foraging further from climbable trees in increased home range areas, resulted in increased levels of intragroup competition and decreased group stability.
In such contexts, including the development of more potent weaponry and its use, even if initially for defense, the capacity for social cognition, and the capacity for increasingly complex language as an instrument for establishing and modulating ever more complicated relationships between our hominin ancestors over almost all of the last 6 million years, larger brain size and the behaviors that would support it would have come under strong selection pressure as a mechanism for learning, communicating and manipulating (shared or private) mental-model representations of the contexts and constraints on action. Aside from object tool use in non-human primates, primatologists repeatedly emphasize the ability of non-human primates to use other individuals as 'social tools' to achieve particular results.]

[1.18 I doubt that with their legs, pelvis and spine not as evolved as ours for moving on the ground, and their small stature (1.1 meters), that australopithecines would have had a snowball's chance in hell of surviving on the ground (which is where their feet (greatly resembling our own) strongly suggest they were spending a great deal of time), without weapons. In my opinion anyone thinking otherwise has a totally inadequate understanding of combat reality and the unarmed defense capabilities of such a creature in the environment it shared not only with large cats and hyenas but also with baboons (though more grass-eating then as they purportedly were) the size of a modern gorilla and very probably other dangerous large animals outside my knowledge.
And I think much the same applies to the larger *Homo erectus* that came later and was more our size – without fangs and claws, or the speed of a gazelle, or the size of an elephant, or the hide and tusk of a rhinoceros, or the reproduction rate of a rabbit, or the taste of a stink-bug, or the camouflage of a stick insect, some other method of survival is required. Weapons were almost certainly it, obviously enough it seems to me, but apparently not to some, as the subject seems generally very poorly dealt with in the literature, notwithstanding that I haven't read or canvassed all of it.]

[1.19 It seems very likely to me that, probably used only in times of great hunger, both scavenging and power scavenging were the only access our earlier hominin ancestors had to the meat of large animals. They probably needed weapons for either method, as scavenging is likely to involve making more sound than the average made for gathering, and moving quietly was probably a central requirement of survival for early hominins other than armed and at the ready with clubs, stones and sharp sticks, and in groups – having such weapons, they would probably have been capable of driving off from carcasses at least sufficiently sated carnivores of some size larger than themselves, as a group.

Raw meat was probably not very palatable to our early (and maybe nearly all) hominin ancestors, who probably preferred raw fruits, nuts, and some vegetative plant parts when available, and probably even legumes and starchy tubers, to raw meat, and almost certainly to raw meat in various stages of rotting.]

[1.20 Various proposed implications of studies of early hominin diets: Cordain; Eaton; Ungar; chapters in Ungar (2007); and Eaton, Shostak and Konner's (1988) book The Paleolithic Prescription, are critically assessed in detail in section 1.20 proper below.]

[My conclusions on chapter 1:

It seems to me that if present-day apes and monkeys can use naturally shaped stones and sticks as tools then even our most distant hominin ancestors would almost certainly have done so too.
And lack of ability to reason using language does not indicate the lack of ability to reason with mental pictures.

Wild animals are, perhaps above all in their behavior, observant – failure to be so is very likely to result in removal from the gene pool at a very early age – I find it very unlikely that even our earliest hominin ancestors were incapable of watching pigs dig up roots on serial occasions and realizing that they could also do so themselves with the use of a stick – in fact, such reasoning in animals has been pretty conclusively demonstrated and reported on by now in 2010.

Likewise, to reason in mental pictures that if nuts could be processed (cracked open) between two hard objects, that it might be worth a try using some variation of that as constitutes grinding i.e. of grains, seems well within the mental capabilities of even our earliest hominin ancestors to me, notwithstanding the frustrating lack of the mental capabilities that one might think reasonable to expect, in many modern and present-day humans…

That the canine teeth of all hominins are/were no longer elongated as they are/were in the apes and monkeys suggests that humans are the least naturally carnivorous among humans and apes (chimpanzees eating meat as a fallback food in seasons of fruit scarcity, and gorillas, orangutans and bonobos eating almost no meat).

That human premolar teeth are molarized (more like the molars), that human molars are flatter, and that the jawbone has been moved to the rear, compared to our ape relatives, suggests an adaptation to more chewing of small seeds like cereal grains, and of other things that require more crushing than shearing, perhaps tubers.

That human/hominin tooth enamel is thicker than that of the thickest-enameled apes, the orangutans, and much thicker than that of the chimpanzees and gorillas is concordant with evolutionary suitedness to crushing hard things, as well as grinding things that grow close to or in the ground and therefore have more abrasive dirt particles on them, such as cereal grains, seeds of small herbaceous plants, and tubers.

It seems very probable to me that even our very earliest hominin ancestors used naturally shaped hammer and anvil stones, naturally shaped mortar and pestle stones, and digging sticks.
The invisibility of this use in the archaeological record due to decay of the sticks and undifferentiability of the stones from other stones clearly make such absence of evidence (which is not the same thing as evidence of absence) not at all a good argument against such use having occurred.

Human digestive tract proportions, and physiological parameters such as stomach digestive juice pH and emptying time, are far closer to those of the apes than they are to more naturally carnivorous animals like the cats and dogs. The slight reduction in relative size of the large intestine and slight increase in relative size of the small intestine are more probably related to increased consumption of macronutrient (protein, fat and carbohydrate) –dense plant foods than of meat (as the nuts, seeds, legumes and grains are much more calorically dense than lean meat and fattier meat of evolutionary relevance).

Processing of foodstuffs by hammer and anvil, mortar and pestle, and grinding stone, and soaking, are more likely to have facilitated the meeting of nutritional requirements than the use of fire (the making on demand of which probably came only very late in human evolution).

If, and I think it definitely had some effect in many places, organized warfare between specialized societies led to preferential selection of larger and stronger men for the armies, who would then frequently kill off large proportions of each other, this could quite easily reduce the population gene pool of genes for largeness, strength and robustness (exactly the opposite of what would happen in combat between random individuals fighting each other, or in small groups where all males took part in fighting – where the smaller are more likely to die).
High population density, itself suggestive of *better* nutrition, favors infectious diseases. Caloric, not protein, shortage is the major cause of stunting, with grain staple foods. Therefore claims that agriculture lead to a reduction in body size due to the supposed unsuitability of agricultural diets for humans are likely to be mostly false, and in any case are made on the basis of a very defective consideration of the evidence.

It may be that the human intelligence was a further evolution in some ways from the relatively high mental functioning continued in evolution into the orangutans – they reportedly remember not only where the different edible-fruited trees in their range are, but also what their fruiting season is, and they use relatively advanced logic to decide on efficient routes between different trees in a day's, or in a series of days', foraging for fruit and the green leaves and insects eaten also in their natural diet. Intelligence is not at all a monopoly of hunters, as some perhaps less capable biologists often state or infer, perhaps in an unconscious attempt to rationalize their own cherished carnivorous dietary habits.

Now, substantial evolutionary changes of long-lived species like humans (of the sort related to evolution of a suitability to at least general dietary food group proportions, and probably also to other, food- or nutrient-specific proportions or amounts), other than the removal of DNA-*degenerative* mutations, probably does not take place over periods of time so short as a thousand or ten thousand years, as the natural mutation rate of our DNA is not fast enough, even allowing for environmental stresses (acting on heat shock proteins) to permit a greater number of mutations to escape correction by cellular DNA

repair mechanisms (assuming that this actually happens outside the test tube, which it might well not…) (as postulated by Maresca and Schwartz, 2006), relative to the length of our generation gap (parent to offspring), to allow it.
This is something many proponents of human dietary suitability who refer to archeological evidence (in itself flawed by vegetable matter being *far* more decomposable than bones…) from the last ten thousand or so years (bones of animals eaten by humans) fail to realize the significance of when they propose that humans have evolved into a substantially naturally (they mean optimally, though they usually seem unaware of the difference…) meat eating animal.

Many proponents of substantially carnivorous diets being optimal for humans, basing their claims on archeological or anthropological evidence, do not go back far enough in evolutionary time in their thinking.
They tend to refer to the most recent cold period (there are warm periods, such as the present) of the most recent ice age (which began 2.6 million years ago, and which we are still in), when hungry humans with their possibly more natural or preferred (in the absence of fire) plant food sources in shorter supply (the early human population explosion a big factor here) probably did eat more other animals than in previous warmer periods. This most recent cold period is only of approximately 10,000 years duration.
Failing firstly to go back far enough in time (tens and hundreds of thousands of years) to discern our very likely evolutionary origins as a very predominantly plant-food eating animal, coming from the same evolutionary origin as the all very predominantly plant-food eating apes (our closest animal relatives), they then secondarily fail to sufficiently address the question of whether most if any human populations have genetically evolved to eat much animal food, as the fraction of time in human history during which we ate much animal food was not long enough for genetic mutations away from our fairly obviously mainly plant-food eating evolutionary origins to occur often enough to have become the standard or most common genes of our species, via being selected for, or perhaps rather, via not being removed as insufficiently functional (more on this below), in the struggle for survival in nature.
Also, they fail to take into account the varying periods of time, in most (and particularly in some major) human populations, where not much (and sometimes very little) animal-food was eaten in the last 7,000 years since the advent of agriculture as a major source of human food, which is likely to have removed much of any DNA-*degenerative* changes (such as vitamin B12 absorption, transport and storage ability decreases) that what (relatively short in evolutionary time) substantial meat-eating that did occur allowed to remain (at some probably relatively low level in any case) in the gene pool unremoved by natural selection.]

Chapter 2. Our closest animal relatives: apes (chimpanzees, bonobos, gorillas, orangutans, and gibbons) and monkeys (briefest summary)

[The distant ancestors of all primates (humans, apes, and monkeys; and lemurs, lorises, galagos and tarsiers) were insectivores – most of their diet was insects.

Our genetically closest animal relatives are the apes, animals that look like monkeys, and are closely related to them, but have no tails. The apes are the chimpanzees, bonobos (sometimes termed pygmy chimpanzees as they are very similar to chimpanzees in appearance), gorillas, orangutans, and gibbons (gibbons are less similar, and they are an ape, being tailless, but not a 'great ape'). Chimpanzees are generally agreed to be genetically the closest to humans, having roughly 97% exactly the same DNA, and it is generally agreed that the chimpanzee (and bonobo) evolutionary line split from the

human evolutionary line about 6 million years ago. The gorilla evolutionary line split a little earlier, and the orangutan line a little earlier still.
All of the four present-day great apes, in their 'natural' habitats, are predominantly frugivorous very predominantly vegetarian omnivores – they all eat some animal foods, but only a little (mostly insects) usually and on average, and among the plant foods, they prefer to eat mostly sweet fruits and do so for most of the year.
They also eat green leaves, legumes, nuts, piths, and seeds including oily seeds and cereal grains (cereal grains not very often).
Chimpanzees like fat-containing foods, whereas gorillas reportedly do not, although how this fits with their insect eating would be interesting, because often the most favored insects for humans and presumably chimpanzees have a high fat content.

Of the apes, chimpanzees eat the most meat, but animal foods are generally agreed to total less than 5 % of total calories, and of this they eat more insects than meat.

The meat hunted by (only adult male) chimpanzees is mostly colobus monkeys (mostly young/infants), but also piglets, small deer called dikdik, and baboons (mostly young/infants). Most female and young chimpanzees get little or no meat.

All chimpanzees eat a significant amount of insects (eggs, larvae/grubs, and adults) (such as would meet their vitamin B12 requirements), but only some adult males eat most of the hunted meat. This is in sharp contrast to naturally more carnivorous or insectivorous animals, where the growing young are fed more of the animal foods, as they (like any other growing young) have a higher protein: energy ratio requirement than the adults.

However, chimpanzees generally do not set out to hunt for meat, but do so, only sometimes (usually not), only in the course of looking for the plant foods that are their preference (sweet fruits), if they happen upon an opportunity.
Furthermore, hunting by chimpanzees increases in the seasons when their preferred foods are in short supply – and it should be noted that at these times they are generally *very* hungry, as it is reported that the growth of young chimpanzees stops during these seasons, which we would term some degree of starvation.

Therefore, it makes more sense to consider meat to be very much only a fallback food of chimpanzees, resorted to mostly in seasons of scarcity of their preferred plant foods.

Bonobos live in a much less seasonal environment than chimpanzees, and they eat much less meat again (nearly none) than chimpanzees.
This reinforces the conclusion that meat is very much only a fallback food of chimpanzees, resorted to mostly in seasons of scarcity of their preferred plant foods.
Bonobos eat enough insects to supply their vitamin B12 requirements.

Orangutans have only very rarely been observed to eat meat (squirrels), and it should be noted that orangutans also experience seasons when they are unable to eat even close to the amount of sweet fruits that they would like to, such that it is no longer the mainstay of their diet that it is in the seasons of plenty. Orangutans eat enough insects to supply their vitamin B12 requirements.

Gorillas have apparently not been observed to eat any meat. They favor sweet fruits, but in some highland regions where such fruits are always in short supply eat mostly other plant matter all year round. Gorillas eat enough insects to supply their vitamin B12 requirements.

There are many more species of monkeys than there are of apes, and some among them, for example capuchin monkeys, are very substantially carnivorous, much more so than chimpanzees. However, taken as one group, the monkeys are predominantly frugivorous or folivorous predominantly vegetarian omnivores.

Apes and humans all have color vision, which was regained in evolution by our earlier common ancestors after being previously lost earlier in evolution (some crustaceans see in 5 or more colors, compared to our three colours). This is probably related to the identification of ripe fruit and young tender leaves.

The human gastrointestinal tract differs from that of the apes in having less large intestine, and more small intestine, such that we are less suited to breaking down cellulose in the large intestine (but keep in mind that we are still suited to digesting some dietary fiber, probably including some amount of cellulose, in the large intestine, because the descending last third of the large intestine prefers the short chain fatty acids that result from that digestion as its energy source, compared to most other body tissues, that prefer glucose, fats, or ketones).
The suggested physiological difference, that humans are suited to a more energy-dense diet, contrary to some authors' theories would very probably be better accounted for by increased dietary energy-dense plant foods like nuts, oily seeds, cereal grains and starchy tubers, than by meat, as raw meat is about 70% water; raw nuts and oily seeds have very little water, the large majority of the weight being taken up by about 20% protein on average, and another 70-odd% being fully fat or carbohydrate – so obviously raw nuts (600 kcal / 100 g) (or oily seeds (600 kcal / 100 g)), and also cereal grains (380 kcal / 100 g) for that matter, are more energy dense than raw meat (190 kcal / 100 g, lean; 230 kcal / 100 g, regular).

The human teeth differ from those of the apes in having canine teeth that are not elongated, which suggests, the reduced need as a combat weapon notwithstanding, that humans are, insofar, less suited for carnivory than any of the apes.
Also, in having flatter molar tooth horizontal surfaces, suggesting that humans are better adapted for grinding things like seeds (including grass seeds i.e., cereals).
Also in having premolar teeth that are more like the molar teeth ('molarized premolars'), which again suggests that humans are better adapted for grinding things like seeds (including grass seeds i.e., cereals).
Also, in having thicker enamel, which again suggests that humans are better adapted for grinding things like seeds (including grass seeds i.e., cereals, which may have abrasive dirt particles in them, being near the ground).
(Humans have thicker enamel than orangutans, which have thicker enamel than chimpanzees, bonobos and gorillas.)
The grinding of tubers, and the resistance to abrasion from dirt remaining on them would also be at least accommodated, and facilitated, respectively, by the premolar and molar surfaces and the thicker enamel.

In many of the places where chimpanzees have been observed, there is a strong possibility that human activity has increased the amount of chimpanzee carnivory, for example, at Gombe National Park.
Note however that (adult male) chimpanzee predation on colobus monkeys has been found in effectively all places in which chimpanzees in their 'natural' environment have been studied – such that some small amount of colobus hunting must reasonably be said to be an evolutionarily natural behavior, even if mostly a fallback one for times of preferred food scarcity - but, in contrast to true carnivores, females and young get little or none of this meat. Also in such contrast, the most fat-rich part, the brain, is eaten first.

Note that over many years, with a probably exponential increase over the last couple of hundred years, chimpanzees have had the opportunity to observe an animal (humans) that they probably recognize (at least to the extent that would lead to a greater tendency to imitation of them) as being more similar to them than, e.g. leopards and lions, hunting, killing and eating other animals. Also, humans have reduced predation on chimpanzees by the cats (hunted for fur), leading to increased food pressure amongst chimpanzees.

Following on from the discussion of factors influencing chimpanzee hunting activity, the study of the 'Pumphouse Gang' baboon troop, which will surely remain for all time a classic of environmentally and socially modified hunting behavior in a higher primate, shows the following:
Baboons (which are not as closely related to humans as apes are) do naturally eat a significant amount of meat.
When human activity (turning the area into a ranch/range farm) reduced the numbers of the animals that naturally prey on them (Lions, in particular), the baboon population increased.
So did the population of ungulates, notably, Thompsons gazelles.
Probably there was population pressure on the baboon's plant foods, and the other (ungulate) species' increased populations probably increased that pressure to some extent, so that the baboons may have been hungrier than they ordinarily would have been.
Active hunting at a long distance from the baboon troop by solitary males, in the absence of the lions, was no longer selected against in natural selection, and so it increased from experimental isolated occurrences by one particularly predisposed male to become fairly prevalent by males.
Then, after experiencing the success of a small number of chance arrangements of baboon hunters to form a cooperatively functional group, cooperative long-distance hunting became prevalent.
Baboons learned to select Thomson's gazelle herds of females and young, not of males, as hunting targets.
The Thomson's gazelles then learnt to regard baboons as dangerous predators and to increase vigilance behavior and to set much greater alarm and flight distances between themselves and the baboons.
Then, following immigration of more powerful males that prevented the successful baboon hunters from actually eating their prey, but who themselves preferred to stay close to the troop as baboons usually do, and the subsequent emigration of a couple of the successful hunter males, the practice of cooperative long-distance hunting ceased.
Focus on hunting in general declined, but increased solitary hunting by immature males, closer to the troop, persevered.
Baboon dismemberment of the prey, in the process of eating it, was often clumsy enough to indicate that it was not as strongly natural an activity as it would be for the dedicated carnivores, who carry out that process in a fairly standard and proficient manner, learning from infancy from the parents notwithstanding.

Tool use by chimpanzees in their 'natural' habitat includes the use of naturally-shaped hammer and anvil stones to crack open nuts, naturally-shaped mortar-and-pestle stones for grinding seeds and tubers, and naturally-shaped pounding logs, as well as the use of twigs to extract termites from termite mounds. Also the use of thrown rocks as weapons in hunting has been reported.
Tool use by capuchin monkeys in their 'natural' habitat includes digging for tubers with stones, cracking open seeds and branches with stones, breaking tubers with stones, and using stones as hammers in combination with wooden anvils to crack seeds.
It is therefore probable that all of these fairly conceptually simple tool uses were carried out also by even our most distant bipedal hominin ancestors.

Among the great apes, our closest animal relatives, chimpanzees (males mostly) have been reported to violently and lethally defend (and acquire or expand) their territories against other groups, gorillas perhaps even more so but maybe less lethally, the savagely solitary orangutans likewise, while the bonobos are reputed to be less violent in general.

It is possible that human encroachment on habitat, impact on large cat predators on apes (i.e., leopards, which I surmise are more populous in the bonobo's habitats (mostly closed forest), which I gather are less impacted by humans), and enactment of imitatable territorial and hunting violence, may have increased the level of territorial violence to some extent, but it does seem likely to be a natural inclination – in the absence of sufficient ape population losses to predators, disease, cannibalism and/or infanticide such behaviour is almost certainly adaptive to survival – in the chimpanzee case combat is most often pursued by a more powerful group, and avoided by a weaker group.

Also, the assumption that chimpanzees habituated to human observation have no reaction to that observation, for example aggression or violence (e.g. as a display), may be false.

The basic relevancies of these to human nutrition and diet should be obvious enough. And it seems rather foolish, to me, and wishful thinking, mostly by those enamoured of human (in particular their own) carnivory, to posit that hunting violence and violence to others of one's own species are not *at all* related – in the human case, surely killing large mammals whose emotional similarity (i.e., in distress) to oneself is obvious enough makes it easier for one to inflict violence on other members of one's own species? And conversely, if humans (in the absence of enough population losses to predation and disease to prevent population stress on i.e. plant food supplies) must kill their own infants or other adults to secure enough territory to get (nearer to) enough plant food, surely this killing makes it more likely for them to kill animals? The former seems likely to be and to have been more in force than the latter, though.]

Chapter 3 Diet-related biochemistry and mutations – indicators of evolutionary adaptations to and for diet (briefest summary)

[The following facts and their derived meanings are in order of the chapter sections:

3.2 Firstly, most genetic mutations involve a loss of function, rather than the development of a new function, for the fairly obvious reason that a single deleterious mutation may suffice to destroy the function of some protein coded for by the (now mutated) DNA, whereas to evolve a new or substantially different function usually requires at least several mutations, and maybe requires these to happen together.
Degenerative loss of function, even if this is loss of repression of a function that therefore allows the function (as in adult retention of lactase), is therefore more likely to occur in evolution, and to require much less time to do so, than the generation of new functions.

3.3 The large majority of animals make their own vitamin C; humans no longer do, and so must get their vitamin C from their foods, indicating persistence of a genetic adaptation to a diet higher in vitamin C than that of most other animals.

3.4 Vitamin B12 intestinal uptake, transport and storage are mostly very well conserved. Saturation of intestinal B12 uptake is said to be at 1.5-2.0 micrograms(ug)/meal, this in the context of a daily requirement of <1ug and the body's ability to store 3,000-5,000 ug. Even western diet-eating omnivorous humans have a low breast milk B12 (1/10 that of cow milk), and their newborn babies have a low liver B12 relative to adults (1/3 to 1/20).

A substantial amount of vegan experience with little or no B12 in the diet for years and no clinically detectable problems (imperfect, patchy and dubious as some of this evidence is) including this author's (my own) experience confirms generally low B12 requirements. These facts are consistent with humans being genetically adapted to a low-vitamin B12 diet with a somehow irregular intake.
Most probably the diet to which humans are evolutionarily suited is one with a regular (seasonally variable) intake of incidentally and deliberately eaten insect matter, larvae and adults, with some meat (incidentally) able to be accommodated in lieu of insects.

3.5 Human dietary protein requirements are very low (20-25g purified amino acids, or, 40g food protein / day / person).
Human breast milk protein/energy ratio is lower than that of all other food types (less than 1/10 that of meat), and much lower (only a fraction) than that of other mammals.

Consideration of human water requirements clearly indicates that if enough fruit and/or root vegetables and leafy vegetables are eaten to meet water requirements, then most of the human protein requirements will be met from those sources, leaving only a small requirement for more concentrated protein sources like legumes, nuts, oily seeds, cereal grains, or animal foods, the first four of which would round out the energy requirement also in the amounts needed to round out the protein requirement, as they are comprised mostly of carbohydrate or fat, compared to meat (and especially lean meat, i.e., wild game other than of very cold climates), which is comprised mostly of water.
These facts clearly indicate that humans are genetically adapted to a diet very low in protein, which strongly argues against the possibility that we have become evolutionarily adapted to a diet with a high or even a moderate proportion of meat.

3.6 Generally, for humans more than 50% of dietary calories as protein is overtly toxic (from the nitrogenous waste resulting), and illness is felt within days.
The excess nitrogenous waste resulting from the breakdown of dietary protein for energy is diuretic – that is, it causes an increased excretion of water as urine. This is undesirable in habitats where water is in short supply, and maybe even in any case, as predators lurk preferentially near groundwater supplies. And that urine will have a stronger smell and therefore be more detectable to predators.
High-protein diets for humans are reported to increase the cost of digestion by about 30% compared to high-fat diets.
Taken together, this information is evidence further suggesting that humans are not genetically adapted to a high meat diet, but rather to a predominantly fruit and vegetable diet.

Claims of meat being more energy-dense than plant foods are very obviously absurd: raw meat is about 70% water; raw legumes, nuts and oily seeds have very little water, the large majority of the weight being taken up by about 20% protein on average, and the another 70-odd% being fully fat or carbohydrate. So raw nuts (600 kcal / 100 g) (or oily seeds (600 kcal / 100 g), and also cereal grains (380 kcal / 100 g dry) for that matter…) are more energy dense than raw meat (190 kcal / 100 g, lean; 230 kcal / 100 g, regular).

3.7 Human starch digestion has been enhanced in evolution by the triplication of salivary amylase after the last common ancestor of chimps and humans. The foods most probably involved in this were starchy tubers and cereal grains.
This information is evidence suggesting that we are evolutionarily genetically adapted to a diet with a higher complex carbohydrate (starch) content than that of our common ancestor with the chimpanzee.
Legumes also have a substantial amount of starch, but consumption of these has probably been relatively constant throughout relevant periods of evolution, as legumes have been

reported to be a dietary mainstay for virtually all relevant non-human primates (see also chapter 2, section 2.5).

3.8 Regarding the supposed human inefficiency in biosynthesis of taurine, as in cats (obligate carnivores), vegan breast milk taurine was reported to be 30 times the level of cow's milk-based infant formula, and to considerably overlap in distribution omnivorous human breast milk taurine, suggesting that humans are just as likely to be evolutionarily genetically adapted to be able to meet their taurine requirements on a very predominantly vegetarian diet as to not be able to.

3.9 Regarding the supposed human inefficiency in elongating and desaturating 18 carbon fatty acids to their product 20 and 22 carbon fatty acids, docosohexaenoic acid (DHA) and arachidonic acid (AA) (similarly to obligate carnivores), comparisons of vegan breast milk with captive and wild apes and monkeys, with similar levels found in vegan breast milk and the milk of wild apes and monkeys, indicates otherwise – that our relatively low fatty acid elongation and desaturation ability is evolutionarily genetically adapted to provide a sufficiency of the DHA and AA products on a very predominantly plant food/ vegetarian diet.

3.10 Preformed Vitamin A from animal food is far in excess of requirements, in western diets. The human ability to synthesize vitamin A from pre-vitamin beta-carotene from plant food sources has been retained through evolution. That humans have not lost the ability to synthesize sufficient vitamin A from the pre-vitamin A (beta carotene), commonly available from many different types of plant foods, despite high-meat diets leading to liver stores (relatively low at birth) of an average 450milligrams, massively in excess of the 1milligram Recommended Daily Allowance (McLaren 1981), is clearly consistent with human high-meat (assuming the liver would have been eaten) diets not having been part of our evolutionary history for long enough to have caused us to evolve to be suited for high-meat diets, and is consistent with humans being evolutionarily a predominantly vegetarian animal.

3.11 Adult lactase persistence in humans is almost certainly the result of only a few or several thousand years of milk consumption after weaning. So, the persistent activation of the genes for lactase has become part of the genome in some small and generally geographically prescribed populations and their emigrants.
Clearly all other mammals on earth drink milk only as infants, then go on to eat the natural (if not always the optimal) foods of their species. The fact that a small number of human groups have evolved by a genetically degenerative (DNA *de*generation) process to be able to digest the milk sugar lactose is not at all evidence that dairy products are the best source of calcium for humans of any age, as other nutrients, and the combinations of foods in meals and diets, and even other factors, need to be addressed in deciding what is the best source of calcium.

People should be skeptical of assumptions that our genome is "adapted for a Paleolithic lifestyle" (Cordain, ch. 19; Eaton, ch. 20), for Paleolithic diets were quite variable in time (short, evolutionarily) and place, and *some* natural selection (i.e. removal of *some* degenerated function) may occur much more quickly than evolution of new function.

3.12 Fructose is a sugar found in fruit, and some vegetables (i.e., sugar beets) where it most often is present in an equal or lesser concentration than glucose.
Inborn errors of fructose metabolism (in Fructokinase, (Fruct)Aldolase B, Aldolase A and Fructose-1,6-diphosphatase) are very low in frequency (1/20,000 or less).
The very common 'fructose malabsorption' seems mostly due to the unnatural modern context of rapid ingestion of some high fructose:glucose ratio foods in the absence of

other meal-balancing protein and fat foods, and to be otherwise beneficial to the digestive tract.
Overall it seems that the human diet must have contained a large proportion of fruit throughout the periods of time that have produced the current rareness of inborn errors of fructose metabolism – the evidence here is for humans being by genetic evolutionary nature very much a fruit-eating animal. However, the evidence is consistent with a modest swing towards suitedness for starchy root vegetables, such that we might be suited for a diet of less fruit and more starchy root vegetables and cereal grains than apes are – perhaps very roughly somewhere between 2:1 and 1:2 of fruits to starchy root vegetables overall.

3.13 Celiac disease is a condition in which an allergic response to the cereal grain (wheat, barley, rye, oats, millet…) protein gluten (its gliadin part) causes structural and functional damage to the surface of the small intestine, causing reduced absorption of nutrients from digested food.
The prevalence of celiac disease, in those countries where reporting can be relied on, is considered to be about 1%.
A 1% prevalence of celiac disease in the context of the predisposing infectious diseases being more prevalent in the higher population densities of the agricultural (and modern urban) era is hardly a very strong argument for humans not being evolutionarily suited to eating cereal grains – particularly as other foodstuffs are also the subject of allergies: peanuts and tree nuts around 1% each in a variety of countries, milk 0.4% in the USA, egg 0.2% in the USA, egg 2% in UK children, 0.1% in adults, shrimp 1% in the USA and shellfish 5% in an Asian country. These figures are only very roughly comparable indeed, as they need to be related to exposure frequency and schedule, and meal context, and corrected for socioeconomic confounders such as affordability, weaning, and related infectious and nutritional deficiency diseases – obviously cereal grains are almost certainly *the* most affordable of these and all foodstuffs, and are therefore very likely confounded in their association with allergy by a more unfavourable collection of socioeconomic confounders such as weaning, and relevant diseases.
Celiac disease's 1% prevalence is only a very weak evidence, or none at all, for humans not being evolutionarily suited to eating cereal grains, nor for cereal grains not having been a substantial part of the diet in various places in the human evolutionary past, as we can safely assume peanuts, tree nuts, eggs and shellfish were also.]

Chapter 4 Food and food group nutrient supply sufficiencies and deficiencies according to Recommended Daily Allowances (RDAs) – indicators of evolutionary adaptations to and for diet food group proportions (briefest summary)

[4.2.1 Calcium (Ca) (RDA 1000 mg / day):
The meaning of the food calcium (Ca) data I interpret to be that it is significant that 500 grams of green leaf, flower and shoot vegetables eaten along with enough fruit to meet caloric requirements would give 850 mg Ca per day, which I strongly suspect in reality would be very adequate for the very large majority of the human population(s), due to the RDA being artificially inflated as discussed above.
This could provide a little too much water for comfort, but replacement of some fraction of the fruit with tubers, legumes, nuts and seeds, and even grains would still very probably provide a sufficient intake of calcium.
4.2.2 Phosphorus (P) (RDA 700 mg / day):
4.2.3 Calcium/Phosphorus ratio (Ca/P):
The meaning of the food Ca/P ratio data I interpret to be that perhaps the most meaningful comment that could be made on this data is that with the exception of dairy milk (which has 4 times the protein/energy of human breast milk and therefore probably

should not make up a large proportion of the diet) the dietary proportionation of food groups that would make the dietary Ca/P ratio most consistent with both the RDA Ca/P and the human breast milk Ca/P would be to obtain most of the calories through fruit, to have also at least 500grams of green leaf, flower or shoot vegetables, and to have a relatively small proportion (all together) of the other foods.

4.2.4 Magnesium (Mg) (RDA 360 mg / day):
The meaning of the food magnesium data I interpret to be that perhaps the most meaningful comment that could be made on this data is that if we take the actual optimal requirements as being probably on average no higher than the RDA (as the breast milk level at 120 mg / 2500 kcal is much lower), then seafoods, cow/goat milk, legumes and seeds (very roughly around 1000 mg / 2500 kcal) should be (all together) no more than a lesser proportion of the diet.

4.2.5 Fiber (RDA 32 g / day):
All of the animal foods; flesh, eggs and milks; are absolutely deficient in fiber, having none.
It is the opinion of many orthodox dieticians and nutritionists that 80 g of fiber / day would be a better target than the RDA here of 32 g, and fruit and green leaf, flower and shoot vegetables seems the best way of meeting or (perhaps to even greater benefit) exceeding that amount.

4.2.6 Vitamin B12 (RDA 2.4 micrograms (ug) / day):
The meaning of the food vitamin B12 data I interpret to be that as the human infant is making many times more the amount of neural tissue (absolutely, not only relative to its bodyweight) and other things, that adult humans are, but needs absolutely less energy (1000 kcal / day for a 10 kg infant, for example), and therefore receives only roughly 1 microgram per day from the breast milk it drinks in meeting its lower absolute energy needs, it seems logical that adult human requirements, at similar asorption plus intestinal production (?), should be (much?) less than 1 microgram of vitamin B12 per day.
This would be met by only a very small amount of animal matter, and could probably all be met by the amount of insect matter, larvae and adults, that apes eat incidentally in plant matter eaten, with or without deliberately focused insect eating; or alternatively by occasional egg and small animal eating.

4.2.7 Folate (RDA 400 micrograms / day):
In considering calcium, calcium/phosphorus ratio, fiber, folate, protein, water and energy together, it seems that any more (as a replacement for any other of the food groups noted here) than the very small amount of animal food required for meeting vitamin B12 requirements would have a detrimental effect (with reference to the RDAs) on a diet composed of mostly (roughly 3kg, such as to roughly meet the average daily water requirements) fruit and starchy root vegetables, having at least 500 grams of green leaf, flower or shoot vegetables, and roughly 200 grams dry weight of combined legumes, seeds, nuts and cereal grains. The same would seem to apply even if these proportions were varied quite a bit.

4.2.8 Zinc (Zn) (RDA 9.5 mg / day):
4.2.9 Iron (Fe) (RDA 8 mg / day for males; 18 mg / day for menstruating females):
The meaning of the food data (for calcium (Ca), phosphorus (P), their Ca/P ratio, magnesium (Mg), fiber, iron (Fe), zinc (Zn), folate and vitamin B12) considered so far in this chapter I interpret to be that in considering calcium, calcium/phosphorus ratio, fiber, folate, protein, water, and energy, to which now are added zinc and iron, all together, it still seems that any more (as a replacement for any other of the food groups noted here) than the very small amount of animal food (and certainly of milk with its great deficiency

of iron) required for meeting vitamin B12 requirements would have a detrimental effect (with reference to the RDAs of the others listed just above) on a diet composed of mostly (roughly 3 kg, such as to roughly meet the average daily water requirements) fruit and starchy root vegetables, having at least 500 g of green leaf, flower or shoot vegetables, and roughly 200 grams dry weight of combined legumes, seeds, nuts and cereal grains. The same would seem to apply even if these proportions were varied quite a bit.

4.2.10 Vitamin C (VitC) (RDA 82 mg / day):

The meaning of the food vitamin C data I interpret to be that:
Taken together with the data named above, it seems safe to say that much more green leaf, flower and shoot vegetables should be eaten than squash and zucchini.
Also it seems safe to say that on the basis of vitamin C content alone, the human diet generally should derive at least half of its calories from fruits and/or starchy root vegetables, and that this does not contradict the optimum proportion of diet of these foods suggested by consideration of the other nutrients done above.

4.2.11 Vitamin A (VitA) (RDA 800 micrograms / day):

The meaning of the food vitamin A data I interpret to be that red and orange/purple fruits and (i.e., fruit-) vegetables should be a regular substantial part of the diet, and that 500g of a good variety of green vegetables must be had to make up shortfalls in red/orange/purple fruit unless it is known that the particular green vegetable(s) eaten are those high in VitA.
However, probably almost any deeply dark green leafy vegetable is likely to be able to do so on its own, our common western cabbage (not the Savoy or purple cabbages, note) being the culprit dragging the averages down, notwithstanding its great usefulness overall as an affordable and readily available green(ish) leaf vegetable staple food.

4.2.12 Vitamin E (VitE) (RDA 15 micrograms (ug) $\propto$ (alpha)-tocopherol / day):

The meaning of the food vitamin E data I interpret to be that to provide a good probability of the diet providing a good supply of vitamin E, on average, a good mixture (on a per kcal basis) of different types of seeds, nuts, tubers and fruit should comprise the large majority of the diet.
These could be substituted (on a per kcal basis) by marine animal foods, or eggs, regards VitE, but this does not consider other factors, which generally suggest that such substitution would be best kept to a lesser proportion by kcal. (see immediately below)

4.2.13 Selenium (Se) (RDA 55 micrograms (ug) /day) (Safe Upper limit 400 ug / day):

The meaning of the food selenium data I interpret to be that the addition of roughly equal amounts of legumes, nuts, seeds and cereal grains to an amount of fruit and/or starchy root vegetables that meets most of the dietary caloric requirement, and 500 g of green leaf, flower and shoot vegetables would provide a nicely sufficient selenium intake well short of the upper safe limit and toxic levels.
Brazil nuts should not constitute more than 1/20 of calories.
Land animal flesh on average should not constitute more than roughly 1/4 of calories.
Marine fish flesh on average should not constitute more than roughly 1/6 of calories.
Marine mollusks and crustaceans on average should not constitute more than roughly 1/10 of calories. Note that consideration of the other nutrients as done above indicates that certainly the 1/4 and quite often the 1/6 fraction given here would be too high.

4.2.14 Thiamin (VitB1) (RDA 1.1 mg / day) ; Riboflavin (VitB2) (RDA 1.2 mg / day) ; Niacin (VitB3) (RDA 15 mg / day) & Pyridoxal/ine/amine (VitB6) (RDA 1.3 mg / day):

The meaning of the food thiamin (VitB1), riboflavin (VitB2), niacin (VitB3) and pyridoxal/ine/amine (VitB6) data I interpret to be that the addition of roughly equal amounts of legumes, nuts, seeds and cereal grains, to an amount of fruit and/or starchy

root vegetables that meets most of the dietary caloric requirement, and 500 g of green leaf, flower and shoot vegetables would provide a very sufficient intake of all four of these vitamins.
It is quite noteworthy that the fruits on their own on average on a per 2,500 kcal basis provide a very sufficient amount of thiamin (VitB1), riboflavin (VitB2), niacin (VitB3) and pyridoxal/ine/amine (VitB6).
Also, that 500 g of green leaf, flower and shoot vegetables on average supply a very significant and useful fraction of the RDAs for thiamin (VitB1), riboflavin (VitB2), niacin (VitB3) and pyridoxal/ine/amine (VitB6).]

Chapter 5. Essential amino acid complementarities among food groups (briefest summary)

[Animal foods require no dietary complementation of their essential amino acid profile by other food groups.

For almost any conceivable practical purpose, legumes require no dietary complementation of their essential amino acid profile by other food groups.

Nuts, seeds and grains are generally mildly or moderately low in lysine. Some nuts are also low in (methionine + cysteine).

Generally, fruit may not be relied on to complement any other food group's essential amino acid profile, as different fruits vary so widely in their amino acid profile.

In a diet made up calorically of mostly fruit or starchy root vegetables in almost any proportion, with the caloric balance coming from almost any proportion of legumes, seeds or grains, with or without nuts, along with 500 grams of green leaf, flower or shoot vegetables, there is almost no chance of there being any significant dietary shortage of any essential amino acid.
The only combination of food groups likely to result in a significant dietary shortage of any essential amino acid would be an unfortunate very unusual exclusive combination of one type of only certain nuts (i.e., macadamia nuts in particular, extremely low ('0/5') in lysine and very low ('1/5') in (methionine + cysteine)) with certain fruits (i.e., pawpaw/papaya, mandarins, apricots, very low ('1/5') in (methionine + cysteine), with or without green leaf, flower or shoot vegetables.]

Chapter 6. The difference between 'natural' and 'optimal' diets: natural statistical distributions of optimality – perfection is unusual; the average is far from perfect (briefest summary)

[Nature is not generally some sort of paradise or utopia for the organisms living in it. It is a somewhat chaotic system (with a Gaussian ('normal' or 'bell curve') or some other such distribution(s) of orderliness in it) in which at any one time a large fraction of the organisms in it is faced with some sort of imperfection in some aspect of its life – predation, starvation, other violence from members of their own or other species, weather, physical accidents, infection, parasitization, cancer, population explosion, extinction or other unfavorable increases or decreases. And, it is logical to assume that the quality or suitability of food that is being eaten has some similar level of variation to it.

So, what happens in nature is that quite often, in various circumstances, the sort of food that an individual or population is eating is not actually what would be the best for their health, it is what they are eating in order to survive. In other words, 'fallback foods', as discussed above.
And population explosions are probably in general the greatest cause of hunger and the eating by a population of foods that are not best for the health of members of their particular species.
Environmental and habitat change would be other such causes.

Foods eaten in natural circumstances are not necessarily the best for the health of the animal eating them, and may be far from it.

Evolution generally selects for organisms to have some flexibility in the foods they can eat for survival and reproduction – this does not indicate that all foods eaten in surviving and reproducing, or the proportions of them, are the best for the long-term health of individuals of that species, only that they will permit survival of that species.]

Chapter 7. Hunter-gatherers and/or gatherer-hunters (briefest summary)

[In the past, and to some extent carrying over to the present in the (mis)conceptions of the public and some academics, there has been a focus on the big-game hunting of those traditional 'hunter-gatherer' (more correctly gatherer-hunter) groups studied, due in no small part to an ethnocentric male-centric investigator base, with notions of manhood tied to savagely dispatching the hapless wildlife rife.

It seems, rather, that other than in the most extreme of habitats such as in the arctic, or in some other particular ones such as some islands and coasts, substantially more plant food (more than 70%) than animal food is generally eaten. Hence gatherer-hunters is usually a more correct term than hunter-gatherers.

And, this needs to be considered in the context that the areas occupied by present-day and historically recent gatherer-hunters and hunter-gatherers are areas that are more marginal (less rich or suited) to supplying the optimal foods for humans, because the richest or most suitable areas have been taken over by agriculture. A particular aspect of this that seems rather obvious once considered, but that receives little mention, is that agriculture is largely based on growing cereal grains, and that therefore, land remaining after agriculture has taken over lands more suitable for the growing of those cereal grains is likely to be less suitable for the growing of cereal grains, and cereal grains are therefore less likely to be substantially represented in the diet of the peoples living in those remaining lands, and therefore their diets will under represent (and maybe greatly) the extent to which cereal grains were a part of our most recent prehistorical gatherer-hunter ancestors' diets. And a similar consideration may apply to the remaining dietary 30% as animal foods – in richer areas, in preagricultural times, the proportion may have been considerably less than this.

Aside from a marginal-rich continuum, some places are more merely *different* from others in the foodstuffs (and therefore the nutrients) they supply, and there is a fair bit of variety in the diets of studied gatherer-hunters and hunter-gathers. This is not at all to be taken to mean that all such diets are equally good for all humans. There are very likely to be significantly beneficial changes that could be made to most of these diets. And, for peoples whose ancestry has a long line in a particular place, eating particular foods such as to give a particular dietary nutrient profile, the optimal diets are likely to be a little dissimilar, though we should be careful not to overestimate the probability of genetic differences arising from such population dietary histories. (see also chapter 3 above)

It seems however, that the substantial use of tubers as a foodstuff has been very present in our most recent prehistorical gatherer-hunter ancestors' diets, as it remains so in studied present-day or recent historical gatherer-hunter groups. The same can be said of fruits, legumes, nuts, oily and other non-cereal seeds, insects and their larvae, and some leaves and other vegetative plant parts (i.e. shoots).

On energetic considerations, it seems that the calories returned from the meat acquired by hunting are less than the calories returned from plant foods acquired by gathering. It seems that there are two possible situations:
1. The plentifulness of plant foods actually allow the nutritionally energetically wasteful hunting to be indulged in (see just below), and when the plant foods are in short supply the men reduce or stop hunting and put more effort into gathering.
2. If plant foods are in such short supply that the whole group together cannot gather enough to meet their needs, then hunting is increased. Here it needs to be kept in mind that more than 50% of calories as protein is overtly toxic to humans and illness is felt – and accordingly, in some conditions of great shortage of plant foods, prey that does not have enough fat is discarded even if the hunters are in a state of great hunger.

Claims of meat being more energy-dense than plant foods are palpably absurd: raw meat is about 70% water; raw nuts and oily seeds have very little water, the large majority of the weight being taken up by about 20% protein on average, and another 70-odd% being fully fat or carbohydrate – so obviously raw nuts (600 kcal / 100 g) (or oily seeds (600 kcal / 100 g), and also cereal grains (380 kcal / 100 g) for that matter…) are more energy dense than raw meat (190 kcal / 100 g, lean; 230 kcal / 100 g, regular).
Furthermore, the proportions of energy/protein in raw nuts are much closer to that optimal/required by apes or humans than is that of meat.

Furthermore, note that high-protein diets are reported to increase the cost of digestion by about 30% compared to high-fat diets.

Also, the excess nitrogenous waste resulting from the breakdown of dietary protein for energy is diuretic – that is, it causes an increased excretion of water as urine. This is undesirable in habitats where water is in short supply.

There are some circumstances where a major focus of gatherer-hunters is on gathering enough water-rich foods, a notable example being melons.
The diets of present-day apes generally supply all of their water requirements.

Humans need very little protein in comparison with their energy needs.
Quantifications of this include that with purified amino acids supplied in the right proportions in experimental diets as little as 20 grams of amino acids per day per average sized male have been sufficient to maintain positive nitrogen balance (to prevent reduction of the body's muscle mass and other protein components), and that generally accepted requirements for providing a good margin of safety to the whole population have been around 0.6 grams of food protein per kg of body weight (about 40 grams of food protein per day for a 60kg human.
Also, that human breast milk protein/energy ratios are a mere fraction of those of other mammals for which it is known (see also section 3.5, chapter 5)

It seems fairly likely to me that present-day/modern gatherer-hunter men (mostly, but also other group members for some things) don't mind that hunting is often nutritionally energetically wasteful, because:

1. They are released from the more mundane and unexciting drudgeries of plant food collecting (of surplus over and above what plant food they eat while looking around for animals to hunt…);

2. They have an outlet for competitive, dominance and aggressive urges;

3. They have a mechanism for proxy displays of martial power to all three of: neighboring group members; other males within their own group; and females; which is safer than actual combat in the context of potentially lethal weaponization.

4. Males with less martial power have a mechanism, via cuckoldry against them deliberately left unknown to them due to them being out hunting, to attempt (or not, to the extent that it was more a psychological displacement activity to remove one from the potential danger of conflict due to food contribution imbalance or cuckoldry prevention or discovery) to balance the imbalance in meat contributions with the more successful (on average more martially powerful by proxy of greater hunting success) males, to exchange sexual access to their wives for reduced risk in potentially fatal weaponized combat with the cuckolders, a share in the food (i.e. meat) greater than that proportional to their own contribution, and a share in the martial power of those better so endowed for the purposes of defence (the last two of which may be related to primitive pimping, a correlate of primitive prostitution).
So perhaps there was not only the fairly well understood female prostitution analogue of exchange of food for sexual access (an evolutionarily ancient behavior), but also a more tacit and less obvious pimping correlate (with evolutionarily ancient origins, but more recently evolved 'refinements') …

But probably these aspects did not apply to earlier hominins because they were physiologically and technologically limited to power scavenging in larger groups, rather than hunting alone or in pairs or small groups, and probably much more importantly, because they were physiologically and technologically limited to defending themselves from large carnivores in larger groups, being unable to do so consistently alone or in pairs or small groups.

It seems that cooked starchy foods have a significantly worse effect in generating dental caries than raw starchy foods, and also that cooked starchy foods have a higher Glycemic Index than raw starchy foods, these two reports being consistent regarding the increased early breakdown of complex carbohydrate into simple sugars.
Also, there is the possibility that the cooking modifies the food fiber such as to reduce its cleaning brushing action.]

Chapter 8. The disappearance of large animals in recent prehistory and The Overkill Hypothesis (briefest summary)

[I interpret the meaning of the disappearance of large animals in recent prehistory to be that human predation (particularly as a proxy display of martial power) probably was involved in their disappearance, but probably as a lesser factor.

Human predation may well have driven certain animal populations already under severe natural stress over the brink of extinction, but that predation probably only either merely accelerated a process that was already destined to culminate in extinction at a not much later point in time, or only just tipped a precarious balance over to the side of progression to extinction.
However, my opinion on this is not a strong one, as I do not have an expert understanding of the data, which is complex – It could be that that hunting may have been a bigger contributory factor in the disappearance of large animals in recent prehistory than Webster (1981) posits as below here and as I tend to agree with, because hunting of large animals is not only a nutritional activity – it also serves as a proxy display of martial power to all three of: members of other groups of humans (with whom territory is contestable and very probably was contested, in the context of an expanding population); male members of one's own group (in establishing prestige, and very probably also some sort of hierarchical position, the reported and ostensibly credible egalitarian nature of gatherer-hunter and hunter-gatherer groups notwithstanding); and females (as a mating display leading to increased and varied sexual activity).

It is therefore possible that many more large animals were killed than were required merely to meet (/complete) nutritional requirements – possibly many times over that amount.

However, as detailed and explored in other chapters of my work here, it does not appear that the extent of meat eating that this contributed to our genetic evolution was great enough for long enough to contribute much change in our nutrition-related genes. Though, there are likely to be some small or very small number of people in some populations who have DNA-degenerative genetic mutations that reduce their ability to absorb, transport, store or use vitamin B12, due to there being an amount of it in the diet (for some thousands of years), of some gene-lines, that was probably far in excess that in the diet of our evolutionary ancestors of the last 6 million years overall.
Notwithstanding my own demonstrated ability to have survived a 20+ year experimental exclusively raw exclusively plant food (vegan) diet with no supplemental vitamin B12, apparently without damage to my nervous system (see also chapter 10), there are very probably some small or very small number of present-day humans who would experience possibly disastrous damage to their nervous system if they were to attempt to replicate my experiment.
Such people are much less likely to be found in populations that have lived for some dozens (or maybe less – in more extreme circumstances only one or two generations might be needed) or hundreds of generations on low-meat diets in historical times (times after the extinction of the large mammals occurred, when agriculture became prominent) because selection of the fittest under various selective stresses would have weeded those DNA-*degenerative* changes reducing the ability to absorb, transport, store or use vitamin B12, back out of the gene pool, preventing generational transfer of such genes along into present-day populations.

In any case, in the absence of a clear family history of some number of generations on both sides of an infant's family having lived and thrived mentally and physically on a very low animal food diet, infants should not be subjected to a purely vegan diet – their diet should have either some substantial animal food B12 content (breast milk from an omnivore or high lacto-ovo vegetarian, but not from a pure unsupplemented vegan) or a vitamin B12 supplement. Infants with nervous system damage from a purely vegan diet have been clearly established in the medical journal literature via reported case histories.

Chapter 9. Examples of human performance: Dave Scott, vegan Hawaii Ironman Triathlon (3.8 km swim, 180 km cycle, 42km run) 6 times world champion; Carl Lewis, vegan 100 m & 200 m sprint and longjump world champion

Dave Scott won the Hawaii Ironman Triathlon (The accepted world championship of Ironman and ironman-distance Triathlons), a 3.8km swim followed by a 180km bicycle ride followed by a 42km marathon run, six times, as a pure vegan, that is, not eating any animal flesh, eggs or dairy products. (Endurance performance)
He is quoted as saying something to the effect that the claim that animal protein was necessary for human health was utter nonsense – and it would seem somehow unreasonable to argue with that, given the evidence of his performance?...
Carl Lewis reached the peak of his performance as a 100m & 200 m sprint and longjump world champion(s) and Olympic gold(s) medalist as a vegan. (Explosive performance)
Also, web-search “vegetarian athlete” and find Ed Moses, Martina Navratilova, etc, etc...

Chapter 10. An example of human performance after 20 years of experimental exclusively raw vegan diet with absolutely no supplemental vitamin B12: the author, in the top 15% of the 2007 Ironman Korea Triathlon (Swim 3.8 km, bicycle 180 km, run 42 km), a lifetime best at age 45 carrying injuries, and nervous system apparently still intact. The details. (briefest summary)

[I have myself carried out a personal *experiment* (*not* a belief system, please note, other than maybe a little very early on…) for more than 20 years (from 1986 to 2008) of eating only a raw vegan (exclusively plant food) diet, eating only 2 meals per day, late brunch and dinner, without any supplementation of vitamin B12 (and this, the lack of any vitamin B12, and very probably this alone is the risky part of the experiment, and I do not advise being without some source of vitamin B12 as an ongoing thing, and I make explicit here that it would be unethical to subject a child to it).

However, I have throughout eaten a large green salad with virtually every meal (only two per day, late brunch and dinner), and green leaves are the main source of cobalt atoms that are not already incorporated in vitamin B12 – therefore it is possible, indeed probable, that my large intestine bacteria did manufacture some vitamin B12 from that, which found its way down a concentration gradient in the moist raw-food-derived large intestinal contents, back up through the ileocecal valve and into the small intestine just on the other side of that valve, where vitamin B12 is absorbed by the human body.

My vitamin B12 levels for most of that 20-plus year experimental period were 40-100 picomolar (normal range 150-450 picomolar) (around 50 picomolar for at least the last 10 years of that period) and my homocysteine levels were 80-100 uM (normal range less than 12 uM).

Regards body muscle and nervous system health:

At the end of that period of 20-plus years of raw vegan diet unsupplemented by any vitamin B12, at the age of 45 years, I finished the 2007 Ironman Korea Triathlon (a 3.8 km swim followed by a 180 km bicycle ride followed by a 42 km marathon run, the same distance as at the famous Hawaii Ironman Triathlon) at the 15th percentile, that is, with 15% of the field in front of me and 85% of the field behind me.
I have never manifested athletic champion genetic ability, and have carried since the age of 18 hamstring and hip joint region damage with adhesions and a partly rehabilitated prolapsed lumbar intervertebral disk. My bicycle was good enough, but nearly the least expensive there. I was in employment, though on less than the normal 30 (~ 24) teaching hours per week, as an ESL English teacher throughout the time of the Ironman training.

Working on farms harvesting fruit in between contiguous years of university study, I have consistently averaged 6 full workdays per week of work, for the 3 month periods, and my reputation as a hard and good worker on those many farms and amongst those I've worked alongside is quite good.
I always had nothing to eat and only water to drink till lunchtime at midday while working on these farms – lunchtime was usually one hour, but sometimes only half an hour – when lunchtime was only half an hour I cut my leafy green consumption to one mouthful at each of the start and finish of the meal, and spent the rest of the short time eating peanuts and rolled oats (chewed dry and properly), and fresh raw mangoes (eaten more quickly than I like to (which is slowly), but not bolted) – my work performance in the afternoons was, like that in the mornings, above average.
Sometimes (say twice a week) I ran for 10-odd km after work, with a 1-hr-+ bicycle ride thrown in maybe once a week after work. Yes, the evening meal included a good gutful of mangoes, but the oily nut and seed intake was no more than 100-odd grams – and, somewhat annoyingly, it was usual for me to have more body fat (though not that much) at the end of the harvest season than at the start of it – mangoes are not as concentrated in energy as bananas, but they are at the high end of the range for fresh raw fruits...

There has throughout the experiment virtually always been mostly daily enthusiastic physical exercising, running and martial arts mostly, with a variable good serving of bicycling (road and MTB) thrown in when convenient.

Regards nervous system i.e., brain health:

In 1996, 10 years into the experimental raw vegan diet, I achieved approximately (the results distribution graph was very small and not lined so as to facilitate reading it very accurately) the (top) 15th percentile of the GAMSAT test that Australian university students must sit to gain entry to postgraduate medical schools to study medicine to be medical doctors – approximately 25% of those who sit the test are, after an interview assessment, admitted to the courses. Had I completed senior high school math's and physics before sitting that exam, which I hadn't, due to being an almost singularly unapplied and rudderless student in high school (as a junk-food eating omnivore, notwithstanding the lack of almost any steering committee...), I would almost certainly have achieved a score higher again than the 15th percentile.

(High school physics and Pure Math's were completed in 2000, 14 years into the experimental raw vegan diet, at the Queensland TAFE's Open Learning Centre, with the highest possible grade category (Honours) of the three passing categories being achieved, with a bare minimum of tutor assistance).

In my most recently completed university degree, Master of Public Health, at one of Australia's most highly ranked universities, the University of Queensland, in 2008, just after 22 years of the experimental diet, and some months into vitamin B12 supplementation (which did not result in any detectable difference in subjective feeling, notwithstanding me being of course non-blinded to my conditions), my grade point average was 6, the possible maximum being 7.

And the state of my brain function may be further assessed by examination of my analyses in this writing here.

I have done most of my university degrees, commenced in 1991 after 5-odd years of raw vegan diet, at an overload of subjects (none repeated subjects), for example 50% overload in the third year of the BSc, 66% overload of PGDip PH&TM and second year psychology (this did have some, but not a full, content overlap in the statistics area), 75%

overload (all subjects passed, at a D or 6/7 gpa, though I did have some prior knowledge) in one semester of Master of Medical Statistics and Master of Nutrition & Dietetics (which degree I was prevented from finishing, during the very *un*objectively assessed practical placements, the final semester of the course, by a most silly little coterie of liars and mental midgets (such as professor(!!...) _oger _ughes, very probably the stupidest PhD candidate that Prof Kerin O'Dea and Dr Shawn Sommerset have ever nursed through that degree (I challenge them to deny it…), and the stupidest blockhead I've ever had as a Head of Department or lecturer anywhere in my university studies, who would not have relished my adversarial and much better qualified and competent presence in the field of public health nutrition that he has elected to network his easiest possible to arrange of paychecks in…) such as are not in short supply in the dietetic profession, but that's another story and I'm sure the intelligent reader could surmise readily enough the general theme of my idiot-inflicted problems, due to my vegetarian"ism", there, for example by fellow student _llison _piegel (previous 'leadership' course attendee somewhere…) who achieved (we corrected each other's exam paper) 6 out of 24 for a multiple choice clinical dietetics exam of 24 questions with 4 answers to choose from per question (The result you'd be most likely to get if you knew nothing *at all* about the examined material and selected answers totally randomly – my result was 22/24, the highest result in the class on that occasion…)).
Of course these academic challenges were far from easy – I submit that very few people would have found them easy… And the regular daily physical exercise was continued alongside that.

Regards the eating/meal schedules:

Since 1986 or so I've virtually only ever eaten two meals a day, late brunch and dinner, with nothing other than water in between (and this water usually mostly in summer, the meals having their own generally nicely sufficient water content), excepting a couple of liters of fruit juice in the middle of the 6-odd hour Ironman training rides and the Ironman race (no-one else does these things without taking in some food along the way either, generally…) and a couple of marathon (42km) runs and one 9 hour running traverse end-to-end of Korea's Baekunsan Provincial Park's mountain ridge system and a 6 hour run followed by 7 hour fastish walk to traverse the even more formidable Jirisan National Park end-to-end in one day (Yes, as a pure run, which was the intention, Jirisan beat me, but the one day full traverse just in itself is nevertheless quite a challenge – Hwaomsa Temple near sea level up to Nogodan, along to Cheonwangbong peak (1,950m), down to Daewonsa Temple, if anyone's interested, and I'd advise not calculating your estimated time of completion solely on the basis of the map distance…).

I do spend some substantial time eating my food, say a total of 3 hours a day (but almost nil time in preparation and washing up…), but I am a slow eater, by choice, because I enjoy what I am eating, and do not like to rush my meals, and I have seen many others eat the foods I linger over in a mere fraction of the time I spend.

Even so, far from always feeling hungry, often enough I choose lower-calorie foods to prevent my tendency to overeat (which I have had since a child, though it has never lead to obesity as formally defined, only to, say, 5kg of excess fat with peak excess maximums not exceeding 10kg) from adding unwanted and physiologically detrimental fat to my carcass – for example, although I like bananas well enough, I usually reserve them for dinners before a race or long bicycle ride (more than two hours) or run (more than one hour) the next morning.
I have only very, very infrequently eaten dried fruit in the 20-plus years of the experimental diet (not at all for most of the first 15 years – I did fall for the dried persimmons in Korea in 2000, and again in 2006-7 and 2009-10, but not heavily), mostly

lately in an effort to be a bit more relaxed about life, but I have stopped this again now (for the present, I will have some more dried persimmons, hopefully, and Chinese dates, in the future…) as only a small amount of dried food at the end of a meal was adding unwanted body fat to me, and was reducing a little my appreciation of raw fresh fruit, which I usually find delicious, not merely tasty.

Furthermore I generally do not eat more than 100-200 grams of nuts and oily seeds plus the dry weight of the soaked legumes with grains per day, as to do so would be a gross excess of protein intake, and also bring my energy intake to an excess – and I can crack brazil nuts, almonds, macadamia nuts etc, with naturally shaped stones, faster than I can eat them – so obviously Professor Wrangham's (of Harvard University…) statements reproduced in chapter 1 section 1.14 here and in his book Catching Fire How Cooking Made us Human (2009), that humans cannot meet their caloric requirements from raw plant foods, are so far wrong as to be laughable, if the possible reinforcement and spreading of this gross ignorance were not a very serious matter regarding the public health.

In short, If I was to spend more than approximately one hour per day eating the more concentrated raw natural plant foods available to me, at the speed that I've seen many people eat them, I would become more and more overweight by the week… if I was to spend two hours per day doing that, I would become morbidly obese, despite my regular exercise, soon enough!...]

Chapter 11. Breast Milk: nearly always the right concentration in nearly all climatic conditions (reproduced from below)

[Most experts agree that breast milk is the best food for infants because it's naturally suited for infants, being naturally evolved to have the right amount of protein, fat, sugar, vitamins, minerals, water and antibodies for infants (and is low in contaminants, generally).

Most experts agree that exclusively breast fed infants should not be given water to drink, even in very hot and dry conditions because the greater amount of milk that the infant drinks in very hot and dry conditions will be appropriately diluted with the mother's own body water, if she is not severely dehydrated (as shown by studies of infant urine osmolality in very hot and dry conditions).]

Chapter 12. Sense of taste: natural taste; habit-formed taste; proper hunger; Sense of smell and visual aesthetics (reproduced from below)

[Regarding the natural taste of humans, the first thing to note is how very flexible our sense of taste is to the forming of habits.

Consider that one's first taste of a cigarette (its smoke, in inhalation) is almost universally revolting. Note that cigarette smoke is very obviously not what we are naturally evolved to breathe, and is unarguably toxic. However, after that first experience, psychological and social factors, in conjunction with the drug action of nicotine, after the forming of the smoking habit cause the taste of cigarettes to become pleasurable.
The drinking of most alcoholic drinks is very similar.

Now, if one can develop, through a habit which is not started until many years of life have gone by, a sense of pleasure from the taste of things which fairly obviously naturally revolt us, then how much easier is it to develop a sense of pleasure from the taste of things that are not outright and obvious poisons, but only second- or third-rate as foods, when the habit has been developed from very early childhood? Much easier, fairly obviously.

I'll begin by considering human taste for the high-protein foods meat, and nuts and seeds (roughly 20% each in protein on average, the nuts and seeds being more variable in this than the meats) (And the considerations apply to legumes also, but for convenience and on the basis of more similarly low carbohydrate contents, I will deal with meat and nuts and seeds.

Now, note that for many of us raised in the west on a diet based on cooked and usually garnished (salt, pepper, chili, garlic, curry, herbs etc) meat, raw nuts and seeds, even encountered later in life and much less frequently than the cooked meat, are much more palatable than raw meat.

We must note that the fair comparison of taste even just in any person's present state of habituation, and not adjusting for that habituation, is raw nuts and seeds compared with raw meat, cooked nuts and seeds compared with cooked meat, cooked and salted nuts and seeds compared with cooked and salted meat, and cooked, salted and garnished nuts and seeds compared with cooked, salted and garnished meat.

Obviously enough, a comparison of raw nuts and seeds with cooked, salted and garnished meat would not be fair, or rather, not be sensible.

If we are naturally predisposed to eat more meat than nuts and seeds, then why doesn't raw meat taste as good as raw nuts and seeds, to people who have been brought up eating far more cooked meat than raw nuts and seeds (And note that this applies also to those people who have been brought up eating far more cooked plant foods than raw plant foods)? Is it just *totally* habit or what they are used to (raw nuts and seeds, and cooked meat), or does this taste preference suggest that our natural predisposition is towards eating much more nuts and seeds than meat?

An alternate suggestion that must be considered is that nuts and seeds might be rarer in nature than meats and more valuable than meat but only as a lesser fraction of the protein food eaten, the greater fraction better being meats – this would be analogous with apes and most monkeys having a preference for fruits over leaves, but the natural situation being that fruits are always in undersupply regards desire, while the leaves are always available, the net result being that for most species of apes and monkeys, more green leaves are eaten than fruit, this being possibly the better diet for them than having an unlimited supply of fruit available. Countering this suggestion is the fact that our anatomy, our teeth, lack of claws or talons, and long digestive tract are similar to those of our very predominantly plant food eating nearest relatives, the apes and monkeys, and not to the sharp pointy teeth and claws or talons, and short digestive tracts of the more naturally predominantly carnivorous animals.

An interesting and relevant phenomenon that I have experienced is that, as a long term (now more than 20 years, but the phenomenon came to be experienced much earlier, as dietary and taste habits were reformed) eater of raw un-garnished foods, when starting meals with some green leafy vegetables (not enough to meet my protein requirements) and following this with high-protein (average 20% protein) raw nuts and seeds, if I do not keep introducing different nuts and seeds for flavor variety, my taste tells me when I have eaten enough protein to meet my requirements – more when exercise has been such that recovery and adaptation would need more protein, and less otherwise – the amount indicated by my taste is in very good agreement with the scientific facts of protein requirement (20grams of protein per day if it has the perfect amino acid complement and is perfectly digested and absorbed, 40grams of protein per day recommended as the

minimum, for a grown adult in a normal, stable physical exercise situation, as those perfections do not usually apply).
My natural taste for protein reaches extinction if and when I have slowly and properly chewed and therefore fully tasted all of the nuts or seeds eaten – I know from my taste that I have eaten enough protein to meet my requirements, and that it is time to move to the foods that will complete my energy requirements, mostly carbohydrate bearing fruits (nuts and seeds having a substantial amount of fat in them will have made a substantial initial contribution to the energy requirement also, as well as some very variable carbohydrate content).
If I keep turning to new nuts and seeds (and there are many tasty nuts and seeds to turn to…) through gluttony, seeking more pleasure than is sensible, through the new flavors that are alongside the taste of protein, then I can easily go on to consume two or three times the amount of protein I really require.

Almost invariably I feel most healthy when I limit my consumption of protein foods to that which meets but not much exceeds my physiological protein requirement, even if consuming several tasty varieties that cause me to want to overeat for the pleasure of it. The phenomenon is not as clearly demarcated when I consume raw soaked grains (dry protein content 10%, the balance being much more carbohydrate than fat).

To those who say that this may be due to individual idiosyncrasy of my biochemistry or physiology, I can only say, do the scientific thing – give it fair trial – get properly lean, rehabilitate your taste buds from the effects of garnishing and flavoring (salt, pepper, garlic, chili, herbs), be physiologically hungry at meals, not just eating for pleasure, overcome any psychological barriers or biases, and see for yourself.
And note that among old sayings, one of the very truest is "Hunger is the best sauce" – without that proper physiological hunger at mealtimes, you'll just be chasing your tail with excessive variety and garnishings and flavorings, seeking pleasure that is not reasonably or rightfully yours to experience…

Relating this to the question of what is the natural food of humans – carnivores almost certainly do not experience protein taste satiation when they have met their protein requirements – they go on to eat until they will meet their caloric (energy) requirements also – 60 kg cats and dogs do not start looking around for carbohydrate sources after eating 50 grams of meat at each of two daily meals. They will meet their energy requirements, as well as from fat (saturated fat, and with cholesterol), from breaking down the protein (amino acids) that is not required for rebuilding proteins into an energy source molecule and a nitrogen-containing waste molecule (ammonia) that will eventually (as a different molecule (urea)) have to be excreted by the kidneys. Some animals (for example sharks) are naturally benefited by high blood nitrogen levels of this source, and others are not. I suggest we are not, and that is why I feel better when I limit my protein intake to what I need for body protein maintenance or building, and why my sense of taste tells me fairly unmistakably when this has occurred in a meal that is eaten in a way that allows my taste buds to properly experience the protein that is being eaten.

The sweetness preference in the human taste, even though some people do not like extremely sweet foods, is still prevalent enough to clearly indicate that we are evolved to seek fruit as a large part of our diet. However, in nature, there is not always the abundance of sweet ripe fruit readily to hand that there is now in modernized societies, and the triplication of the salivary amylase enzyme for the breakdown of starch into glucose and smaller glucose polymers (small starch molecules) is a strong suggestion that we are also evolved to eat starchy foods like starchy root vegetables and cereal grains, where the salivary amylase releases glucose in the mouth to provide a sweet taste. The high resistance to extinction during a meal of the sweetness preference in taste (that most of us can eat quite a lot of sweet food before losing pleasure from doing so) suggests that we are evolved to be suited a diet with a large proportion of fruit, but it should be kept in mind that the high resistance to extinction of the sweetness taste preference may well be evolutionarily due to sweet fruit not being always fully available in the natural environment, such that other things (beneficial by way of contributing to a healthier complement of food in the diet) such as green leaf, flower or shoot vegetables, seeds, nuts, grains and starchy root vegetables, with or without some deliberately eaten animal matter as well as that eaten incidentally along with the plant matter, must be eaten.

The fact that we, as do the chimpanzees, like the taste of fat, in contradistinction to gorillas, who reportedly do not like the taste of fat, suggests that we are also evolved to be suited to a diet with a substantial proportion of fat. Similarly to sweetness preference in taste, if maybe subordinate to it, the high resistance to extinction during a meal of the fat taste pleasure (that most of us can eat a fair bit of fatty food before losing pleasure from doing so) suggests that we are evolved to be suited to a diet with some fat at some times, one having some proportion of oily nuts and seeds, but it should be kept in mind that the high resistance to extinction of the fat taste pleasure may well be evolutionarily due to oily nuts and seeds not being always fully available, at least without some effort, in the natural environment.

In nature a meal may not always contain significant fat or much carbohydrate, but the optimal meal may well contain a certain proportion of protein:fat:carbohydrate.
It seems to me that the best proportion is roughly: only enough protein to meet our protein requirements; the amount of fat that comes with the meeting of those requirements from protein sources that include a good proportion of oily nuts and seeds alongside whatever legumes and cereal grains may also contribute to the protein source mix; and the balance of our caloric (energy) requirements from fruits (which would therefore meet also roughly all our water requirements).

Such a mixture provides a slower uptake of dietary carbohydrate from the digestive tract, and some accompanying fat, and one effect of this is a feeling of well-being due to that steady drawn-out delivery of fuel to the body. Also as detailed below in chapter 20 on weight loss diets, because fat metabolism is activated, when the carbohydrate supplied by the meal is no longer available for ready fuel use, fuel use can switch to depot fat (in our fat stores) more smoothly and with less sense of lethargy and hunger or craving for food than would occur if the meal had had no fat.

It is important to understand that my statements regarding this feeling of wellbeing and its proposed reasons are not negated by reports of a feeling of wellbeing reported by people on the Atkins high protein diet.
The feeling of wellbeing reported by people on the Atkins high protein diet is due to their blood sugar curves previous to the Atkins diet having been wildly fluctuating through (refined or high Glycemic Index) carbohydrate being ingested without being accompanied by any fat to slow down its absorption into the bloodstream, or to prime fat metabolism to make access of depot fat for use as energy easier for the body.
Refined or high Glycemic Index carbohydrate being ingested without being accompanied by any fat to slow down its absorption into the bloodstream, in the context of a low-fat diet, causes initially rearing blood glucose levels, causing rearing insulin levels, which cause subsequent plunging blood glucose levels, in the presence of low blood fat levels, leaving regular periods of time when both blood glucose and blood fat and fat metabolism are low, causing fatigue and hunger, the latter causing overeating in an attempt to regain some short-term blood energy fuel.
Also, the Atkins diet gives a slow and relatively constant feed of fat into the bloodstream that blunts the hunger that occurs when (refined or high Glycemic Index) carbohydrate is

ingested without being accompanied by any fat to slow down its absorption into the bloodstream.
One big problem with the Atkins Diet is that in the long term, atherosclerosis is very likely to occur from the saturated fat, cholesterol, and oxidatively damaged unsaturated fat in which it is high.

Condiments/seasoning such as sugar, salt, pepper, garlic, chili and herbs make food taste good/better and rather than their effect being mostly to provide added nutrients and to be good for health, they pervert the natural sense of taste that should be enough for a properly hungry person to enjoy natural human foods, and often have undesirable and/or unknown pharmacological (drug) activity.
Salt raises blood pressure, which contributes to atherosclerosis, heart disease and stroke. Garlic, chili, curry and pepper irritate the gastrointestinal tract, exacerbating GIT disease, and increasing the chance of cancer there, and of undesirable GIT permeability.
People whose sense of taste has been perverted by habituation to the use of these unnatural, abnormal or undesirable flavourants will find food in its natural state not tasty enough, and will have to endure a period of reduced pleasure from eating before their sense of taste has been rehabilitated and they find food in its natural state tasty (again, or maybe for the first time in their life…).

Our sense of smell is physiologically very closely related to our sense of taste, and therefore very much subject to the force of habit in what we perceive as pleasant in the smell of food – in other words, just as the sense of taste may be very deranged from what it is naturally (by evolution) constituted to be, so may the sense of smell. However, there is one interesting commonality that seems to have prevailed – that as perfumes humans favor those scents that are very closely related to fruit scents, and hardly at all related to animal food scents. Bear in mind that scent preference, like almost all other features of us and any other animal, is selected for in evolution for a reason related to our survival as a species, not for no reason at all or mere artistic flair or caprice – the fairly obvious and very plausible reason for the evolution of that scent preference over other scents seems to be that we are by nature evolved to eat fruit as a large proportion of the diet.

Consider also the reason for the almost universal visual pleasure that humans get from gemstones, with their colours and regular geometric shapes. Again bear in mind that visual aesthetic preferences, like almost all other features of us and any other animal, is selected for in evolution for a reason related to our survival as a species, not for no reason at all or mere artistic flair or caprice – and again a very plausible reason for the evolution of that visual aesthetic preference seems to be that we are by nature evolved to eat fruit, with their colours and regular geometric shapes, as a large proportion of our diet.

Regarding the natural taste of humans for green leaf, flower or shoot vegetables, my own experience is that since my sense of taste has been rehabilitated from the deranged state an early life diet of cooked, garnished and junk foods had rendered it into, I find that if some logistic reason prevents me from eating something like 200 g of green leaf, flower or shoot vegetables (perhaps in some brunchtimes of occupationally restricted length), then at the next meal my sense of taste clearly urges me to eat an amount of green leaf, flower or shoot vegetables such as to make up for the lack at the previous meal.
However, the green leaf, flower or shoot taste pleasure (which may have some element of protein taste pleasure) is subordinate to the sweetness taste preference and the fat taste pleasure (fat usually comes with protein, particularly in plant foods), very probably because evolutionarily these were more readily available than sweet or fatty foods, and therefore needed less drive to ensure their being eaten, as they would be eaten somewhat by default in the natural absence of enough sweet and fatty foods to produce complete satiation.

This aspect (green leaf, flower or shoot) of natural taste pleasure for natural foods was the last to be properly rehabilitated in the course of my excluding cooked and junk foods – replacing sweets/candy and other sweet junk foods with fruits was the first, and experienced as more straightforward and easy to accomplish, and finding the right dietary place for fatty foods was a little more subtle and involved than for the sweet foods. Rehabilitating the taste for and acceptance of green leaf, flower or shoot vegetables requires us to somewhat mimic in our eating habits the natural environment where green leaf, flower or shoot vegetables are immediately available, but the sweet fruits and fatty/oily foods are less available – this can be done as noted elsewhere here, by making the effort to eat some green leaf, flower or shoot vegetables at the start of the meal (which is good for dental health as these foods have no acidity which many fruits have, nor significant carbohydrate for oral bacteria to turn into acid, which most fruits have, and as they have tooth minerals (calcium and magnesium) to provide less of a concentration gradient out of the teeth than some other foods do), and some second in the meal with the smaller amount of more protein-rich foods, and some last in the meal to knock off the teeth some of the remains of the other foods. After a while this practice will no longer require effort, it will have become a beneficial habit that one will reap the benefit of.]

Chapter 13. Dental health, optimization of digestion, and food group eating order (briefest summary)

[13.1 The causes of dental disease:

Genetic and Developmental/Environmental: Due to differences in their genes and in their environment from conception onwards, people develop different tendencies towards dental diseases, including those involving tooth enamel strength, tooth spacing, saliva acidity, saliva volume, mouth versus nose breathing, and taste preferences for carbohydrate versus protein or fat.

Imperfections in dental prosthetics (false teeth plates, caps, bridges, veneers, implants) both in the manufacture and fitting (by dental personnel), and in the use and maintenance (by the person having them) can contribute to further dental disease.

13.3 Nutrition/ diet in dental health management

The details of what nutrients in different foods contribute to tooth decay / caries is included in the following description of the best order in which to eat the foods at a meal. To the extent to which one deviates from the general sequence of four general food types or groups, one is reducing one's resistance to tooth decay. However, some deviations from the sequence are (much) more important than others (see note near end this section). To the extent to which one has dental decay problems, if one's order of eating foods differs from the following, change towards the following will reduce decay problems.

A good sequence/order of eating the foods in a meal is to:

First (perhaps most importantly, for dental health), eat some green leaf/flower vegetables (lettuce, cabbage, broccoli, cauliflower) (alone and without dressings, many of which

contain vinegar which is acidic and therefore damage tooth enamel, all of which contain other food substances best not packed into tooth spaces at the start of a meal)
because,
this gives a similar effect to the pre-meal toothbrushing recommended by dentists in removing some bacteria from the teeth before eating,
and,
when this material packs into any spaces between or in the teeth, it is not acidic to start with, and nor does it have significant carbohydrate for oral bacteria to turn into acid,
and,
leafy greens have substantial calcium and magnesium, and a variable amount of fluoride, so that in general these minerals will be leached more slowly from the tooth enamel, down a more shallow concentration gradient (Such flows are only very small at any one time, but over years, or a lifetime, they are significant), than it would for other foods on average,
and,
the lack of easily digestible carbohydrate will prevent premature triggering of release of the gastric contents into the duodenum (small intestine),
and,
leafy greens do not have a lot of water and so will not dilute the gastric digestive juices very much, so that protein eaten immediately thereafter will still be well digested by concentrated gastric juices that will be allowed time to act properly in breaking down dietary protein.

Second, eat the protein and fat source(s) (legumes (beans, peas, lentils, chick peas etc), nuts, seeds, meats, dairy products (milk, cheese, butter etc)), eggs, and even grains (wheat, oats, rye, barley, rice etc) and grain products such as breads, because although they have more carbohydrate than fat they have roughly 10% protein) (Some leafy greens can be eaten with these),
because,
the protein needs concentrated gastric juices, that will be allowed time to act properly in breaking it down, and fat mixed with carbohydrate before it enters the duodenum (small intestine) slows down the absorption of sugars into the bloodstream, creating a much more stable blood sugar curve, that will have a meal-increased higher blood fat curve alongside it. Also, protein lodging in any remaining spaces between or in the teeth has amino acid parts that buffer acidity, and bacteria take longer to cause a problem by acting on this material than by acting on carbohydrates.

Third, eat the carbohydrate (and/or acid) sources (fruits, and starchy root vegetables such as potatoes and sweet potatoes; carrots having only one half the starch of these),
because,
they are more easily digested and trigger release of the gastric contents into the duodenum (small intestine), and dilute the gastric juices (as fruit and root vegetables have a lot of water), decreasing their ability to break up dietary protein.
and,
this way the carbohydrate is released into the small intestine more slowly than it would be if it was eaten on its own, and it will also be absorbed from the small intestine more

gradually, leading to more stable blood sugar levels, that will be accompanied by beneficial blood fat levels (see also chapter 19 here on weight loss).

After eating the carbohydrate source, particularly if it has any substantial acidity of its own (tastes even slightly sour/tart) (citrus (limes, lemons, grapefruit, oranges, mandarins, lemonades), pineapples, most stonefruit (peaches, nectarines, plums), some grapes and apples, etc), rinse the mouth with water 1-3 times.

Last, finish with more green leaf/flower vegetables (lettuce, cabbage, broccoli, caulifl'r), because,
they are nature's toothbrush; fruits are not as they have substantial carbohydrate and, more often than not, some acidity.

Most important are:
To begin each meal with green leaf vegetables alone for a few well-chewed mouthfuls; thereafter some green leaf vegetables and some root vegetables raw and therefore slowly ingested due to the time taken to chew them properly may be added to the concentrated protein sources (likewise well chewed to maximize digestion).
To not add fresh fruit (and even dried fruit, because the sugars are a signal to the stomach that hastens the release of the stomach contents into the small intestine, possibly before the proteins have been digested as well as possible in the stomach) to the foregoing, rather to eat it after the foregoing, in proper slowness as noted above in order not to cause premature dilution of stomach digestive enzymes and premature passing of the stomach contents into the small intestine.
To end each meal with green leaf vegetables alone for a few well-chewed mouthfuls.

Immediately after each meal, rinse the mouth, brush the teeth, use a proxy brush (preferably) and/or, as suits, dental floss between all teeth, and rinse again to finish.

Use of a high-pressure 'water-pick' will further enhance cleaning between the teeth.

Of course, do not eat between meals, as this disrupts the effects outlined above.

The fiber in natural fruit and vegetables can prevent really thick deposits of almost pure carbohydrates (and of protein and fats/oils) on and in between teeth, by physically brushing some of it off, and by being in amongst the carbohydrate, in effect diluting it.

13.4 Physical exercise in dental health management

Physical exercise contributes to making a healthy physiological hunger, which helps natural foods taste as good as they can taste, and ought to taste. This will help one to enjoy eating natural wholefoods, which on average, and in any case if eaten in the right sequence (as noted above), are less damaging to tooth enamel than most unnatural foods.

A healthy physiological hunger probably enhances salivation (saliva production and release), which is good for tooth enamel health.

It is preferable in general for tooth enamel health to breathe through the nose, rather than through the mouth, because this keeps the mouth and especially exposed tooth surfaces moist, – this is less important for short intervals, such as a few minutes of mouth breathing, regularly, or for the occasional longer race or some-such, but in general try to breathe through the nose where possible, and to moisten the mouth with saliva in between mouth breathing.

Avoid becoming dehydrated during physical exercise, and otherwise, as the dry mouth experienced in association with this doubtless reduces saliva flow onto the tooth enamel surfaces, which reduces tooth enamel health and resistance to decay/ carries.

Chewing of natural foods including a substantial amount of sufficiently fibrous raw fruit and vegetables (grated if need be) can provide the right stimulus to keep/make the teeth solidly placed in their sockets – lack of sufficient chewing can lead to or contribute to the teeth becoming looser in their sockets.

Chewing enhances salivation (saliva release), which is good for tooth enamel health.

13.5 Pharmacy in dental health management.

Fluoride is a necessary mineral, in the right proportion of replacement of hydroxyl in tooth enamel hydroxyapatite, for tooth enamel hardness and resistance to decay. If your local water and soil (and therefore foods) and your imported foods together do not provide sufficient fluoride to augment your genetic and developmentally resultant tooth enamel strength, then for the best resistance to decay you will require some supplementary fluoride (not too much, because this causes the tooth enamel to become too hard, that is, to become brittle, and liable to breakage. And other bones too.). Fluoride painting from your dentist, fluoride mouthwashes (not to be swallowed, as they are somewhat, if cumulatively, poisonous in that sort of dose) and fluoride tablets, are all possible ways to do this.

Chapter 14. Ecology, ethics and animal welfare/ rights

By now in 2010 it seems pretty well agreed that it is better ecologically to consume plant protein sources directly ourselves rather than feed them to animals first and then eat the animals, because only a fraction of the farmland area is necessary that way, and less fertilizer and farm chemical waste pollutes the environment.

Until the stage is reached where all soils are analyzed and their mineral or related deficiencies are corrected by addition of minerals or other (organic, in the C, H, O, N, S sense) soil components, some amount of exchange of foods between geological regions so that individuals' diets may be drawn from them seems desirable.

Until we can get to the stage where we divide up the wealth/resources more or less equally according to need among all people on earth, it seems generally unreasonable (notwithstanding the derangement to individuals' farming that farming for export can sometimes have as a correlate) not to buy food from developing countries whose export economy is centered on food.

Although in many developing countries animal foods are seen as a desirable luxury that was previously unavailable, in some 'developed' countries there is an increasing consideration of the ethics of the unnecessary slaughter of animals for food. Vegetarianism has different motivating factors for different people, but a strong component overall is that vegetarians usually don't eat meat because they don't like people to kill or be cruel to animals, and they don't think eating meat is necessary for good health, or even beneficial to health.
As I think I demonstrate pretty conclusively in this book here, these latter are almost certainly true, other than the need for some small amount of vitamin B12, which can be obtained fairly ethically from free-range eggs or some dairy products, or brainless (and therefore painless) animals like oysters and most other mollusks, or nearly brainless animals like insects and crustaceans (which are very probably only barely aware that they are even alive, and can be (not always are) collected and quickly dispatched without them experiencing emotional stress. (And B12 supplements/ injections are readily available.)

In particular, the cage-farming of animals (particular examples being chickens farmed for eggs, and pigs), seems to many (relevantly qualified experts included), one of the cruelest practices of humankind.
I am still amazed and greatly dismayed if hopefully not rendered utterly hopeless of humans as a species ever learning to overcome their greatest defects, that people can be so mentally deficient as to attempt to justify keeping any animal with a brain and obvious capacity for emotion and need for locomotion, in a small cage for effectively all its life. In this case, and also pretty much for most humans asked about life in general, the quality of life before death matters much more than whether death comes earlier or later.

It is very probably not psychologically healthy for humans to knowingly participate in the cage-farming of animals, whether directly, or indirectly through the paying of others to do this dirtiest of all works by buying the products (cage-farmed eggs, pork etc). Surely psychological health requires us to identify the severest and most easily addressed of these practices and put a stop to our support of them, and then make improvements to our practice over time by ceasing support of the next severest and most easily addressed (it seems to me that cage-farmed eggs and pork are the best places to start – for everybody's sake, people should make a very good start by stopping buying them).

Also, it seems fairly obvious to me that if it's ok to treat animals in the way we do as societies, then it is entirely consistent with that treatment to treat other humans as badly as we can get away with. Those who would argue otherwise should read some very basic psychology about psychological rationalization, the highly prevalent human characteristic of using faulty reasoning in trying to justify our unjustifiable actions or thoughts.

Apart from cow milk being an inferior food for humans, there is also the matter of whether the immediate separation and slaughter of newborn calves, and the problems involved in the genetically oversized udders, are ethically acceptable.

Importantly, school tuckshops and hospitals (their cafeterias, other stores, and vending machines) should not sell unhealthy junk food, because it's unethical/immoral; because unhealthy eating habits acquired during childhood tend to carry on through adult life, and contribute greatly to disease and premature death; and because such sale tends to reinforce the stupidifying and infantilizing effect of junk food advertising on the intellect, as schools and hospitals are information authorities, and school is a pervasive influence in one's formative years, and hospitals are visited for significant health-related reasons.

Chapter 15. The psychology of diet optimization (reproduced from below)

No pain, no gain.
However, we each have different amounts of pain/discomfort in our lives from different sources. So it can be difficult to take on some more discomfort in the short term, even if we know there will be less discomfort/ more pleasure in the long-term.
Think about how to best handle the best amount of short-term pain/discomfort in order to best reduce the amount of long-term pain/discomfort in total (including dental and associated pain/discomfort).
Suitable (not negative/silly) rewards in the short-term may need to be used.

The habits of a lifetime are hard to break, but once broken, they are gone, and you reap the benefits/pleasure.

There is pleasure and satisfaction from breaking bad habits in itself – the feeling of successfully controlling oneself for the better, instead of being a slave to bad habits unfortunately acquired.

Don't punish yourself mentally for your bad habits – perhaps the most fundamental (not fundamentalistic…..) law of science is **"Every action causes an equal and opposite reaction, and every reaction requires an equal and opposite set of preceding causes."** Logically, that is to say that at least nearly everything about us is causally predetermined according to the law of cause and effect – we can't help the way we are. However this does not negate the fact that the subjective feeling of being in control and breaking a bad habit will make us feel more satisfied with that, nor does it negate that being very much a good thing for us, and for others, particularly if we can help ourselves and others to focus on the remedy for bad things rather than looking to place blame for them. Better to say **"I've/you've been a fool in this regard up till now, but what have been the causes of this, and how do I/you improve in this regard from now?"**, than to say **"I've/you've been a fool in this regard, and it's all my/your fault, and I/you deserve to feel bad about it and to be punished for it." Forgive yourself, and forgive others**, it's logical, even if giving unpleasant feedback to ourselves or others can be beneficial in the (very) short term and is natural to some extent.

Two steps forward, one step back, two steps forward; better than none forward at all. **The people who sell junk food** are not really interested in **improving your quality of life**; they're interested in **getting your money** and improving their bank accounts, while actually **reducing the quality of your life**. You can **derive pleasure from not being made a fool of** in this way.

The people who sell junk food use **'clever' psychology** to get you to buy their products, by making you **think and behave more foolishly** than you would otherwise. You can **derive pleasure from not being made a fool of** in this way.
The whole barrage of this sort of advertising has the effect of making the whole population think and behave more foolishly than they would otherwise – given the problems we collectively face on the planet, this is not a good thing, because we need all the brainpower possible.

"Hunger is the best sauce" is an old saying that is very true – when we are physiologically hungry, not just **eating out of boredom or seeking more pleasure**, the taste of food is greatly enhanced. Eat only when you are hungry. If you aren't hungry, don't eat, or don't eat much, or eat mostly salad vegetables.

Seek to identify those areas in your life where you are not gaining the pleasure/ satisfaction you could/should, and to identify those parts of your use of food which are negative counterproductive attempts to compensate for that lack of pleasure/ satisfaction. Then act to remedy those circumstances, or to find a more positive productive way of compensating.

There are many easily accessed ways of using time that are more pleasurable and satisfying than eating unhealthy junk food, or eating excessive amounts of otherwise fairly healthy food. These include quality reading, quality TV/movie viewing, a course of study (TAFE short/long/lifestyle/interest courses/ university), sketching, painting, sculpture, pottery, photography, learning/playing a musical instrument, walking, bicycling, humanitarian activity, etc, etc, etc……. **Don't be using your time so poorly that eating something becomes a way of filling in unoccupied time.**

If you're really more tired than hungry, your body probably really wants rest, and not food - try having the extra rest and not the food, and see if that makes you feel better (It almost certainly will…. You'll probably be quite surprised that going to bed early without any dinner is actually more of a reward than a punishment, in this case.).

When one eats food that has been **treated to make it much more tasty than natural food**, one's **taste bud receptor molecules become down-regulated** (reduced in number) from the excessive stimulation in the same way that a heroin addict's natural endorphin receptor molecules become down-regulated by the excessive stimulation that heroin causes in making its pleasurable effect. And in the same way, that due to that reduction in number of receptor molecules, natural life (without heroin) is unnaturally unpleasant to the heroin addict, and life with heroin eventually not much more pleasant than natural life is to non-addicts, **natural foods become unnaturally unpleasant to those whose taste buds are accustomed to food that has been treated to make it much more tasty than natural food, and the unnaturally treated food eventually tastes more ordinary**.

Given enough time not eating unnaturally tasty food, taste receptor molecules will become up-regulated (increase in number) again to their natural level, and **natural foods will come to be experienced as very nearly as tasty (that is, very tasty)** as the unnaturally tasty foods were when the taste receptors had been down-regulated.

The same (analogous) thing happens with regard to our psychological perception of the different pleasure associated with eating different foods.

Learn how best to say no to well-meaning friends – explain your decision, try to carefully, gently and gradually **raise them up in their lives, rather than letting them bring you down in your life**.

Chapter 16 Natural and synthetic food contaminants; organic food; parasites (reproduced from below)

Some important food (constituents / substances / molecules) other than the major nutrient groups are antioxidants, antinutrients, allergens, toxins, carcinogens, irritant biochemicals, pharmacologically active biochemicals.

Some important unhealthy contaminants of foods are: pesticides (herbicides, fungicides, bactericides, antibiotics, vermicides), farm hormones, funguses[i]/moulds, bacteria, parasites, parasite eggs, viruses, toxins, preservatives, artificial colors, artificial flavors, irritant condiments, pharmacologically active condiments, allergens, carcinogens, leached food container metals and dyes, motor vehicle exhaust substances, industrial pollutants, pollutant metals, and plastics molecules….

If cooked food is insisted on instead of raw food, food is generally better for health stewed, boiled, poached or steamed, because roasted, baked, broiled/grilled, fried and sautéed foods have unhealthy molecules in them, including: carcinogens, such as Polycyclic Aromatic Hydrocarbons (PAH), mutagenic heterocycles, benzene derivatives, and nitrites); causes of atherosclerosis such as oxidatively damaged unsaturated fat; Maillard Products (, which incorporate amino acids, notably lysine, that are thereby not absorbable by the gastrointestinal tract ('GIT')); acrylamides, as yet unknown activity...

It should be borne in mind that raw flesh often carries parasites that can bore into various parts of the body, such as the brain and liver.

Also, raw eggs may carry pathogens, and eating a lot of them can cause a biotin deficiency.

Plant foods that are acquired from near the ground (such as root vegetables and many green leaf, flower or shoot vegetables), such as to be near to animal feces should be washed properly, or peeled, if eaten raw, to remove parasites that can bore into various parts of the body, such as the brain and liver.

Organically grown food is probably better on average than food grown with chemicals, but soil that is deficient in nutrients should have those nutrients enriched. Atoms are atoms and molecules are molecules, regardless of how they are sourced (i.e., metals) or produced (molecules) – it is the surrounding context of other atoms and molecules that may be important, in comparing the different foods that contain them.

Apples and pears are said to be among the worst hit with pesticides and should be peeled.
Strawberries must be suspect if not grown in hothouses.
Peeled food on average is probably safer than unpeeled food.
Food grown in countries where the more dangerous farm chemicals are prohibited or where residues on/in produce are monitored well are generally safer than those grown where they are not prohibited. Exceptions to this generalization may arise as a result of the food being grown for export (more likely to be tested/monitored) or local consumption (less likely to be tested/monitored).

There is a lot more that could be written here on this subject, even though I have not yet been able to find good, comprehensive, detailed information), but that will have to wait for a later edition.

Chapter 17 Balanced meal and balanced diet; vegetarian diet and health; caloric restriction and longevity (reproduced from below)

It seems to me that, in general, on average:

A balanced meal includes a (preferably a mixture of) concentrated protein source(s) such as flesh, eggs, dairy products, nuts, seeds, legumes or grains; a fat source (which is often also a protein source); at least 200 g of green leaf or flower vegetables for an average-sized adult; and root vegetables and fruit (, in the right proportions (, that is, increasing down this list by weight, and such that there is enough root vegetables and fruit (roughly 1-1.5 kg) that the meal roughly meets water requirements (clear urine often enough))).

A balanced diet includes a variety of protein sources, fat sources, green leaf or flower vegetables, root vegetables, fruit, and a source of vitamin B12.
In more detail:
A balanced diet includes a variety of protein sources (such as flesh, eggs, dairy products, nuts, seeds, legumes or grains); a fat source (, which is often also a protein source); an average of at least 500 g / day of green leaf or flower vegetables; root vegetables and fruits; in the right proportions (, that is, increasing down this list by weight, and such that there is enough root vegetables and fruit (roughly 2-3 kg/ day) that the diet roughly meets the water requirements), and in the right schedule (, that is, in balanced meals). Including a source of vitamin B12 (e.g. flesh/insects, eggs, dairy products, vitamin supplement).

Is a vegetarian diet healthier than an omnivorous diet?
A vegetarian diet is not necessarily healthier than an omnivorous diet.
For personal health it is more important that the diet have protein, fat, carbohydrate, vitamins, minerals and antioxidants in the right proportions and combinations, and be not much damaged by any cooking process (see previous page) , than whether or not the diet is vegetarian or omnivorous.
That is to say that it is more important that the diet, and the meals, be composed of a smallish amount of protein and fat foods (such as legumes, nuts, seeds, grains, eggs, dairy products and meats), a largish amount of green leafy vegetables, and the largest amount of fresh fruits and root vegetables (with or without grains, by caloric content not by weight) as the major carbohydrate source.

Maybe the single most important dietary factor for long life is whether one eats and absorbs too many calories worth of food relative to the energy one requires.

The criteria that any member of the general public needs to understand and apply to their assessment of themselves is actually very basic and low-tech - If one cannot see some muscle and ribs under the external body fat, then one is too fat, unless one is expecting a period of unavoidable food shortage sometime soon, or in pregnancy, when a *little* extra body fat may be desirable.

Just very roughly, the difference to lifespan is likely to be at least 10 years in lifespan between good leanness and having 10 kg of excess fat, and more for having more excess fat.

Excessive leanness can be assessed on the basis of physical weakness, especially in walking or running, which are very basic natural things that all humans have evolved to do, and need to do, irrespective of our preferences for other sorts of physical exercise. And by hormonal derangements – in particular growth retardation, and retardation of menarche, or cessation of menstruation.

Chapter 18 Individual health vs. population health - deleterious mutations and the effect of medical treatments and nutritional supplements on their retention and prevalence rate in the gene pool; Individuality within species, populations (reproduced from below)

If a disease is not always fatal, then it is multifactorial in nature.

In infectious disease, for example, not only the inoculation load (the number of infectious disease viruses, bacteria or parasites that make up the incoming load of infectious organisms) but other factors also will contribute to whether the infected person lives or dies.

Some of these other factors will be environmental, for example, stress, nutrition, and chemicals.

Genes are other factors that contribute to the outcome of life or death for the person with the disease. And *many* genes will contribute to the outcome, some more strongly than others.
Deleterious mutations are continuously coming into the human gene pool in a fairly random fashion – all animals/organisms have a particular mutation rate related to the characteristics of their DNA and DNA repair mechanisms.
When disease kills a person before they have reproduced, on average, the death removes a genetic code that overall has more deleterious mutations than the average genetic code among people making up the population. It removes more of the genes more strongly involved in the disease course, but it also removes some of the genes involved less strongly in the course of the disease.
So the net amount of deleteriously mutated DNA in the population is determined by the balance between the rate that random mutations come into the gene pool, and the rate at which death (i.e., from disease) before reproduction removes deleterious mutations from the gene pool.

When any intervention, be it medical or nutritional, results in preventing disease from killing people before they are able to reproduce, at a population level the result is that the gene pool comes to include more deleterious genetic mutations.
The more such interventions are successful, and the longer they go on for, the more deleterious mutations come to be included in the gene pool.
If we go on generally intervening to prevent disease killing people before they are able to reproduce, without genetic engineering of humans, eventually the human genetic code will become so riddled with deleterious mutations that it will become a major and eventually uniformly fatal disease(s) in itself.
My personal preference, and I doubt cessation of intervention to prevent disease killing people before they are able to reproduce will become the popular preference, is for ongoing intervention plus genetic engineering of humans.

Now, different people have different geological and dietetic evolutionary backgrounds. We should not make too much of this, but it probably applies more to humans than it does to other animals.
As dealt with elsewhere in this book, some lines of humans had a high vitamin B12 intake in those (evolutionarily relatively short) periods of our history when we ate a lot of animal food and less plant food (including the colder portions of the most recent ice age). So, even though later on animal food consumption was significantly decreased back down from those evolutionarily uncommon highs, some lines of the human race will have retained through natural selection some deleterious mutations decreasing the ability of some few individuals to absorb, transport or/and store vitamin B12 (In the same way that

we totally lost the ability to synthesize vitamin C, which most other animals still have, because evolutionarily our diet contained such a high amount of vitamin C).
So, have the Swiss, or people of Swiss ethnic background, from living in the iodine-deficient mountain regions, on average, less iodine absorption-, transport- and metabolism-related deleterious mutations in their gene pool, because they were more severely selected against in natural selection than in populations elsewhere? Have they, from eating the dairy foods that genetically prolonged their lactase function into adulthood, more calcium absorption-, transport- and metabolism-related deleterious mutations in their gene pool, because they were less severely selected against in natural selection than in populations elsewhere?
What about the coastal peoples eating marine mollusks and crustaceans with their high load of nutrient minerals – would nutrient metal absorption-, transport- and metabolism-related 'deleterious' mutations in their gene pool, actually be protective from the excessive metal nutrient levels found in these foods?
Would people with an evolutionary background in the selenium deficient Keshan area of China (intake 3-22 ug Se / day) fare extra poorly if living in the selenium-richest Enshi area of China (intake 3,200-6,700 ug Se /day) or at the coast and eating a lot of marine mollusks and crustaceans, from selenium toxicity? And vice-versa?

It seems that many of these questions may soon be answerable through genetic epidemiology, and answerable at the individual level with genetic testing, such that people will be able to know what changes in their nutrient intake they might benefit from.

In the meantime, on the basis that deficiency is more likely to be a problem than toxicity from excess, I think that on average a vitamin and nutrient supplement of very roughly 1/2 of the RDAs might be a good idea for all people to consume, once a day or two, with a meal, to account for unknown gene differences and differences in produce nutrients due to geographic regions of origin, and that individuals should take their genetic evolutionary history into account where feasible in adjusting their intakes of particular nutrients.
However, the vitamin and nutrient supplement should not be seen as a replacement for healthy whole foods, because many studies using supplements of isolated nutrients have found that the benefits of the nutrients in that context have been nowhere near as great as the benefits associated with the whole food sources of the same amount of the nutrient.

Chapter 19 Eating for weight (surplus body fat) loss (reproduced from below)

In nature a meal may not always contain significant fat or much carbohydrate, but the optimal meal may well contain a certain proportion of protein:fat:carbohydrate.
It seems to me that the best proportion is roughly: only enough protein to meet our protein requirements; the amount of fat that comes with the meeting of those requirements from protein sources that include a good proportion of oily nuts and seeds alongside whatever legumes and cereal grains may also contribute to the protein source mix; and the balance of our caloric (energy) requirements from (raw) starchy root vegetables and fruits (which would therefore meet also roughly all our water requirements).

Such a mixture provides a slower uptake of dietary carbohydrate from the digestive tract, and some accompanying fat, and one effect of this is a feeling of well-being due to that steady drawn-out delivery of fuel to the body. Also, because fat metabolism is activated, when the carbohydrate supplied by the meal is no longer available for ready fuel use, fuel use can switch to depot fat (in our fat stores) more smoothly and with less sense of lethargy and hunger or craving of food than would occur if the meal had had no fat.

It is important to understand that my statements regarding this feeling of wellbeing and its proposed reasons are not negated by reports of a feeling of wellbeing reported by people on the Atkins high protein diet.
The feeling of wellbeing reported by people on the Atkins high protein diet is due to their blood sugar curves previous to the Atkins diet having been wildly fluctuating through (refined or high Glycemic Index) carbohydrate being ingested without being accompanied by any fat to slow down its absorption into the bloodstream, or to prime fat metabolism to make access of depot fat for use as energy easier for the body.
Refined or high Glycemic Index carbohydrate being ingested without being accompanied by any fat to slow down its absorption into the bloodstream, in the context of a low-fat diet, causes initially rearing blood glucose levels, causing rearing insulin levels, which cause subsequent plunging blood glucose levels, in the presence of low blood fat levels, leaving regular periods of time when both blood glucose and blood fat and fat metabolism are low, causing fatigue and hunger, the latter causing overeating in an attempt to regain some short-term blood energy fuel.
Also, the Atkins diet gives a slow and relatively constant feed of fat into the bloodstream that blunts the hunger that occurs when (refined or high Glycemic Index) carbohydrate is ingested without being accompanied by any fat to slow down its absorption into the bloodstream.
One big problem with the Atkins Diet is that in the long term, atherosclerosis is very likely to occur from the saturated fat, cholesterol, and oxidatively damaged unsaturated fat in which it is high.

The one valuable piece of information that can be derived from the Atkins Diet experience is that having a *totally* fat free diet is *not* the best way to shed excess body fat, because fat metabolism is reduced with every new load of dietary carbohydrate, and when the carbohydrate is used up the blood will be low in not only sugar but also fat, such that before depot fat is accessed for energy, enough time goes by with the person in a state of both low blood sugar and low blood fat/ fat metabolism that they feel substantially weak, irritable and hungry.

So my suggestion for diet modification for the loss of excess body fat is to reduce the dietary carbohydrate, and to keep the protein and fat content the same as I recommend for the general diet (per day): 100-200 grams combined weight of nuts, seeds, cereal grains and legumes (with or without animal foods: flesh, eggs, dairy products), at least 500 g of green leaf, flower or shoot vegetables, and enough fruit and starchy root vegetables to not quite make the total energy/calorie intake equal to the total energy/calorie use.
In general, the food should be raw wholefoods.

The foods in a meal should be eaten generally in the order described in chapter 12 here, to optimize the delivery of sugar and fat from the digestive system into the bloodstream, to optimize dental health, and to optimize digestion of proteins:
Some green leaf, flower or shoot vegetables first for dental health, high protein and fat foods next (allowably interspersed with more green leaf, flower or shoot vegetables) for optimal digestion by concentrated digestive juices over a sufficient time, then starchy root vegetables and fruit, and a little more green leaf, flower or shoot vegetables last to knock the residue from the other foods off the teeth for dental health (followed by rinsing, brushing and picking the teeth after every meal, of course).

In general, for adults, there should be only two meals per day, brunch as close to noon as feasible and tolerable, and dinner in the evening after some session of physical exercise. Three meals otherwise.
Nothing but water (if necessary) in between meals.

Chapter 20 Homocysteine, vitamin B12, and folate (reproduced from below)

Homocysteine at levels of 200-500 uM in a few severe genetic metabolic diseases is associated with premature atherosclerosis, and likely to be a major part of the cause of it in patients with these diseases, due to interference with disulfide bonding in (particularly the formation of) structural proteins in tissues, themselves already (in most cases, the main disease type) compromised by the reduced cysteine levels caused by the block in metabolism of homocysteine to cysteine, one of three available metabolic pathways for converting ostensibly useless homocysteine back into something useful (cysteine or methionine).

Various authorities have posited that with a population normal range given as 8-12 uM, levels of 12-20 uM, which are associated with atherosclerosis, are a contributory cause of that atherosclerosis. Other authorities, myself included, think it more likely that the homocysteine at this level is only a marker for other things that are the real causes of this atherosclerosis – lack of folate and fruit & vegetables – and disease process components.

Folate, found in higher levels in plant foods, and vitamin B12, found virtually only in animal foods, are equally involved in reducing homocysteine levels through a particular metabolic pathway. Vitamin B6 is involved in another pathway, but it is plentifully present in both plant and animal foods.
Exclusive vegans with no vitamin B12 supplementation may have homocysteine levels as high as 100uM.
There have been no reports in the scientific journal literature of vegans and vegetarians having more atherosclerosis, usually, in fact, the opposite, of them having less atherosclerosis.
Studies to date have incompetently failed to assess whether a case of raised homocysteine level was due to low vitamin B12, from eating less animal food, or due to low folate, from eating less plant food. Given the large amount of reference there has been in the mass media to animal fat and fruit and vegetables regards atherosclerosis, I will not insult the reader's intelligence by spelling out why it would be crucial to determine whether raised homocysteine levels in atherosclerosis cases were due to low vitamin B12 or due to low folate. I will however mention that animal foods are themselves a marker for greater oxidative cooking damage to foods (i.e., their unsaturated fat) and plant foods are themselves a marker for less oxidative cooking damage to foods.

My advice to vegetarians is that if your homocysteine levels (say 12 – 20 or 30 uM) are due to low vitamin B12 levels, and your diet is composed mostly of raw or not very heat-damaged plant foods, that level of homocysteine is very unlikely to contribute anything other than maybe a tiny, tiny, utterly insignificant part of any atherosclerosis that you may develop (or have had already). I recommend not worrying about it at all.
If such levels were due to low consumption of plant foods and therefore low folate, then worry about atherosclerosis is entirely appropriate and dietary change (*not* merely folate supplementation) is very much called for.
Exclusive vegans with higher levels of homocysteine, up to 100 uM are very probably much more at risk of damage to their nervous system, and maybe of getting some cancers related to global DNA hypomethylation, than they are of serious atherosclerosis, but given some other bad factors (genetic or environmental) atherosclerosis could become problematic despite the high levels of protective factors in diets with a lot of raw fruit and vegetables.
Unless a person wants to participate in a somewhat risky personal experiment and contribute some more data, I recommend that they should have enough vitamin B12 supplementation to bring blood levels into the low 'normal' range, and to bring their homocysteine levels roughly under 20 uM.

Chapter 21 Afterthoughts: The China Study; Ornish's and Esselstyn's heart disease reversal diets: Inflammation Index and other potential refinements of diet

The China Study (Campbell & Campbell, 2006, Benbella Books) warrants a much more detailed treatment than that I provide here, which will be forthcoming. In very brief, a major epidemiological study comparing many different regions of China (constituting a 'natural experiment' due to the little immigration into and emigration out of the different regions, with their different diets) found that even at lower levels of animal-food consumption, lower consumption of animal-food was associated with lower heart disease and cancer. There is doubtless much of great value in this study. However, it does not consider the oxidative damage to unsaturated fat that various methods of cooking cause. (The result is that the C=C double bonds open up to include oxygen atoms in epoxides and peroxides, which are the same type of chemicals that macrophages cause when they lyse things with the 'oxidative burst' of hydrogen peroxide that is the mainstay of their activity, these chemicals then attracting other macrophages to the site, which is a main part of the process of the generation of atherosclerosis (arterial hardening and plaqueing)) Nor does it consider that to some extent animal products are actually a marker for more oxidatively damaging cooking methods (such as frying, roasting, grilling, sautéing).

Both Ornish's (1990, Dr Dean Ornish's Program for Reversing Heart Disease, Ballantine Books) and Esselstyn's (2007, Prevent and Reverse Heart Disease, Penguin Books) warrant a more detailed treatment than that I provide here, which will be forthcoming. While they offer much of value in improving on the traditional western diet, unfortunately they too do not consider the oxidative damage to unsaturated fat that various methods of cooking cause, and that to some extent animal products are actually a marker for more oxidatively damaging cooking methods. Consequently they have un unreasonably severe and inadequately differentiated negative view of *all* dietary fats.

I only make a bare mention of an interesting development/ possibility currently part of the SelfNutritionData website: the Inflammation Index, derived from an algorithm I have not acquired the details of, which incorporates various food-related factors to produce an overall number quantifying the potential contribution of the food to the body's general state/level of inflammation. My general recommendations for diet would give a very favourable Inflammation Index... Something for us all to find out some more about...

Marine versus land polyunsaturated fats are another issue, and, other than having read some studies that did not convince me due to methodological shortfalls/ publication bias (studies finding some difference in their results get published more often than studies finding no difference in their results, at least for smaller, investigatory studies), and I will say no more for now other than that we are a land animal, not an aquatic or marine animal. And that the question of the suitability of a food for humans must be resolved through accounting for multiple factors in the particular food, in the meal of which it is to be a part, and in the diet of which it is to be a part – which is far and away most often not at all done... Something for us all to find out some more about...

Would eating only the minimum necessary amount of protein, and ensuring it was as fully digested as possible by chewing it fully (and mechanically grinding/pasting it before eating if necessary), and by eating it near the beginning of a meal, and eating the meal slowly, result in less polypeptides able to affect the function of the immune system getting through the always somewhat permeable intestinal walls into the bloodstream, and might this be to the benefit of health conditions involving overactivation of the immune system? (some inflammatory diseases respond quite well to fasting...)

END OF WHOLE BOOK SUMMARY

Chapter 1. Evolutionary Origins, Directions, Milestones and Timeframes:

1.1 Summary of Chapter 1, by chapter sections, and conclusions p66

1.2 Human and ape genetic and evolutionary relationship; ape-like human origin p88

1.3 Earlier insectivore origin, persistence of insectivory, vitamin B12 p89

1.4 Colour vision, sweetness preference and fruit p90

1.5 Archaeological remains biases p90

1.6 Archaeological interpretation and attribution biases due to lack of appropriate quantification and epidemiological consideration p90

1.7 Archeological site discovery biases p95

1.8 Archaeological burial soil contamination of remains (diagenesis), Sr/Ca; Ba/Ca; $^{13}C/^{12}C$ p96

1.9 Some archaeological sites, their evidence, and interpretations p98

1.10 Feet, bipedalism, arboreality p134

1.11 Teeth, the supra-pharyngeal vocal tract and language p139

1.12 The gastrointestinal/ digestive tract p159

1.13 Tooth isotopic ^{13}C from C_4 plant origins p160

1.14 Cooking p164

1.15 Energetic models p176

1.16 The brain and intelligence p179

1.17 Cracking and grinding tools, digging tools p183

1.18 Defense needs and weapons p184

1.19 Scavenging, power scavenging, and hunting p184

1.20 Proposed implications of studies of early hominin diets: Cordain; Eaton; Ungar; in Ungar (2007) Eaton, Shostak and Konner (1988) (includes sodium:potassium ratios (p192) and acid/base contributions (p191) of foods) p185

References for chapter 1 p207

1.1 Summary of chapter one, and conclusions
(by section, followed by conclusions)

[1.2 Human and ape genetic and evolutionary relationship; ape-like human origin (summary)

Genetically we remain very similar to the anthropoid apes that are our closest relatives: chimpanzees, bonobos, gorillas, orangutans and gibbons - DNA studies show approximately 97 % exact similarity in DNA between humans and apes.

It seems reasonable to think that the diet of our distant ancestors in the times when our line diverged from the apes (6 million years ago) was more like that of present-day apes than like that of present-day humans. This is because humans have an essentially unnaturally radical course of behavioural evolution in general, and more radical in later periods than in earlier ones.]

[1.3 Earlier insectivore origin, persistence of insectivory, vitamin B12 (summary)

Many more millions of years before the evolutionary divergence of humans and apes, all of us, and the monkeys, came from evolutionary origins in primates that were very strongly insectivorous – all present-day apes and monkeys, in fact all or very nearly all primates, continue to have a substantial, regular, daily deliberate or incidental (in plant foods) ingestion of insects and their larvae (or pupae) and eggs.

This has a particular relevance for vitamin B12, which is generally taken to be available to humans only (other than from supplemental vitamins) from foods of animal origin.

It is possible that a vegan diet (having only foods of plant origin) that contained a sufficient quantity of green leafy vegetables (the best plant food source of cobalt, the metal atom in the vitamin B12 molecule) that had been grown in soil that was not deficient in cobalt might provide the bacteria in a human large intestine, if the bacterial population was of a healthy species composition and population level and in a conducive environment, with enough cobalt to manufacture some vitamin B12; and if the contents of the big intestine were sufficiently continuously moist enough due to the intake of moisture in the diet being optimal in regard to being part of the vegetables and fruit consumed, then a useful if potentially insufficient quantity of vitamin B12 might flow down its concentration gradient to pass through the ileocaecal valve back to the last part of the small intestine where vitamin B12 is taken up.

For some few or very few humans, particularly among those from some population evolutionary course, who have received deleterious mutations in their DNA coding for vitamin B12 uptake, transport and storage proteins, because these deleterious mutations were allowed to remain in the genetic code due to there having been for long and continuously and sufficiently recently enough in their bloodlines' recent evolutionary past a relatively high consumption of animal foods, this amount would probably be too small to prevent damage to the nervous system from vitamin B12 deficiency.

Even if such deleterious mutation-carrying humans are in the minority, and probably a very small minority in many populations, it would not be ethical to subject an infant to the risk of permanent brain damage through a purely vegan diet – that is a risk for adults to undertake, for whatever reasons of esthetics or scientific curiosity, properly equipped with the knowledge and understanding of their risk, by their own informed choice.

Some seaweed makes biologically active vitamin B12, but more often seaweed makes a vitamin B12-like molecule that is not biologically active and is worse than useless because it occupies molecular vitamin B12 sites in the human body (notwithstanding some physiological mechanisms for screening it out existing, reportedly) thereby preventing what biologically active vitamin B12 is present from binding to those sites and performing its biochemical function.]

[1.4 Colour vision, sweetness preference and fruit (duplication)

Although many animals have a more complex colour vision system than humans, for example many invertebrates (i.e., butterflies, stomatopods (i.e., shrimps with a dozen or more different colour vision receptor types)), tropical fish and birds, and reptiles and amphibians, in the evolution of mammals, segmenta of colour vision were lost, and then for a few species of primates, regained by gene duplication. Placental mammals other than those primates (i.e., cats, dogs and most farmed animals) have dichromatic vision with blue and yellow, and therefore green, but not red vision. Various authors have suggested that this evolutionary change was related to a dietary shift to greater consumption of fruits and young leaves with a red colour.]

[1.5 Archaeological remains biases (summary)

The archaeological record is highly biased; some materials (such as stone and bone) preserve very well and are routinely found during excavation, while others (organic material such as plant material (i.e., food remains and digging sticks) and non-bone body parts) either rot entirely away or leave only such subtle traces as are missed.
Much dietary theorizing, in particular that done without properly considering not only what biased remains there are (as well as how they could or are likely to be biased), but also the other relevant facts known from fields of knowledge such as nutrition, physiology, biochemistry, chemistry, genetics, epidemiology, primatology, and anthropology, produces, accordingly, very faulty theories of what the diet(s) of our hominin ancestors were.
In particular, theories grossly overestimating the proportion of animal food, in particular the meat of hunted animals, and underestimating the proportion of plant foods such as fruits, leaves, legumes, nuts, oily seeds, cereal grains and tubers and corms.]

[1.6 Archaeological interpretation and attribution biases due to lack of appropriate quantification and epidemiological consideration (summary)

Archaeological interpretation and attribution biases due to lack of appropriate quantification and epidemiological consideration are very common, perhaps even usual.

Almost invariably, in studies of animal bones left in concentrated deposits, with stone tool cutmarks on some but not all bones, by our hominin ancestors, there is no attempt made to perform what calculations are able to be performed to estimate what the amount of meat thus gained per hominin per unit of time might be, and so to estimate what proportion of the diet that meat might have been. Almost invariably, large deposits of bones such as those at the Solutre site (see section 1.6 below), are taken to indicate a most ferocious level of carnivory – but when the period of time, thousands or tens of thousands of years, and the number of humans involved, some smallish group, are taken into consideration such as to be used in calculating an estimate of what the amount of meat thus gained per hominin per unit of time might be, such calculation indicates that the meat gained through such serial slaughtering was very likely to have been only a very small proportion of the diet, or ~ nil if obtaining skins for warm clothing and tools was the main purpose of the slaughter, which plausibly roughly fits quantitatively.

Regarding the supposed larger size of preagricultural humans compared to agricultural humans, attributed for example to the diets, commonly in what literature I have read, the theorists fail to consider the possibility that in general, organized warfare between specialized societies led to preferential selection of larger and stronger men for the armies, who would then frequently kill off large proportions of each other, such that this could quite easily reduce the population gene pool of genes for largeness, strength and robustness, which is exactly the opposite of what would happen in combat between random individuals fighting each other, or in small groups where all males took part in fighting – where the smaller, weaker and less robust are more likely to die, increasing the population gene pool of genes for largeness, strength and robustness.]

[1.7 Archeological site discovery biases (summary)

Various habitats are not equally conducive to fossilization – factors that affect the ability of fossils to survive the forces of decomposition include rainfall, temperature, humidity, type of vegetative cover, amount of biological activity in the topsoil, and the speed of the covering of remains. For example, forest floors are said to be generally very poor places for fossilization.

Now, we may accept that Africa was the place of origin of humans, on the basis of relatively recent DNA analyses showing more genetic diversity among African hominins than elsewhere on earth (reference required).
However, if the preferred foods (such as fruits) of hominids in Africa were in forests, and the savannas were a fallback place with fallback foods for seasons of shortage of the forest foods, or places that competitively or otherwise dispossessed or dispersed hominids lived out of necessity or habituation, then the assemblages (grouping of all sorts of remains: bones, what little if any plant remains, buildings, excavations, tools and footsteps) of fossil remains found will be biased so as to be not properly representative of what the average nor the variety of prehistoric diets were. In other words the sites found and studied may represent the fringes of the hominid environments rather than the main hominid environments.

Also, as modern gatherer/hunters, and baboons and chimpanzees, exploit a number of habitats during a yearly (i.e., seasonal) round, early hominins may likewise have seasonally lived in various habitats, where they ate different diets.

Also, note that very often it is the initial finding of collections of animal bones and manufactured stone tools, not the finding of hominid bone remains, that is more likely to be the trigger for archaeologists to excavate a site – this again leads to a bias to finding evidence of carnivory rather than of plant eating.]

[1.8 Archaeological burial soil contamination of remains (diagenesis), Sr/Ca; Ba/Ca; $^{13}C/^{12}C$ (summary)

Archaeological burial involves long periods of time, and soil contamination of fossil remains (diagenesis) has been found to be significant/substantial for all chemical constituents commonly assayed (i.e., Sr/Ca; Ba/Ca; $^{13}C/^{12}C$).

Even with the scrupulous application of commonly accepted cleaning methods, comparisons between bone samples may only be done if the relevant benchmarking or comparator samples are available for assay and incorporation into the analysis. And even then, as diagenesis can vary from part to part of the one fossil specimen, statistical significance must be sought in the analysis before proceeding to drawing conclusions. The ramifications of diagenesis are complex enough to be difficult to fully detail in brief so as to be understandable to readers other than the small number of professionals in the area, and so I recommend that readers of this book read section 1.7 and 1.8 in order to gain an understanding of what needs to be considered in deciding whether to accept as plausible or not the conclusions drawn by authors using such assay comparisons as the bases for them.]

[1.9 Some archaeological sites, their evidence, and interpretations (summary)

There have been enough archaeological sites examined, and the interpretation of the evidence they provide is reasonable enough for the purpose (if not for the quantification of the meat eating as a proportion of diet), to be able to state as a matter of virtual fact that our hominin ancestors did eat some meat at some times.
And, that at some evolutionarily very recent times (for example within the last 40,000 years), probably under climatic and population stress, in a substantial number of areas, meat did constitute a substantial proportion of their diet.

However, the theorizing on the meaning of the findings at the various sites is plagued with defects of the sort I have mentioned above – readers may read through my dealings with a substantial number of various studies and their author(s)' theorizing, for more detailed examples, in section 1.6 and 1.9 below, on which I have critically and perforce all too often destructively commented.

A few more enlightened commentators have noted the strong tendency for theorists to interpret the data (incomplete as I note that the limited subset of it available that they have chosen to include in their analyses and theorizings is…) from a position strongly biased by their own cultural and personal dietetic positions.

Some have stood back and considered the larger picture of the various possible environments that our hominin ancestors would be likely to have lived in, and the importance of realizing how their diets may have differed in those places and in various climatic periods. Also, the importance of understanding which type of areas were likely to have been the more suitable, hominin population-producing (more reproduction than death) ones, and which areas were likely to be the less suitable, population-absorbing (more death than reproduction) ones.]

[1.10 Feet, bipedalism, and arboreality (summary)

It has been said that *Australopithecus afarensis*, of approximately 4 million years ago, probably of our ancestral line, had a pelvis and lower limbs suggesting that it was capable of walking bipedally but probably only for short distances – I take the point of view that A. afarensis having feet much as our own, was capable of walking for *more* than short distances, as chimpanzees with their feet much less well adapted for walking (they still have fully opposable big toes, like hands do) walk for approximately all of their travel, climbing a particular tree on arrival, and human-type feet are obviously *much* better adapted for walking than ape feet.

Some of the 3% DNA difference between humans and chimpanzees is in the feet (humans not having opposable big toes and therefore prehensile feet) – note that humans still climb trees *quite* well, just not *as* well as our nearest animal relatives do. Certainly modern humans and all of our hominin ancestors climbed trees well enough to be harvesting and eating fruit from them as a daily staple.

We humans are better evolved to travel longer distances between trees because we can cover open ground better than the apes, and our use of weapons to defend ourselves (strangely overlooked, as against for killing things, as a possibility, by many archeological and anthropological theorists) offers us a defense while on the ground that the other apes don't have, and that the gorillas don't need due to their larger size. It may well be that the evolution of the use of defensive weapons enabled the evolution of the human feet.

Modern humans need remarkably little muscle activity to stand in a relaxed posture, and our evolutionary ancestors would have been able to be erect for long enough at a time to carry out the visual vigilance against predators required for survival, even when not walking.

Activities possibly involved in the inauguration and development of bipedal posture and locomotion in hominins include:
carrying infants, food and tools (including weapons);
throwing of tools (including weapons) and other missiles;
threat displays;
aggression;
evasion;
vigilance and sentinel behaviors;
sexual display and nuptial gifts;
arboreal gathering of food;
terrestrial gathering of food; aquatic gathering of food;
terrestrial predation (i.e. hunting);
terrestrial scavenging (of carcasses);
efficiency in long-distance travel/foraging;
thermoregulation during high activity in hot environments;
walking on snow or ice;
iodine deficiency (reduced metabolism and activity capacity?);
none at all due to bipedalism being the primitive mode of locomotion from which all quadrupeds evolved (seems very unlikely?)

Almost certainly our early hominin ancestors would have been able to carry clubs, babies, some (i.e. missile and hammer) useful stones, and some food, maybe in bags made of hide. Modern chimpanzees are reported to do all of these except the last.

Consideration of nut gathering and cracking with hammer and anvil stones, grain gathering and grinding with useable natural stones, and root vegetable finding, digging up, transport, and pounding/grinding, as well as fruit tree location, access and climbing, is necessary for sensible theorization about the evolution of bipedalism in hominins.

Consideration of highly proficient weapon use, which requires fast, precise, stable movements, and driving force from the feet on the ground, through the body, to the weapon, as appropriate, is essential in the question of the evolutionary drives for bipedalism.]

[1.11 Teeth, the supra-pharyngeal vocal tract and language (summary)

The teeth of the apes are generally similar to those of our own, and are presumably adapted to their predominantly frugivorous very predominantly vegetarian diets. Human teeth, and those of even our earliest hominin ancestors to a very substantial extent, differ from ape teeth in the following ways:

The canine teeth of all hominins are/were no longer elongated – this suggests not only that they were no longer needed for the purpose of threat or defense displays such as remain in the apes, but also that they were no longer needed for (/ nor became useful for) the purposes of killing prey and tearing its flesh into lumps of an ingestible size, a lesser behavior of chimpanzees that is virtually absent in gorillas, orangutans and bonobos, and that is the obvious dominant feeding behaviour of the natural carnivores.

The premolar teeth are molarized (more like the molars), and the molars are flatter – this suggests an adaptation to more chewing of small seeds like cereal grains (and the enamel prism attitude of our australopith ancestors is reported as being suited for roll-crush and mastication, which is consistent with eating smaller seeds and grains), and of other things that require more crushing than shearing, perhaps tubers.

The jawbone has been moved to the rear, which distributes a stronger chewing force over the premolars while retaining the strong force over the molars, chewing musculature being equal (chewing musculature is less in modern man, but probably was equal in our earlier hominin ancestors) – this is concordant with more chewing of small seeds like cereal grains, and of tubers.

The tooth enamel is thicker than that of the thickest-enameled apes, the orangutans (who are reported to crush more hard seeds and nuts with their teeth than the other apes), and much thicker than that of the chimpanzees and gorillas – this is concordant with crushing hard things, as well as grinding things that grow close to or in the ground and therefore have more abrasive dirt particles on them, such as cereal grains, seeds of small herbaceous plants, and tubers.

The teeth of modern humans are smaller, though the enamel is still close to as thick as it has ever been in our hominin ancestors. The jawbone is smaller, but the rearward movement of the jaw has retained good force delivery to the molars and premolars. The jaw musculature is substantially reduced – suggesting either or both of:
The use of food processing by the use of stone and other tools (cracking nuts with hammer and anvil stones, and grinding seeds and tubers, and quite possibly soaking foods in water), which I think is far more likely to be the large majority of the cause of this evolutionary change, as these tools and methods were very probably used by even our earliest hominin ancestors, as modern chimpanzees are reported to still do (other than the soaking); and
Cooking, which also makes food easier to chew, I think is far less likely to be the large majority of the cause of these evolutionary changes, as the ability to make fire on demand for daily use was very probably generally possessed only by our very recent hominin ancestors (maybe 200,000 years ago) and not by our hominin ancestors of the preceding 6 million years.

Regards the final changes in the human mouth and dental structures, it needs to be borne in mind that in more modern humans, in a process only finalized sometime in the last 100,000 years, maybe as recently as 40,000 years ago, the roof of the mouth and the lower jaw have been reduced compared with non-human primates and archaic hominins. The reduced chewing ability is offset by the increased phonetic range of the supralaryngeal airway, such that the modern human vocal tract can seal off the air pathway that leads through the mouth from the nose, so yielding non-nasal sounds, resulting in clearer, more readily identified sounds.

The comparative anatomy of living primates and of hominid fossils suggests that the evolution of the human supralaryngeal vocal tract probably started in early (about 1.6 million years ago) African populations of *Homo erectus* and was not completed until the appearance of fully modern humans.]

[1.12 The gastrointestinal/ digestive tract (summary)

It seems generally agreed that the modern human digestive tract has less large intestine, and more small intestine, than that of modern apes – quantification of these differences is generally absent from literature mentioning it, but from examination of pictures, these differences must be relatively small, and the digestive tract of humans is *much* more like that of chimpanzees than those of the more natural carnivores like the cats and dogs.
A related issue illustrating the great difference between the human digestive tract and those of the more naturally highly carnivorous animals is that although gut passage rates are similar between humans and dogs, dogs retain food in their stomachs for much longer (around 4 to 12 hours) than humans (around 1 to 2 hours).

Also, the digestive juices of the human stomach seem to be agreed to be clearly not as acidic (roughly only one tenth as acidic - keep in mind the pH scale is logarithmic) as those of the carnivores i.e., the cats and dogs (which tend to bolt their meat with little chewing – the teeth of the cats and dogs are not well adapted for chewing, but are well adapted for tearing).

It seems that modern humans are less well adapted for digesting cellulose than the apes – it should be borne in mind, however, that the last third of the modern human large intestine favors for its energy supply the short chain fatty acids that result from the breakdown of cellulose, in contradistinction to all other human body tissues, which use glucose, fatty acids or ketones for their energy supply.

Claims that the evolutionary changes leading to the slight reduction in the large intestine and slight increase in the small intestine (that plausibly took place since the human line diverged from the chimpanzee line) were based on a large intake of meat in our ancestral diet are generally quite defective in that they fail to consider that the oily nuts and seeds are actually clearly *more* energy-dense than meat is (as meat is approximately 70% water), undried cereal grains roughly as energy-dense dense as meat (so obviously raw nuts (600 kcal / 100 g) (or oily seeds (600 kcal / 100 g), and also cereal grains (380 kcal / 100 g) for that matter…) are more energy dense than raw meat (190 kcal / 100 g, lean; 230 kcal / 100 g, regular)), and starchy tubers energy-dense enough to supply human energy needs without recourse to much cellulose digestion. And that these foods are fairly likely to have been used by our ancestors for the very large majority of the 6 million years since our line diverged from that of the chimpanzees. And that the taste of raw nuts, seeds and tubers (as well as fruits) was probably preferred to the taste of raw meat by our evolutionary ancestors.]

[1.13 Tooth isotopic ^{13}C from C_4 plant origins (summary)

In tropical savannas, trees, bushes, and forbs [broad-leafed non-grass herbaceous plants especially of prairie or meadow] use the C_3 photosynthetic pathway, while tropical grasses and some sedges use the C_4 pathway. ……
Because of both anatomical and biochemical differences, C_3 plants discriminate more heavily against ^{13}C during photosynthesis than do C_4 plants.
As a result C_3 plants have highly depleted $^{13}C/^{12}C$ ratios, which are expressed as delta values in parts per thousand relative to a Cretaceous fossil cephalopod which is highly enriched in ^{13}C, resulting in most terrestrial plants and animals having negative carbon isotope ratios (more correctly, negative delta ^{13}C values).
C_4 plants, on the other hand, are much less depleted in ^{13}C (Smith and Epstein, 1971; Vogel, Fuls and Ellis, 1978).
(C_3 forbs, minus 27; C_3 trees, minus 27; C_4 grasses minus 12 (delta^{13}C values), with virtually no overlap at all of statistical distributions). ……

These stable carbon isotopes in plants are passed down into the tissues of animals that eat them, and on into the tissues of animals that eat those animals.

The carbon isotopes in bone and tooth enamel reliably provide dietary information in modern animals, but only tooth enamel levels have proved reliable with fossils, as chemical exchange with the surrounding soil alters bone levels substantially and variably.

Thirty-seven australopith specimens (dental specimens, judging from the reference titles) from Swartkrans, Kromdraai, Sterkfontein, and Makapansgat have been analysed, the data for which are:

60 x C_4 feeders, delta ^{13}C = minus 1;
19 x *Australopithecus*, delta ^{13}C = minus 7;
18 x *Paranthropus*, delta ^{13}C = minus 8;
61 x C_3 feeders, delta ^{13}C = minus 11

The data also show that both *Australopithecus africanus* and *Paranthropus robustus* are highly distinct from the C_3 browsers and the C_4 grazers with which they are associated (Scheffe test, $p < .00001$), although they are not different from one another.
This indicates that both australopiths consumed foods enriched in ^{13}C, such as C_4 grasses, C_4 sedges, or animals that ate these foods. ……

However, the australopith delta ^{13}C is highly variable, far more so than most modern and fossil species in South Africa. Two or three of the 37 australopiths analyzed had nearly pure C_3 diets (lowest ^{13}C enrichment), while several were extremely enriched in ^{13}C, with delta ^{13}C values within the range of the grass-eating baboon *Theropithecus oswaldi*.

Modern chimpanzees, living in either heavily forested environments where few if any C_4 resources are available, or in semiarid woodland environments where C_4 foods are extremely abundant, do not usually eat detectable quantities of C_4 foods.

Because many australopiths probably inhabited environments similar to those of today's woodland chimpanzees (though we might not have remains from such environments), this suggests that different dietary adaptations, rather than environmental differences, might be responsible for the difference in their consumption of ^{13}C-enriched foods.

Baboons (*Papio ursinus*) living in marginal environments may eat up to 50% ^{13}C-enriched foods such as C_4 grasses and CAM (a third, but minor, type of photosynthetic pathway) plants, while those in preferred richer woodland environments eat only about 10% C_4 vegetation. Likewise, some baboon fossils from the period of the australopith(ecine)s. Indicating not necessarily consumption of the same food items by our early hominin ancestors and modern/early baboons (though I think this is fairly probable), but certainly some dietary flexibility and willingness to use C_4 resources.

Baboons are known to eat grass-eating grass-hoppers (*Acrididae*) almost exclusively during temporary gluts, and these may be feeding on C_4 plants.

Regarding our early hominin ancestors, grass-eating termites are another dietary possibility, particularly given the possibility that australopithecine bone tools were used to extract termites from mounds, as are larvae of other insects, in contributing to ^{13}C-enrichment via feeding on C_4 plants – and many present-day gatherer-hunters eat larvae and/or insects.

It is impossible to use stable carbon isotopes to distinguish between diets as different as frugivory, folivory, and carnivory in forest ecosystems, because all are ultimately based on C_3 plants.

Claims that if australopiths had a grass-based diet similar to the modern gelada baboon (*Theropithecus gelada*), this would indicate that their diets were less nutrient-dense than those of modern apes, possibly placing important limitations on burgeoning hominin brains and sociality, are quite mistaken – if a lesser but substantial amount of nuts, oily seeds and starchy tubers were eaten alongside the grass (particularly seed and corm) base, particularly if the nutrient-denser parts of the grass (seeds and richer corms) were selected, but even if not, there would be no such limitation – the same basic mistake in thinking applies to much theorizing about the small evolutionary decrease (reasonably inferred from comparison with present-day apes) in human large intestine and increase in human small intestine.

Stable isotope data can only indicate an individual's "average" diet, and it is quite possible that C_4 resources were largely fallback foods that made up the bulk of australopiths diets during the dry season when preferred resources were no longer available. Alternatively, australopiths might have eaten ^{13}C-enriched foods throughout the year, perhaps C_4 grass and sedge underground storage organs (USOs) during the dry season and grass seeds during the rains.

It is essential to understand that there are many diets that can produce the same stable carbon isotope ratio ($^{13}C/^{12}C$), and that it is impossible, for instance, to use the ratio from fossil teeth to distinguish between, for example, a purely vegetarian human ancestor that ate 50% C_3 fruits and 50% C_4 grass and a highly predaceous human ancestor that ate nothing but C_3- and C_4-derived mammal flesh in roughly equal proportions.]

[1.14 Cooking (summary)

Our ancestors very probably ate very nearly all their food raw (uncooked) for almost all (roughly 97%) of the 6 million years since our evolutionary line split from the chimpanzee line. That is, for all but the last 200,000 years, in which time observable (i.e. skeletal and dental) or inferable (i.e. from muscle attachment points on bones and other bone anatomy) evolutionary changes are relatively quite trivial. There are no actual fossil digestive tracts remaining preserved available to provide direct evidence of when the small changes from the common ancestor of humans and chimpanzees, a slightly increased small intestine and a slightly decreased large intestine, took place.

It has been stated that fire was controlled in several sites in southern Europe during the Middle Paleolithic back to at least 250,000 years ago, and probably as early as 300,000 – 500,000 years ago.
Firstly, for this to have much meaning it needs also to be known what percentage of all sites from a particular time period have evidence for the control of fire. Then the biases in the finding of sites (see sections 1.5, 1.6, 1.7 below) need to be accounted and adjusted for, or at least kept in mind.
Secondly, there needs to be differentiation among: using fire fortuitously or opportunistically obtained; *controlling* fire fortuitously obtained for relatively long periods of time; and *making* fire as required, which are three very different things…

It seems very doubtful to me, from what I have read in the literature, that there was a general ability of our ancestors to *make* fire as required until 200,000 years ago at the earliest.

Theories that cooking was not merely one of many extra-oral food-processing techniques (such as cracking, pounding, grinding, drying, soaking and sprouting, at least the first three of which were probably used by even our earlier hominin ancestors, and all following them…) that can raise food quality, and therefore unlikely to have caused much evolutionary change leading to the modern human digestive tract, but that it was a decisive factor in the evolution of the modern human digestive tract, alongside whatever survival benefit it had in some circumstances through allowing the eating of a wider range of foods, are very faulty regards the following:

The Basal solution theory, the hypothesis that cooking was adopted around the origin of *Homo erectus* (~2 million years ago) and was responsible for many of the features that characterize human evolutionary changes from australopithecines, such as the reduced jaw and teeth, evidence of smaller gut, and yet higher energy expenditure, fails to consider whether, as it seems to me, these evolutionary changes are much more likely to be due to early tool use of digging sticks in digging for tubers, pounding rams in breaking up tubers, naturally formed hammer and anvil stones in i.e. nut cracking, naturally formed mortar and pestle stones in i.e. seed grinding, and naturally formed sharp-edged stones in tuber peeling, along with soaking of various foods.

More conservative theories that cooking was limited in its evolutionary effects to eventually causing a reduction in tooth size, beginning around 100,000 years ago, are nearly all propounded without mention of the fact that selection for a supralaryngeal vocal tract able to make all of the sounds of modern humans, which was not finalized in evolution until 'fully modern man' evolved 35,000-odd years ago (see section 1.11), need to be considered in the final reduction of human tooth size and final jaw shaping.

Theorists often fail to consider that cooking also requires time; and that the smoke may attract unwanted attention that then has to be guarded against; and that the food being cooked is less immediately portable than it was before, should leaving the area quickly become necessary.

There have been some theorists, whose gross ignorance must put them in the category of the ignorami, who have made statements such as the following:
"It is also known that a diet of raw plant food creates substantial energetic problems for humans under even the best conditions. ……
Second, even under the most favorable conditions people who attempt to restrict their diets to raw foods do not thrive. …… In sum, the sparse [Indeed, and it must be *very* sparse!...] current evidence suggests that raw-food diets produce inadequate energy for humans even under excellent conditions, at least when the diet is dominated by plant items. ……
Third, even the most committed raw-foodists find it difficult to keep to their regime because they are consistently hungry even when they eat as much as possible. ……
Evidence from raw-foodists indicates that under subsistence conditions humans would not survive long on the kinds of raw foods that are available in the wild. ……
Humans are not known to be able to survive on raw food."

Statements such as these are the most arrant nonsense!

(See also more complete versions of the following in sections 1.14 and Chapt. 10 below.) I have myself carried out a personal *experiment* (*not* a belief system, please note, other than maybe a little very early on…) for more than 20 years (from 1986 to 2008) of eating only a raw vegan (exclusively plant food) diet, eating only 2 meals per day, late brunch and dinner, without any supplementation of vitamin B12 (longterm lack of any dietary or supplemental vitamin B12 is risky for the nervous system, and some people would incur damage, and I make explicit here that it would be unethical to subject a child to it). Although some very very small amount of vitamin B12 probably was supplied by vinegar fly or other insect eggs and larvae, it would certainly have been orders of magnitude below that amount that apes get incidentally in nature from only their plant food, without even considering what they get from adult insects or larvae deliberately sought.

Regards body muscle and nervous system health:

Near the end of that period of 20-plus years of raw vegan diet unsupplemented by any vitamin B12, at the age of 45 years, I finished the 2007 Ironman Korea Triathlon (a 3.8 km swim followed by a 180 km bicycle ride followed by a 42km marathon run, the same distance as at the famous Hawaii Ironman Triathlon) at the 15th percentile, that is, with 15% of the other participants in front of me and 85% of them behind me.

In between contiguous years of university study, I worked on farms harvesting fruit, consistently averaging 6 full workdays per week, for the 3 month periods, having always nothing to eat and only water to drink till lunchtime at midday while working on these farms – lunchtime was usually one hour, but sometimes only half an hour – my work performance in the afternoons was, like that in the mornings, above average.
Sometimes (say twice a week) I ran for 10-odd km after work, with a bicycle ride thrown in maybe once a week after work.
Yes, the evening meal included a good gutful of mangoes, but the oily nut and seed intake was no more than 100-odd grams – and, somewhat annoyingly, it was usual for me to have more body fat (though not that much) at the end of the harvest season than at the start of it – mangoes are not as concentrated in energy as bananas, but they are at the high end of the range for fresh raw fruits…

Regards nervous system i.e., brain health:

In 1996, 10 years into the experimental raw vegan diet, I achieved approximately (the results distribution graph was very small and not lined so as to facilitate reading it very accurately) the 15th percentile of the GAMSAT test that Australian university students must sit to gain entry to postgraduate medical schools to study medicine to be medical doctors.
I have done most of my university degrees, commenced in 1991 after 5-odd years of raw vegan diet, at an overload of subjects (none repeated subjects), for example 50% overload in the third year of the BSc, 66% overload of PGDip PH&TM and second year psychology (this did have some, but not a full, content overlap in the statistics area), 75% overload (all subjects passed, at a D or 6/7 gpa, though I did have some prior knowledge) in one semester of Master of Medical Statistics and Master of Nutrition & Dietetics.

Regards the eating/meal schedules:

Since 1986 or so I've virtually only ever eaten two meals a day, late brunch and dinner, with nothing other than water in between (and this water usually mostly in summer, the meals having their own generally nicely sufficient water content),
I do spend some substantial time eating my food, say a total of 3 hours a day (but almost nil time in preparation and washing up…), but I am a slow eater, by choice, because I enjoy what I am eating, and do not like to rush my meals, and I have seen many others eat the foods I linger over in a mere fraction of the time I spend.
Even so, far from always feeling hungry, often enough I choose lower-calorie foods to prevent my tendency to overeat (which I have had since a child) from fattening me up.

Furthermore, I generally do not eat more than 100-200 grams of nuts and oily seeds per day, as to do so would be a gross excess of protein intake, and also bring my energy intake to an excess – and I can crack brazil nuts, almonds, macadamia nuts etc, with naturally shaped stones, faster than I can eat them – so obviously statements such as those quoted above and addressed here are so far wrong as to be laughable, if the possible reinforcement and spreading of this gross ignorance were not a very serious matter regarding the public health.

In short, If I was to spend more than approximately one hour per day eating the more concentrated raw natural plant foods available to me, at the speed that I've seen many people eat them, I would become more and more overweight by the week… if I was to spend two hours per day doing that, I would become morbidly obese, despite my regular exercise, soon enough!...

An old saying: "It is better to remain silent and be thought a fool, than to speak and remove all doubt." …

In the much the same vein of ignorance are statements such as:
"First, although raw plant food is evidently a poor diet for humans, a sufficient inclusion of raw meat, in theory, can create an energetically adequate diet."
The basic absurdity of this sort of statement is obvious from a nutritional content analysis at a childish level of mental difficulty: raw meat is about 70% water; raw nuts and oily seeds have very little water, the large majority of the weight being taken up by about 20% protein on average, and the another 70-odd% being fully fat or carbohydrate - so obviously raw nuts (600 kcal / 100 g) (or oily seeds (600 kcal / 100 g), and also cereal grains (380 kcal / 100 g) for that matter…) are more energy dense than raw meat (190 kcal / 100 g, lean; 230 kcal / 100 g, regular) (the maths here is at about a primary/ elementary school level – one need only open up a standard food nutrient content tables and do the elementary school maths!...)

Furthermore note that high-protein diets are reported to increase the cost of digestion by about 30% compared to high-fat diets.

Statements to the effect that cooking results in a large improvement in the diet are generally not well considered.
Firstly, improving the palatability of some foods might actually decrease the quality of the diet if more suitable alternative foods were then not sought.
Secondly, changes to fiber in foods, to break it down and render it more digestible, might not be beneficial – there has been quite a lot of association of dental caries with cooked carbohydrate rich foods in the literature, as well as extracted uncooked carbohydrate – such material might be more likely to stick in tooth-tooth and tooth-gum crevices than such foods uncooked and whole.
Related to this, the increase of the speed at which the carbohydrate in a food is able to be absorbed into the bloodstream, a measure of which is the familiar 'glycemic index', by cooking, is very probably not a good thing.
Making it possible to eat a meal much more quickly than if the food were raw might not be good for digestion (see chapter 13, section 13.3), nor for psychological satisfaction and social aspects.
The formation of 'Maillard Products' in which the essential amino acid lysine is bound to a sugar and therefore not available for absorption from the digestive tract is not a good thing.
The oxidative damage of unsaturated fat subjected to high temperatures in the presence of oxygen to form peroxides and epoxides, reactive biochemicals which play a role in

damage to artery surfaces and stimulation of macrophages to enter artery tissue and progress atherosclerosis (hardening of the arteries) is not a good thing.
The conversion of nuts into more allergenic foods by roasting is not a good thing.
On the other hand there is the beneficial inactivation of some toxic legume lectins, but this needs to be considered in the context of raw legumes having been reliably reported to be one of the very main foods of a wide range of primates, allowing their ecological development as a group – is it possible that only a lesser number of legumes have harmful lectins, and that these lectins are less harmful if eaten in the context of a whole raw meal, perhaps because there are other chemical moieties in the intestinal residue that they can bind to instead of the exposed carbohydrate chains of the glycoproteins in the human cell wall membranes?
These considerations need to be made before thrusting forward one's theory as a definitive or even potentially plausible statement of the state of affairs…]

[1.15 Energetic models (summary)

At walking speeds of between 2.4 and 6.0 km/hour, humans expend significantly less energy than most quadrupeds, whereas at running speeds bipedalism is less economical, as humans expend significantly more energy (per body weight per distance travelled) than quadrupeds. Over the entire range of walking speeds, human males expend 13% less energy than their quadrupedal counterparts, whereas human females expend 25% less (presumably as they carry less upper body musculature than males).

In contrast to humans, chimpanzees expend 35% to 40% more energy for quadrupedal locomotion than an "average" mammalian quadruped of the same size. ……

Modern apes move only about 1.8 km/day, whereas human foragers move an average of 13.1 km/day.
For an animal the size (and presumably the shape) of an early australopithecine, for the smaller day ranges of 2.0 and 5.0 km/day, the energetic savings due to bipedalism accounts for about 1% to 3% of estimated resting metabolic rate per day; however, for day ranges similar to those of modern human foragers (presumably for them) the savings is 12% to 16% of RMR, a substantial portion of the daily energy budget.
Selection for energetically efficient locomotion is likely to be stronger among animals that must travel longer distances because they have the most to gain by being more efficient.

In humans, brain metabolism accounts for 20% to 25% of Resting Metabolic Rate (RMR), as compared to 8% to 10% in other primate species, and 3% to 5% in nonprimate mammals.

So, modern humans require more energy per unit of total body weight than chimpanzees to sustain their proportionally larger brain, but on the other hand they require less energy to travel from place to place per kilometer but more energy absolutely due to the much greater distances travelled.
However, if chimpanzees are spending more time climbing trees then this is likely to significantly affect the comparison, and often this aspect is not properly incorporated in analyses and theorizings. Likewise, the weight of ingested food to be carried, and the weight of gathered food to be carried, and the effort involved in travelling to and carrying drinking water if sufficient water is not taken in as part of foods like fruit and vegetables. Most energetic models do not come close to properly considering these factors.

Perhaps it would suffice at this point to say that the net meaning of this evidence, relevant to the questions addressed in this book here, is that the gastrointestinal/ digestive tract of modern humans is adapted to a slightly different diet to those of the chimpanzees and other apes – slightly more small intestine, slightly less large intestine, less digestion of cellulose. And that this signifies a slightly more energy/macronutrient-dense and slightly less fibrous diet.

Even among chimpanzees, the most predatory of the great apes, only about 5% of their calories are derived from animal foods, including insects.
Again, in the area of scientific specialty that is primatology, as in some others covered here in other sections, there are some grossly fallacious assumptions apparently made without the little effort that opening up standard food nutrient tables and applying primary/ elementary school maths would entail: that meat and other animal foods are much more dense in calories and nutrients than most of the plant foods typically eaten by large-bodied primates, and that eating of these "higher quality" animal foods means that humans need to eat less volume of food to get the energy and nutrients that we require. The basic flaws in such thinking: raw meat is about 70% water; raw nuts and oily seeds have very little water, the large majority of the weight being taken up by about 20% protein on average, and another 70-odd% being fully fat or carbohydrate – so obviously raw nuts (600 kcal / 100 g) (or oily seeds (600kcal/100g), and also cereal grains (380 kcal / 100 g, dry) for that matter…) are more energy dense than raw meat (190 kcal / 100 g, lean; 230 kcal / 100 g, regular) - the maths here is at about a primary/ elementary school level – just open up a standard food nutrient content tables and do the elementary school maths!
Furthermore, the proportions of energy/protein in raw nuts/seeds are much closer to that optimal/required by apes or humans than is that of meat.

Now, a certain amount of water is needed every day – if it does not come in the food, then it must come from elsewhere – why wouldn't food be a good place to be getting most of the required water from?
It is reported that the natural diet of apes and monkeys supplies them with sufficient water.
Water taken in as part of fruit and vegetables takes longer to be absorbed from the space in the stomach and intestines because osmotic force and other chemical forces act to keep it in the bulk of food (/remains) as that bulk/solution/suspension/emulsion is being moved through them towards its eventual evacuation as feces some 24-plus hours later – the result is that water is more likely to be delivered from the intestinal tract as required, rather than to be excreted as surplus to immediate requirements in the urine, which is more likely to happen if the same amount of water was taken in as water and mixed with little or no food in the intestines.
And, in nature in general, drinking water sites are more risky sites for visiting unless one is invulnerable to predation from large carnivores, because large carnivores are aware that if animals (must) go to such rivers and streams, then they may be hunted there with considerable success. Which is to say nothing of crocodilians, which can lunge suddenly and viciously from the water, to take the hapless prey by surprise in their ghastly jaws, drag it back into the water, apply the dreaded 'death roll', and devour the resultant corpse. Also infectious bacteria and parasites are fairly likely to be encountered in many bodies of water (probably the more so the larger the body of water, on average), whereas they are much less likely to be encountered in the water inside fruit and vegetables, although they may be encountered on their surfaces – cooking has the benefit of killing infectious agents in food, but unless the food is cooked in water, which is clearly an evolutionarily later practice than merely bringing the food into contact or next to fire, water will then have to be acquired from some other source, and more of it, because cooking outside of water clearly must dehydrate the food to some extent.
And, without detailing of just *which* "nutrients" are better provided by animal foods (which is not at all the same thing as a nutrient being more present in a food, because

there are other very important factors to consider, including other nutrients, and other foods in combination in the same meals and in the diet), isn't that part of their statements here just another motherhood saying about the diet they are habituated to and cherish? … the only nutrient of any real concern here on average is vitamin B12, and we need *very* little ((less than) 2 to 3 *micro*grams, not *mille*grams, per day (RDI)) of that…

Now humans have a relatively enlarged small intestine, and a reduced colon, compared to those of the apes.
Statements that this human anatomical proportion is more similar to that of carnivores are quite incorrect - the morphology is in fact more (and apparently much more) similar to that of apes than it is to carnivores! It is changed only a *small* amount in the direction of the carnivores.
And, evolutionarily, this change could easily be accounted for by an increase in more macronutrient- (carbohydrate, fat, and protein, in order of optimal proportions, from most to least) dense plant foods such as nuts, oily seeds, cereal seeds, starchy tubers and legumes, with or without tool use for processing, without any theoretical need for the invoking of substantial meat eating, particularly in the absence of any evidence of it being generally the case other than in some areas in archaeologically very very recent areas, such that the relatively very very recent increases in meat eating would not have been the driver or facilitator of the gut changes, being mostly another fallback strategy, as well as availing escape of male gathering drudgery into hunting excitement, displays of martial power to neighboring groups, males within one's own group, and females within one's own group, as detailed in chapter 7, section 7.3 here.

Claims that consumption of more animal foods was necessary for the evolution of the larger human brain on the basis that the composition of all mammalian brain tissue is similarly rich in two fatty acids, docosohexaenoic acid, DHA, and arachidonic acid, AA, and that these two fatty acids are richer in animal foods than in plant foods, might best be considered along with the fact that the elephant has the biggest brain on the planet, twice the size that of the blue whale and four times the size that of humans - and elephants are not generally known to eat meat, though they would ingest some insect matter incidentally in the plant matter they eat. And, that the apes are all relatively big-brained, yet eat generally very little meat, though again some insect matter, deliberately as well as incidentally.
Furthermore, frugivorous bat species' brain size, relative to body size, is typically almost twice as large as in aerial insectivore bat species, with no overlap b'tween the two groups. Nectarivorous bats have relatively large brains and generally overlap with the frugivores, whereas the remaining bats (carnivores, piscivores, sanguivores and foliage gleaners) typically have brain sizes intermediate between frugivores and insectivores.

Claims that in their raw form, the starch in tubers is not absorbed in the small intestine and is passed through the body are very incorrect - if one actually chews the stuff properly first, instead of bolting it unchewed like a dog, then the large majority of it will in fact be digested and absorbed – an easy enough experiment to try by oneself (a radical suggestion, apparently, for some armchair theorists and parrotic vocalizers in nutrition and dietetics, but there we go…), even without the simplest of blood sugar assay equipment…]

[1.16 The brain and intelligence (summary)

In our ancestral line, brains much larger than those of the apes only arrived in evolution with the arrival of the genus *Homo*, but it should be kept in mind that there was a gradual increase in brain size in the period of time from when our line split from the line of the chimpanzees up until the beginning of the emergence of the genus Homo, where the rate of brain size increase began to significantly increase – overall an 'exponential' growth curve.

A positive correlation between comparative brain size and home range size in primates, after effects for body size had been taken into account, has been reported, and note the context is that folivores generally need a smaller range than frugivores to find their food in.
It is probably crucial to note that the orangutan is a savagely solitary animal, and its large brain (much the same size as that of the chimpanzee) and high intelligence must therefore be related to the complexity of its diet, where it remembers what fruits, nuts, seeds, and legumes are in season, and where in its home range they are, and then makes a decision on an efficient foraging route through them all.
Furthermore, frugivorous bat species' brain size, relative to body size, is typically almost twice as large as in aerial insectivore bat species, with no overlap between the two groups.
Nectarivorous bats have relatively large brains and generally overlap with the frugivores, whereas the remaining bats (carnivores, piscivores, sanguivores and foliage gleaners) typically have brain sizes intermediate between frugivores and insectivores.

Overall the energetic costs of brain maintenance for modern humans (20% of human basal (resting) energy expenditure) are about three times those of a chimpanzee. Growth costs will also be commensurately larger. However it must be borne in mind that the context of how the locomotory energy requirement of humans (more walking) and chimpanzees (more climbing) differ, needs to be known, before a conclusion on how overall energy requirements might differ can be made. As well, any differences in the basal metabolic rate itself need be known.

The slower rates of development and longer period of immaturity among the genus *Homo* suggests that human mothers were under energetic constraints which could not be compensated for entirely by a shift to a more macronutrient-dense diet (that differences in the human digestive tract from that of the ape digestive tracts suggest), and that as a result there was evolutionary selection for a shift towards slower development. This probably involves some reduction of basal metabolic rate, and also of the protein/energy requirement ratio of the growing human.

Note that the ratio of protein/energy in human breast milk is very low, only a fraction of what it is for all of the several other mammals I've found data for, indicating that we remain after evolution a naturally physiologically and biochemically low-protein diet consuming animal.

Accordingly, most often protein will not be the limiting macronutrient factor in the diet, including that of the breast-feeding mother - energy will most often be the limiting factor. So it seems quite possible that the use of starchy root vegetables, added to the diet of the common ancestor of the chimpanzees and humans (which was probably a generally very ape-like diet), and increased nut, seed and grain consumption, which were all very likely to have been increased as a proportion of the diet of our hominin ancestors, taken together, could meet any increased energy requirement. And bear in mind that even in the absence of any increase in energy requirement, an increase in the macronutrient-density of the diet could result in the slightly increased small intestine and slightly decreased large intestine of humans in comparison with chimpanzees – bear in mind that for chimpanzees sweet juicy fruits are the preferred foods and nuts requiring cracking more often fallback foods for times (i.e. seasons) of scarcity of sweet juicy fruits – it is possible that our hominin ancestors ate more nuts, seeds and grains because that is what was more available, not necessarily because they needed a more energy-dense diet.

Alternatively, increased returns for food foraging and processing effort may be an important pre-requisite for evolving a larger brain, and a larger brain may be necessary to organize human food foraging and processing behavior, particularly in view of a diet of 'imbedded foods' (foods that are neither visually obvious nor available without expending considerable time and energy in processing them, being extracted from a secondary source) probably requiring more complex cognitive skills.

It seems obviously correct to note that tool use increases the efficiency of extraction of such foods from their coverings, and widens the range of resources available, and as such could enable an increase of net energy such as would contribute to evolution of a larger brain.
However, not enough consideration has been given to the possibility of the use of spears as a defensive weapon for ground foraging, and the use of (accurately thrown) spears and stones to defend tree fruit from competitors such as monkeys, apes and birds, and the cognitive requirements of this as a strategy requiring an evolutionary step up, but a plausibly achievable step up, in planning for the near future.
Furthermore, that the full exploitation of nuts ripe on the ground, that only some chimpanzee groups have learned to use anvil and hammer rocks to start doing, would be facilitated by spears, as hominids, with their feet differentiated from the apes and monkeys, while still able to take shelter in trees as unencumbered adults with sufficient speed to escape predation, would not be as fast at doing so as apes and monkeys, particularly if encumbered with infants and very young children, particularly as nut-cracking is likely to be a noisy process, potentially attracting predators, and our earlier hominin ancestors very probably relied on quiet movement and activity for survival.

Statements and theories to the effect that meat represents a high quality resource which could enable an increase of net energy such as would contribute to evolution of a larger brain should be viewed in the context that a food's being denser in energy and/or protein, is not the same thing as its being of higher quality, because there are other very important factors to consider, including other nutrients, and other foods in combination in the same meals and in the diet (something that a large number of more simplistic vocalisers and writers seem utterly unaware of in making statements that some food is a good source of some nutrient… *good* source and *rich* source are two very different things…).
Furthermore, to repeat from the summary of section 1.15 above: raw meat is about 70% water; raw nuts and oily seeds have very little water, the large majority of the weight being taken up by about 20% protein on average, and another 70-odd% being fully fat or carbohydrate – so obviously raw nuts (600 kcal / 100 g) (or oily seeds (600 kcal / 100 g), and also cereal grains (380 kcal / 100 g) for that matter…) are more energy dense than raw meat (190 kcal / 100 g, lean; 230 kcal / 100 g, regular) - the maths here is at about a primary/ elementary school level – just open up a standard food nutrient content tables and do the elementary school maths!
Furthermore, the proportions of energy/protein in raw nuts are much closer to that optimal/required by apes or humans than is that of meat.

The evidence for claims of (even seasonally-varying) heavy consumption of meat by Homo erectus is very poor - nearly always this claim is pure assumption.

Also, taste preferences have a bearing on diet (surprise, surprise…).
If a more energy-dense diet was required for the evolution of the larger human brain, then an increase in intake of nuts would very certainly meet that criterion, as would an increase of smaller such things generally termed seeds, likewise grains (These are in order of decreasing ease of extraction, large nuts being very readily extracted with the correct use of hammer and anvil stones, seeds less so (though not all may require

extraction prior to eating), and grains maybe less so again (though not all may require extraction prior to eating at all stages i.e. sprouting or soaked or ground up.)).
If the taste of these was preferred to the taste of raw (including rotting) meat, then meat may have been only a fallback food in times of scarcity of the preferred foods.

Aside from the requirements for a larger brain in more complex food gathering and processing, the other main factor is, very probably, increasing social complexity and its cognitive demands, including the following (and please see the full section 1.16 below for the authorship citation) in ordering of ascending difficulty, with apes able to do, and doing, more of the complex things than monkeys:
discovery behaviours (curiosity and exploration of physical objects, and social play for gathering information about social relationships);
innovation or tactical flexibility as responses to novel associations of events (spatio-temporal fluctuations in food availability, or in social interactions the abstraction of "conditional social rules" as a basis for social actions, in relation to specific actions by other con-sociates whose past social-interactive patterns can be stored in short-term and long-term memory);
cognitive representations of the intentionality and attentional states of others, perhaps involving attributions of several levels of intentions ("X thinks Y knows that X wants Y to do such-and-such", or in more complex tripartite interactions "X thinks Y knows that Z wants Y to do such-and-such", etc);
and (perhaps only in humans) *self-reflective monitoring* as a means of abstracting an explanatory model of human action.

Non-human primate grouping patterns and social systems are determined by three major factors: dietary characteristics, predation risk, and social competition.

Perhaps an increase in body-size, coupled with larger group sizes than were optimal simply for foraging due to an increased predation risk from foraging further from climbable trees in increased home range areas, resulted in increased levels of intragroup competition and decreased group stability.
In such contexts, including the development of more potent weaponry and its use, even if initially for defense, the capacity for social cognition, and the capacity for increasingly complex language as an instrument for establishing and modulating ever more complicated relationships between our hominin ancestors over almost all of the last 6 million years, larger brain size and the behaviors that would support it would have come under strong selection pressure as a mechanism for learning, communicating and manipulating (shared or private) mental-model representations of the contexts and constraints on action. Aside from object tool use in non-human primates, primatologists repeatedly emphasize the ability of non-human primates to use other individuals as 'social tools' to achieve particular results.]

[1.17 Cracking and grinding tools, digging tools (summary)

Modern chimpanzees have been well documented to use, in their 'natural' habitat, naturally shaped hammer and anvil stones, naturally shaped mortar-and-pestle stones, naturally shaped grinding stones, and sticks/twigs for termite fishing …(see chapter 2 below), and capuchin monkeys have been documented to use, in their 'natural' habitat, naturally shaped hammer and anvil stones, naturally shaped mortar-and-pestle stones, and naturally shaped grinding stones (see section 2.16 of chapter 2 below).

Such naturally shaped tools are invisible in the archaeological record because the twigs, sticks and branches decay fully into dust, and the naturally shaped stone tools look, for all practical purposes in the present state of technology and method/practice, like just like stones among other stones.

However, it seems to me that if present-day apes and monkeys can use naturally shaped stones and sticks as tools then even our most distant hominin ancestors would almost certainly have done so too. And lack of ability to reason using language does not indicate the lack of ability to reason with mental pictures.

Wild animals are, perhaps above all in their behavior, observant – failure to be so is very likely to result in removal from the gene pool at a very early age – I find it very unlikely that even our earliest hominin ancestors were incapable of watching a pig dig up roots on serial occasions and not deciding that they could also do so themselves with the use of a stick – in fact, such reasoning in animals has been pretty conclusively demonstrated and reported on by now in 2010, though I will not provide any references in this present version of this book.
Therefore, to reason in mental pictures that if nuts could be processed (cracked open) between two hard objects, that it might be worth a try using some variation of that as constitutes grinding i.e. of grains, seems well within the mental capabilities of even our earliest hominin ancestors to me, notwithstanding the frustrating lack of the mental capabilities that one might think reasonable to expect, in many modern humans…

It seems very probable to me that even our very earliest hominin ancestors used naturally shaped hammer and anvil stones, naturally shaped mortar and pestle stones, and digging sticks. The invisibility of this use in the archaeological record due to decay of the sticks and undifferentiability of the stones from other stones clearly make such absence of evidence (which is not the same thing as evidence of absence) not at all a good argument against such use having occurred.]

[1.18 Defense needs and weapons (summary)

I doubt that with their legs, pelvis and spine not as evolved as ours for moving on the ground, and their small stature (1.1 meters), that australopithecines would have had a snowball's chance in hell of surviving on the ground (which is where their feet (greatly resembling our own) strongly suggest they were spending a great deal of time), without weapons. In my opinion anyone thinking otherwise has a totally inadequate understanding of combat reality and the unarmed defense capabilities of such a creature in the environment it shared not only with large cats and hyenas but also with baboons (though more grass-eating then as they purportedly were) the size of a modern gorilla and very probably other dangerous large animals outside my knowledge.
And I think much the same applies to the larger *Homo erectus* that came later and was more our size – without fangs and claws, or the speed of a gazelle, or the size of an elephant, or the hide and tusk of a rhinoceros, or the reproduction rate of a rabbit, or the taste of a stink-bug, or the camouflage of a stick insect, some other method of survival is required. Weapons were almost certainly it, obviously enough it seems to me, but apparently not to some, as the subject seems generally very poorly dealt with in the literature, notwithstanding that I haven't read or canvassed all of it.]

[1.19 Scavenging, power scavenging, and hunting (summary)

There is a side debate which is somewhat peripheral to the main questions addressed by my book here – that debate is about whether our earlier hominin ancestors acquired what meat they did eat by scavenging, power scavenging, or hunting. Power scavenging involves driving off whatever animal is feeding on a carcass and taking it over for eating.

It seems to me that by the time our earliest ancestors had evolved feet like ours they, although still very proficient climbers of trees for the purposes of getting fruit, would not have been good enough climbers to hunt monkeys in the way that chimpanzees sometimes do, though use of thrown weapons might have improved their ability a little.

However, they probably were able to hunt on the ground in a manner resembling that of chimpanzees observed hunting piglets and baby deer. However, in the more open country that it seems likely that our early hominin ancestors spent a lot of time in, there is no great certainty that this would have been as feasible as it was for the chimpanzees.

Furthermore, our early hominin ancestors already had lost the elongated canine teeth such as chimpanzees (though mostly in combat, not predation) and carnivores use in killing and tearing – there are other ways to kill, such as by battering or beating, but this lack of elongated canine teeth would reduce the size of animals able to be killed.

I doubt that with their legs, pelvis and spine not as evolved as ours for moving on ground, their shoulders probably not as well evolved for throwing things and maybe not even as evolved for club use as ours, their smaller stature than ours, their (at least earlier) lack of quality spears, and their lack of bows and arrows, that our earlier hominin ancestors, the australopithecines, were able to hunt anything other than very small animals, and these probably were not taken by active hunting and chasing, but seized or cornered more by chance as baboons do with small animals and young animals who use freezing as a defence, that they come upon more by chance. Baboons have been recorded as chasing down hares, but it seems very unlikely to me that an australopithecine could run at the speed that would require – I also doubt that a chimpanzee would be capable of it, and it seems unlikely to me that pure running down would be the method that chimpanzees use in their recorded occasional predation on young baboons.

It seems very likely to me that, probably used only in times of great hunger, both scavenging and power scavenging were the only access our earlier hominin ancestors had to the meat of large animals. They probably needed weapons for either method, as scavenging is likely to involve making more sound than the average made for gathering, and moving quietly was probably a central requirement of survival for early hominins other than armed and at the ready with clubs, stones and sharp sticks, and in groups – having such weapons, they would probably have been capable, as a group, of driving off from carcasses at least sufficiently sated carnivores of some size larger than themselves.

Raw meat was probably not very palatable to our early (and maybe nearly all) hominin ancestors, who probably preferred raw fruits, nuts, and some vegetative plant parts when available, and probably even legumes and starchy tubers, to raw meat, and almost certainly to raw meat in various stages of rotting.

This is of course conjecture, but we are all limited here to trying to think sensibly about the capabilities and taste preferences of our hominin ancestors – conclusive evidence is unlikely to ever be available.]

[1.20 Proposed implications of studies of early hominin diets: Cordain; Eaton; Ungar; in Ungar (2007) Eaton, Shostak and Konner (1988) (summary not given - see section 1.20 proper below)

The following works are critically reviewed below in section 1.20, and as I cannot shorten them substantially while still giving readers sufficient insight on the extent and nature of the mistakes in the authors theorizing and conclusions, and the process of how such work is critically assessed for such mistakes by a sufficiently competent scientist (for example, I posit, myself), all I will do here is to list the full references, and urge the reader to read section 1.20 below in full, so as to gain something more of the proper understanding of the extent of the flaws very common but variably distributed in work by supposed experts (i.e., university professors) in the fields relevant to the main questions addressed by me here.

Cordain L. Chapter 19 Implications of Plio-Pleistocene hominin diets for modern humans. In, Ungar PS. (Ed) (2007) Evolution of the human diet The known, the unknown, and the unknowable. Oxford University Press.

Eaton SB. Chapter 20 Preagricultural diets and evolutionary health promotion. In, Ungar PS. (Ed) (2007) Evolution of the human diet The known, the unknown, and the unknowable. Oxford University Press.

Ungar PS. Chapter 21 Limits to knowledge on the evolution of hominin diet. In, Ungar PS. (Ed) (2007) Evolution of the human diet The known, the unknown, and the unknowable. Oxford University Press.

Eaton SB, Shostak M, Konner M (1988) The Paleolithic Prescription. A program of diet & exercise and a design for living. Harper and Row, New York.]

[My conclusions on chapter 1

It seems to me that if present-day apes and monkeys can use naturally shaped stones and sticks as tools, then even our most distant hominin ancestors would almost certainly have done so too.
And lack of ability to reason using language does not indicate the lack of ability to reason with mental pictures.

Wild animals are, perhaps above all in their behavior, observant – failure to be so is very likely to result in removal from the gene pool at a very early age – I find it very unlikely that even our earliest hominin ancestors were incapable of watching pigs dig up roots on serial occasions and realizing that they could also do so themselves with the use of a stick – in fact, such reasoning in animals has been pretty conclusively demonstrated and reported on by now in 2010.

Likewise, to reason in mental pictures that if nuts could be processed (cracked open) between two hard objects, that it might be worth a try using some variation of that as constitutes grinding i.e. of grains, seems well within the mental capabilities of even our earliest hominin ancestors to me, notwithstanding the frustrating lack of the mental capabilities that one might think reasonable to expect, in many modern and present-day humans…

That the canine teeth of all hominins are/were no longer elongated as they are/were in the apes and monkeys suggests that humans are the least naturally carnivorous of humans and apes (chimpanzees eating meat as a fallback food in seasons of fruit scarcity, and gorillas, orangutans and bonobos eating almost no meat).

That human premolar teeth are molarized (more like the molars), that human molars are flatter, and that the jawbone has been moved to the rear, compared to our ape relatives, suggests an adaptation to more chewing of small seeds like cereal grains, and of other things that require more crushing than shearing, perhaps tubers.

That human/hominin tooth enamel is thicker than that of the thickest-enameled apes, the orangutans, and much thicker than that of the chimpanzees and gorillas is concordant with evolutionary suitedness to crushing hard things, as well as grinding things that grow close to or in the ground and therefore have more abrasive dirt particles on them, such as cereal grains, seeds of small herbaceous plants, and tubers.

It seems very probable to me that even our very earliest hominin ancestors used naturally shaped hammer and anvil stones, naturally shaped mortar and pestle stones, and digging sticks.
The invisibility of this use in the archaeological record due to decay of the sticks and undifferentiability of the stones from other stones clearly make such absence of evidence (which is not the same thing as evidence of absence) not at all a good argument against such use having occurred.

Human digestive tract proportions, and physiological parameters such as stomach digestive juice pH and emptying time, are far closer to those of the apes than they are to more naturally carnivorous animals like the cats and dogs. The slight reduction in relative size of the large intestine and slight increase in relative size of the small intestine are more probably related to increased consumption of macronutrient (protein, fat and carbohydrate) –dense plant foods than of meat (as the nuts, seeds, legumes and grains are more calorically dense than lean meat).

Processing of foodstuffs by hammer and anvil, mortar and pestle, and grinding stone, and soaking, are more likely to have facilitated the meeting of nutritional requirements than the use of fire (the making on demand of which probably came only very late in human evolution.

If, and I think it definitely had some effect in many places, organized warfare between specialized societies led to preferential selection of larger and stronger men for the armies, who would then frequently kill off large proportions of each other, this could quite easily reduce the population gene pool of genes for largeness, strength and robustness (exactly the opposite of what would happen in combat between random individuals fighting each other, or in small groups where all males took part in fighting – where the smaller are more likely to die).
Therefore claims that agriculture lead to a reduction in body size due to the supposed unsuitability of agricultural diets for humans are likely to be mostly false, and in any case are made on the basis of a very defective consideration of the evidence.

It may be that the human intelligence was a further evolution in some ways from the relatively high mental functioning continued in evolution into the orangutans – they remember not only where the different edible-fruited trees in their range are, but also what their fruiting season is, and they use relatively advanced logic to decide on efficient routes between different trees in a day's, or in a series of days', foraging for fruit and the green leaves and insects eaten also in their natural diet. Intelligence is not at all a monopoly of hunters, as some perhaps less capable biologists often state or infer, perhaps in an unconscious attempt to rationalize their own cherished carnivorous dietary habits.

Now, substantial evolutionary changes of long-lived species like humans (of the sort related to evolution of a suitability to at least general dietary food group proportions, and probably also to other, food- or nutrient-specific proportions or amounts), other than removal of DNA-*degenerative* mutations, probably does not take place over periods of time so short as a thousand or ten thousand years, as the natural mutation rate of our DNA is not fast enough, even allowing for environmental stresses (acting on heat shock proteins) to permit a greater number of mutations to escape correction by cellular DNA

repair mechanisms (assuming that this actually happens outside the test tube, which it might well not…) (as postulated by Maresca and Schwartz, 2006), relative to the length of our generation gap (parent to offspring), to allow it.
This is something many proponents of human dietary suitability who refer to archeological evidence (in itself flawed by vegetable matter being *far* more decomposable than bones…) from the last ten thousand or so years (bones of animals eaten by humans) fail to realize the significance of when they propose that humans have evolved into a substantially naturally (they mean optimally, though they usually seem unaware of the difference…) meat eating animal.

Many proponents of substantially carnivorous diets being optimal for humans, basing their claims on archeological or anthropological evidence, do not go back far enough in evolutionary time in their thinking.
They tend to refer to the most recent cold period (there are warm periods, such as the present) of the most recent ice age, when hungry humans with their possibly more natural or preferred (in the absence of fire) plant food sources in shorter supply probably did eat more other animals than in previous warmer periods. This most recent cold period is only of approximately 10,000 years duration.
Failing firstly to go back far enough in time (tens and hundreds of thousands of years) to discern our very likely evolutionary origins as a very predominantly plant-food eating animal, coming from the same evolutionary origin as the all very predominantly plant-food eating apes (our closest animal relatives), they then secondarily fail to sufficiently address the question of whether most if any human populations have genetically evolved to eat much animal food, as the fraction of time in human history during which we ate much animal food was not long enough for genetic mutations away from our fairly obviously mainly plant-food eating evolutionary origins to occur often enough to have become the standard or most common genes of our species, via being selected for, or perhaps rather, via not being removed as insufficiently functional (more on this below), in the struggle for survival in nature.)
Also, they fail to take into account the varying periods of time, in most (and particularly in some major) human populations, where not much (and sometimes very little) animal-food was eaten in the last 7,000 years since the advent of agriculture as a major source of human food, which is likely to have removed much of any DNA-*degenerative* changes (such as vitamin B12 absorption, transport and storage ability decreases) that what (relatively short in evolutionary time) substantial meat-eating that did occur allowed to remain in the gene pool unremoved by natural selection.]

1.2 Human and ape genetic and evolutionary relationship; ape-like human origin

[Genetically we remain very similar to the anthropoid apes that are our closest relatives, chimpanzees, bonobos, gorillas, orangutans and gibbons - DNA studies showing approximately 97% exact similarity in DNA between humans and apes (I think the accepted figure is less than that now, with individual humans on average having only 97% exact similarity of DNA) make this as much a fact of science as anything can be.

Kelley J Section 6.4 Evolution of apes. In, Chapter 6 The primate fossil record. In Jones S, Martin R, Pilbeam D, Bunney S (1992) The Cambridge Encyclopedia of Human Evolution. Cambridge University Press.

While there is some question about whether chimpanzees are more closely related to gorillas than to humans, it is generally agreed that all of these three are more closely related to each other than to orangutans and gibbons.

Although the apes too have evolved over the six-odd million years since our evolutionary lines diverged, and, like humans and hominids, it is not possible to tell exactly or for some things even closely how their dietary and diet-related behaviour may have changed, it is reasonable to think that the diet of our distant ancestors in the times when our line diverged from the apes was more like present-day apes than like present-day humans. This is because humans have an essentially unnaturally radical course of behavioural evolution in general, and more radical in later periods than in earlier ones.]

1.3 Earlier insectivore origins, persistence of insectivory, vitamin B12

[Many more millions of years before the evolutionary divergence of humans and apes, all of us, and the monkeys, came from evolutionary origins in primates that were very strongly insectivorous – all present-day apes and monkeys, in fact all or very nearly all primates, continue to have a substantial, regular, daily deliberate or incidental ingestion of insects and their larvae (or pupae) and eggs.
Many humans likewise continue to deliberately eat insects and their larvae, and it would be unusual for any person to not incidentally, unknowingly, eat some insect matter in the food they consume.
The alternative might be to ingest in one's food an unacceptably high dose of farming insecticide residue.

This has a particular relevance for vitamin B12, which is generally taken to be available to humans only (other than from supplemental vitamins) from foods of animal origin.

It is possible that a vegan diet (having only foods of plant origin) that contained a sufficient quantity of green leafy vegetables (the best plant food source of cobalt, the metal atom in the vitamin B12 molecule) that had been grown in soil that was not deficient in cobalt might provide the bacteria in a human large intestine, if the bacterial population was of a healthy species composition and population level and in a conducive environment, with enough cobalt to manufacture some vitamin B12; and if the contents of the big intestine were sufficiently continuously moist enough due to the intake of moisture in the diet being optimal in regard to being part of vegetables and fruit consumed, then a useful if potentially insufficient quantity of vitamin B12 might flow down its concentration gradient to pass through the ileocaecal valve back to the last part of the small intestine where vitamin B12 is taken up.
For some humans, particularly those from some population evolutionary course, who have received deleterious mutations in their DNA coding for vitamin B12 uptake, transport and storage proteins because these deleterious mutations were allowed to remain in the genetic code due to there having been for long and continuously and recently enough in their bloodlines' evolutionary past a relatively high consumption of animal foods, this amount would probably be too small to prevent damage to the nervous system from vitamin B12 deficiency.
Even if such deleterious mutation-carrying humans are in the minority, and probably a very small minority in many populations, it would not be ethical to subject an infant to the risk of permanent brain damage through a purely vegan diet – that is a risk for adults to undertake, for whatever reasons of esthetics or scientific curiosity, properly equipped with the knowledge and understanding to do so in the context of an informed choice.

Some seaweed makes biologically active vitamin B12, but more often seaweed makes a vitamin B12-like molecule that is not biologically active and is worse than useless because (some physiological screening mechanisms notwithstanding) it occupies molecular vitamin B12 sites in the human body thereby preventing what biologically active vitamin B12 is present from binding to those sites and performing its biochemical function.]

1.4 Colour vision, sweetness preference and fruit

[Although many animals have a more complex colour vision system than humans, for example many invertebrates (i.e., butterflies, stomatopods (i.e., shrimps with a dozen or more different colour vision receptor types)), tropical fish and birds, and reptiles and amphibians, in the evolution of mammals, segmenta of colour vision (i.e. red) were lost, and then for a few species of primates, regained by gene duplication. Placental mammals other than those primates (i.e., cats, dogs and most farmed animals) have dichromatic vision with blue and yellow, and therefore green, but not red vision. Various authors have suggested that this evolutionary change was related to a dietary shift to greater consumption of fruits and young leaves with a red colour.]

1.5 Archaeological remains biases

[Mann (1981) notes] "The archaeological record is a highly selected information source; some materials preserve very well and are routinely found during excavation, while others either are not preserved or leave such subtle traces that they are missed.
Many dietary reconstructions have been based exclusively on the recovered evidence, ignoring items that might originally have been present but that have disintegrated long since. Because vertebrate bones preserve better than plant materials or invertebrate exoskeletons, the basic diet during much of human evolution has been viewed in terms of an emphasis on a single food resource: mammals that are hunted."

[In the entry for Paleodietary Analysis in Delson et al (Eds) (2000):] "Moreover, it is notoriously difficult to calculate the proportional importance of different classes of foods consumed from archeological residues. For example, the archeological record tends to skew the evidence of prehistoric diets toward meat eating because the bones of edible animals are more likely to survive than the remains of plant foods."

1.6 Archaeological interpretation and attribution biases due to lack of appropriate quantification

Potts R. Section 9.1 The hominid way of life. In, Chapter 9 Early human behaviour and ecology. In Jones S, Martin R, Pilbeam D, Bunney S (1992) The Cambridge Encyclopedia of Human Evolution. Cambridge University Press.

"The current view is that throughout the long australopithecine phase, which lasted until less than 2 million years ago, hominids were basically ape-like in their levels of behavioural complexity and social organization. They mainly depended, as do all primates including humans, on plants for food, and there is as yet little direct evidence that they ate more meat than do some chimpanzee populations today."
[Thereafter he speculates on the significance of archaeological findings of manufactured stone tools and animal bones with human fossils regards an increase in the consumption of meat by hominids, somewhat cautiously, but without recognizing the discovery bias in archaeological sites due to the facts that bones survive where plant matter does not, and, that manufactured stone tools, although able (as many writers seem to forget) to be used to advantage in dealing with plant material as well as dealing with animal carcasses to obtain hides for clothing and bags, and to obtain sinews and bones for manufacture materials, are more likely to be used in butchery than are the naturally shaped (and

therefore archaeologically invisible) stone tools almost certainly used for cracking nuts, grinding seeds and grains, and grinding, pulping or mashing root vegetables – which makes discovery of hominid or human remains in the presence of bones and manufactured stone tools more likely than discovering them in the absence of these things because often it is the sighting of the animal bones and manufactured stone tools that triggers the excavation and discovery of the associated hominid or human remains.
He also does not consider whether the climatic conditions (dryness or coldness) that are conducive to survival of hominid or human remains are also those that might have caused an increased reliance on meat as a fallback food in the absence of naturally more preferred foods, adding to discovery bias – in this regard there seems to be some consensus that over the last 5 million years there have been multiple alternating periods of temperature difference, for example the 9 degrees C difference between the glacial and interglacial periods of the most recent Ice Age.
He also does not explicitly recognize the very general lack of proper attempts to quantify the rate of consumption of meat per human per unit of time, in all the reports (on remains from any time period) of supposedly substantial meat eating, though he does note ", that specific artifacts and other refuse are not produced in proportions to the frequency of a particular activity, " (see immediately below).]

[However, the following is an example very typical of the almost universal failure to do what tentative quantifying is readily able to be done (the maths is very simple):
Although Olsen (1989) does a good job of analyzing the probabilities that various hunting methods were used in the generating of the great pile of horse remains in the Upper Paleolithic (from at least 30,000 years before the present up to 12,000 years before the present) big game kill site in Solutre in what is now France, she unfortunately did not carry out the mathematically childish calculation of horse/human/year – most readers are unfortunately (both on her, and on their, accounts….) likely to form the notion that the accumulation of the remains of "32,000 to 100,000" horse carcasses at the site is indicative of a most ferociously carnivorous existence of the humans involved.
The calculation proceeds as follows:
Taking the middle value of 60,000 horse carcasses, over the least number of years (20,000 years), gives 3 horses killed per year.
Now, unlike the much less numerous reindeer remains at the site where only a cursory examination of 100 bones "yielded numerous examples of reliable butcher marks", only 17 of the 4,483 horse bones examined bore signs of butchery – this is 0.38%.
However, to allow for the possibility that the bones with butcher marks were much larger (10 x) than the bones that didn't have butcher marks, 0.38% is changed to 3.8%, which might well lead to a large exaggeration of the consumption of horse per person per year, but is unlikely to underestimate it.
So, 3.8% of 3 horses killed per year is 0.11 horse butchered/eaten per year.
And, this one tenth of a horse was eaten, in a year, by the group of people involved, say a group of 10 people, which is probably a minimum, the number quite possibly being twice or more that.
This gives one hundredth of a horse (the meat, not all of the carcass) eaten per person per year.
For an average horse of 400kg, very roughly 200kg of meat, one hundredth of that is 2kg of meat.
That is, 2kg of horse meat per year per person.
Or, 5 or 6 grams (.005 or .006kg) of horse meat per day per person.
And maybe it wasn't all eaten.
Maybe the hides were what they really wanted, to make warm clothes or even dwellings or tools/ accoutrements (being French as they were…) from.

And the real figure could be *much* less than this estimate which has been deliberately made conservative towards the orthodoxy's claims of humans having a ferociously carnivorous past and away from a more predominantly vegetarian past.
Which hardly constitutes a most ferociously carnivorous ex(subs)istence.
So, it seems likely, every now and then (maybe every year or few years), probably as much for the excitement, display of martial power, and ego-trip of it as for anything else other than to get hides for clothing, and/or for some ceremonial/religious reason(ing) (the horses were taken in summer, when other food was plentiful, the reindeers were taken in winter, when other food was not plentiful) they drove some horses up the canyon to the dead end, killed them (maybe released some), ate little, took hides, and left the rest to rot. Violent, yes, carnivorous, much less so, it seems, but without this little bit of basic maths to calculate the likely actual contribution to *daily* diet of this activity carried out over tens of thousands of years the image very falsely given is one of predominant carnivory....]

[And another example of biased conjecture not based on a sufficiently sensible critical analysis of the evidence available:]

Richards MP (2002) Review A brief review of the archeological evidence for Paleolithic and Neolithic subsistence. Eur J Clin Nutr 56:1-9.

"... the first tools that can be demonstrably shown to have been purposefully created by hominids are the simple Oldowan choppers and scrapers that appear in Africa ca 2.5-1.8 million years ago (Jurmain et al, 2000) ......
With the appearance of Homo erectus we find an associated more complex stone tool industry, the Acheulian, characterized primarily by the ovate bifaces of various sizes that have been found throughout the old world.
The uses of the tools in these two traditions are still not known definitively.
Experimental archeology, where modern replicas are made and used in a variety of ways, has shown that these tools, such as the handaxe, are remarkably diverse, and could be used effectively to butcher animals as well as cut up plant foods.
Microwear analysis (Keeley, 1980), where the pits and scrapes on tools are recorded with a microscope, has been used in a number of locations to infer the use of some of these early stone tools. These studies generally give mixed signals, indicating that most tools were used for a variety of purposes, or that the method is flawed, and can be confounded by the addition of new microwear after deposition. ......
What is missing from the archeological record, even up to more recent time periods such as the Upper Paleolithic, is plant remains. They only rarely survive, and as they may not have required sophisticated stone tools to collect and process, we may not recognize the stone tools that may have been used to collect and process them.
[For example, nut- and seed-cracking hammer and anvil stones, and grass seed-grinding stones, all of which can be comprised of natural rocks of suitable shape requiring no modifications, that are therefore virtually invisible archeologically, and tuber-digging sticks which, like plant food refuse, decompose quite away, leaving nothing to evidence them.]
To summarize, the archeological record, consisting of stone tools and animal bones, is most likely biased and unrepresentative of the daily subsistence patterns of hominids. Our inferences, from what we do have, is of a strongly meat-based subsistence, starting with the appearance of our line, Homo, in Africa, as evidenced by the remains of butchered animals and increasingly sophisticated stone tools.
[This is sheer nonsense – no attempt whatsoever has been made to even roughly calculate what rate of consumption per hominin is suggested by the fossil evidence – nor is the concept that such calculation is required even mentioned – nor is there any mention of the possibility of scavenging or hunting for hides or bones not necessarily including

consumption of the meat available, which it might not have been had more palatable foodstuffs been readily available…..] ……
The earliest hominids that have been studied are Australopithecines from South Africa (Sponheimer and Lee-Thorp 1999, Lee-Thorp et al 1994) …… It is unclear why they argue that the presence of C_4 signal necessarily means that the hominids were carnivorous [They didn't argue that, see section 1.13 below].
If the majority of C_4 plants were grasses, and therefore inedible to humans (which is not known),
[and unlikely, it being a small step from cracking nuts with hammer and anvil stones to roughly grinding seeds with appropriately shaped natural stones; as well as eating (sufficiently chewing) undried seeds (mature and immature) being quite feasible without grinding; and eating dried seeds after soaking being quite feasible.]
the C_4 signal must have come from consuming animals (or insects) that consumed C_4 plants. However, they do indicate the probable presence of tubers of C_4 plants that could have been consumed by the Australopithecus afarensis."

[And another example of biased conjecture not based on a sufficiently sensible critical analysis of the evidence available:]

Eaton SB, Konner M, Shostak M (1988) Stone agers in the fast lane: chronic degenerative diseases in evolutionary perspective. American J Med 84: 739-749.

"The rapid cultural changes that have occurred during the past 10,000 years have far outpaced any possible genetic adaptation
[This is not correct with respect to disadvantageous genetic variations being removed (for example susceptibilities to some infectious diseases); nor is it correct with respect to allowing deleterious mutations to build up in the gene pool (for example through medical preventive, remedial and curative treatment allowing carriers to reproduce at a greater rate than they would have otherwise);
it is more correct with respect to more than one mutation coming together to make a new function in the same way that we understand that very often something can be destroyed much more easily and quickly than it can be created (For example in the acquisition of lactose tolerance by adults in some populations, which has come into the gene pool(s) probably inside the last 10,000 years, it is the destruction of a mechanism previously switching infant/child lactose digestion off that occurred, rather than that a new mechanism was created for adults)] ……
The late Paleolithic era, from 35,000 to 20,000 B.P., may be considered the last time period during which the collective human gene pool interacted with bioenvironmental circumstances typical of those for which it had been originally selected.
[It probably wasn't, because the climate at this time was that of one of the glacial periods of the most recent ice age, and glacial period bioenvironmental circumstances may not be taken to be typical of ice ages (as they also include warmer (by 9 C) interglacial periods), and ice age bioenvironmental circumstances may not be taken to be typical of non-ice age bioenvironmental circumstances, there having been several ice ages and non-ice ages, as well as of the glacial and interglacial periods within each ice age, and moreover, the last glacial period of the last ice age (in which the late Paleolithic period was, and in which we still are), is, as a matter of fact, not "typical" of the bioenvironmental circumstances in which most of our gene pool was originally determined]
It is because of this that the diet, exercise patterns, and social adaptations of that time have continuing relevance for us today.
[Rather, it is because of these climatic aspects I note here that we must look at a much more complete complement of our archeological and genetic past, not just at the period

noted here by Eaton et al which does not at all provide the necessary context for answering the question of just what sort of a diet is best for what we have evolved to be]. Nutrition.

The diets of Paleolithic humans must have varied greatly with latitude and season, just as do those of recently studied hunter-gatherers; undoubtedly there were periods of relative plenty and others of shortage; certainly there was no one universal subsistence pattern. However, the dietary requirements of all Stone Age men and women had to be met by uncultivated vegetables and wild game exclusively;

[fruits, nuts, oily seeds, cereal grains, root vegetables and green leaf vegetables, all need to be enumerated and addressed to be able to have a sensible discussion on this subject!] from this starting point, a number of logically defensible nutritional generalizations can be extrapolated [4 [cites Eaton & Konner 1985]]. (1) The amount of protein, especially animal protein, was very great

[The evidence provided here does not support this claim – this claim is very wildly assumptive, and probably tends from the well-documented great tendency of humans to psychologically rationalize whatever they are habituated to like to do, in this case, eating lots of meat….].

The mean, median and modal protein intake for 58 hunter-gatherer groups studied in this century was 34 percent

[cites Eaton & Konner 1985, but, as they note there, "In comparison with the majority of Paleolithic human beings, existing hunter-gatherers occupy marginal habitats, and their lives differ in many ways from those of people living before the advent of agriculture". Also they fail to note that as the human population expanded exponentially (very probably with the evolution of language, and the supra-laryngeal (mouth cavity) structures required for it, providing the final degree of immunity from predation by large carnivores, and/or reduction in infanticide, and/or reduction in warring, that had previously held some if not quite sufficient check on population growth), with the population curve starting to steepen up notably through the Late Paleolithic and preferred foods probably therefore becoming in shorter supply, proportionately more (and maybe much more) non-preferred foods were eaten in this period – and on top of this it was a glacial period of an ice age – so that what was eaten may not logically be assumed to be what humans were evolved to best eat….] ……

(10) Because they had no domesticated animals, they had no dairy foods; despite this, their calcium intake, in most cases, would have far exceeded that generally consumed in the 20th century.

[Purportedly the large amount of high calcium plant matter consumed, with or without ground/chewed bone, provided a large amount (~1,800 mg/day) of calcium – see also review of Eaton, Shostak and Konner (1988) section 1.20 below]. ……

Skeletal remains can be used for estimation of strength and muscularity. The prominence of muscular insertion sites, the area of articular surfaces, and the cortical thickness and cross-sectional shape of long bone shafts all reflect the forces exerted by the muscles acting on them. Analyses of these features consistently show that preagricultural humans were more robust than their descendants, including the average inhabitants of today's western nations. This pattern holds whether the population being studied underwent the shift to agriculture 10,000 or only 1,000 years ago, so it clearly represents the results of habitual activity rather than genetic evolution.

[This is very incorrect – if for example, and I think it definitely had some effect in many places, organized warfare between specialized societies led to preferential selection of larger and stronger men for the armies, who would then often enough kill off large proportions of each other, this could quite easily reduce the population gene pool of genes for largeness, strength and robustness (exactly the opposite of what would happen in combat between random individuals fighting each other, or in small groups where all males took part in fighting – where the smaller are more likely to die) – and this is a fairly basic consideration for anyone even moderately competent in human genetics and anthropology to make, the absence of which consideration does not suggest a very good analysis, as the paper is, after all, about the evolution of humans….]

1.7 Archaeological site discovery biases

[(Mann 1981) discusses the bias that stems from many sorts of habitat environments not being conducive to fossils remaining, but rather to decomposition:]
"However, conditions in various habitats do not equally favor fossilization, but vary according to such factors as rainfall, temperature, humidity, type of major vegetative cover, amount of biological activity in the topsoil, depositional conditions favoring rapid covering of remains, and other features (Brouwer, 1966). In general, forest floors are very poor places for fossilization. Unless a forest animal falls into a context where rapid deposition of sediments is occurring – such as a stream – and the bones can be quickly removed from destructive agencies, its remains rapidly disappear. Only a small number of habitats offer optimal conditions for deposition of bony remains
[And note how much less likely remains of plant foods would be at such places].
Therefore, most fossil samples may derive from those environments most conducive to long-term preservation, even though these environments may have been exploited only marginally by hominids, or occupied for only brief intervals during the year."

[Now, we may accept that Africa was the place of origin of humans, on the basis of relatively recent DNA analyses showing more genetic diversity among African hominins than elsewhere on earth (reference required ???). However, if the preferred foods (such as fruits) of hominids in Africa were in forests, and the savannas were a fallback place with fallback foods for seasons of shortage of the forest foods, or places that competitively or otherwise dispossessed or dispersed hominids lived out of necessity or habituation, then the assemblages (grouping of all sorts of remains: bones, what little if any plant remains, buildings, excavations, tools and footsteps) of fossil remains found will be biased so as to be not properly representative of what the average nor the variety of prehistoric diets were. In other words the sites found and studied may represent the fringes of the hominid environments rather than the main hominid environments.
As Mann (1981) notes:]
"In view of the ability of modern gatherer/hunters to exploit a number of habitats during a yearly round and the similar patterns observed in baboons and chimpanzees, early hominids may have seasonally exploited various habitats, with groups specializing in different food resources in separate geographic areas."

[A cursory look at some of the sites in Delson et al's (Eds) (2000) Encyclopedia of human evolution and prehistory suggests that the hominid fossils that have been found in Africa are those preserved by volcanic ash coverings ('tuff' layers) and alluvial sedimentation. Isaac and Isaac (1977) note just how rarely there have been environmental conditions such as to leave high quality archaeological remains:]
"Olorgesailie was suitable as a testing ground because artifacts could be recovered from fine-grained sediments that had preserved detailed evidences of archeological associations. These could be related to stratigraphy and to features of the original environment. This preservation was in marked contrast to the sites that have yielded most of the evidence on which classic interpretations of Middle Pleistocene culture had previously depended. Until recently, most hand axes found in Africa, Asia, and Europe had been recovered from relatively coarse alluvial or littoral deposits. Even those that were extracted from more promising contexts generally had provenance records concerning only position in a generalized vertical stratigraphic sequence."

[Also, it is fairly obvious that even if sites for exploratory investigation were chosen randomly (and they almost certainly weren't, because fairly early in the history of archaeology, certainly of early hominid Archaeology, archaeologists gained an understanding of what areas would be more likely to have preserved remains available to be found, and chose to look in those sorts of areas), within the area looked over, very often it is the initial finding of collections of animal bones and manufactured stone tools, not the finding of hominid bone remains, that is more likely to be the trigger for them to excavate a site – this again leads to a bias to finding evidence of carnivory rather than of plant eating. One of the few examples readily available in the literature, of the way in which archaeological sites are located is this (partial) one by Isaac and Isaac (1977):]
"During our journey from Naivasha to Magadi Mr. Hobley and I found independently near our camp at the Ol Keju Nyiro, at the northern foot of Mount Ol Gasalik, some large roughly chipped axes…"

1.8 Archaeological burial soil contamination of remains (diagenesis), Sr/Ca; Ba/Ca; $^{13}C/^{12}C$

[Fabig and Herrmann (2002) provide a report of work indicating the contamination of bone remains Strontium/Calcium (Sr/Ca) ratio by burial soil:]

[Regarding the use of Sr/Ca and Ba/Ca ratios to impute diet of fossil human/hominids:]
"After the cleaning process, bone samples histiomorphologically determined as poorly preserved [(S3)] did not differ from well preserved bones [(S1)] in respect of the values of the essential mineral chemical control criterion, the Ca/P ratio. [i.e. Kerscheim/Teck site: S1 2.22 S3 2.18 nss (nss = not statistically significantly different);
Wenigumstadt site: S1 2.34 S3 2.30 nss ;
Neresheim site: S1 2.23 S3 2.24 nss ;
Weingarten site: S1 2.22 S3 2.17 nss]
Therefore, the sample cleaning procedure was regarded as successful. ……
the histiomorphological subsets of poorly preserved (S3) and well preserved samples[bones] (S1) always differ significantly in their [Sr/Ca] concentration[ratio]s (see Table 1);
[i.e. Kerscheim/Teck site: S1 0.44 S3 0.49 ss $p < .0041$ (.05/12 = .0041Bonferroni adj. for 12 compar. grps.)
Wenigumstadt site: S1 0.73 S3 0.90 ss $p < .0041$ (.05/12 = .0041Bonferroni adj. for 12 compar. grps.)
Neresheim site: S1 0.53 S3 0.81 ss $p < .0041$ (.05/12 = .0041Bonferroni adj. for 12 compar. grps.)
Weingarten site: S1 0.88 S3 1.40 ss $p < .0041$ (.05/12 = .0041Bonferroni adj. for 12 compar. grps.)]
…… In detail ["particular" better translation], Sr/Ca and Ba/Ca
[Ba/Ca ratio differences were very similar to the Sr/Ca in all four cases, and all were likewise ss (statistically significant)]
ratios increase with decreasing degree of histiomorphological preservation in all sites where analyses were possible
[Bones of intermediate (S2) preservation were not always of higher Sr/Ca ratio than their respective well preserved S1 counterparts, but were always of lower Sr/Ca ratio than their respective poorly preserved S3 counterparts.] ……
This finding is explained in terms of parsimony [the simplest explanation] by diagenetic changes in bone mineral. The obvious systematic character [consistent: found at all four sites examined] of diagenetic influence on the trace element concentrations we observed is in conflict with the general assumption that trace element concentration estimates yield dietary information if the accepted sample procedures [cleaning procedures] are used.

Even if the direction of diagenetic concentration change is systematic, the indirect results of our analysis confirm that the extent of the diagenetic processes is not consistent.
First, the varying state of the histiomorphological preservation, or the degree of diagenetic change in samples from one site, is caused by the variable soil properties within this site (see Hanson and Buikstra 1987).
Second it is noteworthy that the four early medieval populations under investigation here are dated to the fifth-eighth centuries AD. Therefore, all four samples are basically coeval [of the same time period] ...... do not depict a gradient of time. They only show varying degrees of the common characteristic pattern: dissimilar site-specific properties are bound to cause different degrees of diagenetical alterations.
[Note: there is the possibility that confounding might occur through diet being a factor in living bone porosity, further complicated by the probability that in the times of history and prehistory of interest bone porosity was more likely to be due to anemia due to blood loss from parasitization.]
...... Therefore, intra-population grouping structures based on trace element concentrations might only be interpreted very tentatively as dietary differences. ...... A diet-related interpretation is only permissible if subsets of concentration do not correspond positively with subsets of histomorphological or other markers for diagenetic alterations [but see my note above...]. ......
The second, and by far the most critical, consequence of our study is concluded from the fact that diagenetic interference on element concentrations could be observed, although the chemical cleaning of the bone specimens was carried out prior to the element analyses. All samples were proclaimed to be sufficiently cleaned of diagenetic components when simply abiding to the control criterion of the Ca/P ratio. ......
Repeated criticism regarding the suitability of the Ca/P ratio as a criterion for biogenic bone mineral therefore seems to be warranted (see Radosevich 1993) ......
There are several explanations for this dilemma: cationic substitutions of calcium, especially by strontium as well as by barium, during the burial period without an increase of solubility may occur (Radosevich 1993).
Furthermore, the anionic incorporation of fluorine may lead to an increased state of crystalinity, decreasing mineral solubility (see review by Sillen 1989).
The modification of the biogenic mineral structure after the death of an organism can be seen as a highly complex multidirectional chemical process (Pate and Brown 1985).
Thus, the removal of the more soluble compounds might not be the method of choice for sufficiently eliminating all traces of diagenetic concentration alterations (see implications recently pointed out by Budd et al. (2000)). ......
It does not seem possible at present to clearly distinguish between biogen and diagenetic signal using current analytical procedures."

[Therefore, both the statistical significance (having enough bones or (better, usually) individuals), and the characteristics of the soil both surrounding and near to the bones, should be taken into consideration in interpreting bone mineral content for indications of diet. If there are bones of other animals with acceptably well known or assumable diets nearby in (apparently) similar soil types, then use of these for comparison should be made, but the limitations of the lack of a desirable quantity of data, and of the uncertainty of the comparability of soil types (including what moisture there has been in the soils in times past) surrounding and next to the bones being compared, must be borne in mind as always indicating some degree of uncertainty in the comparison and drawing of dietary inferences.]

Schoeninger MJ. Chapter 9 Reconstructing early hominin diets Evaluating tooth chemistry and macronutrient composition. In, Ungar PS. (Ed) (2007) Evolution of the human diet The known, the unknown, and the unknowable. Oxford University Press.

"In a series of studies, my colleagues and I analyzed fossil teeth from a single excavation in Locality 261-1 (M. Leakey, personal communication, 1996) at the Pliocene site of Allia Bay, east Lake Turkana, Northern Kenya, which produced several specimens of the early biped, Australopithecus anamensis (Coffing et al., 1964; Leakey et al., 1995). …… These results indicate that use of strontium [Sr] and/or barium [Ba] concentration for diet reconstruction would produce inaccurate results at this site (contre Sillen, 1992, for fossils recovered in South Africa). ……

All of the browsing fauna had exterior enamel that has a higher delta ^{13}C value than their interior enamel. In contrast, all of the grazing fauna and mixed feeders, with the exception of two bovid samples, had exterior enamel that has lower delta ^{13}C values than their interior enamel.

In other words, the enamel exteriors of browsing and grazing fauna are more similar to each other [in delta ^{13}C values] than are the enamel interiors. The pattern of alteration suggested that the surface of the enamels was altered in the direction of sediments that surrounded the enamel during fossilization (see Lee-Thorp, 2000, for another example). The magnitude of the change, however, is much larger in some cases than previously recognized, and it varies between fragments.

In some cases, the alteration is great enough to misinterpret the animal's feeding strategy where the external enamel indicated a mixed feeder, while the interior enamel indicated a browser or a grazer. ……

When we removed the altered enamels from the fossils from Allia Bay, most of the fauna have lower delta ^{13}C values in their enamel interiors than is true of similar fauna today and of fauna from the later site of Swartkrans in South Africa (see fig. 9.2; Schoeninger, Reeser and Hallin, 2003). This suggests that the animals fed in regions that were more closed (i.e., forested) than today. In contrast, if we had used data from the enamel exteriors, we would have concluded that the fossil environments were similar to today. ……

We could use the diagenetic alteration in fossil teeth to reconstruct the ecological history of a fossil site (see Karkanas et al., 2000, for a similar approach). Once we identify particular environments and/or histories, we could focus our analyses on faunal groups where all the fossilized animals experienced the same extent and type of diagenetic process. Such similarity would suggest that they had been buried at similar times and hence had been alive at the same or nearly the same time. ……

The element compositions of teeth from the single excavation at Allia Bay, northern Kenya, clearly show variable diagenetic alteration. This argues strongly against the common assumptions that fossils taken from single deposits within sites represent a single assemblage. Rather, it points to the necessity for dealing with each fossil sample individually, demonstrating through screening methods like cathode-luminescence that they really are comparable before doing isotopic or trace element analyses. …… At this point, I choose to err on the side of caution whenever diagenesis is as likely as a biological process. At the same time, I recognize that the alternative explanation cannot be rejected completely."

1.9 Some archaeological sites, their evidence, and interpretations

Harding RSO, Teleki G. Chapter 1 Introduction. In, Harding RSO and Teleki G (Eds) (1981) Omnivorous Primates Gathering and hunting in human evolution. Columbia University Press.

"He [(Dart, 1925)] reasoned that the abundance of food in the forests had offered no challenge to the apes, and thus they had remained apes; the open veldt, by contrast, required "higher manifestations of intellect (p. 199)." ...... Dart, for instance relied on the masses of broken bone found in the breccias [rock made of solidified rubble] of Makapansgat and other South African sites for evidence that the australopithecines were carnivores, ... "Man's predecessors differed from living apes in being confirmed killers: carnivorous creatures, that seized living quarries by violence, battered them to death, tore apart their broken bodies, dismembered them limb from limb, slaking their ravenous thirst with the hot blood of victims and greedily devouring livid writhing flesh." ......
Having established the australopithecines' carnivorous habits to his own satisfaction, Dart proceeded, in a circular argument repeated by later writers, to establish that these creatures were hominids by comparing their purported behavior with his own rather dismal view of history. ......
By contrast, there was little disagreement that hominids had been involved in creating some of the patches of bone debris uncovered at Olduvai gorge, for these were associated with clearly utilized stone tools (Leakey, 1971) ...... Circumstantial though it was, this evidence was especially appealing because an association had long existed in the minds of scientists between the emergence of the hominids, their use of stone tools as weapons, and an evolutionary reduction in the size of canine teeth in the hominid line. This association bore all the authority of Charles Darwin, who had stated: {The free use of the arms and hands, partly the cause and partly the result of man's erect position, appears to have led in an indirect manner to other modifications of structure. The early male forefathers of man were, as previously stated, probably furnished with great canine teeth; but as they gradually acquired the habit of using stones, clubs, or other weapons, for fighting with their enemies or rivals, they would use the jaws and teeth less and less. In this case, the jaws, together with the teeth, would become reduced in size.} (1871:435)
[It might be just as probable that early hominids ate *less* meat than even the very low proportion of diet eaten by the common ancestor of apes and humans, due to the increased difficulty of catching it before the use of throwing spears was well enough developed – they would not have climbed well enough to catch monkeys, nor ran well enough to catch most ungatherable terrestrial animals in open country – and that this contributed to the reduction in size of the canine teeth – even the venerable C. Darwin not perfect in his thinking...]
Thus from Darwin's time onward, tools came to be thought of as weapons, even though no analysis of these mute stones had provided any clue to their original use. A typical example of this bias can be found in Ardrey's African Genesis (1961), where the reader will search the index in vain for references to tools; instead there is only the brief notation, "see Weapons." From the use of tools as weapons, it was but a short step to a hypothesis in which tools made possible a hunting way of life.
The paleoanthropological preoccupation with hunting may have stemmed in part from the fact that the first prehistoric excavations were carried out in Europe, in deposits laid down in glacial times when the severe climate probably did make human populations especially dependent on animals for food. As Mann has suggested (personal communication), the abundant evidence for hominid exploitation of animals in the periglacial environments of the Upper Paleolithic in Europe may have colored the way in which many subsequent sites were interpreted, even if these newer sites were located in Africa and at much earlier time levels. ......

In fact, any primate living in an equatorial habitat would find it virtually impossible to avoid consuming substantial amounts of animal matter along with plants; nearly every plant, and certainly every ripe fruit, is probably infested with invertebrates. ......

the collection of insects, for example, may involve motor patterns exactly like those performed while collecting grass seeds. ......

Were it not for the immense power of labels in regulating thoughts and forming concepts, often fixing inaccurate images in our minds well before solid evidence has been obtained ……

…… In this line of reasoning, the hunting aspect of resource exploitation has been heavily emphasized while gathering has received little attention, to the point where many earlier writers omitted gathering completely when referring to these people as "hunters." …… It is important to stress here that all these approaches [(various theorizings of carnivorous behavior types and their predominance)] are founded on the tacit assumption that the obtaining and eating of meat was the decisive adaptation in early hominid evolution: thus they lack any explanatory power if this assumption cannot be demonstrated. ……

We are convinced that most of the evidence and argument advanced in support of hunting as the "master behavior pattern" has never been rigorously tested on the basis of all the available information.

In fact, the issue may never be resolved so long as those writing on either side of the question continue to rely on subsets of facts that support one particular position or prejudice. ……

The primate evidence in particular suggests that subsistence activities wax and wane over time in different populations, so that a pattern may become preeminent under certain environmental conditions while remaining relatively unimportant under other circumstances. ……

Primates are indeed omnivores; from prosimians and monkeys to apes and humans, we share an adaptation which, in some members of the order, has developed further into dietary specialization."

[Harding and Teleki do not go far enough here, in their not noting that 'omnivore' is not a particularly useful term as it does not indicate on its own where along the continuum of omnivory the diet lies, and that humans are probably optimally very predominantly vegetarian omnivores, eating a fair variety of the plant food types but with a favoring of fruit: fruit, legumes, nuts, oily seeds, root vegetables, green leaves, and cereal grains; and that insect consumption be greater than meat consumption, meat not being necessary, nor evidenced to be beneficial to health, given just a small consumption of insects.]

Mann AE. Chapter 2 Diet and human evolution. In, Harding RSO and Teleki G (Eds) (1981) Omnivorous Primates Gathering and hunting in human evolution. Columbia University Press.

"The archaeological record is a highly selected information source; some materials preserve very well and are routinely found during excavation, while others are not preserved or leave such subtle traces that they are missed.

Many dietary reconstructions have been based exclusively on the recovered evidence, ignoring items that might originally have been present but that have disintegrated long since.

Because vertebrate bones preserve better than plant materials or invertebrate exoskeletons, the basic diet during much of human evolution has been viewed in terms of an emphasis on a single food resource: mammals that are hunted ……

Even if an archaeological site preserves an unusually complete record of the foods eaten at that spot, an appreciable percentage of the diet may have been consumed elsewhere. ……

The major problem with an analogical [(with living hunter/gatherers / gatherer/hunters)] approach to reconstructing dietary patterns is the cultural and temporal distance between earlier hominids and modern gatherer/hunters. …… there is some dispute about how

representative of earlier gatherer/hunter hominids these groups may be (Perper and Schrire, 1977).
Contemporary gatherer/hunters exist for the most part in marginal environments (but see Silberbauer, in this volume), perhaps forced there by more powerful, agriculturally based neighbors.
The earlier richness of the gatherer/hunters' subsistence may never be fully appreciated, given the context of their currently impoverished refuge areas.
It has also been observed that living gatherer/hunters use a number of very sophisticated weapons, such as poison darts, bows and arrows, snares, and traps, which may have been beyond the technological capabilities of all but our most recent ancestors. Can the subsistence patterns of these well-equipped humans be compared to those of earlier hominids who had Oldowan, Acheulian, or Mousterian tool kits? ……

Wolpoff (1973a,b, 1976) has emphasized the large size of the back teeth in all early hominids; the summed area of the posterior teeth, when compared with rough estimates of body size, show that surface chewing areas were very large in proportion to body weight. He suggests that this large chewing area with its characteristic thick occlusal enamel is indicative of very heavy mastication, perhaps of seeds but also of roots and other items that are hard to chew.
[But it is the *bulk*, not the difficulty, which is related to area, and this could include grass seeds – it is the *hardness* that relates to enamel *thickness*, though a greater span will require a greater thickness, even though this span is well supported underneath.] ……

However, it is unfortunate that the last items to be eaten before the sample is collected tend to leave the most obvious markings on the occlusal enamel (Walker, Hoeck, and Perez, 1978, and Walker, personal communication).
Thus if early hominids had a diverse diet, a great many would have to be sampled to obtain an accurate picture.
Even then, conclusions may be biased if the hominids from one site represent a particular seasonal feeding time or a particular specialization. ……

We need to know more about the exchange of chemicals between buried bones and the surrounding soil, ……

Finally we must consider how the culturally influenced food preferences of Western societies have shaped our view of earlier dietary adaptations.
For instance, few discussions of the evolution of hominid dietary patterns mention insects as a possible source of high-quality protein (with adequate amounts of all the essential amino acids), yet insects have been and continue to be extremely important in the diets of many peoples (Bodenheimer, 1951; Meyer-Rochow, 1973; Oliviera et al., 1976).
Similarly, it is possible to ask whether the widely held view that the early hominid diet was predominantly meat may also derive in part from modern Western food preferences. ……

It is also possible that some human populations have specific nutritional requirements not found among the majority of living humans. ……

As Scrimshaw and young point out (1976), deficiency diseases will result from the absence or low intake of certain nutrients, but good health and well being, the other end of the scale, are very hard to define. ……

Animal tissue, eggs, and dairy products contain enough of the eight essential amino acids to permit the body to synthesize protein. All other protein sources lack one or more of these essential amino acids.

[This is gross ignorance of the most basic sort! The sort of statement made by people who have not taken the little time required to look at a tabulation of food amino acid contents – for example the qualified dietician who managed to be allowed to give me a failing grade in a public health nutrition course at JCU – whose error in making just that statement I had pointed out in an assignment, in a very appropriately unflattering tone, though without naming her exactly. The fact of the matter is that the proteinaceous plant foods are only a *little* low, relative to the proportion to other amino acids required, of some amino acids – most notably cereal grains and nuts and seeds being generally a *little* low in lysine, and a *little* high in methionine, and legumes being generally a *little* low in methionine, and a *little* high in lysine. (see chapter 5 below)
Furthermore, it is not "very important" to combine these two groups in a meal to optimize amino acid ingestion, it merely improves (and that only potentially, as other foods eaten in the meal also contribute their, lesser, amounts of various amino acids to the total complement) it a little. Almost invariably we eat more protein than we need, because we need so little of it – and because the limiting amino acid is only limiting to a relatively small extent, most often a meal with legumes but no grains, nuts or seeds, or a meal with grains, nuts or seeds but no legumes, will provide more than enough of each and all of the essential amino acids needed.
Furthermore other than supposedly eggs (which by the way have purportedly only 1/6 the vitamin B12 absorbability of other common animal foods), all the animal foods have their own limiting amino acid too – and in the case of cooking at the usual temperatures other than boiling, Maillard Products (Lysine-sugar) will be formed, reducing the amount of lysine that can be absorbed by the gastrointestinal tract from ingested food, rendering lysine even more so the limiting amino acid that it commonly (slightly) is in animal foods to start with…] ……

Scrimshaw and Young (1976) discovered that a group of American undergraduate males achieved protein balance with the consumption of 0.57g of protein per kg of body weight per day. While both the FAO tables and most recent NRC standards now recommend between 0.5 and 0.6 g of protein per kg of body weight per day, it is still unclear whether this amount is needed by all populations. As noted earlier, some studies suggest that some modern human groups may remain fit with lower daily intakes of protein. ……

Linoleic acid [an essential fatty acid] is found primarily in vegetable fats, especially in nuts and seeds, and is also well represented in the fats of some insects (Oliviera et al., 1976). …… The fats of some amphibians (frogs, tortoises [reptiles, not amphibians]), some reptiles (lizards, pythons, crocodiles), and some birds contain relatively high proportions of linoleic acid. Most of the limited number of animals examined, however, contain only small amounts of linoleic acid. ……

Most mammals possess the biochemical machinery to synthesize vitamin C (ascorbic acid), but this ability has been lost in humans and many other primates tested (Stone, 1965). Pauling (1970) hypothesized that the enzyme which is responsible for production of ascorbic acid and is found in the liver of most other mammals was lost during primate evolution because ample supplies of this vitamin were present in the diet. ……
Vitamin C is found widely in fruits and vegetables, but only small amounts are present in animal tissue. ……
Draper believes that "The Eskimo practice of eating their food in the raw, frozen, or lightly cooked state was a critical factor in preserving the small amounts of vitamin C necessary to prevent scurvy (less than 10mg/day)" (1977:310). This observation is interesting because it suggests that the discovery of fire, usually considered to have been a great advantage to hominids because it broadened the diet to include items that needed to be cooked to soften or detoxify them, may also have had some serious disadvantages for any hominids subsisting largely on meat. …… Nickerson, Rowe, and Richter (1973)

discuss the methods used by some North Alaskan Eskimos to exploit the sometimes abundant, but short-lived, appearance of plant foods. They report that six different varieties of berries and the shoots, buds and leaves of willow, various roots, and wild spinach are gathered in season, with some being frozen or preserved in seal oil for consumption during the winter. ……

There is little agreement, however, about what the hominids might have eaten in this habitat [the savannah].
Leopold and Ardrey (1974) have argued that there would have been only a limited number of vegetable foods available on the savannah to a mammal with few dental and digestive specializations, and thus early hominids are best viewed as predators.
In contrast C.J.Jolly (1970, 1972), who also sees the early hominids as open-country forms, feels their diet was composed of "the seeds and rhizomes of grasses and the tough, dry fruits of savannah trees and shrubs, with the addition of many insects and leaves, occasional fleshy fruits from riverine forests, and some vertebrate meat. In other words … foods requiring fine comminution and hard grinding" (1872:10).
Szalay, however, feels that the early hominid dental system is better seen as an adaptation to hunting and scavenging, with the cheek teeth with their thickened enamel being used for "cracking ribs, metapodials, joints, or for increased longevity or both" (1975:428) ……
However, conditions in various habitats do not equally favor fossilization, but vary according to such factors as rainfall, temperature, humidity, type of major vegetative cover, amount of biological activity in the topsoil, depositional conditions favoring rapid covering of remains, and other features (Brouwer, 1966).
In general, forest floors are very poor places for fossilization. Unless a forest animal falls into a context where rapid deposition of sediments is occurring – such as a stream – and the bones can be quickly removed from destructive agencies, its remains rapidly disappear.
Only a small number of habitats offer optimal conditions for preservation of bony remains.
Therefore, most fossil samples may derive from those environments most conducive to long-term preservation, even though these environments may have been exploited only marginally by hominids, or occupied for only brief intervals during the year. ……

In view of the ability of modern gatherer/hunters to exploit a number of habitats during a yearly round and the similar patterns observed in baboons and chimpanzees, early hominids may have seasonally exploited various habitats, with groups specializing in different food resources in separate geographic areas. ……

The archaeological sites at Olduvai (Isaac and Crader, in this volume; M.D. Leakey, 1971), east of Lake Turkana (Isaac, 1978; Isaac and Crader, in this volume), and the Omo Basin (Merrick, 1976), are the remains of hominid living floors. They contain scattered animal bones (sometimes thoroughly smashed), stone tools, and occasionally hominid bones and teeth. ……

Chimpanzees, after all, have undergone a great deal of evolutionary change since they last shared a common ancestor with hominids and therefore may not be valid models for them."
["may have", not "have", and therefore one needs to look for commonalities among all the apes, and if one does, one will find some… notably, predominantly frugivorous very predominantly vegetarian omnivory with insectivory predominating over carnivory]

Isaac GLI, Crader, DC. Chapter 3 To what extent were early hominids carnivorous? An archaeological perspective. In, Harding RSO and Teleki G (Eds) (1981) Omnivorous Primates Gathering and hunting in human evolution. Columbia University Press.

"The adaptive strategies of recent humans who live without agriculture are outside the range of those found among other primates, or indeed among other mammals.
[These peoples may not be representative of recent gatherer/hunters living on less marginal lands, nor of the earlier hominids] ......

The foregoing [mention of division of labor between the sexes, and of food sharing] is a cursory sketch of some characteristics of human socioeconomics in which our species differs from other primates. Even without consideration of phylogeny or the fossil record, this scheme inevitably suggests that the adoption of meat-eating by our protohuman ancestors may at some stage have had a crucial formative and facultative influence on the course of human evolution. This belief is highly plausible and is widely held by anthropologists on the basis of synchronic comparisons between modern humans and their living primate relatives. ......
In the dynamics of human evolution, there are two intersecting dimensions of variation: (1) the degree to which meat was a component of the diet, and (2) if meat was a component, the proportion acquired by hunting relative to scavenging or "gathering."
In fact, in considering any given stage of human evolution, the model must be elaborated immediately, as we need to allow for variation between contemporary populations in different circumstances, whether geographic, environmental, seasonal, or even "cultural."
Most scientists concerned with the study of the last few million years of human evolution assume that the lineage ancestral to mankind was to a degree "carnivorous" and that this dietary involvement had a significant influence on the trajectory of change. We regard this view, when not pushed to extremes, as very reasonable and plausible.
However, it is beginning to be incorporated into the literature as established knowledge in a way that it should not be. "Early man, the hunter" is in danger of becoming a fact by default. ......
Given that recent humans are far more intensive flesh-eaters than any other living primates, the question can be rephrased: When did hominid meat-eating start to intensify, and how much interpopulation variation has there been at each successive stage?
[These are crucially important questions, and the answers I posit are: relatively very late, in an exponential fashion, with a fair bit of interpopulation variation.]

As we go back in time, for each major phase we ought to ask if there is clear evidence that at least some hominid subsistence patterns involved the eating of significant amounts of meat. The first step back in time takes us to the nonagricultural societies of the Upper Pleistocene, that is, the last 100,000 years or so. For these societies, the answers to our questions are unambiguous.
[They are in fact far from unambiguous with regard to the actual percent of diet constituted by this supposedly/assumedly "significant" and "substantial" carnivory – My own calculations here on pages 91-92 above fairly strongly suggest the contrary; that the carnivory was a quite low percent of diet until the very last preagricultural times.]
In many but not all regions, human artifacts of this age are very often incorporated in middens with impressive amounts of broken, burnt bone. The case for substantial meat-eating by some humans in this time range is well proven (see Klein, in this volume).

If we move farther back in time to the early Pleistocene and late Pliocene, to the span covering 700,000 to 3,000,000 years BP, the evidence is relatively scant. ......

For example, the flat-crowned cheek teeth of all hominids have thickened enamel and are quite unlike those of any of the carnivores, except for superficial resemblances to the

teeth of omnivorous or non-meat-eating members of the order, such as bears and the aardwolf (Proteles cristatus) (Dorst and Dandelot, 1970).
Given the morphology of hominid and human teeth, then, it is unlikely that our phylogeny includes any prolonged period of exclusively carnivorous diet, Ardrey (1976) notwithstanding. ……
However, we should recognize in advance that although scratch and chippage patterns may well indicate vegetable components in diet, the eating of meat itself is unlikely to leave particularly distinctive traces.
Tooth wear patterns may show that meat was not the exclusive food of a given fossil hominid, but they are unlikely to help us determine whether meat comprised 0%, 5% or 20% of its food. (See also Mann, in this volume, for discussion of these questions.) ……

Almost all direct evidence of protohuman life in the period before 700,000 years ago comes from a very small number of geographic areas, namely Java, northwest Africa, the East African Rift Valley, and the Transvaal in South Africa. …… At present Java contributes little or nothing to the solution of the problem at hand. …… In the Transvaal, hominid fossils have been found during the excavation of deposits that accumulated in caves and fissures. They occur jumbled with numerous bones of other animals. …… The question is; were the nonhominid bones at these sites accumulated in whole or in part by the hominids as food refuse? …… Many of the bone assemblages in question are unlikely to be middens accumulated by hominids (Brain, 1958, 1976a). Along the Eastern Rift, hominid fossils, numerous nonhominid fossil bones, and artefacts occur together in the deposits of large sedimentary basins. ……

Concentrations of broken bone in which more than 15-20% is unidentifiable bone splinters seem to be distinctive features of (1) human food refuse; (2) carnivore lairs at which bone is recurrently chewed; (3) some cave deposits, which may well be fossil examples of 1 or 2; (4) deposits where weathering and soil pressures have broken up the bone (Klein, 1977a). The degree of bone fragmentation observed at the early East African archaeological sites is consistent with the hypothesis that the bone was food refuse, but other agencies may also have been responsible in whole or in part. ……
Species composition does not by itself establish the role of hominids in creating the accumulations. ……

In summary, we find that even after ultracritical review, there is good prima facie evidence for the involvement of early tool-making hominids in meat-eating.
First, there are a number of instances in which anomalously high densities of bones coincide with high densities of discarded artefacts, under circumstances where water flow seems unlikely to have washed the stones and bones together.
Second, in many instances the aggregations of bone show a degree of fragmentation that is unusual in the natural bone aggregates observed thus far in open terrain. The co-occurrence of artifacts and fragmented bones is a reliable indicator of the habitations of later prehistoric and recent humans, and the notion that many of the Plio-Pleistocene sites represent early instances of analogous protohuman behavior seems highly plausible to us.
However, … alternative explanations …
1. Tool-making hominids may have frequented the sites, repeatedly bringing, using, and discarding stones and artifacts. They may also have brought bones encased in meat to these sites, and broken the bones in the process of eating meat and marrow.
2. At slightly different times both tool-making hominids and carnivores may have frequented the same restricted locality – for instance a grove of shady trees. The hominids may have discarded stone tools, while the carnivores brought bones and broke them up by chewing.

3. Tool-making hominids may have been attracted to places such as waterholes where quantities of bones were lying around, and the hominids recurrent or prolonged activity in the vicinity may have created a coincident scatter of artifacts.
4. Tool-making hominids may have habitually collected and broken dry bones for some reason other than meat-eating, perhaps for tool use.
5. Both artifacts and bones may have been swept together from initially dispersed and separate locations by stream action or flash floods. This alternative applies only to those sites associated with channels or depressions.
6. Given the formation of both stone artifact concentrations and bone concentrations in ancient landscapes, some could come to coincide purely by chance. ......
While we await the results of more rigorous tests, we tentatively conclude that meat-eating by hominids is the best of several possible explanations for the Plio-Pleistocene sites. ......

in turn raises other questions: Did the early tool-makers acquire meat by hunting or by scavenging? If both, can we determine the relative proportion of each? Assuming that gathered foods such as plants, insects and eggs were as important in prehistoric diets as in current human diets, can we assess the proportion of meat intake relative to these other foods?

What may have been the ecological, sociological, and evolutionary implications of meat-eating in protohuman subsistence economy? ......

However, it appears that even the nonhuman primates that are sometimes active predators either do not scavenge (chimpanzees: Teleki, 1973a) or do so very rarely (baboons: Harding, personal communication; Strum, in this volume)
[They all have been documented to 'power scavenge' taking by force freshly killed prey from the possessor – it seems most likely to me that early hominids probably power scavenged with weapons, due to a great distaste for decomposing meat, and this more in times of shortage of preferred plant foods, because lack of speed on the ground, and climbing ability in the trees, made it the most feasible way of getting fresh meat.] ......

Outside of the Arctic and cold temperate regions, gatherer/hunters who do not depend heavily on fishing have a diet consisting of more than 50% gathered foods, of which the bulk are plants (Hiatt, 1970; Lee 1968; Teleki, 1975).
Models of the early hominid diet that contain smaller proportions of plant food seem inherently much less probable than those that recognize the importance of foods other than meat.

At present archaeology can provide very little direct evidence. ......

In summary, it can be said that while the archaeological evidence is insufficient, it is at least consistent with modern hunter/gatherer [gatherer/hunter] dietary patterns, in which varied fruit, vegetable and insect foods make up more than 50% of the total food intake.
......

Equipment is involved in yet another aspect of the modern human adaptive mode, namely gathering, by which we mean acquisition and postponed consumption. Collection of tubers, fruits, nuts, insects, fungi, and foliage [and seeds and grains] for later meals is really feasible only with the aid of containers such as trays, bags, baskets, or pouches. Lee has called this set of subsistence aids "the basic human invention." ......
Only stone artifacts and very few ad hoc bone tools have been recovered from early hominid sites, but it is highly probable that among the most important functions of stone artifacts was the manufacture of simple wooden equipment.

The core tools (choppers etc) can be used to hack off fresh hardwood branches, which can then be trimmed and sharpened into digging sticks, spears, and so on, by using sharp flakes with or without secondary retouch. ......

Binford (1972) has observed that an abundance of handaxes tends not to be associated with abundant bone, implying that these ubiquitous tools were used mainly in activities not associated [directly] with meat consumption (see Freeman, in this volume; Isaac, 1971, 1975). ......

Most of the evidence for hominid subsistence activities from 3-4 and 0.7 million years BP comes from regions that today are neither forest nor extreme desert. In general, the early sites are found in the varied habitats that we may conveniently, but vaguely, classify as savannah – regions with a heterogeneous cover of grassland and woodland, interspersed with ribbons and patches of riverine or mountain forests.
The available paleo-environmental data strongly suggest that the early sites were formed in similar but slightly moister savannah, or grassland-woodland, habitats (see Bonnefille, 1972; Butzer, 1971a, 1978; Isaac, 1975).
[This is debatable as there have been marked variations over time in climate, and drought cycles are common enough in the present…] ......

Archaeologists have long been interested in the possibility that early hominids exhibited seasonal patterns of range utilization (e.g., Clark, 197:84-86; Howell and Clark, 1963; Kleindienst, 1961).
Speth and David (1976) have recently suggested that most of the lakeside sites at Olduvai were occupied during the dry season, based on the presence of significant amounts of tortoise remains at only 4 out of 22 hominid sites. They argue, in part by analogy with data for the !Kung of Botswana, for whom tortoises are a wet-season dietary item, that the ratio of dry- to wet-season sites at Olduvai is about 18 to 4. ...... 4 wet months and 8 dry months of the year … $p>0.10$. The data as they stand do not seem to preclude year-round hominid usage of the basin floor, with each site being occupied only briefly. ......

… the archaeological evidence emphatically does *not* indicate, contrary to what Ardrey (1976) so strenuously maintains, that early hominids were critically dependant on meat as the main item of diet.
Thus, although we suspect that the importance of meat per se has probably been exaggerated, we would argue that increased meat consumption may have been more important in human evolution than we can assess simply from the caloric value of the meat consumed.
Increased involvement in meat-eating may well have established an asymmetry in early hominid social life that in turn made division of labor and the development of reciprocal social obligations an advantageous adaptive trend. In many current hunter/gatherer [/ gatherer/hunter] societies meat acquisition and consumption seem to have quasi-symbolic overtones and to affect status as well as nutrition (Hayden, in this volume). ......

These authors [Morgan, 1972; Tanner and Zihlman, 1976; Zihlman and Tanner, in press] propose that a crucial step in the development of cumulative behavioral differences between evolving protohumans and the hominoid common ancestor of apes and humans was the inception of the practice of gathering plant foods for delayed consumption. ......
Why would a predominantly vegetarian primate find it advantageous to postpone consumption of fruit or roots instead of consuming them as they were obtained? We are skeptical about Zihlman and Tanner's suggestion that it may have been safer to acquire such foods in haste and then to take cover so as to minimize exposure to predators.
[Why be skeptical of this? Early hominids almost certainly lacked the speed on the ground to escape by flight across the ground, which is what other smaller prey animals

rely on to escape, as well as a fecundity and rapid developmental growth, that hominids also did not have.]
However, it is possible that if hominids fed extensively on items that had to be prepared with equipment, such as nuts needing to be opened with nut-cracking stones
[or seeds or tubers that were ground using naturally shaped mortar and pestle stones], then it might be easier to collect the food and go at intervals to the apparatus.
[And bear in mind that these processes would involve making noise, a dangerous practice regards attracting predators, for a small and not yet well terrestrially-adapted hominid whose method of subsistence very probably included quietness in movement, even if armed with the earliest evolutionary form of the spear and club.]
Alternatively, feeding in the shade may have been important.
[And all the more so if the tree were climbable for escape from predation.] ……

We thus favor a model in which the active delivery of some meat to fellow members of a social group developed in a reciprocal relationship with the practice of transporting and sharing some surplus plant foods. We see the model as representing a functionally integrated behavioral complex, in which any attempt to isolate one or another component as an initial or prime mover is probably misleading.
[I disagree, on the grounds that there was probably very little meat-eating for most of human/hominid prehistory, it having grown in an exponential manner mostly in very recent history, in particular when the use of fire made meat palatable, even if as a habit. Furthermore, the division of labor could well have been the division of the danger of gathering different items, which possibility seems to have been surprisingly unexplored in the literature, though it seems to me a rather obviously quite plausible, and fairly simple thing in practice for hominids to do.] ……

Our findings can be summarized as follows:
1. The most convincing positive evidence [of early hominid carnivory] consists of the patches of discarded artifacts and broken bones that have been excavated at Olduvai and in the Koobi Fora formation. ……
5. Ethnographic, primatological, and ecological considerations lead us to surmise that foods such as tubers, roots, fruits, nuts, eggs, insects and other small animals were quantitatively the main items of early hominid diet.
6. Some sites that lack bone, in spite of apparently favorable conditions of preservation, may attest to situations in protohuman life where meat was not a significant part of the diet, owing to seasonal, locational, or accidental factors. We cannot yet test this proposition rigorously."

Freeman LG. Chapter 4 The fat of the land: notes on Paleolithic diet in Iberia. In, Harding RSO and Teleki G (Eds) (1981) Omnivorous Primates Gathering and hunting in human evolution. Columbia University Press.

[Iberia is in Spain]

"Evidence about food, fuel, and raw material acquisition (hunting/gathering operations, butchering techniques) is commonly reported as though it were direct evidence for consumption, which, of course, it is not. ……
Even if we could recover an unbiased sample of all diet-related imperishables produced by an extinct human society, it would not give a complete idea of the diet of the times, since a large proportion of any past meal may have consisted of foodstuffs that do not leave anything we now recognize as a durable material trace (beverages, boned meat, greens, ground meal, and so on). ……

Among some societies today, much of the total dietary intake is consumed on the spot (Hayden, Mann, in this volume); this is especially true for small, perishable items such as berries or shellfish.
[The point is probably more the immediate edibility, rather than the perishability, the latter pertaining more to storage.]

In some groups where there is a pronounced division of food-acquisition activities by age and sex, during certain seasons children and women satisfy their major dietary requirements for days at a time in this way. ……

Opinions about palatability vary widely from group to group, of course, and prehistorians must keep their own ethnocentric biases from influencing their interpretations in this context. ……

In addition, from at least Solutrean times, fires were used to make flint more workable in the tool-making process, especially when pressure-flaking was involved. Thus caution should be exercised in interpreting the remains of fireplaces as indicating that food was cooked. ……
The approach taken by the "site catchment analysts" (who seem to assume that whenever a resource is available, it will be exploited) glosses over both the known complexities of hominid behavior and the great difficulties involved in reconstructing prehistoric environments in useful detail. ……
…, the Acheulian complex (a set of assemblages usually including characteristic tools called bifaces) is found in Africa, Europe and part of Asia for more than a million years, whereas the later Solutrean complex is known in the strict sense only from France and Spain, and lasts only a few thousand years. ……

The Acheulian complex (prior to ca. 100,000 BP)
The Torralba/Ambrona sites are sometimes characterized as "specialized big-game hunting" or "specialized elephant-hunting" camps. ……
[Now in 2010 it seems agreed by archaeological burial science (taphonomy) experts that the animal bones at Torralba were in fact deposited by water flow, and not by humans.]
The animals butchered at Torralba represent an immense quantity of edible meat. ……
There is no way to prove that most or any of the meat represented by the Torralba kills was eaten. ……
However, with one possible exception, nothing suggesting a deliberate fireplace was recovered. ……
The location and elevation of the Torralba and Ambrona sites are such that, today, the climate can be bitterly cold in winter, and nights are quite cool even in summer. During the full glacial period when the sites were utilized, the mean annual temperature of the region was about 10 degrees C lower than it is now, which would make fire, good shelter, and heavy clothing essential to survival in the winter months. …… If Torralba and Ambrona were warm-season camps …… In any case, it seems clear that the kinds of food acquisition activities we can detect at Torralba and Ambrona may have taken place only periodically, and perhaps on a seasonal basis. ……

Middle Paleolithic Patterns (from ca. 100,000 to 35,000 BP)
[Abric Agut, middle Paleolithic] The lagomorphs [rabbits] "average medium" and "large medium" animals are evidently human food refuse. Each category contains at least some bones showing butchering marks or traces of burning. ……
A number of samples from the Agut Mousterian were subjected to flotation, in the hopes of recovering macrobotanical remains. Twelve whole or fragmentary seeds were recovered from two very localized patches in adjacent 1-m squares in one level. ……
Their occurrence in restricted patches suggests to me that they were deposited in human

feces which subsequently disintegrated and disappeared. ...... At some time during the 24-hour period before visiting the site, an individual apparently ate a meal including legumes and grains of several species, some of which were accidentally charred before they were eaten. [This seems unlikely?] The charred seeds were deposited, probably in feces, on the site. ......
Several large, flattened circular grindstones (18-20-cm diameter) were recovered from one Mousterian horizon at Castillo. Smaller grindstones are known from other occupation levels. Apparently, such objects may occur in assemblages of any facies [parts/locations of a site]. We do not know what material may have been processed with these implements. Dried meat, vegetable foods, and coloring matter have all been suggested. ......
There is some evidence of fire in all facies, and in the upper Denticulate Mousterian horizon at Cueva Morin we found a single large hearth that had contained an intense fire... ......
The fauna from that level is extremely rich (Altuna, 1972), but I do not believe that it is a reliable indication of hominid diet. The exceptional number of cave bears and other carnivores suggests that the site served as a den for these animals and that hominid occupation residues probably contributed only a small part of the faunal remains. ......
I have suggested elsewhere that it is a specialized (perhaps seasonal) hide-working area, and that many of the bone fragments found in the occupation level are hide-working tools or blanks for manufacturing such tools (Gonzalez Echegaray and Freeman, 1978; see also Semenov, 1968:156-71). ......
Most other Mousterian occupations at Morin have little in the way of faunal remains.

Synthetic Overview: Lower and Middle Paleolithic
...... We know nothing of the vegetable foods that probably were utilized, perhaps in great quantity, by Iberian Acheulian hunters ......
Nor are the species lists [of animals reasonably thought to have been eaten] as diverse as they became in the Upper Paleolithic. Mousterians seem to have neglected many potential food resources in their environments, even when they were apparently aware that those resources existed. ......
It is no surprise that regional differences in lifeways became much more marked during the Mousterian than they ever had been before. ......

Diet in the Earlier Upper Paleolithic (from ca. 35,000 to 18,000 BC[(BP?)])
As representative of the available information concerning Upper Perigordian (Gravettian) horizons, I have selected two other levels at Morin: upper level 5, deposited under cold climatic conditions dating to 18,760(+/- 340) BC, and level 4, in temperate deposits. ...... hearths have been described ......charred, checked and sometimes calcined bone fragments ...... Two irregular "storage pits" (whose functions are not as clear as their name implies) were found in another Aurignacian level at Morin. The major dimension of one was 40cm, that of the other 1m, and both were about 0.5m deep and absolutely jammed with fragmentary bones and stone tools. The fill they contained was clearly not stored food remains but "garbage." However, true food-storage pits are sometimes jammed with debris just before a campsite is abandoned. In a fill deliberately packed between two of the graves in the Morin structural level, we found several fragmentary hazelnut shells next to a pair of large double-pitted hammer/grindstones ... some of the nutshells had been gnawed by a rodent ... possible ... rodent meals. ......
The species lists [of animals eaten by hominins] (table 4.10) are more diverse than is usual for reliably excavated Mousterian horizons in Cantabria, and the body size of animals appears to become more diversified over time. ...... beef 40-50%; deer 25-35%; horsemeat 15%; other mammals 10%. ......
It is suggestive that local populations of cave-dwelling carnivores, in direct competition with humans for food and living space [which are two very different things, note...],

declined or vanished by the earliest Upper Paleolithic, and elephants and rhinos suffered a similar decline soon after.
[If this sequence is correct, it is more consistent with human hunting having been a significant contributory factor to the disappearance of the elephants and rhinos, because the reduction of large carnivores (due to their displacement from necessary shelter) should have led to an increase in numbers of their prey species, not a decrease – unless an increase of some preferred prey species numbers led to them out-competing other non-preferred little-preyed-on species by eating more of their common foods – I am not aware that this aspect has been properly examined in the literature, but I am not fully familiar with that literature.] ……

Diet During Later Phases of the Upper Paleolithic (from ca. 18,000 to 8,500 BC) ……
Apparently, concentration on rabbit is a recurrent feature of Paleolithic subsistence strategies in the Mediterranean region. …… Some levels, like that at Aitzbitarte, have a good representation of burrow-dwelling and nocturnal carnivores. …… They may have been sought primarily as sources of pelts rather than food. ……
No suitable containers have survived, but liquids can be boiled over an open fire in a skin or gut bag, a wooden trough, or a bark bucket, all of which are extremely perishable. It is also possible to boil liquids without subjecting the container to direct heat; heated stones are simply placed into the liquid. ……
The most reasonable explanation for the food acquisition practices documented for the later Upper Paleolithic in Cantabria is that resources were becoming scarce, owing to increases in the size and density of human populations in the region, with some of the apparent increase possibly due to immigration from other regions. …… Where … and emigration is denied or checked by the presence of other populations in neighboring regions, intrinsic human population growth over a few generations might easily suffice to put pressure on the local resources. ……

Summary and Conclusions ……
Until the Upper Pleistocene, humans were probably very rare animals in the Iberian landscape. ……
Most important of all, the technique of periodic (probably seasonal) cropping of especially productive resources (wild-harvesting) was introduced [Documented with respect to shellfish, no mention made of plant foods] ……
Dietary evidence from the Iberian Peninsula is scanty and circumstantial, and after examining it we are left with more questions than answers. Nevertheless, the outlines of Pleistocene dietary practices have begun to appear. The most striking conclusion they suggest is that subsistence-related behavior during the Iberian Paleolithic was extremely variable from place to place and time to time, and became progressively more so. In retrospect this conclusion seems almost tautological. However, it directly contradicts the now-popular assertion that man's long history as a hunter immutably fixed the direction of his social, psychological and cultural evolution in a narrowly circumscribed rut, from which escape is impossible. …… "Hunting" is not a single set of behavioral patterns and attitudes, but thousands of different sets, many of which have little in common. Food-gathering demanded from our hominid forebears' intelligence, inventiveness, cooperation, information exchange, and adaptability. These are the hallmarks of the modern human condition."

Klein RG. Chapter 5 Later stone age subsistence at Byeneskranskop Cave, South Africa. In, Harding RSO and Teleki G (Eds) (1981) Omnivorous Primates Gathering and hunting in human evolution. Columbia University Press.

"… much more often the only preserved food wastes are animal bones. ……

My own research, devoted primarily to faunal remains from sites in the southern Cape, has lead to several conclusions (Klein, 1974a, 1975a, 1977a):
(1) During the last glacial (approximately 70,000-10,000BP), the southern Cape contained substantially more grassland than existed during the preceding interglacial (roughly 130,000-70,000BP) or during the Holocene (the last 10,000 years);
(2) Middle Stone Age people were less effective at exploiting local animal resources than their later Stone Age successors;
(3) Later Stone Age peoples were at least partly responsible for the extinction of several large mammalian species in the southern cape roughly 10,000 years ago;
(4) The introduction of domestic sheep to the southern Cape between 2,000 and 1,600BP led to differences in the way Later Stone Age peoples dealt with the indigenous fauna. ...... The environmental change (a reduction in grassland, leading to fewer animals with smaller ranges) cannot have been the only cause of these extinctions, because the same animals had survived a similar environmental change that occurred about 130,000BP, in the transition from the penultimate glacial to the last interglacial. ...... The Pre-Wilton occupants depended more on large, migratory plains game, the Wilton ones more on small, nonmigratory ground game and probably also on plants (especially geophytes [= a perennial plant, propagated by buds on underground bulbs, tubers, or corms, these latter being eaten]). Both the Pre-Wilton and Wilton inhabitants ate tortoises in large numbers, but tortoises clearly were more important as a reliable dietary staple in Wilton times. [Klein makes no attempt to calculate what percent of diet might be constituted by this carnivory.]

Blumenschine RJ, Pobiner, BL. Chapter 10 Zooarcheology and the ecology of Oldowan hominin carnivory. In, Ungar PS. (Ed) (2007) Evolution of the human diet The known, the unknown, and the unknowable. Oxford University Press.

“Larger-mammal carnivory signals a major niche expansion within the hominin lineage. The likely frugivorous ancestral diet (e.g., Andrews and Martin, 1991) that probably included small animals (<5kg body weight) was supplemented by foods from animals of similar and far greater body size than hominins. The novel stone-tool slicing technology of the Oldowan Industry enabled this niche shift (e.g., Isaac, 1983). ......
The suite of carnivores associated with landscapes supporting patchy tree cover theoretically afforded [exposed] hominins varying degrees of predator encounter risk and competition for carcass foods. ......
Hominin dietary interpretations based on assemblage-level traces, such as species, skeletal element, and age profiles, are subject to interpretive ambiguity arising from equifinalities in process-trace relationship (e.g., Marean et al., 1992; Marean, 1995, 1998; Bartram and Marean, 1999; Marean and Assefa, 1999; Pobiner and Blumenschine, 2003). ...... Such taphonomic singularity would pertain only if hominins maintained exclusive access to carcass-processing locales during the time that their food debris remained nutritionally attractive to other carnivores and if no other agents subsequently added or subtracted bones from the locale. These conditions appear unlikely to apply to Oldowan localities [Olduvai Gorge in Africa]. Deemphasis of the effect of nonhuman agencies, a common occurrence in zooarchaeology, suppresses the ecological information encoded in taphonomically complex assemblages. ......
Otherwise, butchery marking [on bones] is uncommon in the other Oldowan assemblages that have been examined taphonomically (table 10.1). The paucity of evidence for Oldowan hominin butchery may be real, leaving FLK 22 as an unexplained behavioral anomaly. ......
Assuming that the butchery marking signals extraction of edible tissues (as opposed to only skinning, disarticulation, or bone breakage for nondietary reasons) hominins consumed foods from animals ranging in adult body size from approximately 10kg to

more than 2,500kg and representing grazing and browsing species associated with a range of vegetation settings (table 10.2). ……
Butchery marking has not been reported on birds, amphibians, reptiles, or fish (Stewart, 1994) in any Oldowan setting. [No remains left to find?] ……
Oldowan hominins may have eaten grease but, lacking boiling technology, this may have been accomplished by gnawing softer cancellous bone portions of small animals or by chewing pounded, pulverized [Making a lot of noise, attracting predators?...] cancellous bone from larger animals. …… Unknowable aspects of the larger mammal composition of Oldowan hominin diets include the species consumed only in nondepositional settings and consumption of viscera and blood, which would not be expected to leave distinctive traces on bone. ……
Use of carcass soft tissues for nonfood purposes [i.e., clothing] is probably not knowable. ……
The contribution of larger mammal foods to Oldowan hominin diets is a long-standing issue (e.g., Isaac and Crader, 1981) with a number of components. These include the frequency and quantity of nutrients consumed from different carcass tissue types and species relative to other foods, as well as variability in these parameters by season, landscape setting, and consumer sex and age. While many of these aspects of diet are unknown, with some likely remaining unknowable, the zooarchaeological record provides some reliable and relevant information on the topic. ……
Following optimal diet breadth models (e.g., MacArthur and Pianka, 1966), food items in the optimal set are expected to be consumed whenever encountered. The question of the regularity of animal food intake by Oldowan hominins therefore is about the prehistoric availability of larger mammal carcasses,
[Foolish to omit considerations of taste preferences, which is a major factor…] …
Regardless, the absolute high yield of marrow alone suggests that hominins would exploit it whenever encountered
[Not if they didn't prefer the taste of it, perhaps analogously to chimpanzees not exploiting prey opportunity whenever they encounter it, particularly not in the seasons of plenty of preferred foods…],
assuming low risk
[This is probably a very incorrect assumption, as the hammering sounds involved in breaking open bones would be very likely to attract predators from relatively far and wide, particularly at this early stage of evolution when size and weaponization were less than later on…] ……

A current debate centers on whether the frequency and anatomical patterning of cut marks on larger mammal bones indicates defleshing of large muscle masses or only small scraps of flesh. …… The presumption is that large quantities of flesh were consumed. …… Binford (1986, 1988) argues, however, that the cut marks at FLK 22 were inflicted while extracting relatively small amounts of near-bone flesh scraps abandoned by nonhuman carnivores. ……

If hunting accounts for the full size and ecological range of butchery-marked animals (table 10.2), capabilities comparable to today's top African predators (lions, crocodiles) would be indicated (cf. Schaller, 1972; Pooley, 1989). Such capabilities imply that hominins possessed effective weaponry and used cooperative tactics to gain regular access to and a major dietary contribution from larger mammal foods.
[This is quite erroneous – defense uses much the same type of weaponry and cooperative action, and was very probably the origin of weaponization, rather than for hunting – if other foods were preferred, as by evolution they were likely to be, particularly before the advent of fire-making and cooking, then hunting would have occurred mostly only in times of shortage of those other, preferred, foods, and a "major dietary contribution" would not have occurred, other than in the context of a relatively very small amount of

hunted meat being life-saving in times of shortage of other foods, and even this latter case did not necessarily apply, as many predominantly vegetarian animals have fall-back plant foods, that they use in times of shortage of more preferred foods, such as is well documented for the present-day great apes…]

Early emergence of dominant parasitizing capabilities within the guild [of carnivores] is implied by the hypothesis that Oldowan hominins obtained larger animal carcasses by driving predators from kills. Such confrontational, or "power" (Bunn and Ezzo, 1993; Dominguez-Rodrigo, 1997, 2002; Bunn, 2001) …
[That (however infrequent) carnivory of large mammals that did occur seems much more likely to have been by power scavenging than by hunting, as the taste of decomposing flesh was probably genetically unpleasant to even very hungry hominids (scavenging of old carcasses by apes or monkeys is a relatively very rare event, despite opportunities for it…), and as the ability of the small (1.1 m high) early hominids with their relatively short (and not yet well evolved for running) legs and long arms, to hunt would probably have been significantly inferior to evolutionarily later hominins like modern-day gatherer-hunters, but in a large enough group armed with stone projectiles, clubs and sharpened (by abrasion on rocks, etc) spears-come-digging sticks, they could probably have driven off at least some types of (half-sated?) carnivores.] ……

Oldowan hominins also remained a prey species for larger carnivores. Adult hominins were likely vulnerable to all large felid, canid, hyaenid, and crocodilian predators, and young hominins were additionally vulnerable to smaller predators. ……
[Almost certainly necessitating some degree of weaponization, obviously enough?...]

Knifelike stone flakes were used to remove flesh, but the amount of flesh Oldowan hominins consumed is uncertain.
[Very uncertain, on the basis of the evidence provided here, such that there is no reason provided to lean towards assumptions of "major dietary contribution" from flesh]

Bunn HT. Chapter 11 Meat made us human. In, Ungar PS. (Ed) (2007) Evolution of the human diet The known, the unknown, and the unknowable. Oxford University Press.

"The FLK Zinj archaeological site at Olduvai Gorge [Africa] is one of the largest excavations (~300m^2) and apparently one of the most completely excavated archaeological sites known from the Plio-Pleistocene (Leakey, 1971).
Bones and stone tools have been found in a discrete, four-inch-thick layer in sediments, which accumulated within a kilometer of the margin of a fluctuating, saline lake. An air-fall ash abruptly buried the stone tools and bones, some of which actually extended into the ash layer (Leakey, 1971; Hay, 1976; R.L. Hay, personal communication, 1991). Leakey, (1971) reported a total of 2,470 Oldowan stone artifacts (more than 90% were sharp-edged flakes and flake fragments and the remainder were a mix of core tools [(what the flakes were knocked off)] and a total of 3,510 bone specimens of which 1,090 were considered identifiable to taxon or skeletal part; some had been gnawed by carnivores. …… The preservation of surface detail on the original bone surfaces is unparalleled
[And therefore the site may not be representative of the average Plio-Pleistocene hominid subsistence, on this basis as well as on the other bases discussed in sections 1.5, 1.6 and 1.7]
for the Plio-Pleistocene and so good that one can readily recognize and confidently and accurately distinguish, with unaided eyes, the telltale signatures of artifact-induced cut marks, carnivore gnaw marks, rodent gnaw marks, and root marks. ……

The MNI [(minimum number of individuals)] estimate for larger mammals is at least 48: 29 bovids, 5 equids, five suids, 2 giraffes, one hyenid, 2 small carnivores, two hominins, one hippo (tooth fragment), and one elephant (tooth fragment).

...... Paleontological work on selected subsets of the assemblage and its derivatives (Binford, 1981; Potts, 1988) obscure, rather than reveal, the actual skeletal pattern. ......

A high degree of fragmentation is a common feature of humanly broken bones at more recent archaeological sites, in contrast to the relatively more complete bone elements characteristic of many bone accumulations generated by large carnivores.
[Including those generated by bone-crunching hyenas?...] ......

In the FLK Zinj bone assemblage, abundant percussion notches and large bone flakes, which were caused by hominins using hammerstones to break open marrow-rich limbs and hemimandibles, establish hominins as the dominant taphonomic agent causing bone fragmentation. ...... Quantification of bone fragmentation (Bunn, 1989), and particularly quantification of fracture morphology (Capaldo and Blumenschine, 1994), strengthens the conclusion that hominins caused most of the breakage of marrow-rich bones.
[The quantifications should have been given here...]
Marks inflicted in bone surfaces are abundant and varied in the FLK Zinj assemblage. The mere presence of cut marks (fine slicing grooves) produced by hominins using sharp stone flakes in butchery puts to rest the concern that for taphonomic reasons, the bone assemblage might have no functional association with hominin behavior.
But investigating what that hominin behavioral role might have been requires more evidence.
Documentation of more than 200 bone specimens with cut marks in the assemblage provides a more secure foundation.
The total number of cut-marked bone specimens is actually closer to 300, but discussion here refers to the 200 or so specimens that have been identified by both macroscopic and microscopic criteria to meet the requirements of all researchers involved. ......
Most of the cut marks, however, are clustered on once-meaty limb bone shafts near major muscle attachments.
[Which does not exclude the possibility that most of those major muscles had already been removed by nonhuman carnivores]
The composite pattern of defleshing cut marks on the FLK Zinj limb elements indicates that meat was a primary target of hominin butchers ......
Judging from the multiple taphonomic agents involved with the bone assemblage, FLK Zinj was a popular location on the Olduvai landscape during formation of the site.
[I don't think that it is possible to logically draw such a conclusion, and certainly no reasoning is provided to support the statement – alternatively it was a fallback site for hard times such as of shortage of preferred foods.] ......
Transport of stones and bones by hominins to the site is strongly implicated. ......
Yet, the vast majority of the bone surfaces are fresh to lightly weathered (i.e., weathering stages 0 to early stage 2 of Behrensmeyer, 1978), indicating a maximum of several years of subaerial exposure and weathering, rather than decades, centuries or millennia (Bunn and Kroll, 1987). The concentration of at least 48 individual larger mammals
[Only some parts of them...]
in this temporally restricted bone assemblage is anomalous relative to the distribution of bones on the modern and (samples of) the ancient landscape and indicates the carcass remains were transported to the location (Behrensmeyer, 1983; Blumenschine and Masao, 1991; Bunn, Kroll and Bartram, 1991; Tappen, 1995). ......
The abundance and strong patterning of defleshing cut marks on limb elements from FLK Zinj documents a systematic, thorough butchery for meat by knowledgeable butchers.
[This is very probably a great exaggeration of their skills and by implication experience

and involvement – the marks don't seem to me to indicate anything more than them just cutting off what meat there was to see…] ……

Scavenging hyenas and other nocturnal carnivores probably had unhindered nocturnal access to the day's freshly discarded bones, which they exploited, thereby producing the documented depletion of limb epiphyses in the bone assemblages at FLK Zinj.
[And therefore producing as well a significant fraction of the bone shaft fragmentation?...] ……

The bone assemblage as a whole, accumulated through repeated daytime visits by hominins during the several-year time interval preceding the abrupt burial of the site.
[What about the years before that interval – most "popular" locations of that open-air type would be expected to have some remains of that also, as middens?...]
["repeated daytime visits" – (some lesser, and maybe much lesser, part of) 48 larger mammals over i.e. 5 years is once every 5 weeks, among perhaps very roughly a dozen hominins, which does not indicate anything more than a fairly small fraction of meat intake as a fraction of total food intake, at this dubiously representative site…] ……

The most parsimonious taphonomic sequence forming the FLK Zinj bone assemblage is thus:
1. Repeated acquisition by hominins of intact carcasses through hunting of smaller carcasses
[Unlikely - given the small size of the early hominins, and their anatomically poor development for running chasing, and the generally high speed of smaller prey animals, this does not seem probable…]
and power scavenging
[This is more probable, for all sizes of prey, though probably as an infrequent subsistence activity in harder times of shortages of more preferred foods]
(an/or less likely hunting) of larger carcasses (i.e., carnivore exploitation strategy).
2. Transport of field butchered portions to the FLK Zinj location.
3. Further butchery, sharing, and consumption of substantial quantities of meat and fat at FLK Zinj during the day.
4. Hyena removal of fresh limb epiphyses and some hyena gnawing [And crushing?...] of butchered bones; more intensive gnawing by small carnivores (jackals, viverrids) and by small rodents of butchered bones. ……

A recent reexamination of FLK limb specimens concludes that most of the linear grooves identified as carnivore gnaw marks by Blumenschine and colleagues actually result from biochemical decay, not carnivore gnawing
[If the bone epiphyses were destroyed by hyenids, as stated to be the case by Bunn here, it seems unlikely that there were not also substantial gnawing of the bone shafts by hyenids – also, would not something that started as a gnaw mark be a more likely place for biochemical decay to start from?...] ……

Third, studies sampling bone distributions on modern landscapes in ecological settings comparable to Olduvai indicate that associations of more than approximately three animals in an area the size of FLK Zinj result from transport or from detectable catastrophic events (Behrensmeyer, 1983; Bunn, Kroll and Bartram, 1991; Tappen, 1995).

The MNI estimate is more than 15 times greater than the sample likely to result from in situ death, unless considerably more time for site formation was involved.
Both taphonomic and geological evidence, however, preclude that, which yields the reconstruction that carcass portions were transported to FLK Zinj. ……

The salient point is that carcass size, fat content, and adhering meat would certainly influence bone selection by hominins and hyenids for consumption and for transport.
……
How important was meat in hominin diet, and how did that vary from place to place?
……

Tubers and/or Meat.

Wrangham and colleagues (1999) argue that because of butchery evidence at 2.5 Mya [million years ago] (de Heinzelin et al., 1999), a dietary change to increased meat consumption cannot have been the principal cause in the appearance of *H. erectus* more than one-half million years later.
[This reasoning is not correct, as an uncommon event can grow to become a common event, which is very probably the course of meat-eating in human evolution – the questions remain as to how common and frequent, at what time, was meat-eating, and what thresholds may have been reached in the characteristics of the meat-eating and -acquisition (regardless of its probably being a lesser, and much lesser, fraction of the diet). Change of some aspect is not always constant or regular.]
Instead Wrangham and colleagues claim that a dietary change to increased tuber consumption
[It seems to me much more likely that tuber consumption had been increasing and/or substantial for a long time before the evolution of *H. erectus* – the early hominin can see food acquisition of tubers by other animals, and it can readily evolve the use of the digging stick to facilitate their acquisition for itself….]
enabled the evolution to *H. erectus*.
Claims of supporting evidence have been questioned elsewhere (Bunn, 1999).

The postulated shift to tuber consumption needs to be considered in relation to documented physical and metabolic changes in the hominins themselves, such as increased brain and body size and decreased cheek tooth and gut size.
Foraging for tubers would not have selected for most of these, at least not if realistic estimates of the nutritional yields of wild tubers are considered (Schoeninger et al., 2001a, 2001b). Only by doubling or tripling the actual energetic value of wild tubers can O'Connell, Hawkes and Blurton Jones (2002) describe them as an attractive energy source relative to meat.
[A claim like this must be supported by evidence of energy content, including without the thick skin (that can be removed) of most wild tubers, and with the spitting out of fiber residual after juice extraction by chewing; energy involved in acquisition; risks involved in acquisition, and taste preferences, none of which has been provided here, nor probably in the original works…]

So, does it follow that increased meat and marrow in hominin diet made us human independently of other components of hominin lifeways, that other aspects of foraging strategy, social organization, biology, and diet immediately melted away in favor of a foraging strategy obsessed over meat? Not at all. I contend that in the context of existing foraging adaptations for a plant-based diet, meat is what changed first and what altered the dynamics toward a more humanlike foraging adaptation and diet in H. erectus.
Time spent foraging specifically for meat is time not spent accessing and consuming plant foods (In practice, of course, an individual can do both simultaneously but probably less efficiently;
[Not less efficiently, if the foraging behavior of chimpanzees, well documented as taking hunted meat only opportunistically when encountered while looking for their (apparently preferred over meat) preferred plant foods, and then only sometimes, mostly in seasons of shortages of those plant foods, is anything to go by, which it seems logically to be…];

this is meant as a heuristic discussion).
Individuals devoting significant time and energy foraging for animal carcasses, particularly when unsuccessful, which is most of the time judging from ethnography, [that is, studies of modern gatherer-hunters – and so, how much more likely for the much less run-chase physiologically adapted early hominins with their relatively short legs and long arms, and their less effective weaponization?]
come up short, with a deficit in the energy budget
[But maybe the present-day/modern gatherer-hunter men don't mind this deficit, because:
1. They are released from the more mundane and unexciting drudgeries of (surplus, over what they eat while looking around for animals to hunt…) plant food collecting;
2. They have an outlet for competitive, dominance and aggressive urges;
3. They have a mechanism for proxy displays of martial power to all three of: neighboring group members, other males within their own group, and females; which is safer than actual combat in the context of potentially lethal weaponization.
4. Males with less martial power have a mechanism, via cuckoldry against them deliberately left unknown to them due to them being out hunting, to attempt (or not, to the extent that it was more a psychological displacement activity to remove one from the potential danger of conflict due to contribution imbalance or cuckoldry prevention or discovery) to balance the imbalance in meat contributions with the more successful (more martially powerful by proxy of greater hunting success) males, to exchange sexual access to their wives for reduced risk in potentially fatal weaponized combat with the cuckolders, a share in the food (i.e. meat) greater than that proportional to their own contribution, and a share in the martial power of those better so endowed for the purposes of defence. So perhaps there was not only the fairly well understood female prostitution analogue of exchange of food for sexual access (an evolutionarily ancient behavior), but also a more tacit and less obvious pimping analogue (with evolutionarily ancient origins, but more recently evolved refinements) …
But probably these aspects did not apply to earlier hominins because they were physiologically and technologically limited to power scavenging in larger groups, rather than hunting alone or in pairs or small groups, and perhaps much more importantly, because they were physiologically and technologically limited to defending themselves from large carnivores in larger groups, being unable to do so consistently alone or in pairs or small groups.]
much of the time, unless foraging dynamics change.
[Indeed, but purely nutritional bioenergetics are only a fraction of the overall efficiency equation or picture, as I have elaborated some reasonably plausible possible aspects of just above, and, indeed foraging dynamics would have changed with the evolution of physiology and weaponization, perhaps roughly equally for defensive as for offensive considerations.]

What are some of the possibilities? One possibility is increasing foraging time, which is constrained by day length for a diurnal hominin.
Another is increasing emphasis on higher-ranked food items, such as nuts and seeds rich in lipids and protein (Schoeninger et al., 2001b).
Yet another is discovering and adding a new plant food, such as tubers, or increasing use of other carbohydrate sources, for example, honey, to the existing diet.
[Indeed, and taking all of this expansion of food sources together, and adding it to the use of naturally formed and the most primitive of artificially formed tools (the sharpened digging stick might well have been this), and adding also the needs for evolving defensive strategies, including the use of the earliest of weapons, for foraging further afield from the safety of climbable trees, this is very much the point, given that the very solitary orangutan apparently evolved or kept its considerable mental faculties (equal to the very social chimpanzees) through their requirement for the memorization and planning of efficient complex food acquisition routes, even in the apparent absence of

significant tool use. Meeting the requirements for these behavioral changes alone would logically very probably account for the very large fraction of the evolutionary increase in the size of the brain to that of *H. erectus* (and also, for the (small, actually) evolutionary changes noted in the intestines (increase of the small intestine, decrease of the large)), without any need to look to the supposed/ assumed substantial increase in hunting behavior as the driving/facilitating reason other than as a (probably very) minor one. The causes of, and things facilitating, subsequent changes in brain size remaining unanswered and a different though related question.]

Given that female chimps use sexual access as leverage to induce males to share meat (e.g., Mitani and Watts, 2001; Stanford, 2001), how big an intellectual leap would it have been to collect and then use surplus plant foods as leverage to induce meat sharing?
[This very much misses the point that for chimpanzees meat seems to be more of a fallback food than a preferred food – *hungry* female chimpanzees trade sexual access for raw meat - would a well-fed early hominin want to trade raw hazel nuts, almonds, or fruit for raw meat? Not in general if its taste preferences weren't evolved to, it seems to me – it might not even want to trade starchy tubers peeled and ground, or oily or cereal grains ground or soaked, for raw meat, other than to be social…
And maybe just as much the point, the female would probably surrender surplus food as required to the male on pain of receiving a beating and having the foods taken by force, if documentation of the behavior of apes and baboons is anything to go by…
Also, and maybe at least as important regarding leverage, ultimately, the females needed the males for protection of them and their young, otherwise they probably wouldn't have evolved to tolerate the subjugation…]

Encephalized Homo was adapted for more complex decision making than australopithecines or chimps. This is not a particularly complex decision for early H. erectus to make. It is called division of labor.
[Division of labor is generally accepted to have its roots in the division of foraging risk, which is far from certainly related as closely to hunting for early hominids as it is for chimpanzees, and may be more (taking food type diet proportions into account) related to the environmental position and noise required for the acquisition of other foods.]

The escalation [To what extent is moot, probably only slightly] of meat eating, which appears at the beginning of the Pleistocene, may well have been supported by the addition of other energy sources to the diet (Aiello and Wells, 2002), but tubers did not initiate or support the beginning of meat eating in the hominin adaptation.
[Bunn's reasoning here in support of this claim is fatally flawed, as detailed below, and the claim itself is likely enough to be incorrect.]

On the basis of a simplistic view of Hadza foraging, a two- to threefold exaggeration of the energetic yields of Hadza tubers,
[This is quite incorrect – the energy content would be quite sufficient, as the thick skin that most wild tubers have can be easily enough removed against a sharp rock edge or by breaking up the tuber with a stone, and the fiber residual after juice extraction by chewing can be spat out as required, as some gatherer-hunters have been documented to do.]
and an incautious, distorted reading of the Plio-Pleistocene archaeological evidence,
[This seems to me, as I have detailed thoroughly enough in my comments interspersed here, to be at least as much a characteristic of Bunn here as of others, though the same could indeed be said of many others, as detailed here elsewhere in my comments…]
O'Connell, Hawkes, and Blurton Jones (1999, 2002) imagine that foraging for tubers by female *H. erectus* enabled male hunting and scavenging.
For the Hadza, recognizing some relationship between foraging for tubers and for meat seems reasonable, given that those foods are reciprocally shared.

[At least as likely the most important reciprocation is the exchange of food (and sexual access), from the females, for defense and protection, from the males. The meat is just as likely to have its value to the females as a proxy indicator of martial power and skill, rather than as a nutritional benefit.]

But characterizing that relationship evolutionarily as O'Connell, Hawkes and Blurton Jones do confuses the dynamic by reversing cause and effect and thereby losing a sense of time depth in evolution. How did a tuber-meat relationship develop for the Hadza, or, hypothetically, for ancestral Plio-Pleistocene hominins?
No plausible evidence or claim of tuber use and cooking technology, the combined adaptation envisioned by Wrangham and others as an efficient substitute for meat, extends back into the Pliocene.
A recent hypothesis by Laden and Wrangham (2005), however, proposes tubers as a pivotal fallback food in the early evolution of australopithecines but without considering how the earliest hominins of the Pliocene and even Miocene would have acquired deeply buried tubers in the first place.
[Not all tubers are deeply buried, and if chimpanzees naturally use sticks as tools for getting foods out of interior locations, and early hominins can see other animals digging up and eating tubers, it's pretty obvious that they very probably would have used sticks to dig up tubers at the earliest bipedal stages of evolution.]
Evidence of butchery for meat and marrow, however, extends all the way back to 2.6 Mya (de Heinzelin et al., 1999; Semaw et al., 2003; Dominguez-Rodrigo et al, 2005). Thus, consumption of meat and marrow of large animals greatly preceded any likely tuber foraging.
[Obviously enough, this is very probably very incorrect.]

Meat provided the selective pressure. Meat was the evolutionary dynamic driving the adaptation and defining the cause-and-effect relationship. Tubers (the effect), or honey, more likely, may have compensated for foraging time and food value lost in unsuccessful searches for meat (the cause). [Probably not, as detailed…] ……
If tubers were a significant new dietary component at the beginning of the Pleistocene, then an even more pronounced response in foraging effort for meat and marrow would be predicted.
[Not unless taste preferences favored raw meat, which they probably didn't, if the foraging behavior of chimpanzees, well documented as taking hunted meat only opportunistically when encountered while looking for their (apparently preferred over meat) preferred plant foods, and then only sometimes, mostly in seasons of shortages of those plant foods, is anything to go by, which it seems logically to be.
And, one should also consider that most often, even modern humans raised on cooked meat-based diets seem to prefer the taste of raw almonds, cashews, brazil nuts, hazelnuts etc. to the taste of raw meat, exceptions to this needing to be considered in the light of an understanding of the human propensity to acquire through repeated use a taste for things initially and perhaps naturally found quite distasteful (cigarettes and many alcoholic drinks for example) …]

[The next page of Bunn's work here (p205), which I will not transcribe here, suffers from the following collectively fatal flaws: failure to consider the collective role of nuts, oily and cereal seeds, and tubers; failure to understand the likelihood of processing of these foods by pounding, mashing, grinding, cracking and soaking, using naturally shaped stone and wooden tools; failure to consider the discovery biases of archaeological sites (as I've detailed in sections 1.5, 1.6 and 1.7 above) including the very obvious inability of wooden digging sticks to remain in the archaeological record, and of naturally shaped stones to be visible in the archaeological record, whereas bone remains to be found, and moreover, often triggers the finding of archaeological sites; and failure to note that the human gut is only slightly different from that of the apes, remaining nearly as different as the guts of the apes are from the guts of natural carnivores, and that this slight difference (slightly longer small intestine, slightly smaller large intestine) from the guts of apes is easily enough accountable for by an increase in the consumption of the more macronutrient dense foods that nuts, and oily and cereal seeds are, and that tubers can be easily enough processed to be]

[The final pages of Bunn's work here (p206-7), are correct in noting that there is increasing and extensive evidence over archaeological time of meat-eating, but very incorrect in jumping from that to the assumption that this equates to evidence of extensive meat-eating, which it does not, not least due to the biases in archaeological site discovery and interpretation I've elaborated on above in sections 1.5, 1.6 and 1.7 – as is common in much writing on the topic, no attempt whatsoever is made to use the evidence available, such as it nevertheless imperfectly is, to quantify what percent of diet meat might have been at those various times in human evolution – Bunn just jumps to the assumption that it increased to be a substantial percent of diet by the time of *Homo erectus* – obviously much more on the basis of vehemence than of evidence….]

Egeland CP, Dominguez-Rodrigo M (2008) Taphonomic perspectives on hominid site use and foraging strategies during Bed II times at Olduvai Gorge, Tanzania. Journal of Human Evolution 55: 1031-1052.

From the abstract:
"A reexamination of all of Leakey's Bed I sites has shown that hominids played little or no role in the formation of all but one of these faunal assemblages [(collection of fossil animal bones)], a finding that prompted the reanalysis of the Bed II sites presented here. …… Our data contradict previous interpretations of MNK Main as a hominid accumulation but uphold the contention that BK represents a primarily hominid accumulation reflecting early access to carcasses.
The small and poorly preserved assemblages from FC West and TK are difficult to link unambiguously to either hominids or carnivores.
Site MNK Main and HWK East Levels 3-5 appear to be death arenas where carcasses accumulated via natural deaths and/or serial predation.
Site SHK is severely biased by selective retention and therefore little can be said about its formational history. Nevertheless, no hominid modifications were documented in this assemblage.
Comparisons with other Olduvai sites indicate a more conspicuous hyena taphonomic signal during Bed II times than Bed I times, which appears to mirror the changing configuration of the large carnivore guild.
These findings also beg the question of what activities were being carried out by hominids with the stone tools discarded at these sites.
Although it seems clear that hominids were utilizing stone tools to carry out subsistence activities unrelated to carcass butchery, more excavation and techniques such as phytolith analysis should be employed to explore alternative explanations."

Shea JJ. Chapter 12 Lithic archaeology, or, what stone tools can (and can't) tell us about early hominin diets. In, Ungar PS. (Ed) (2007) Evolution of the human diet The known, the unknown, and the unknowable. Oxford University Press.

"Stone tools are the most durable residues of hominin behavior. ……
Stone tools are still used as subsistence aids by chimpanzees, our nearest primate relatives. ……

In seeking these clues (to hominin dietary shifts), however, we need to be alert to the complicating effects of behavioral variability. Recent stone-tool-using humans exhibit considerable variability, not only in the kinds of tasks for which stone tools are used but also in the choice of technological strategies they deploy in their land-use strategies and subsistence adaptations. ......

Juxtaposed concentrations of animal bones featuring cut marks and percussion fractures together with stone tools have been discovered at numerous sites in Africa and western Asia, dating to more than 1 Myr, including (to name a few) Olduvai Gorge, Tanzania ... East Turkana, Kenya ... Olorgesailie, Kenya ... and Ubeidiya, Israel ... .

The question of hunting versus scavenging as the principal mode of meat/fat procurement at these and other sites is still debated (Bunn, 2001; Lupo and O'Connell, 2002; Dominguez-Rodrigo and Pickering, 2003).

Nevertheless, the inferred link between early hominins' making stone tools and meat eating/ animal fat procurement retains near-universal support among contemporary archaeologists (Schick and Toth, 1993; Stanford and Bunn, 2001). ......

Microwear- and residue-based functional analyses of early Paleolithic tools from East Turkana and other sites indicate stone tool use for working wood and processing soft plant matter (Keeley and Toth, 1981; Dominguez-Rodrigo et al., 2001).

The ethnographic record is also rich in accounts of simple stone tools like those found in early Paleolithic assemblages being used as a "secondary technology," as tools for making other tools of wood, bone, and other perishable media and as aids to the collection and processing of plant foods (Gould, 1980).
Replicative experiments affirm Early Paleolithic tools would have been effective butchery tools (Jones, 1980), but experiments suggest that they would have been effective in many other tasks as well (Schick and Toth, 1993). ......

The technological simplicity of Early Paleolithic stone-tool technology has been viewed as a product of relatively simple early hominin cognitive capacities. I argue here that this simplicity is deceptive and that it is better explained as a strategic response to the intrinsic variability of early hominin subsistence activities and other tasks requiring stone tools.
It is tempting to see the inception of stone-tool use by hominins as a revolutionary development, but the cognitive capacities underlying stone-tool production are probably evolutionarily primitive among hominins and at least some African hominoids (chimpanzees and bonobos) (Toth et al., 1993; Wynn, 2002). ......
Chimpanzee nut-cracking is a plausible evolutionary precursor to stone tool production. Cracking nuts with stone and wood hammers is a widespread chimpanzee subsistence activity (McGrew, 1992), and one that requires a degree of control over fracture properties similar to that involved in flintknapping [(making stone tools by knocking off pieces of a stone)]. ......
Most lithic assemblages formally recognized as Oldowan date to more than 1.5 Myr [Million years ago], but younger assemblages are sometimes likened to the Oldowan as well (Plummer, 2004). ......
Who made these Early Paleolithic industries? It is generally supposed that most Early Paleolithic stone-tool assemblages were produced by species of the genus *Homo*. Stone-tool assemblages become more consistent features of the African paleolandscape after the emergence of *Homo habilis* circa 2.4-1.6 Myr.
Yet, earlier occurrences of stone tools are associated with australopithecines and paranthropines. ......

Though the Acheulian is often seen as the material culture of *Homo erectus*, it would be a mistake to reflexively equate this industry solely with this hominin, or vice versa. ……
If the link [stone tool design to particular subsistence tasks] is weak or equivocal, then we should expect a weak correlation between lithic industries and hominin taxa. I would argue that the latter is precisely what we see from the earliest archaeological sites, through the Early and Middle Paleolithic and into the most recent phases of Stone Age prehistory. …… Particular stone tools from recent contexts may resemble Early Paleolithic tools morphologically, but they may have been integrated into their respective users' technological adaptations in fundamentally different ways. …… One limitation for microwear and residue analysis is that both are only preserved under exceptional conditions.
Consequently, generalizations about Early Paleolithic stone tool use based on microwear/residue analysis are that both are only preserved under exceptional conditions.
Consequently, generalizations about Early Paleolithic stone tool use based on microwear/residue analysis are often extrapolated from small and nonprobabilistically selected samples
[Which is an overgeneralization methodologically, and therefore quite likely to be incorrect in fact]. ……

…, demonstrating that one can butcher a horse with a replica of an Acheulean handaxe does not prove that a particular handaxe, much less handaxes as a group, were used to butcher horses. ……
Flaked stone tools are unquestionably useful in subsistence-related tasks. The sharp edges of stone tools offer considerable advantages in butchering animal carcasses, and cut marks on animal bones from numerous Early Paleolithic sites clearly affirm this mode of tool use. ……

Rather, it is to suggest that particular qualities of Early Paleolithic stone-tool assemblages, predominantly local raw materials and relatively little discrimination of raw material quality, reflect technological strategies emphasizing the provisioning of places and "events" with useful lithic materials. These events might be particular short-lived resources (an animal carcass) at which tools were abandoned or strategic caches of raw material in areas where resources requiring stone tools were available on a predictable schedule (Potts, 1988).
Such a strategy contrasts with the increased strategic emphasis on provisioning people with mobile toolkits that one sees in the ethnographic [Of studies of living human groups] record (Binford, 1979) and in late Paleolithic assemblages (Kuhn, 1992). ……

There is thus far little evidence to suggest considerable investment of postproduction effort in stone tools, such as attaching them to handles.
Most Early Paleolithic stone tools are relatively thick in cross section, which is an obstacle for hafting. The kind of stone-tool-shape standardization usually associated with hafting does not become a regular feature of the African record until after 0.2-0.3 Myr (Ambrose, 2001). ……

Microwear and residue studies of retouched Early and Middle Paleolithic stone tool types consistently point to multiple modes of use (Keeley and Toth, 1981; Anderson-Gerfaud, 1980; Shea, 1981. Ethnographic observations of flake tool use (e.g., Gould, Koster and Sontz, 1971) and experiments using replicas of Early Paleolithic stone tools (e.g., Toth, 1997) both point to versatility as an important feature of Early Paleolithic stone-tool design. ……
Several researchers have noted the superior qualities of Acheulean handaxes in butchery tasks, particularly in skin removal and disarticulation of larger mammals. …… large flakes work just as well as handaxes of similar size in most tasks, including butchery.

Kohn and Mithen argue that the extensive retouch and imposition of symmetry involved in handaxe design reflect male hominins signaling their fitness as potential mates by demonstrating complex cognitive and technological skills in tasks that do not directly influence foraging success. Such "costly signaling is not uncommon in the annals of ethology and human behavioral ecology (Bird, Smith and Bird, 2001) ……

We know that Early Paleolithic stone tools were used from about 2.7 Myr onward to aid early hominins in butchering animal carcasses.
We know this, however, more from stone-tool cut marks on bone than we do from the analysis of stone tools themselves.
We also know, from wear pattern analyses and residue studies, that Early Paleolithic stone tools were used for other purposes not directly related to subsistence, such as woodworking.
Consequently, we know that we should not automatically assume that accumulations of stone tools necessarily reflect subsistence activities, much less carnivory, specifically.
…… A third prominent "known unknown" for the Early Paleolithic is the extent to which stone tools were used to create a secondary technology, wooden tools, for example, that had a more direct role in subsistence. ……

No matter how careful you are, if you knapp stone, you will get cut … injuries … infections … ……

If, as studies of chimpanzee stone-tool use suggest (Boesch and Boesch, 1984), the capacity for procuring useful stones and transporting them to places where they are needed is evolutionarily primitive among hominins, there may be a long prologue to the Early Paleolithic.
[In my opinion this is very obviously very highly probable…] ……
Rather, we need to be aware that the recognizably flaked stone tools that populate the oldest Early Paleolithic assemblages may not so much reflect the inception of stone-tool use but rather the increasing need [provision, more the point, other than need due to population increase…] for sharp stone tools on parts of the landscape where they did not naturally occur.

Peters CR. Chapter 13 Theoretical and actualistic ecobotanical perspectives on early hominin diets and paleoecology. In, Ungar PS. (Ed) (2007) Evolution of the human diet The known, the unknown, and the unknowable. Oxford University Press.

Limitations in our paleoanthropological knowledge sometimes invite overly rich interpretations and speculations uninformed in critical ways. ……
[Indeed!!]

What plant and animal foods early hominins actually ate is virtually unknown (perhaps, in almost all cases, empirically unknowable).
What they were capable of eating and probably ate are more fundamental, theoretically approachable questions. ……

Historically, humans and baboons also exploited a taxa-rich variety of wild plants for their rootstocks (underground storage organs), indicating that this was probably an additional food-item type exploited by early hominins.
A similar analysis could be made of the small wild animals eaten by these three extant primates [human, chimpanzee and baboon], and some of those small animals could be added to the fundamental trophic niche(s) of the early hominins.

One overall result of this method is a general characterization of their trophic niche(s) as omnivorous.
[Ok, but this is not very informative, as omnivory covers the broadest of spectrums from almost exclusively leaf eating, to almost exclusively fruit eating, to almost exclusively carnivorous, to almost exclusively insectivorous, at the extreme peripheries, as well as more mixed compositions of diet… in other words, it says nothing about proportions, which is very much the crucial point…]

Another general inference is that the abstract and fundamental overlap in diet of these three extant primates implies high potential competition in closely related extinct species of hominins, hominids [homin*oids* (apes)?], and large-bodied monkeys (also see Peters and O'Brien, 1982).
The easy to exploit fruits are theoretically the plant-food type on which interspecies competition would be focused the most (Peters and O'Brien, 1981). ……

Stahl was not familiar with Africa's wild-plant foods, so she was not aware of the nutlike oil seeds, their significant role in traditional plant-food diets, and their theoretical importance for early hominin diets
[What a grossly ignorant position to be venturing theory from!...]
(Peters and Maguire, 1981; Peters and O'Brien, 1984; Peters, 1987a, 1987b, 1988). ……

The most challenging potentially high-ranked foods are the nutlike oil seeds (they have very strong endocarps) and the rootstocks (they are often not easily dug/chopped out of the soil and water). ……
All but the most resistant nutlike oil seeds theoretically were probably within the masticatory capabilities of adult male paranthropines [Heavy-jawed early hominids not of our ancestral line] (Peters and Maguire, 1981; Peters, 1978a; Demes and Creel, 1988; Lucas, Peters and Arrandale, 1994; Lucas and Peters, 2000) but not weaning-age infants or young subadults.
However, judging from their jaws and cranial architecture, weaning-age paranthropines probably had a significant advantage over "gracile" australopithecines and early Homo infants, masticating a variety of the relatively less-challenging foods (Peters, 1979, 1993).

…… Such discomfort [with (theoretical, in this case) morphological specialization being part of a broadened trophic niche] may be caused by the confusion stemming from what Robinson and Wilson (1998) have called Liem's paradox: species that are highly specialized in phenotype (e.g., morphology, physiology, and behavior) may act as ecologic generalists. ……
Robinson and Wilson (1998) show that theoretically the most extreme phenotypic specializations should occur in the virtual absence of trade-off and point out that a community of such "specialists" can coexist with or even replace phenotypically generalized consumers that are slightly better at using easy-to-use resources but have no specialized abilities to exploit less-easy-to-use potential resources. In a sense, the specialists create new niches.
They also note that key evolutionary innovations may allow exceptional degrees of both feeding versatility and phenotypic specialization. …… In our case, the ecophenotypic tool-use specializations of early *Homo* (and possibly *Paranthropus* [*Australopithecus* (of our line), more the point?...]) recognized in table 13.1 can readily be added to this picture of potential trophic versatility, especially considering their morphological correlates in the hands and bipedalism (Markze, 1996). ……
[Table 13.1 (p238-241) includes a variety of potential C4 foods.] …
Theoretically, environmental ecostructure dominates locomotor requirements (e.g., Youlatos, 2004). ……

Theoretically, it appears there is room for amphibious locomotor repertoires in the earliest hominins, within the settings of herbaceous-wetland and woodland mosaics. ……

For the summer rain climate region [in Africa] south of the equator, there is a well-known, pronounced seasonal change in the quality as well as the quantity of types of food items from the full spectrum of food types provided by several different types of plants during the rainy season, to a limited fare in the dry season, consisting mainly of fruits and nut-like oil seeds (relatively scarce) provided by trees, dry beans from the Leguminosae provided by trees (especially, miombo species, e.g., Suzuki, 1969) and rootstocks provided primarily by herbaceous plants.
Wild-plant foods from rootstocks are generally available year-round, but those of the geophytes [plant parts growing in soil] often become bitter in the dry season from water loss and increased concentrations of secondary compounds.
Wetland forms do not experience the same degree of seasonal drought stress, and those of perennial wetlands remain fully succulent year-round. ……

Theoretical examples [of specializations that provide a broader niche given seasonal differences] include consumption of nutlike oil seeds at the end of the rainy season to beginning of the dry season, putting on fat reserves that buffer the height of the dry-season [food] stress (e.g., Peters, 1987a).
Another is the use of perennial wetlands for edible rootstocks, for example, those from Phragmites (Peters, 1990, 1994), Typha (Peters, 1994), and Cyperus papyrus (Peters, 1999). ……
Some of the nut-species trees and dry-fruit species shrubs are dry-climate forms and are able to produce good crops in drought years. ……

… within this mosaic some landscapes/habitats are demographic sources [where hominins reproduce more than they die], while others are demographic sinks [where hominins die more than they reproduce] (Pulliam, 1988) ……
In terms of landscape and habitat, for hominins source areas would be characterized by mosaics of extensive woodlands, with local groundwater forest and tall bushland; riparian forest and medium-to small-sized freshwater wetlands; numerous small streams with high quality water and relatively few crocodiles; high productivity and species richness of edible plants; and safe sleeping sites.
[More detail is required on just what is a safe sleeping area, for what hominins in what stage of physiological evolution, weaponization and fire use]
Areas functioning as demographic sinks for hominins would be characterized by extensive montane and lowland rainforest; extensive semiarid bushland; hot, dry broad valleys; saline and alkaline wetlands; coastal forest with very limited potable water; and increased competition for and/or lower productivity and lower species richness of edible wild plants. ……

For Gaborone [semiarid Botswana], 33 woody species provide fleshy fruits, compared with 81 species for Harare [subhumid Zimbabwe].
This difference in species richness is also reflected in dry-season fruit species: 6 and 2 versus 16 and 13 species for the early and late dry season, respectively.
The Gaborone area has 1 nutlike oil seed species, compared with 2 species for the Harare area.
Theoretically, without wetlands, semiarid climate areas were probably marginal habitats for early hominins, sometimes occupied but functioning as demographic sinks. ……
Even reliably fruiting tree species can vary dramatically in individual and stand productivity from year to year. Species richness in edible fruit taxa is important insurance for relatively high annual habitat productivity, even in well-watered summer rain environments. ……

O'Brien and Peters (1999) hypothesized that the East African Plateau was a demographic sink for hominin species in the Pliocene and that south-central Africa was the hominin demographic source area for both eastern and southern Africa. ......
At a finer scale, some of these [favorable] characteristics are present on the East African Plateau, closer to the fossil sites that provide the paleoanthropological record, and they increase the likelihood that a hierarchy of geographic and ecosystem relationships shaped the early course of hominin evolution. ......
The footslopes to toeslopes of well-watered highlands in the subhumid tropic and subtropics theoretically offer all of the characteristics in the profile of the ideal early hominin demographic source area, previously summarized, except perhaps for the wetlands. The water-quality factor afforded by footslope streams is a very important consideration in the face of waterborne zoonoses (especially debilitating to infants and young children) associated with contaminated drinking water (see appendix in Peters and Blumenschine, 1996).
[It may be possible that Peters here has, as most people do, underestimated the ability of the natural diet to provide the fluid requirements, or how it may relate to ideal source areas including as a function of climate, even if it is a finer point of idealness of source area...] ......

Dambos are seasonally waterlogged (for the most part), ill-defined floodplains, channelless (unincised) drainageways forming the headwaters of incipient valleys in areas of low relief. ...... Under the current subhumid to semiarid climatic regimes grasses make the greatest contribution to the biomass of most dambos; sedges are second. Sedges become co-dominant to dominant in wetter conditions. Some of these species provide edible plant parts: seeds, stem/leaf bases, and especially shallow rootstocks.
A variety of rootstock-food plants belonging to the petaliferous monocots are also present, including terrestrial orchids in wetter environments. ......

During the late Cenozoic, intensification of seasonality in rainfall and decreasing overall precipitation probably resulted in grasses increasingly replacing marsh in dambos (also along lake margins, in riverine floodplains and in pans). ......

Before their [recent historical] destruction, they [dambos] constituted the major natural grasslands of tropical/subtropical Africa. If early hominins exploited grasslands in any major way, dambos were probably the primary grasslands to which they were adapted.
......
Mt Elgon is an ancient massif (fig. 13.5). Because of its size, fertile soils, and surrounding wetlands and woodland mosaic, this ecosystem is hypothesized here to have been the most important demographic source area for hominins in east Africa during the (Mio-) Pliocene and early Pleistocene. ......

First, we need to be able to picture the land systems and vegetation ecostructure to postulate patterns of selection for common and divergent hominin forms of locomotion and food acquisition. ......
Perhaps the most striking observation that can be made about the plain geometry of these [Kitale and Eldoret land systems of Mt Elgon] wetlands is that they are pan-like or, more often, irregularly linear, and usually about 1 km or less in breadth, while commonly 3-5 km or more in length. ......
They are individually much smaller as open areas than those of the Serengeti plains, which cover distances of 15-25 km or more, but those volcanic ash edaphic grasslands are probably later Pleistocene in origin (Peters et al., n.d.). ......
It may turn out that the fossil sites are not located in the landscapes that drove hominin evolution, either with regard to origins or divergence in the speciation process. ......

Hominin scavenging from lion kills is in part the business of risk management. The broader the open landscape facet, the greater the distance to the safety of trees, that is, the greater the risk of death from encounters with large carnivores …
The fine-scale dambo-wetland mosaic minimizes this risk and takes a good deal of the danger out of the problematic open-habitat adaptation concept applied to early hominins. …… This might also open up the possibility not just of confrontational (interference [power]) scavenging but of hunting larger game as well.
[But why give preferential consideration to scavenging/hunting over travel between trees and rootstock and seed gathering?... And more needs to be said on just what is the activity of all other relevant animals, large carnivores and prey animals of any sizes, in this sort of landscape…]

Reed KE, Rector AL. Chapter 14 African Pliocene paleoecology Hominin habitats, resources and diets. In, Ungar PS. (Ed) (2007) Evolution of the human diet The known, the unknown, and the unknowable. Oxford University Press.

"… we are interested in the ecological behavior of our ancestors and as such in combining as much data as we can to arrive at the best possible guess for foods consumed. ……
It is expected that African Pliocene habitats and extant African habitats under similar [that is, *when* they were similar…] climatic regimes would have virtually the same vegetation structure. ……

In Africa, habitats that receive less MAR [Mean Annual Rainfall] than in forests and more MAR than occurs in deserts fall under the rubric of "savanna" (White, 1983). The term *savanna* thus covers a multitude of habitats, such as woodland, bushland, shrubland, and grassland, and is therefore somewhat useless in discussion of early hominin habitats. Instead, more precise terms and habitat definitions can be applied.
The structure of woodlands is different from forests in two major ways. First, the ground cover in woodlands consists of grasses, whereas in forests epiphytes and ferns abound (White, 1983). Second, no matter how dense the woodland, there are no canopy levels, and the crowns of trees are not consistently interlaced. Density of woodlands (i.e., open to closed) is based on the percentage of ground covered by tree canopies if viewed from above (Pratt and Gwynne, 1977). ……
Bushlands are characterized as receiving from 250 mm to 600 mm MAR. …… Sept (1994) has reported, however, that bushlands have more fruiting and likely edible plant species than woodlands. ……
Moving along the habitat gradient to drier realms, shrublands usually have the least amount of MAR combined with a lack of nutrients in soils, especially contrasted with woodlands and bushlands. ……
Finally, grasslands are obviously characterized by the presence of few trees. In Africa, there are essentially two types of grasslands. Edaphic grasslands are exemplified by floodplain grasslands and/or wetlands like the Kafue Flats in Zambia, or by the unique Serengeti Plains, which is grassland on volcanic soils (Archibold, 1995). These types of grasslands may receive abundant water in some months of the year either from rainfall or from flooding rivers. [But why are there no trees there? The flooding?]
Secondary grasslands are derived and are caused by overgrazing, regular burning, or both.

Much of Africa that is covered by secondary grassland today would revert to some type of woodland if they were not either consistently burned or overgrazed (Pratt and Gwynne, 1977).
This may mean that the African Pliocene, and indeed much of the Pleistocene, had dominant habitats unlike the prevailing habitats of eastern and southern Africa today.

Habitats across the landscape occur in mosaic patterns because of changes in soil types, the presence of subterranean water, and other factors.
This is important to consider when reconstructing past environments because river courses and lakes provide much of the mosaicism, and almost all of the East African hominin localities were accumulated under these depositional environments [River courses and lakes]. ……

In the tropical and subtropical belt of Africa, however, the differences in seasons are dependent on when and how much rain is received instead of fluctuations in temperature. …… In southern Africa today, there is a unimodal rainfall pattern resulting, in general, in six months of rain and six dry months. In eastern Africa, the rainfall tends to be bimodal with two wet and two dry seasons per year, none of which lasts six months. ……
The numbers of wet- and dry-season months as well as plant productivity are important to consider in reconstructing both the paleohabitats and diets of early hominins. …… … hominins, and indeed other primates, that existed in these habitats may have also needed to switch to different and possibly less-nourishing food resources for the lean season. ……
Unfortunately hominin fossil localities often provide little direct evidence of the plants that previously existed. ……
However, most of these localities are mosaic in nature and assignments to one overall habitat type are problematical. ……

Short-term climate changes are seasonal and in any time period are usually somewhat predictable on a year-to-year basis.
Long-term climate change is caused by a variety of factors, but in Africa over the last 10 million years the major influence has been glacial cycling. The cooling caused by this cycling has caused overall drier and more open habitats to dominate in Africa today, contrasting the forests and deciduous woodlands that are reconstructed for the early to middle Miocene (Crowley and North, 1991).
These forested and woodland habitats of the Miocene indicate higher MAR than seen today in the same regions of Africa (Foley, 1987; Andrews and Humphreys, 1999; Jacobs, 2002).
After the middle Miocene, from about 10 Ma [million years ago] to about 7.0 Ma, the continuous glaciations at both the north and south poles caused the first aridification and cooling trend of the climate across Africa.
However, at the end of the Miocene (~6.0 – 5.4 Ma) this drying trend was reversed as only the South Pole maintained an ice cap.
The early Pliocene (5.4 Ma until ~3.0 Ma) was consequently characterized by a wetter, warmer climate in Africa than the previous 5 million years (Marlow et al., 2000).
Another glacial cooling cycle evinced by low sea surface temperatures (SSTs) began at about 3.2 Ma and continued through to about 2.2 Ma (Marlow et al., 2000). Accordingly, terrestrial temperatures and MAR must have been reduced through the same time period. deMenocal and Bloemendal (1995) note increased levels of eolian dust in sea cores at circa 2.8 Ma, which indicate terrestrial xerification [drying out].
SSTs recorded from 2.1 to 1.9 Ma suggest fairly swift cooling of both sea and land temperatures, resulting in further terrestrial aridity by 1.8 Ma (Marlow et al., 2000).
Each of these events is associated with discernable vegetation fluctuations and resulted in likely dietary changes of early hominins.

(continued below)

Hominin Habitats

Before the drying trend of the late middle Miocene, African habitats were characterized as more forested than any habitat associated with hominins of the Pliocene.
Work by Jacobs (1999a, 1999b, 2002) on fossil plant leaves shows changing seasonality across 6 million years at Tugen Hills. At this site, the length of the dry season changed from about 0 to 4 months at circa 12.8 Ma to about 3 to 7 months at 9 Ma and then reverted to the original 0 to 4 months at circa 5.5 Ma. This corresponds to the climatic aridification patterns iterated earlier and indicates that dry seasons in parts of eastern Africa were likely less than 4 months long at the beginning of the Pliocene.

The climatic data would also suggest that the habitat of Toros-Menalla in Chad where *Sahelanthropus* has been recovered (Brunet et al., 2002) was more arid and perhaps more open than the later Pliocene habitats. However, other fauna recovered with this taxon indicate the presence of water, which may have locally altered the habitat to be wetter and more closed (Vignaud et al., 2002).

The earliest east African hominins (*Ardipithecus* spp.) from sites dating from 5.8 Ma to 4.3 Ma are suggested to have lived in fairly closed woodlands or possibly forested environments based on the mammals found in the same strata with the hominin fossils (WoldeGabriel et al., 1994, 2001; Haile-Selassie, 2001).
The Gona *Ardipithecus* sites, however, appear to be more varied and include some wooded grasslands (Semaw et al., 2005). We do not include these sites in our analyses because the fossil species found are far enough removed from extant groups that trophic assignments based on uniformitarianism might be problematic.

Hominins recovered from sites from about 4.2 Ma to 3.0 Ma include *Australopithecus anamensis, Australopithecus afarensis, Kenyanthropus platyops*, and possibly *Australopithecus africanus*. These hominins lived in regions that were often mosaic and included, in various proportions, closed to open woodlands, bushlands, riverine forests, and seasonal floodplains …
However, some localities such as Allia Bay (*A. anamensis*) and the Tulu Bor Member at Koobi Fora (*A. afarensis*) are suggested to have been fairly dry and open (Reed, 1997; Wynn, 2000).
The *A. africanus* sites (Makapansgat and Sterkfontein) are also mosaic, but the Sterkfontein 4 deposit, from which the majority of *A. africanus* specimens come, reconstructs as drier than the other localities.

As the climate in the earliest Pliocene was comparatively wet and warm, extended rainy seasons that supplied preferred food items to early hominins for long periods of each year are expected. ……

Australopithecus afarensis, which has the most data on varying habitats at different sites (Laetoli, Tulu Bor, Maka, Hadar), seemed to have existed in fairly different habitats across space and through time. ……

Sometime between 3.0 and 2.5 Ma two distinct lineages of hominins appeared, *Paranthropus* and *Homo*.
Habitats such as open woodlands, bushlands, and shrublands with wetlands have been constructed for these hominids before circ 2.0 Ma (Reed, 1997; Spencer, 1997; Bobe and Eck, 2001). There appears to be general trend toward more open habitats in this time period, although this depends on the particular regions in which fossils are recovered.

...... Olduvai Bed 1 from this time period appears extremely open and arid in our analysis, as do the Swartkrans sites from South Africa. ......

Seasonality in the 3.0 – 2.0 Ma time period seems to be more in line with current seasonal regimes in eastern Africa.
Whether this was unimodal or bimodal is probably an unknowable.
However, these habitats had large river and lake systems that would have supplied water in dry seasons, perhaps to a greater extent than what we see today.

After about 2.0 Ma, more coverage of grasslands is indicated by carbon isotopes in soils and increased numbers of grazing mammals at about 1.8 Ma (Cerling, 1992; Reed, 1997; Spencer, 1997; Marlow et al., 2000).
More open, arid, and thus seasonal environments typified the existence of hominins recovered from African sites from circa 2.0 to circa 1.0 Ma.

Grasslands did not come to stay, however, and Cerling (1992) has shown two extensive grassland peaks at about 1.7 Ma and 1.2 Ma with subsequent returns to bushlands and woodlands. Sites in this time range from all of the localities are in the open and dry range of the graph (fig. 14.4), indicating a fairly broad change in climate and habitat. This is the climate in which *Homo erectus* [of our line] appears and in which *Paranthropus* [not of our line] continued to exist.

The trend from the middle Miocene to the late Pleistocene was thus from wet, warm climates in east and southern Africa to decreased rainfall and increased seasonality caused by glaciations of one or both poles (Marlow et al., 2000).
However, many differences have been noted in the rates of species turnover and habitat change at different localities, indicating that the trend was not uniform across time and space (fig. 14.4; Behrensmeyer et al., 1997; Bobe and Eck, 2001; Alemseged, 2003; Reed, Lockwood and Arrowsmith, 2003). ......

Many of the fossil localities have more suid [pig, porcine] species than are present in any extant site. Irrespective of whether one classifies them trophically as grazers or omnivores, they cause the fossil sites to have different mammalian community structures than any extant locality.
This suggests that the habitats might have been different in some way from any that exist today, and/or there was more plant productivity. ......
Fossil suids have been shown to have eaten C_4 plants (Harris and Leakey, 2003b) even if their dentition did not have the shearing crests that are hallmarks of grass and leaf diets (Janis, Damuth and Theodor, 2002). ......
... the earliest fossil localities in particular have higher percentages and numbers of browsing ungulates in their communities than many of the extant sites. ......
In fact, numbers of browsing mammals appear to decrease at about 1.88 Ma, about the time of one of the grassland expansions described by Cerling (1992). ......

Finally, can we relate these reconstructed habitats with hominin diets? ......

Ergo, it is possible that early hominins were eating soft and small hard fruits, flowers, seeds, and so on, that simply required thicker enamel and larger teeth because they were foraged from bushes and trees closer to the ground in these dusty habitats.
However, if fruit production was slightly diminished because of more leafing in plants, early hominins may have had to shift to other resources in the short dry season.
These resources might have included underground storage organs (USOs), which may have been quite plentiful if the number of suids in these fossil sites is an indication of the presence of roots and tubers. ......

Dry-season months increased as habitats became drier and there might have been more reliance by some populations of hominins on USOs.
This situation would be reflected in hominin dentition as increased crushing and more contact with grit would require stronger and more robust dental morphology.
[Fails to properly consider nut-like oil seeds, grass seeds (cereals), oily herbaceous seeds (e.g. sesame), starchy non-cereal herbaceous seeds (e.g. buckwheat), and legumes]
[Also, significant periodic variation of climate seems to have occurred more often than detailed here (see section 1.6, p91 above)]

Sept J. Chapter 15 Modeling the significance of paleoenvironmental context for early hominin diets. In, Ungar PS. (Ed) (2007) Evolution of the human diet The known, the unknown, and the unknowable. Oxford University Press.

"All models make different trade-offs between the three dimensions of generality, realism and precision and are always simplifications of a complex world, in effect, a balancing of the known against the unknown … ……

Elisabeth Vrba has argued in her "turnover pulse hypothesis" that climatic shifts that caused the expansion of African grassland habitats generated broad patterns of bovid speciation and extinction and that such significant habitat shifts were also correlated with hominin evolutionary patterns (Vrba, 1985, 1988, 1995a, 1995b).

Potts (1996a, 1996b, 1998, 2004) has developed an alternative paleoenvironmental model of hominin evolution called the "variability selection hypothesis" that emphasizes adaptive responses to environmental instability, rather than habitat-specific adaptations, as the key selection factor. ……

But when framed as conceptual arguments in this way [speculations about different types of keystone foods framed with references to the dietary preferences of living (nonhuman) primate or human populations, and the relative nutritional quality of the food] food models tend to define the possible but not the probable. ……

However, how these meaty meals were typically acquired, the extent to which they were eaten or shared in relation to other types of food, and how frequently or dependably hunting or scavenging opportunities were accessed in different environmental contexts remain matters of significant debate, as discussed later in this chapter and in other contributions to this volume. ……

The Grandmother Hypothesis posits that *Homo ergaster* evolved in response to climate-driven changes in female-foraging and food-sharing practices. In brief, they [Hawkes, O'Connell, Blurton Jones] argue that climate change decreased the availability of food sources easily available to children, such as soft fruits, making *Homo* more dependent on adult-access foods with high processing costs (tubers, nuts, hard seeds), especially during the dry season. They argue that under such conditions there would have been selective advantages to behaviors such as tool use and social strategies such as food sharing. They suggest that selection pressures would also have favored longevity in postmenopausal females because, by analogy with Hadza and other human foragers today, females with reduced maternal burdens could have shared food with the group and raised inclusive fitness. ……

Could sedges have been the main C_4 foods contributing to the relatively positive ^{13}C isotopic signatures of different South African hominin species, or was it more likely to have been grass seeds, or termites, or the meat from grazing herbivores? Evaluations of

C_4 food distribution in the context of regional vegetation patterns have helped frame the food debate (Peters and Vogel, 2005), but because an almost infinite number of dietary ways exist to create an average isotopic signature (the principle of equifinality), … ……
[The claim here by Sept that Sponheimer at al. (2005) have used the isotopic variability of sedges and termites across landscapes to realistically argue that significant omnivory was quite probable for both Australopithecines and Paranthropines requires, particularly in the context of the foregoing discussion of equifinality, more detail for it to be other than just another motherhood statement for yet another pet theory. Sponheimer at al., in their focus on present-day C_4 sedge underground storage organ and termites in Kruger National Park, extrapolated to *Australopithecus africanus* and *Paranthropus robustus* (of 3 Sterkfontein Valley caves within 3 km of each other) ^{13}C levels, fail to address the issues of possible C_4 sedge *seed* consumption, and C_4 *grass seed* (for example maize, millet and sorghum) consumption, as well as the possible role of C_3 foods such as nuts, oily seeds and starchy tubers in food energy density. Also they fail to differentiate the energy density of grass *seeds* from that of the rest of the plant, in their reference to gelada baboon diets and their energy density. Also, regarding the lack of well-developed rhizomes and tubers on the C_4 sedges that they sampled, the general representativity of this is in question as others have reported nutritionally useful C_4 sedge underground storage organs.]

…… Other elements of their [Peters and Blumenschine, 1995, 1996] [scavenging] model, such as plant food availability, are based on informal, subjective assessments, pending systematic field study (Copeland, 1999) and not calibrated methodologically, so their realistic accuracy if difficult to evaluate. [Just fitted to their pet theory, as they went along?...] …..

… it should be evident from this brief review that all good models are explicitly derived from an empirical base but that they vary in scale and emphasis with the scope of the questions they are designed to explore. …… … in other words, how sensitive is their analysis to the assumptions of their methodology?
[Which is what 'sensitivity studies' are designed to address, should modelers and theorists be so competent as to include them in their studies as appropriate…] ……

Their [Wood and Strait, 2004] critique reminds us that generalized models assuming simplistic contrasts between the diets and subsistence strategies of fossil taxa run the risk of neglecting ecological realism in our efforts to chart evolutionary trends and challenge us to make the assumptions of each component of a model as explicit as possible, so that they can be formally evaluated and adjusted to compare alternatives. ……

… some archaeological analyses that might appear to be "precise" to nonspecialists, such as the SEM [Scanning Electron Microscope] analysis of cut marks on faunal remains [fossil animal bones] from archaeological sites, are not. [Due to subjectivity in overall interpretation of the meaning of the marks] ……

But because the faunal exploitation studies were done during slightly different periods in different areas with different subgroups [of present-day African gatherer-hunters, the Hadza], the studies produced different results. Opportunities to work out how these differences could be understood in terms of a model of Hadza faunal exploitation were passed by in the authors' eagerness to claim that their results could be generalized to interpret definitively different aspects of the early archaeological record.
[All scientists hope to arrive at the definitive answer to some important question – that is to say, that they answer the question completely for all intents and purposes in the real world – in order to make their work seem this way, which can generate both kudos (and the related hubris…) and grant money - it is quite a common occurrence that they omit

(consciously, as well as in ignorance) to mention important caveats – that is, they omit to mention the ways in which their answer is uncertain or incomplete in its general usefulness to answering the question and solving the problem to which it is addressed – and in a presentation to peers or public they may hope that no-one is present in the audience with enough competence to be able to discern these shortcomings at short notice and point them out (which is often not easy to do, even for scientists fairly competent in the (related) things that they themselves are focally involved with…)] ……

From a botanical perspective, so little is preserved of the behavioral interactions between hominins and plants that there is not much empirical data to generalize from. ……

Because of the trade-offs between generality, realism, and precision, any individual model will always be constructed with large elements of "unknowable" built right in.

…… Sometimes those traits [derived from existing data sets] may seem overly exaggerated because of taphonomic [burial science related] or theoretical biases.
[For example, carnivory evidenced by bones remaining where plant food refuse does not remain, and the biases of those who cherish and psychologically rationalize their acculturated carnivorous dietary habits, and the theories they have expounded in the past even when further considerations of the evidence should logically have reduced the vehemence with which those convictions are held…]

Copeland SR (2007) Vegetation and plant food reconstruction of lowermost Bed II, Olduvai Gorge, using modern analogues. Journal of Human Evolution. 53: 146-175.

"Within the 50-80 kyr deposition of lowermost Bed II, periods of drier climate were characterized by low lake stands and a broad eastern lacustrine plain containing a mosaic of springs, marsh, woodland, and edaphic grassland. Based on results of this study, plant food diversity in each of those habitats was relatively low, but the mosaic nature of the area meant that hominins could reach several different habitat types within short distances, with access to potential plant foods including marsh plants, grass grains, roots, shrub fruits, edible parts from palms, leafy herbaceous plants and Acacia pods, flowers and gum. Based on Manyara analogues, a greater variety of plant foods, such as tree fruits (e.g., Ficus, Trichilia) and the roots and fruits of shrubs (e.g., Cordia, Salvadora) would be expected further east along the rivers in the lacustrine terrace and the alluvial fans." [(the interfluves of the fans having seasonal edible grasses in some (drier) periods)]

1.10 Feet, bipedalism, arboreality

Henry AG, Wood W. Chapter 2 Whose diet? An introduction to the hominin fossil record. In, Ungar PS. (Ed) (2007) Evolution of the human diet The known, the unknown, and the unknowable. Oxford University Press.

"Fossils from sites in Kenya and Ethiopia that date from 4.2 to 3.9 Myr [million years ago] have provided evidence of *Australopithecus anamensis*, a hominin that might be the immediate ancestor of *Australopithecus afarensis*. …… The size and shape of the pelvis and lower limb remains suggest that *Au. afarensis* was capable of walking bipedally but probably only for short distances."
[Some of the 2%(>3%?) DNA difference is in the feet (humans not having opposable big toes and therefore prehensile feet) – note that this does not mean humans cannot climb trees well, just not *as* well as our nearest animal relatives. I myself (only very average athletically) coming into tree climbing in commercial fruit harvesting at the age of 30

quite readily developed enough facility at it to see clearly that this is a very natural activity for humans – and those many native peoples living a less modernized existence, whose population average tree-climbing ability well exceeds my own fairly good ability, are not peculiarities or aberrations of nature in that regard, they are merely doing something we humans have evolved to (still) do quite well.
The smaller apes are better adapted than we are to actually living much of the time in trees, which is necessary for protection from predators, the gorilla being able to defend itself (themselves) on the ground on the basis of size and elongated canine teeth.
We humans are better evolved to travel *between* trees because we can cover open ground better than the apes, and our use of weapons to defend ourselves (strangely overlooked, as against for killing things, as a possibility, by many archeological and anthropological theorists) offers us a defense while on the ground that the other apes don't have, and that the gorillas don't need due to their larger size.
It may well be that the evolution of the use of defensive weapons enabled the evolution of the human feet.

Fleagle GJ. Section 2.8 Primate locomotion and posture. In, Chapter 2 The life of primates. In Jones S, Martin R, Pilbeam D, Bunney S (1992) The Cambridge Encyclopedia of Human Evolution. Cambridge University Press.

"Humans…… Our hindlimbs are long, providing a relatively long stride; and our foot lacks the grasping abilities of all other living primates. It has become a stiff propulsive lever with a long heel and short toes, and with the big toe aligned with the other toes."

"Reconstructing the ecological factors that lead to bipedalism is a fascinating and controversial issue in the study of human origins. There are many theories of the emergence of this unique pattern of locomotion.

Some argue that bipedalism evolved in conjunction with the use of stone tools. This is not supported by the fossil record, which indicates that human ancestors were bipedal well before they made such artifacts.
[But probably *not* before they were using *naturally-formed* stone hammers, anvils, mortars and pestles!…]

Others suggest that the evolution of bipedalism is associated with the freeing of hands for transporting food and water from distant foraging sites to offspring and other family members, so that bipedalism is related to reproductive success. ……

The evolution of bipedalism has also been linked with the need to forage in open woodland with widely scattered food resources. Bipedal walking seems to be an energetically efficient way for a large primate to travel long distances over flat country."

[Unfortunately, there is no mention here by Fleagle of defensive vigilance and weapon use, nor of low branch fruit picking]

Alexander RM. Section 2.9 Human locomotion. In, Chapter 2 The life of primates. In Jones S, Martin R, Pilbeam D, Bunney S (1992) The Cambridge Encyclopedia of Human Evolution. Cambridge University Press.

"People need remarkably little muscle activity to stand in a relaxed posture."

Jablonski NG, Chaplin G (2004) Becoming bipedal: how do theories of bipedalization stand up to anatomical scrutiny. Chapter 13 in Anapol F, German RZ, Jablonski NG Shaping primate evolution Form, function and behavior. Cambridge University Press.

"Many different selective agents have been singled out over the years for their role in inaugurating and developing bipedal posture and locomotion in hominins. ......
carrying infants, food and tools [including weapons];
throwing of tools [including weapons] and other missiles;
threat displays;
aggression;
evasion;
vigilance and sentinel behaviors;
sexual display and nuptial gifts;
arboreal gathering of food;
terrestrial gathering of food; aquatic gathering of food;
terrestrial predation [i.e. hunting];
terrestrial scavenging [of carcasses];
efficiency in long-distance travel/foraging, thermoregulation during high activity in hot environments;
walking on snow or ice;
iodine deficiency [reduced metabolism and activity capacity?];
none at all due to bipedalism being the primitive mode of locomotion from which all quadrupeds evolved [seems very unlikely?] ......

Evaluating the hypotheses.

Food carrying.

The lifting of heavy objects in humans is made possible by the very large weight-bearing surfaces of the vertebrae, especially in the lower lumbar region, and the recruitment of a large hip extensor musculature, the glutei (Shapiro 1993; Farfan, 1995).
These anatomical provisions are absent in the African apes, hence their limited abilities to carry external loads without placing a hand on the ground (Farfan 1985).

The anatomical situation in early hominins with respect to the existence and degree of the lumbar lordosis thus becomes critical. ......
When the vertebrae of this specimen and others with relatively complete thoracolumbar vertebral series such as Stw-H8/H41 are examined, however, it appears that the australopithecine vertebral column was adapted to human-like lordosis and stable balance of the trunk over the pelvis (Sanders, 1998).
The relatively small size of the lumbar vertebral centra in these specimens, however, indicates that vertical forces through the vertebral column are transmitted differently than in modern humans or that smaller increments of compression passed across their centra (Sanders, 1998).
The logical deduction in terms of function is that a vertebral column with this configuration was not adapted for carrying anything but a very light load [No weight given...].

A similar conclusion was made from the anatomical observations of the early *Homo* skeleton, KNM-WT 15000 (Latimer and Ward, 1993).
[It seems to me unlikely that there was not a significant difference between australopithecine and early *Homo*, even if early *Homo* differed from later *Homo*...]

The paleontological evidence thus suggests that australopithecines engaged in bipedal walking, but that they could not carry heavy loads at the same time.
[Almost certainly they would have been able to carry clubs, babies, some (i.e. missile and hammer) useful stones, and some food, maybe in bags made of hide]

Whether they could engage in "consistent, extensive walking" is also somewhat questionable.
[The point is that they probably gradually evolved to be able to do more and more consistent extensive walking than i.e. extant apes are able to].

The relatively short legs of Australopithecus did not preclude a modern form of bipedalism (Kramer, 1999), but they did not guarantee it either. .... Thus, at the very best, early bipeds were walking far but empty handed, or at worst weren't walking far at all.
[Their putative abilities need to be assessed in the context of the correct comparisons with i.e. chimpanzees and the most active of modern humans – it seems more likely that australopithecines were able to walk a fair distance, carrying light objects – these are, after all, wild animals conditioned to a large amount of physical activity on a daily basis, and as such their abilities need to be compared not with the modern human 'couch potato' or the more shiftless of academics, but with modern elite troops that can walk 40-plus kilometers with a 60kg load on their back and carrying a rifle, ultra-marathoners who run 80-100km more or less in one go, and long-distance trekkers – and, maybe moreover, australopithecines didn't have feet like modern humans for nothing – it is hardly likely that such feet would have evolved from the more general primate pattern for anything other than substantial walking (even though they are still very useful for climbing trees)].

Arboreal feeding.

...... One of the important consequences of bipedal feeding postures in which the forelimbs are raised above the level of the shoulder joint is that this activity shifts the body's center of gravity and imposes high physical stresses on the thoracic and lumbar vertebrae.
In bipedal feeding postures, animals with relatively long forelimbs such as modern apes and, presumably, the earliest hominins, are carrying the equivalent of a heavy parcel away [How far away is much the point] from their bodies at the level of the shoulders or above, thus greatly increasing the moment of the load. This is still a problem even if only one forelimb is used in food harvesting, while the other supports part of the body weight [The point, importantly, is that it is then much *less* a problem, and maybe not really a "problem", just a matter of a natural level of exertion]. ......

Clambering bipedally on arboreal substrates [tree branches] (or their equivalents) is energetically very costly (Elton et al 1998). Simply standing bipedally on a branch is also relatively expensive in terms of energy because such a posture requires adoption of a compliant posture with flexed hip and knee (and possibly foot) joints, with its attendant metabolic costs.
[This is much less the case if one arm is used to negate the need for balancing.]

A further issue to be considered here is how females with dependant infants would have managed in this context.
With their higher basal metabolic costs (if weaning) and at least one forelimb occupied in stabilizing their infant, they would have been hard pressed to harvest sufficient high-quality food on a regular basis to make bipedal arboreal feeding a worthwhile activity.
[This might not have been a problem at all - if the infant could have been put down on the ground near the feeding mother, or handed to a carer on the ground or in a tree, then the

mother might have been able to feed, with her child in safety, on the ground or in the tree].

We can only conclude, therefore, that the energetic penalties of this posture and mode of locomotion in food harvesting argue against feeding as the primary agent of selection in bipedalization.
[Jablonski and Chaplin have not considered nut gathering and cracking with hammer and anvil stones, nor grain gathering and grinding with useable natural stones, nor root vegetable finding, digging up, transport, and pounding/grinding – and without these considerations, sensible theorization is impossible, and conclusions could be correct only by chance…].
In fact, feeding in low trees or shrubs favors a morphology in which long, powerful forelimbs operate freely from a trunk that is stabilized in sitting or squatting on the ground, such as in giant ground sloths or – much closer to home – living gorillas. In such postures, long forelimbs served by powerful retractor musculature operate from an erect trunk and are used to pull food-laden branches closer to the animal's center of gravity, where they are consumed (Jablonski and Chaplin, 2000).
[This is incorrect, as anyone who has worked in commercial fruit harvesting will be able to attest – here one person harvests (manually or near enough) enough fruit, growing in various average positions relative to the picker and the ground, (notwithstanding the artificial density of it) to feed hundreds of people…].

Threat displays.

…… Therefore, we suggest that bipedalization was driven by a different, relatively rare behavior – the bipedal threat display and appeasement behavior complex – that enhanced the reproductive success of particular individuals and ultimately had dramatic consequences for species demography.
[It seems more likely that they would have done a lot better to consider weapon use both in defense from predation and in defense from and offense to other hominid groups or individuals, together with vigilance vision.]

Conclusions.

…… We recognize the importance of a multiplicity of factors in the evolution of modern bipedalism, as emphasized by Rose (1984, 1991), but the primacy of one in initiating the process of becoming bipedal.
[It is always helpful to identify a major factor(s) in the causation of something, but it is always simplistic, and often or usually detrimental, to neglect other factors that are contributing either in a simple additive way or in a complex multiplicative (interaction) way…]

Behavioral hypotheses for the evolution of hominin bipedalism are often eschewed because they are viewed as untestable, umparsimonious, and unscientific (e.g., Lewin, 1999) … . This ignores the critically important role of behavioral innovation in primate evolution, especially as a way of creating new ecological opportunities (Lee, 1991).
[Indeed, and in a sense this is in agreement with my comment immediately above, but in that sense they do not take their own advice…]

1.11 Teeth, the supra-pharyngeal vocal tract and language

Henry AG, Wood W. Chapter 2 Whose diet? An introduction to the hominin fossil record. In, Ungar PS. (Ed) (2007) Evolution of the human diet The known, the unknown, and the unknowable. Oxford University Press.

"Fossils from sites in Kenya and Ethiopia that date from 4.2 to 3.9 Myr [million years ago] have provided evidence of *Australopithecus anamensis*, a hominin that might be the immediate ancestor of *Australopithecus afarensis*.
The canines of *Au. Anamensis* are more chimplike than those of *Au. afarensis*; yet the chewing teeth are very different from those of chimps. ......
The incisor teeth of *Au. afarensis* are smaller than those of chimpanzees, but the postcanine teeth are larger.
This suggests that the diet [of both *Au. Anamensis* and *Au. afarensis*] included more hard-to-chew items than does the diet of chimps. ......

Homo ergaster [early African Homo erectus] is the earliest hominid with a body whose size and shape is more like that of modern humans than any of the archaic or transitional hominin taxa.
In relation to the size of its body, the teeth and jaws of *H. ergaster* are smaller than those of the archaic and transitional hominins. This means *H. ergaster* either had a different diet than that of archaic and transitional hominins or it was eating the same sort of foods but processing them outside instead of [/as well as] the mouth. [i.e. by grinding or mashing with mortar and pestle] ......

By 600 Kyr [thousand years ago] we begin to see at sites in Africa, such as Bodo in Ethiopia and Kabwe in Zambia, evidence of hominins attributed to *Homo heidelbergensis* because they lack the distinctive features associated with *H. erectus*. ...... There is also a further reduction in the size of the jaws and chewing teeth. ......

It is possible that the species (or two groups of species, *P. aethiopicus* and *P. boisei* [not in our line] vs. *H. habilis*, *H. ergaster*, *H. erectus*, and *H. rudolfensis* [in our line]) differentiated by adopting different dietary strategies."

Lucas PW. Chapter 3 The evolution of the hominin diet from a dental functional perspective. In, Ungar PS. (Ed) (2007) Evolution of the human diet The known, the unknown, and the unknowable. Oxford University Press.

"Cooking greatly affects texture, and because individuals can both detect and select (within the limits of what particular modes of cooking can do) desired textures, this allows very rapid evolutionary changes in the dentition in response.
[Likewise grinding, mashing, and soaking of uncooked foods would allow dental changes, and it is very deficient that these things are not addressed in this chapter...] ......

Thus broader incisors remove more of the outer peel of a fruit with each action than narrower ones.
Analogously, larger canines are more likely to penetrate the skin of male conspecifics.
......

I hypothesize that tool use in human evolution has its major influence on the size of anterior teeth by reducing input particle size.

[It seems to me just as likely that it had just as much influence on the posterior teeth, by performing the function of grinding that would otherwise have to be done by the posterior teeth – and Lucas provides neither evidence nor plausible theory to the contrary here…]

Cooking influences the size of the postcanine teeth by modifying foods such that their fracture is made easier. ……
[It also makes it easier to bite pieces of the food off, which is the function of the anterior (front) teeth, which it is defective of Lucas not to mention here…]

Thus, a potential food like meat, which normally obstructs free-running fracture via notch insensitivity, becomes much easier to consume with the relatively blunt human dentition when cooked (Lucas, 2004).
[Other than by bolting unchewed or relatively unchewed lumps whole, which is what the carnivores do once they have detached a small enough lump, because they have the stronger gastric (stomach) digestive juices to deal with meat swallowed fairly unchewed]
……
The gape of the human mouth is controlled by the need to ingest food particles because, with reduced canines, there is little to gain by threatening social competitors. If canines were still involved in threat gestures, then no reduction would be possible. However, without this factor, jaw lengths and incisal sizes can be thought of as tied together in this argument and would reduce together if ingested food particle size were reduced by tool use. ……

Cooking is a universal practice in modern human groups and may be almost as old as stone tool use (Wrangham et al, 1999).
[This latter suggestion is extremely unlikely to be correct – the use of naturally formed hammer and anvil stones is well documented in chimpanzees in nature, and the use of naturally formed hammer and anvil, and mortar and pestle, and other grinding stones is therefore very likely to go right back to the earliest hominids, antedating the controlled use of fire (*making* fire, in particular), which is archeologically a very recent thing, by millions of years.] ……

At the very simplest level, the linear dimensions of the cheek teeth vary as the cube root of change in food toughness.
A general result of cooking is often reduction of food toughness.
If so, then tooth size would decline in accord.
If so, it does not take much effort to see that tool use could reduce food particle size (to which jaw length and anterior tooth size are proportional) much more rapidly than cooking could change the ratio of cooked-to-raw food toughness.
[This (relative rapidity) is not apparent from the data and reasoning provided – what is obvious enough is that both tool use and cooking are quite likely to be capable of having an effect.] ……

Australopithecines have relatively large postcanine teeth, suggesting, among other things, that they were not cooking food (Teaford and Ungar, 2000). ……

Cooking is much more critical for understanding tooth size trends in humans than tool use.
[Very unlikely! See my immediately previous comments.]

(section continues below)

Ungar PS. Chapter 4 Dental functional morphology The known, the unknown, and the unknowable. In, Ungar PS. (Ed) (2007) Evolution of the human diet The known, the unknown, and the unknowable. Oxford University Press.

"... insectivorous primates have reciprocally concave blades well suited to shearing tough insect chitin between the leading edges of the crown crests, whereas frugivore molars possess cusp tips oriented more parallel to the occlusal plane [like human molars, the surfaces of them being flatter] for crushing and grinding three-dimensional fruit flesh and seeds (fig 4.1). ......

Comparative studies have shown that folivores and insectivores have higher SQs than do closely related frugivores and that, among frugivores, soft-fruit flesh feeders have higher SQs [Shearing Quotient, an index of the extent to which molar cusps are pointed rather than flat] than hard-object feeders.
[So humans would be at the hard-object feeder side of the frugivores.] ......

Results suggest that the australopiths both had relatively flat and blunt molar teeth compared with extant frugivorous hominoids ["homin<u>oids</u>" means animals like humans, i.e., apes] but that *A. africanus* [of our line] had higher SQ values (more pointy molars) on average than did *P. robustus* [not of our line].
This is consistent with Grine's (1981) observation that "gracile" australopiths [like *A. africanus* of our line] had more occlusal relief than did "robust" forms and suggests that while *A. africanus* molars may have been better able to shear than could those of *P. robustus*, neither could have had a very tough diet (Kay, 1985); Teaford, Ungar, and Grine, 2002).
[Note here that the term "tough" is used in the sense that the science of physics has it, which is explicitly different from being "hard" – a leather would be very tough to chew (or try to...) but not very hard, (dried ripe) hazelnut shells are fairly hard, but not very tough (they are fairly brittle, readily breaking up when a certain pressure is reached] ......

First, most teeth, whether they are in museum collections or still in the mouth, are in fact worn. ......

All three of these great apes [chimpanzees, gorillas and orangutans] prefer soft, succulent fruits when available.

The central African chimpanzee and western lowland gorillas take many of the same foods when these apes are sympatric [live in the same area], but they also differ in their diets, especially at times of fruit scarcity.
At such times gorillas fall back on more tough, fibrous foods than do chimpanzees (Tutin et al., 1991; Remis, 1997).
Average annual food-type proportions reported for chimpanzees include 70-80 percent fruit flesh, as compared with 45-55 percent fruit flesh for gorillas (Williamson et al., 1990; Kuroda, 1992; Nishihara, 1992; Tutin et al., 1997). ......
average fruit percentage of 55-65 percent for the orangutans (Rodman, 1977; MacKinnon, 1977). ......

At any given stage of wear, however, gorillas had the steepest slopes followed by orangutans.
Chimpanzees had the shallowest molar cusps (fig 4.4).
This example suggests several things.
First, tooth shape changes with wear. As teeth wear down, they become flatter. This is not rocket science!

Second, cusp slope values mirror leaf-to-fruit ratios reported in the literature for these taxa, suggesting that we are picking up functional aspects of form.
Finally, and perhaps most important, the differences among the species are of about the same magnitude at each of the recorded wear stages.
This suggests that differences between species remain consistent through the wear sequence, so that we can compare chimps, gorillas and orangutans at any given wear stage and get the same results. ……

Results indicate that at any given stage of wear, *A. afarensis* shows flatter molar teeth with less occlusal relief than do early *Homo* specimens.
This suggests that early Homo would have been capable of more efficiently fracturing tough foods than would their australopith predecessors.
Indeed, differences between these two taxa are of about the same magnitude as those separating chimpanzees and gorillas. This may imply similar differences in their diets.
If so, perhaps both ate many of the same sorts of foods. Early members of our own genus may have occasionally consumed more tough foods, especially at times of resource stress, whereas the australopiths “fell back” on more hard, brittle foods.

This brings up another important point. Apes have a penchant for succulent, sugar-rich foods – a legacy of the ancestral catarrhine dietary adaptation (Milton, 1993; Ross, 2000; Ungar, 2005). The same was likely true of early hominins.

Differences in diet among living catarrhines often hinge on a seasonal shift to fallback foods taken when preferred resources are less available (Rogers et al,. 1992; Lambert et al., 2004).

Fecal studies for gorillas and chimpanzees at Lope in the Gabon, for example, show 60 percent to 80 percent plant species overlap (Williamson et al., 1990; Tutin and Fernandez, 1993). These apes both eat fruits much of the year but diverge at “crunch times” when preferred fruits are scarce. At such times gorillas fall back more on leaves and other fibrous plant parts. ……
It is not surprising then that differences in occlusal morphology between living great apes relate more to fallback foods than to preferences per se. ……

While Hartmann’s (1988) analysis of hominoid [apes] molars suggests that tooth form differences among apes more reflect diet than evolutionary propinquity, phylogenetic inertia clearly plays an important role in how adaptations manifest themselves.
[Meaning that changes from two different original habits to the same final habit may arrive at that final habit with body structures or functions that are still different, and maybe even a lot like the originals, with more time being necessary for more change to make the two final structures/functions more similar, and even then they may never become closely alike because they are in the context of being attached to other body structures or functions that are different and remain different.] ……
No matter how many times a given tooth form is associated with a specific diet, we cannot prove that this association has held or will hold in all cases.
[For example, do naturally more carnivorous animals tend more to bolt their (raw) meat after little chewing (they are poorly equipped with teeth suited for chewing) because they have stronger, more acidic gastric digestive juices, whereas a hominid might chew the meat more before swallowing it, which would be in character with how it deals with some of the plant foods that naturally comprise the majority of its diet?] ……

In the end, we have no choice but to use the present to interpret the past, or we risk leaving paleobiological interpretation in “indecipherable chaos” (Martin, 1991).

[But of course there is a requirement to point out the known, unknown, unknowable, possible, and probable in a fully rigorous manner, for the science to be acceptably good instead of just another attempt at a half-baked justification for some cherished belief system or an attempt to make some academic work look like a definitive seminal work instead of the often much less so something it really is…]

We know, for example, that early *Homo* was better able to process tough elastic foods than could their Australopithecus predecessors.
[So what were these tough elastic foods? Leaves? Drier legumes more integrated with their more leaf-like pods? Tougher fruit? Corms/roots/tubers? Meat? Soaked or greener grains?]
[But how much better? Our own teeth have little facility, or substantially less facility, for this compared to the apes. Was there really much difference? Was it due to the grinding function being carried out with stone tools, leaving the shearing function free to be selected for again, rather than more tough elastic foods being actually eaten?]
[It is unfortunate that Ungar did not here better at least lay out these questions clearly, even if he chose not to elaborate what answering of them there may be available…]

Teaford MF. Chapter 5 What do we know and not know about diet and enamel structure? In, Ungar PS. (Ed) (2007) Evolution of the human diet The known, the unknown, and the unknowable. Oxford University Press.

"For instance, primates characterized as hard-object feeders (e.g., Cebus apella) tended to have thicker molar enamel than did soft-fruit eaters (e.g., Ateles belzebuth).
However, humans also seemed to have relatively thick molar enamel, raising the possibility that we, or our ancestors, were also somehow adapted for eating hard objects.

…… There [modern humans] investigators have shown that not only do distal molars tend to have thicker enamel than more proximal ones (Macho and Berner 1993, 1994; Spears and Macho, 1975, 1998; …; Schwartz, 2000a, 2000b; Gantt et al., 2001; Grine, 2005). ……
From this perspective, thicker enamel would then help the tooth withstand higher occlusal forces thought to occur on either more distal molars, or the so-called functional cusps of molars. ……

If one were to use a summary "measure" to characterize molar enamel thickness in modern hominoids, it would probably be something as follows: *Gorilla* and *Pan* [chimpanzees and bonobos] have relatively thin enamel, *Pongo* [orangutans] has medium-thick enamel, depending on which teeth are examined, and *Homo* has thick enamel (Shellis et al., 1998; Schwartz, 2000b). ……

However, as in humans, these summaries may also mask important functional differences across the tooth crown. For instance, while *Pongo* [orangutans] may have slightly thinner enamel than *Homo*, the distribution of enamel across their molar crowns is also different with *Pongo* having relatively thick enamel in areas associated with crushing (Schwartz, 2000b; Kono, 2004) and *Homo* having thicker enamel across the entire crown. ……
[Which would be consistent with Homo chewing more grains (plus some grit), and orangutans cracking more nuts?...]

Initially it was felt that the earliest Miocene apes had relatively thin enamel, and thus, this might be a primitive trait for apes and humans (Martin, 1983, 1985; Andrews and Martin, 1991). More detailed analyses have subsequently shown that Miocene apes show a range of molar enamel thicknesses, from relatively thin in *Proconsul major* and *P. africanus* to relatively thick in *P. nyanzae* and various sivapithecids (Beynon, Dean et al., 1998; Smith, Martin and Leakey, 2003). ……

… there is a general consensus that the early hominins had relatively thick enamel compared with modern humans and most other living primates.
[This needs to be quantified here!] ……

Finally, while it may be difficult to separate early *Homo* from some australopiths based on isolated measures of enamel thickness (Tobias, 1991; Ramirez-Rozzi, 1998), *Homo erectus* appears to have the thinnest enamel of those Plio-Pleistocene hominins analyzed by Beynon and Wood (1986).
This raises the possibility that, with the arrival of *H. erectus* and probably a heavier emphasis on tool use, enamel thickness may have started a gradual decline to this day. Of course this raises questions about the selective mechanisms behind such a trend. So why have thicker, or thinner, molar enamel? Two adaptive explanations are usually given in the literature:
(1) thick enamel may have evolved to prolong the functional life of teeth exposed to an abrasive diet (Macho and Spears, 1999), or
(2) it evolved to strengthen teeth given high occlusal forces generated by a diet that includes hard foods. ……
However, most researchers have recently viewed the second explanation as more acceptable for primates because primate hard-object specialists have been shown to have thicker tooth enamel than closely related taxa that habitually eat softer foods (Kay, 1981; Dumont, 1995).
[But these aren't operating near the ground, where dust and grit contamination are greater than up in the trees…]
Thin tooth enamel, however, may be better suited to a tough food diet, as dentin exposure yields sharp edges on an occlusal surface that can improve shearing and slicing efficiency (Delson, 1973; …). ……

As it turns out, many seeds eaten by primates are not hard but instead are pliant and tough (Lucas and Teaford, 1994), and even those seeds that are encased in hard pericarp are often opened with the front teeth, leaving the molars to process seeds that are again surprisingly soft (Kinsey and Norconk, 1990). ……

So it may indeed be that the early hominins had relatively thick, decussated [made up of interwoven prisms that therefore resist crack propagation and breakage better] enamel, which left them equipped to deal with a variety of foods (Teaford and Ungar, 2000); and it may also be that members of our genus had progressively thinner enamel, allowed (in part) by the advantages of tool use (Teaford, Ungar and Grine, 2002). ……

But as we come to know more about the intricacies of dental development (Weiss, Stock and Zhao, 1998; Salazar-Cuidad, Jernvall, and Newman, 2003), could there be epigenetic effects that we are overlooking? Could deciduous tooth use be influencing permanent tooth structure? [Sounds Plausible.] If so, what would be the driving force of such influence, foods most frequently processed, or foods of critical importance to survival at certain periods in life (Kay, 1975; Kinzey, 1978)?

(section continued below)

Daegling DJ, Grine FE. Chapter 6 Mandibular biomechanics and the paleontological evidence for the evolution of human diet. In, Ungar PS. (Ed) (2007) Evolution of the human diet The known, the unknown, and the unknowable. Oxford University Press.

"A thin section of cortical bone from the base of the mandibular corpus of a single specimen of the Plio-Pleistocene hominin *Paranthropus robustus* [not in our evolutionary line, a dead-end sideline] has been examined for evidence of Haversian remodeling (fig 6.1). This particular fossil reveals a strikingly low incidence of secondary osteons and is more similar to those recorded for *Alouatta* and *Ateles* [soft-food eaters] than for *Cebus* and *Chiropotes* [hard-food eaters].
[However, *P. robustus* has a relatively massive jaw – so massive as to never be sufficiently stressed as to develop secondary osteons?...] ……

Attempts to invoke contemporary analogs for feeding behavior and diet in the two genera [*Australopithecus*, of our line; *Paranthropus*, not of our line] have resulted in wildly divergent interpretations of trophic adaptations (cf, Jolly, 1970 [seed-eating]; Szalay, 1972 [bone-crushing]; Cachel, 1975; DuBrul, 1977), presumably due to unwarranted (and often unstated) assumptions about paleoecological variables. ……

… proposed that the gracile *Australopithecus africanus* was omnivorous, while the robust *P. robustus* was a more dedicated herbivore.
[A good example of "unwarranted assumptions"? – Just as likely, if not more so, to be due to a difference within plant foods choices, and/or to differences in stone tool use for food processing…]
The implication of divergent dietary adaptations for biomechanical comparisons is straightforward, if one assumes that bone tissue models and remodels itself to maintain stress levels within some optimal interval. Specifically, given that mastication of fibrous and tough foods engenders higher strains in the mandible (Hylander, 1981, 1988), then a more demanding diet for *Paranthropus* would require a mandible that was structurally suited to accommodate greater masticatory forces while maintaining stress magnitudes comparable to Australopithecus. In the vernacular, *Paranthropus* should have a more robust jaw than *Australopithecus*. ……
…both fossil species are functionally distinct from the great apes in having relatively stronger mandibles (Daegling, 1990).
With respect to each other, however, they are not distinguishable biomechanically. In other words there is no basis for assuming that *P. robustus* jaws could accommodate larger bending forces than *A. africanus* mandibles and experience similar stress magnitudes at the same time (fig 6.5)
[Maybe, but are the specimens properly representative of the averages for their kinds? Only very small numbers are available for use, and even less used in analysis…]

…… Resistance to shear is essentially a function of cross-sectional area.
Because the overall corpus dimensions of both *A. africanus* and *P. robustus* are typically larger than those of modern hominoids [this includes modern apes], and because the amount of bone per unit of subperiosteal area is comparable to living forms (Daegling, 1990), it follows that the fossil hominins were better equipped to deal with large shearing forces. …… An alternative measure of torsional strength that accounts for this asymmetry [of bone] (a variable known as Bredt's formula in the engineering literature) suggests that *A. africanus* and *P. robustus* jaws were, in fact, absolutely stronger than those of extant hominoids [this includes modern apes] (fig 6.6).
But this finding of absolute strength differences is not terribly informative because it does not correct for the different magnitudes of twisting moments that may (or may not) operate in the different species. ……

An interesting finding in regression of mechanical rigidity and strength variables against moment arm surrogates is that modern human mandibles tend to fall above empirical best-fit and forced isometric lines (figs 6.5 and 6.6), thus prompting the interpretation that, by an allometric criterion, modern humans have relatively strong jaws (Daegling, 1990, 1993).
But the difference between modern human and early hominin jaws indicates that strength is achieved in fundamentally distinct ways.
Modern human jaws are strong because they are short (i.e. corpus size and proportions are unremarkable for a hominoid but the moment arms of the associated masticatory loads are much smaller).
This contrasts with the early hominin case, in which massive cross-sectional dimensions are incorporated in a mandible that has an essentially apelike configuration.
Whether this means that stress levels are comparable in these cases is arguable: masticatory muscle hypertrophy is reasonably inferred from cranial form (especially in *Paranthropus*), whereas it is generally believed that the increasing gracility of *Homo* crania through time is linked to a reduction of adductor muscle mass.
If both suppositions are true then the evolution of the homininae may be marked by a temporal trend toward reduction in masticatory stresses.
Under this interpretation, the mandibles of Homo sapiens are not functionally diminutive, they are actually overdesigned (i.e. needlessly strong) with respect to masticatory loads.

What do these biomechanical properties of *A. africanus* and *P. robustus* jaws tell us about diet in these early hominins?
An important consideration is that postcanine megadontia [having big teeth behind the canine teeth] is a feature of both taxa [species].
It has been argued that functional requirements for supporting these large teeth account for the large corpus proportions (Wolpoff, 1975), but subsequent analysis suggests that mandibular dimensions are not determined by the size of the dentition (Daegling and Grine, 1991; Plavcan and Daegling, in press). Instead, the most parsimonious interpretation of early hominin jaws is that the large corpus dimensions were resisting elevated adductor muscle forces used to create high-magnitude bite forces. Because these bite forces were distributed across an enlarged postcanine dental battery, it is likely that the bite pressures across the occlusal surfaces were comparable to living hominoids (Demes and Creel, 1988).

The combination of elevated masticatory forces and postcanine megadontia could be interpreted as indicating selection for processing greater amounts of food, while not necessarily indicating that the food items themselves were qualitatively different from those processed by the great apes today (e.g., see Walker, 1981).
The concept of equivalent bite pressures is crucial to the interpretation because it underlies the interpretation that these early hominins were simply processing relatively more of the same types of foods that apes eat today. In other words, it is the quantity, rather than the quality, of the diet, that accounts for the morphology of the fossils.
Such an assumption is flawed, however. Even though the concept of equivalent bite pressures is sound in theory, it does not follow that early hominins [/hominids ?] were no more mechanically efficient than their modern counterparts among the hominidea [homininea ?].

If ingestion of small, hard objects was part of the behavioral repertoire of early hominins (Grine, 1981), the application of bite force during the initial phases of food breakdown would have been localized to the object borders (assuming a three-dimensional geometry). In this case, the stress imparted to the object would be much higher than implied by bite pressures because the adductor force would be applied as a point load (or for multiple objects a series of point loads, with local stresses many multiples of a hypothetical bite

pressure imparted over the entire occlusal area). Thus, postcanine megadontia does not necessarily imply that bite pressures applied to food items could not have been very large. Instead megadontia could be seen as the result of selection for mitigating the affects of attrition imposed by the repeated application of high, localized occlusal stresses. The thick enamel characteristic of *A. africanus* and *P. robustus* molars is consistent with this interpretation.
[I suggest that the broadness of the teeth, in particular the occlusal surface being broader than the base of the teeth, is more consistent with chewing multiple small harder things like seeds all at the same time, rather than the crushing of bones, which is better done by a narrower occlusal surface supported by a triangulation down to a broader base, such as more truly natural bone-crushing animals like hyenas and dogs have. The purpose of bone-crushing is to expose the marrow for acquisition and digestion, rather than to chew the bone itself to fine particles for digestion. The thickness of the enamel could be consistent with either the chewing of multiple hard things or the crushing of bone, but is not required for the consumption of the fruits and leaves that the general shape of the human teeth is suited for in the same way that the general shape of the apes' teeth suits them for the consumption of fruits and leaves.
It would certainly help for the consumption of those tubers that were tougher for them to be ground or beaten down into a rough mash using available natural stones, and this would be easily, and plausibly enough have been, done.
The thickness of the enamel is also consistent with protection from dirt particles attached to food from near or in the ground, even if that was not its evolutionary origin.]

By a strictly biomechanical criterion, the mandibles of *P. robustus* and *A. africanus* are indistinguishable (given the comparative variables chosen).
This functional similarity stands in contrast to the cranial and dental differences that distinguish the two taxa.
These differences have traditionally, and sensibly, in our view, been related to distinctive trophic adaptations.
The question naturally arises as to why the mandibles of these forms – elements so clearly functionally linked to mastication – do not appear to reflect the dietary differences to which the teeth and crania seem to point.
We view the most profitable avenue of investigation to be the reassessment of how mechanical competence is measured and quantified. ……
Biomechanically speaking, *A. africanus* and *P. robustus* have mandibles that do not seem to be functionally distinct from one another.
Cranial form, dental proportions, and patterns of dental microwear, however, clearly distinguish these early hominins.
Their diets were probably different, but it does not follow that their jaws must therefore be morphologically distinct for them to have been mechanically competent for their respective functional environments."
[The diets would probably have differed mostly in proportions of the food types eaten, rather than more absolutely as in the types of foods eaten, and that maybe in fall-back foods, in the same way that chimpanzees and lowland gorillas mostly differ in their diets. And perhaps *A. africanus* [of our line] used tools more, but the literature does not seem to provide resolution of what the dietary ramifications of this might have been.]

Teaford MF. Chapter 7 What do we know and not know about dental microwear and diet? In, Ungar PS. (Ed) (2007) Evolution of the human diet The known, the unknown, and the unknowable. Oxford University Press.

"So, while the shape of unworn teeth can tell us a great deal about what a tooth is capable of processing, tooth wear can give us insights into how a tooth is actually used. ……

In fact, very few foods in our diets can scratch our teeth because our foods are so clean, cooked and processed.

Still, without such processing, many foods (e.g. certain leaves) include abrasives that can cause striations on teeth (Lucas and Teaford, 1995; Gegel et al., 2001; Teaford et al., 2006; fig 7.3).
Others include acids, which can etch the teeth (fig 7.4; Puech, 1984b; Puech, Cianfarani and Albertini, 1986; Teaford, 1988a, 1994; Ungar, 1994; King, Andrews and Boz, 1999b).
......
... the incisors [front teeth] and canines are used mainly in *ingesting* food, while the premolars and molars are used primarily in *chewing* food once it has been ingested.

Analyses of incisor microwear have yielded two basic conclusions.
First, animals that use their incisors very heavily in the ingestion of food show higher densities of incisal microwear features (Ryan, 1981; Kelley, 1986, 1990; Ungar, 1990, 1994, 1996b).
Second, the orientation of striations on the incisors reflects the direction of preferred movement of food (or other items) across the incisors (Walker, 1976; Rose, Walker and Jacobs, 1981; Ryan, 1981; Ungar, 1994). ......

... primates that eat hard objects usually show large pits on their molars, while leaf eaters tend to have relatively more scratches than pits on their molar enamel (Teaford and Walker, 1984; Teaford, 1988a; fig 7.5).
Those "hard objects" can evidently include hard nuts but also smaller items such as insect exoskeletons (Strait, 1993; Silcox and Teaford, 2002).

Microwear is also occasionally found on the buccal or lingual ("nonocclusal") surfaces of molars, which may give additional indications of the abrasiveness of the diet, the size of food items, or even the degree of terrestriality (Puech, 1977; Lalueza and Fox, 1992; Lalueza, Fox and Perez-Perez, 1993; Ungar and Teaford, 1996). ......

Finally, the laboratory study also showed that the turnover in dental microwear could be quite rapid in animals with an abrasive diet, as all of the microwear features in an area sampled by an SEM [Scanning Electron Microscope] micrograph would change in one to two weeks, depending on whether the animal was raised on the hard or soft diet. ......

For instance, the amount of molar microwear [In Costa Rican Howling monkeys] may vary from season to season and between riverine and nonriverine microhabitats (Teaford and Glander, 1996).
The studies at La Pacifica have also shown that tooth wear generally proceeds at a rapid pace in the wild – in fact, at about eight to ten times the pace of that in U.S. dental patients (Teaford and Glander, 1991).
This has led to the idea of the "Last Supper" phenomenon (Grine, 1986); that is, in some situations, dental microwear may only record the effects of the most recently eaten foods on the teeth, although some investigators feel that microwear on the sides of the teeth may show far slower turnover (Perz-Perez, Lalueza and Turbon, 1994). ......

The precise timing of dental microwear analyses in this case (before the heavy onset of new leaves and grasses in this seasonal environment), allowed Nystrom, Phillips-Conroy, and Jolly (2004) to implicate "small-caliber environmental grit" as the main cause of the observed microwear patterns, which included no significant differences between the sexes, age groups, or different troops [of Ethiopian baboons, which are terrestrial]. ......

Peters (1982) used standard physical property-testing equipment while examining the effects of a range of African foods on dental microwear, ultimately showing that few foods could actually scratch enamel; instead extraneous abrasives were one of the prime culprits (see also Puech, Cianfarani and Albertini, 1986).

Only with more detailed analyses did subsequent work (e.g., Gugel, Grupe and Kunzelmann, 2001) demonstrate the effects of specific foods on microwear patterns (e.g., "cereal-specific" microwear related to phytolith content in certain grains; fig 7.6 [(in-vitro, in lab experiments) enamel losses: modern spelt, 7um; modern and historical wheat, 22um; millet, 33um]). ……

For instance, certain cooking procedures may introduce abrasives into foods, causing a high incidence of microscopic scratches on teeth, scratches not caused by the foods themselves but by the methods with which they were prepared (Pastor, 1992, 1993; Teaford and Lytle, 1996; fig 7.7) [i.e. grindstone particles].

Similarly, animals may also eat soft foods and still show many scratches on their teeth, if the food is coated with abrasives (e.g., earthworms coated with dirt or leaves and ivy coated with sand; Silcox and Teaford, 2002; Merceron, Viriot and Blondel, 2004a). ……

However, the incisors of the prehistoric populations [North American] do show more variable scratch orientations, suggesting they were used for a greater variety of tasks than in the historic populations (Teaford et al., 2001).
As far as molars are concerned, early quantitative studies of Native American populations noted correlations between the amount of microwear and the abrasiveness [From what?!] of the diet and the presence of large pits and the consumption of hard objects [= What?!] (e.g., Shkurkin et al., 1975; Rose and Marks, 1985; Harmon and Rose, 1988). ……

In central North America, there seems to be a transition in preagricultural populations to a slightly harder diet before the actual switch to agriculture (Schmidt, 2001), with subsequent agricultural populations having a softer, less variable diet (Bullington, 1991). Along the southeastern coast of the United States [of America…], however, microwear evidence for that dietary transition is complicated by significant local differences in microhabitat, most notably between coastal and inland sites, with the former evidently including significant amounts of large-grained abrasives in foods (Teaford, 1991, 2002; Teaford et al., 2001).

In other regions of the world (e.g., the Indian subcontinent), interpretations are further complicated by changes in food-processing techniques, as stone grinding tools introduced significant abrasives into what would have otherwise been a fairly nonabrasive diet (Pastor, 1992; Pastor and Johnston, 1992). ……

Some changes in food processing, most notably the boiling of foods, led to a marked decrease in the amount of microwear at some sites (Molleson, Jones and Jones, 1993).

However, these changes also left microwear researchers with a bit of a dilemma: how to distinguish the microwear signatures of meat eating versus the consumption of cooked food (Molleson, Jones and Jones, 1993; Organ et al., 2005)? ……

Molar microwear analyses have also helped to document the effects of phylogenetic constraints in fossil apes by documenting similar functions in taxa that have undergone shifts in molar morphology through time (Ungar, Teaford and Kay, 2004).

As might have been expected, analyses of human ancestors have focused on whichever fossils are available.

For the anterior teeth, qualitative studies have suggested similarities between early hominin incisor wear and that observed on modern primates that routinely employ a great deal of incisal preparation (Puech and Albertini, 1984).
Quantitative analyses of *Australopithecus afarensis* suggested incisal microwear similarities with those documented for lowland gorillas or perhaps savanna baboons (Ryan and Johanson, 1989).
More detailed analyses of *Paranthropus robustus* and *Australopithecus africanus* (Ungar and Grine, 1991) showed great variability within each species in most standard microwear measurements.
However, the greater density of features on the incisors of A. africanus helped to show that A. africanus had a heavier emphasis on incisal preparation than in P. robustus.

In the molar region, qualitative analyses have raised many possibilities that have been often repeated in the literature. For instance, the robust australopithecines [not of our line] have been characterized as indistinguishable from modern chimpanzees or orangutans (Walker, 1981), perhaps with more abrasive molar wear than in the gracile australopithecines [of our line] (Puech et al., 1985; Puech, Cianfarani and Albertini, 1986; Puech, 1986b).
By contrast, *Homo habilis* [of our line, later on, closer to us, 2.5 to 1.5 million years ago] has been characterized as using high occlusal pressures but on foods that can chemically etch the enamel [Tough acidic fruits? Acidic fruits eaten with tough leaves?] (Puech, Serratrice and Leek, 1983; Puech, 1986b). ......

What is the magnitude of seasonal, annual, geographic, and interspecific differences in dental microwear for other species elsewhere in the world? ......
As primate diets are normally quite variable, and as dental microwear features can change quite quickly (Teaford and Oyen, 1989a, 1989b; Teaford and Glander, 1991), what are the effects of different food items on overall microwear patterns within a specific diet? Will items that are abrasive, hard, or acidic effectively swamp other microwear patterns? Will the Last Supper phenomenon vary between species and populations? In essence we must not lose sight of the fact that the microwear pattern on a given tooth is essentially a complex summary of past events, often with multiple signals superimposed on one another. Some wear processes, like acid etching, leave fairly distinctive patterns but can have varying effects on teeth, depending on the length of exposure of the tooth to the acid (Teaford, 1994). ......

When this [differences in study methods] is coupled with the fact that most analyses have used relatively small [numbers of] samples (even for species with variable diets), the net effect is that dental microwear analyses have barely begun to live up to their potential. ...... As a result, the consumption of certain soft foods may remain invisible, and the relative importance of different foods within overall diets may also be indecipherable because certain foods, or abrasives on foods, effectively overwhelm other evidence. ......

Also, for past populations, we are limited to working on only those specimens preserved in the fossil or archaeological record. Does that mean that we will never obtain samples from certain environments? Probably. Does it mean that we will never obtain samples from certain seasons? Possibly. Does it mean that we will never accurately bracket the range of variation of dietary habits of some species? Probably. ...... For the earliest hominins, we probably cannot come very close; for more recent hominins, perhaps we can come much closer."

Dean C. Section 2.4. Jaws and teeth. In, Chapter 2 The life of primates. In Jones S, Martin R, Pilbeam D, Bunney S (1992) The Cambridge Encyclopedia of Human Evolution. Cambridge University Press.

"Some early fossil hominids were adapted to eating tough and fibrous plant foods (such as certain fruits and tubers [And maybe certain seeds and grains]) and had premolars with expanded biting surfaces, or even extra cusps. These molarized premolars resemble molar teeth. Fossil hominids such as *Australopithecus africanus*, *Paranthropus* (or *Australopithecus*) *robustus* and *Paranthropus* (or *Australopithecus*) *boisei* have bigger premolar teeth than do *Australopithecus afarensis* or later fossil hominids in the genus *Homo*.

The teeth of gorillas and chimpanzees have a thinner layer of enamel, the hardest biological substance known, than those of fossil hominids, modern humans, orangutans or even fossil apes such as *Sivapithecus*.

Modern humans have shorter jaws and smaller teeth than apes."

Andrews P, Martin L (1991) Hominoid dietary evolution. Phil Trans R Soc London B 334: 199-209.

"Summary

During the later Paleocene and early Miocene, catarrhine primates and the evolving hominoids had adaptations for frugivorous diets, with the emphasis on soft foods.
Early in the middle Miocene the hominoids underwent a major shift, both in morphology and in habitat, with the morphology characterized by thickened enamel on the molars, enlarged incisors and massive jaws.
The diet indicated by this morphology is interpreted as still mainly frugivorous but with changed emphasis, possibly towards harder objects.

The thick-enameled hominoids are found associated with more open forest habitats, and the distribution of food resources in equivalent habitats today is discontinuous both in time and in space, leading to evolutionary pressures particularly affecting locomotion, brain size and social behavior .

The earliest known hominid fossils differed little in dental and mandibular morphology from the middle Miocene apes, and the implied dietary similarity…….

Hard food includes seeds, nuts and unripe fruit ….. hard-carapaced insects. [Andrews and Martin fail to mention grains]

Frugivores have relatively small molars and large incisors, whereas in folivores the proportions are reversed.

Relative enamel thicknesses [The last two figures are obviously not significant]:

Gorilla	10.04;
Chimp	10.10;
Orangutan	15.93;
Heliopithecus leakeyi	17.35 (an early ape)
Australopithecus africanus	22.17; (of our line)
Homo sapiens	22.35

Early Miocene

.... the combination of characters seen in Afropithecus and Heliopithecus leaves little doubt that fruits were a major part of their diet. ....

Middle Miocene.

Thick-enameled hominoids, such as *Kenyapithecus* from East Africa and '*Sivapithecus*' from Pasalar and Candir in Turkey, both dating to about 15 Ma BP, first appeared in the middle Miocene.
They are seen as a continuation of the early Miocene trend towards thickening enamel initiated in the progression from *Proconsul* to *Afropithecus* and *Heliopithecus*.
They were associated with more open habitats than were the early Miocene hominids: tropical to subtropical deciduous forest is indicated for both ((Andrews et al 1979; Andrews 1990), with a strongly seasonal climate and hence supplies of fruit .....

... thick enamel is present also in the hominoids that are phylogenetically linked with the orangutan, such as *Sivapithecus* and its included genus *Ramapithecus*. New data presented here suggest the presence of even thicker enamel in *Graecopithecus*.
The presence of thick enamel at this stage would appear to be a retained character from a thick-enameled ancestor of the middle Miocene of Kenya or Turkey, and thick enamel may have either retained its adaptive function in *Sivapithecus* or be a heritage character whose function has changed.
There is, however, some additional evidence for *Sivapithecus* from the microwear on the molars. The relative proportions of pits and scratches, as well as lengths and breadths of defects, on *Sivapithecus* molars match the soft-fruit pattern and are similar to that observed today in chimpanzees (Teaford and Walker 1984). It is significantly different (Teaford 1988) from the pattern observed in animals that eat hard fruit, which is the function assigned by Kay (1981) to thick enamel.

Teaford and Walker (1984) conclude from this that the function of thick enamel is uncertain, but it is also possible that its function when first evolved was, as Kay suggested, for grinding hard objects, but later hominoids changed diet to softer fruits as in chimpanzees without at first losing thick enamel.
[Note that the number of individuals that the teeth came from that were examined for scratches were not provided in this paper and it is possible that seasonal or longer-cyclical (such as the glacial-interglacial cycles, which change temperature through approximately 9 degrees C) variation plays a role in dietary and tooth wear variation in found remains.]

Late Miocene and Pliocene.

..... The proportions of the incisors of *Australopithecus afarensis* appear primitive, and the molars appear to retain primitively thick enamel.
There is no evidence from the molar or mandibular morphology of any major dietary change from the pattern observed for thick-enameled apes during the preceding 10Ma.

Most interpretations of Laetoli and Hadar paleo-environments, where *A. afarensis* has been discovered, suggest either open savanna habitats (Leakey and Harris 1987) or more wooded habitats with seasonally deciduous woodland or forest (Andrews 1989).

There are also indications that the hominids retained a considerable arboreal capacity (Stern and Susman 1983) [Note: As we still do......]

Later species of australopithecine have extensive modification of their teeth, particularly the robust australopithecines (Grine 1986; Kay and Grine 1988; Grine and Martin 1988), but these will not be discussed here. [Note: probably because these homin(o?)ids are not ancestors of modern humans, being evolutionary dead-end side shoots.]

4. HOMINOID DIETARY TRENDS

Several trends have become apparent from the foregoing descriptions of hominoid tooth morphology.
The most notable of these was the increase in enamel thickness during the early to middle Miocene, coupled with the development of flattened molar occlusal morphology, and these are often accompanied by premolar enlargement and increase in mandibular robusticity, as seen in the sequence: *Proconsul-Heliopithecus-Kenyapithecus*.
The dietary significance of these changes is far from certain, as the functional significance of thick enamel is still being debated.
It has been claimed to be associated with terrestrial foraging (Jolly 1970) [Note: The paper's title refers to seed-eating] as an adaptation for hard abrasive foods found on the ground, but Kay (1981) rejected the terrestrial aspect of this idea, claiming that hard and brittle foods alone may have provided the selective impetus for thick enamel, regardless of where they occurred. ……

Small but hard objects make up a large part of the diet of the orangutan, which also has moderately thick enamel, whereas the folivorous gorilla has the thinnest enamel and the longest shearing crests seen in extant apes (Kay 1977; Martin 1985). Similarly, with monkeys, the species with diets of hard fruit have the thickest enamel, whereas the folivorous species have the thinnest enamel (Kay 1981, 1985).

The evidence of microwear on later Miocene thick-enameled hominoids is most similar to that seen today on soft-fruit eaters (Teaford and Walker 1984), which suggests no change from their thin-enameled ancestors, but no work has yet been done on such fossils as *Heliopithecus* and *Kenyapithecus* where the trend of increasing enamel thickness was first observed.
Diets of the later sivapithecines [genus of early apes, 12.5 to 8.5 million years ago; India, Nepal, Pakistan; possible orangutan ancestor] may have changed, as indicated by the microwear, without apparent change of the enamel, and it is on the teeth of the fossil hominoids where thick enamel is first observed that the significance of thick enamel is most likely to be understood.

Another interpretation of thick enamel is that it was an adaptation for a more varied diet. This would have been the product of increasingly seasonal climates and habitats, the seasonality being in part the result of topographic change within the tropics, producing more open and seasonally varied habitats at *Kenyapithecus* localities such as Fort Ternan and Makobo Islands (Andrews et al 1979) and in part latitudinal, as hominoids spread northwards beyond the tropics (Andrews 1990).

Chimpanzees living today in seasonal habitats have been observed to have more varied diets than forest chimpanzees (Collins and McGrew 1988), and during the dry season they eat dry fruits, hard seeds and hard-shelled fruits.

….. A further possibility is that thick enamel did not evolve in response to dietary shifts at all but was coopted for diet after its appearance [Note: This seems (much?) less likely, but is reproduced here for completeness.]

….. Thick enamel appears to have been retained in the ancestors of both the orangutan and the hominid clades.

After the branching of the orangutan clade, the limited fossil evidence available indicates retention of the thick enamel and robust jaw adaptations in the ancestors of the African ape and human clade. The early australopithecines maintained this ancestral pattern, with the inference of no dietary change from ape ancestors, despite the evidence for increasingly seasonal habitats associated with hominid fossils.
This conclusion is supported by microwear studies (Grine 1986), and similar conclusions emerge from other lines of inquiry, for example the retained adaptations for ape-like arboreality in the postcrania of *A. afarensis* (Stern and Susman 1983) and the short developmental period of tooth crown formation in australopithecines, again a retained ape characteristic (Beynon and Dean 1988).
The lack of change in the teeth and jaws of the earliest hominids suggests that the dietary heritage from thick-enameled apes was sufficient for dietary needs during the early stages of human evolution.

……. He compares this with the transport of nuts and hammer stones to suitable anvils by chimpanzees, or alternatively an anvil and hammer stones carried to the source of nuts (references in Potts (1988)).

….. Fruit is a discontinuous resource, both in space and in time (Milton 1988), and being more widely dispersed in the environment than are leaves or grass, frugivores and predators tend to have larger home ranges and therefore larger brains relative to body size than do herbivores and grazers (Clutton-Brock & Harvey 1980).
Fruit is also a predictable resource in that fruiting occurs regularly from one year to the next, and this leads to an emphasis on learning and memory together with greater flexibility in behavior (Milton 1988).
It is proposed by Milton, therefore, that larger brains, and by implication greater intelligence, arose initially as a dietary response rather than to facilitate social relations.

…..australopithecines show little or no change from fossil apes in brain size, dental development, type of diet, overall body size or even, perhaps, in cultural development and tool use, and as a result it may be concluded that they were the same evolutionary grade as the Miocene and recent apes.
Subsequent brain enlargement, dental reduction and changes in the gut proportions in the genus *Homo* (Milton 1988) suggest a higher quality diet and food processing, which may be the result of increasingly sophisticated use of tools.

Discussion

P. Andrews.
….. It is shown here that the functional adaptation of the main battery of chewing teeth is essentially unchanged in the early australopithecines from their ape ancestors, although the later australopithecines show some modifications to this pattern, such as the enlargement of the molars and the development of two-cusped third premolars.
[Is the significance of this an increase in food grinding?]
The incisors of most australopithecines also show no reduction from the ancestral ape condition."

[Note: But is there any seasonal or cyclical effect on hominoid remains surviving destruction by predators and scavengers, or by weather, and how might this bias the implications of the findings?]

Demes B (1988) Bite force, diet, and cranial morphology of fossil hominids. J Human Evolution 17: 657-70.

"In modern Homo, weak bite force is coupled with small molar teeth.
[Tool use, as in hammer and anvil stones to crack nuts and seeds, naturally- or manually-shaped grinding stones for grass seeds and tubers, and cutting implements for hide acquisition and carrying bag manufacture, lessen the requirement for bite force, and, the evolution of the supra-laryngeal vocal tract for speech may have had some role in this too. Also, the *chewing* force is not weak, due to the evolutionary rearward movement of the jaw (see p146 here, Daegling and Grine (2007))] ......

Among the hominid fossils, only the robust australopithecines seem to have been capable of generating bite forces that were exceptionally high compared to extant species.
Because their molar teeth were so large, however, the occlusal pressures applied between their molars would not have been exceptional if distributed over the entire molar surfaces.
If the molarization and considerable dimensions of their premolars had been considered here, their bite forces relative to occlusal area may well have proven quite modest.
Bite force equivalents for the gracile australopithecines and early members of the genus *Homo* are of average size compared to extant hominoid primates.
However, as the molar teeth are larger than average, force per unit of molar crowns is low.

Walker's interpretation of the australopithecine chewing apparatus as adapted to low-energy food that had to be processed in great quantities is fully compatible with our results.
Dietary hypotheses that postulate the processing of hard objects can be maintained only under the assumption that the area of contact between teeth and food was small. This implies that food objects were small or hard [objects could be small, hard, and many] and round[ed] in shape (seeds: Jolly, 1970; nuts: Peters, 1987), and that only a small mouthful of food was triturated [broken up by the teeth] at any one time (Lucas et al., 1985).
Large molar surfaces make sense for both menus.
[This is all compatible with some amount of gra(i)nivory]
Whereas they offer sufficient area for triturating a great quantity of food (Walker, 1981), they also increase the probability of fracture for small, hard objects (Lucas et al., 1985).

Evidence from other sources, especially dental microwear (Grine, 1986), supports a hard object diet of the robust australopithecines.

Our data also support theories on dietary divergence of gracile and robust australopithecines (DuBrul, 1977; Hylander, 1979; Shea, 1985; Grine, 1986).
Bite force equivalents for both *Homo erectus* skulls are higher than that for modern Homo sapiens.
Bite force reduction seems to have been a late trend in hominid evolution and to have paralleled a reduction in dental occlusal area [tooth surfaces between which food is broken up].

...... Incisor uses that presume high incisal [of the front teeth] bite forces have been suggested for the robust australopithecines by Szalay (1972: bone cracking)
[This plainly silly suggestion (Demes and Greel provide the mechanical details for it being unlikely) is included here only as a more extreme example of a plainly silly suggestion – contemplate yourself choosing whether to crack a bone with your front teeth or your back teeth....] ......

In facial configurations such as those of non-human primates, crown area enlargement and molarization of the premolars must result in a relatively longer bite force load arm. In hominid evolution, however, the jaw was shifted posteriorly to a position beneath midface and brain case; a favorable chewing muscle lever arm to load ratio is maintained.

One could even argue that the molarization of the premolars was only possible with these accompanying changes in facial configuration.

In robust australopithecines [not of our line], the mechanical advantage of the masseter muscle was improved further by extending the insertion of this muscle anteriorly and greatly enlarging facial height (see also DuBrul, 1977; Ward and Molnar, 1980; Rak, 1983; Shea, 1985; Preuschoft, in press).

Tall mandibular rami also favor simultaneous tooth contacts (Ward and Molnar, 1980) and tooth loading in axial direction.

The increase in facial height, as well as the position of the zygomatic arch above the tooth row, affect the lever length of the masseter muscle while shifting the face beneath the brain case primarily improves the leverage of the temporalis muscle.

[All of this seems probably compatible with doing more chewing, such as of grains…]

Among mammals, this cranial configuration is unique to hominoid primates.

As a secondary effect of forebrain enlargement, the anterior part of the brain case offers abundant space for the insertion of muscle fibers.
With a large and vertically oriented temporalis muscle, the facial retraction effects a considerable increase of bite [as distinct from chewing] force magnitude.
[This could be useful in biting harder or bigger fruit, or tubers, or opponents – removal of meat from carcasses without tools remains best dealt with by canines. Do elongated canine teeth impair chewing in any way?...] ……

The shift of the dental arcade to a position beneath the midface and brain case in australopithecines is not confined to the molar region. Because the hands are used for food preparation and uptake, an anteriorly positioned front dentition is no longer necessary.

The evolution of the hominid facial profile is thus determined by the need to develop high forces to the molar region and a loss of function of the front teeth for food preparation.
……

The hominid face with its typical profile first appeared in the fossil record with the australopithecines.

As discussed above, this typical configuration may be interpreted as a series of adaptations that help to generate and dissipate the high bite forces necessary in forms with large dental occlusion areas [i.e., those suited for chewing].
It was retained in all subsequent stages of human evolution.

In modern man, who because of his greatly reduced dentition and the use of fire and tools to process his food has no need to generate high bite forces, this development has culminated in a dramatic reduction of masticatory musculature."

Lucas PW (2004) Plant mechanics and primate dental adaptations: an overview. Chapter 9 in Anapol F, German RZ, Jablonski NG Shaping primate evolution Form, function and behavior. Cambridge University Press.

[From this work unfortunately not dealing with tooth enamel thickness, gelada baboon grass seed-eating, or tuber eating, nor relating any of what was dealt with to the hominid and human dentition available for comparison, I reproduce the following parts:]

"When plant toughness is low [fruit flesh and the mesophyll of leaves] …, cusps can propagate cracks easily. Note that a frugivorous primate with cusped molars can chew very young leaves, which have relatively undifferentiated cells. For most such tissues, fragmentation of the tissue seems less important than opening the maximum amount of cells per chew.
Blunt bulbous cusps have a distinct advantage here in that their "zone of action," the region around the cusp tip that is capable of fracturing cells, is enlarged.

When toughness is higher … crack growth [from eating/mastication] starts to be impeded … fruit pods and some seed coats … . Bladed features, i.e. extensions of a point contact on molars, are the optimal solution to driving these cracks, the proximity of the blade edge to the crack tip acting to maintain crack direction and so keep the crack moving.

Additionally, most leaves, except the youngest, are toughened by sclerenchyma fibers that are usually arranged around veins in the form of a sheath … approximately in the optimal range for toughness. Even a few such fibers resist leaf fragmentation very effectively, and, being elongated cells, they frustrate the propagation of cracks in all but one direction – along the side of cells. Since venation in most dicotyledonous angiosperm leaves is in the form of a complex net, primate herbivores [here in the usual sense of an eater of green leaves as distinct from other plant parts] need to direct fractures across veins and so need bladed molars. ……

Most seed-eating primates appear to have cusped molars, superficially resembling those of frugivorous primates.

Similarity in the fracture resistance of their typical foods could explain this. A frugivorous primate could eat very young leaves without the need for blades. For all but the youngest leaves, however, bladed molars would be needed because woody fibers would be differentiating and starting to lay down their secondary cell walls. ……

As has been recognized by countless authors, incisal [of the front teeth] design is critical for most primate frugivores, the fruit flesh having generally little resistance to mastication.
The limit to the defense [of the fruit from being eaten] lies in the capacity of the target disperser to remove the peel.
In contrast, leaves have anti-mastication defenses [possibly mostly evolved to insects] that because of their thinness are controlled essentially by their toughness.
Accordingly, it is molar design that is critical in adaptations to folivory on all but the youngest leaves."

(section continued below)

Lieberman P Section 3.5 Human speech and language. In, Chapter 3 The brain and language. In Jones S, Martin R, Pilbeam D, Bunney S (1992) The Cambridge Encyclopedia of Human Evolution. Cambridge University Press.

"Human adults are less efficient at chewing because the roof of the mouth and the lower jaw have been reduced compared with non-human primates and archaic hominids. ......
These deficiencies in the mouth of an adult human are offset by the increased phonetic range of the supralaryngeal airway.
The human vocal tract can seal off the air pathway that leads through the mouth from the nose, so yielding non-nasal sounds. This results in clearer, more readily identified formant frequency patterns.
Nasalized speech is inherently subject to more errors in interpretation than is unnasalised speech – for example, 30-50 per cent error rates for nasal vowels as against 5 per cent for unnasalised vowels under similar conditions. ......

The comparative anatomy of living primates and of hominid fossils suggests that the evolution of the human supralaryngeal vocal tract probably started in early (about 1.6 million years ago) African populations of *Homo erectus* and was not completed until the appearance of fully modern humans.

The later, or classic, Neanderthals of Europe were the last to retain the ancient non-human supralaryngeal vocal tract.
But predating the late Neanderthals, around 100,000 years ago, were other hominids – for example those from the Israeli sites of Skhul and Jebel Qafzey – with human supralaryngeal airways. ......

Speech may even have served as an isolating mechanism, hence promoting speciation in the evolution that led to anatomically modern *Homo sapiens*.
[And strain differentiation leading to the human races, more the point?]

Great apes (which lack a functional equivalent of Broca's area and the connecting pathways)
[This is no longer considered fully true.]
cannot be taught to control their supralaryngeal airways to produce sounds in any way similar to those of human speech.
[This seems to be still considered very largely true.]
Although the pongid airway is inherently unable to produce all of the sounds of human speech, it does have the mechanical potential to produce at least some of them.
However, non-human primates lacking the brain mechanisms for controlling the voluntary articulation maneuvers that underlie human speech are unable intentionally to produce even these sounds."
[By definition, almost tautologically so, this must be true. Which non-human primates fit the description, and to what extent, is in question, and not explored further here by me.]

Macho GA, Shimizu D (2009) Dietary adaptations of South African australopiths: inference from enamel prism attitude. Journal of Human Evolution 57: 241-247.

"The South African hominins differ significantly in their relationships between wear facets and prism angulations. Teeth of *P. robustus* [(not of our line)] are better adapted to more vertical loads during mastication (Phase I), whereas those of *A. africanus* [(of our line)] are better adapted to cope with more laterally-directed loads (Phase II) commonly associated with roll-crush and mastication."

[This seems very consistent with eating smaller seeds and grains being a part of our evolutionary line/heritage.]
"Overall, teeth of *P. robustus* appear stiffer, while those of *A. africanus* seem more wear resistant."
[This is consistent with eating foods growing closer to the ground, such as starchy root vegetables/ underground storage organs and grains and shrub/ford seeds, being a part of our evolutionary line/heritage.]

1.12 The gastrointestinal/ digestive tract

[It seems generally agreed that the modern human digestive tract has less large intestine, and more small intestine, than that of modern apes. I have not found quantification of these differences in any of the works referenced here. Looking at some drawings of various animal digestive tracts, it seems that these differences must be relatively small. It seems therefore that modern humans are less well adapted for digesting cellulose than the apes – it should be borne in mind, however, that the last third of the modern human large intestine favors for its energy supply the short chain fatty acids that result from the breakdown of cellulose, in contradistinction to all other human body tissues, which use glucose, fatty acids or ketones for their energy supply.

Also, the digestive juices of the human stomach seem to be agreed to be clearly not as acidic (roughly only one tenth as acidic - keep in mind the pH scale is logarithmic) as those of the carnivores i.e., the cats and dogs (which tend to bolt their meat with little chewing – the teeth of the cats and dogs are not well adapted for chewing, but are well adapted for tearing).
Related to this, although gut passage rates are similar between humans and dogs, dogs retain food in their stomachs for much longer (around 4 to 12 hours) than humans (around 1 to 2 hours; Ragir, Rosenberg and Tierno, 2000). The apparent paradox is explained by the dog's lack of chewing to preliminarily break up the meat, which it instead tears off a swallowable lump of, swallowing it with little or no chewing. Humans are obviously dentally evolved to do much chewing.

Claims that the evolutionary changes leading to the slight reduction in the large intestine and slight increase in the small intestine (that plausibly took place since the human line diverged from the chimpanzee line) were based on a large intake of meat in our ancestral diet are generally quite defective in that they fail to consider that the oily nuts and seeds are actually clearly *more* energy-dense than meat is - raw meat is about 70% water; raw nuts and oily seeds have very little water, the large majority of the weight being taken up by about 20% protein on average, and another 70-odd% being fully fat or carbohydrate – so obviously raw nuts (600 kcal / 100 g) (or oily seeds (600 kcal / 100 g), and also cereal grains (380 kcal / 100 g) for that matter…) are more energy dense than raw meat (190 kcal / 100 g, lean; 230 kcal / 100 g, regular).
Also, in failing to understand that starchy tubers are energy-dense enough to supply human energy needs, certainly with the minimal processing available even to our earliest hominid ancestors, without recourse to much cellulose digestion. And that these foods are fairly likely to have been used by our ancestors for the very large majority of the 6 million years since our line diverged from that of the chimpanzees.
And, in failing to take into account that the taste of raw nuts, seeds and tubers (as well as fruits) was probably preferred to the taste of raw meat by our evolutionary ancestors.]

1.13 Tooth isotopic ^{13}C from C_4 plant origins

Sponheimer M, Lee-Thorp J, De Ruiter D. Chapter 8 Icarus, isotopes, and Australopith diets. In, Ungar PS. (Ed) (2007) Evolution of the human diet The known, the unknown, and the unknowable. Oxford University Press.

In tropical savannas, trees, bushes, and forbs [broad-leafed non-grass herbaceous plants especially of prairie or meadow] use the C_3 photosynthetic pathway, while tropical grasses and some sedges use the C_4 pathway. ……
Because of both anatomical and biochemical differences, C_3 plants discriminate more heavily against ^{13}C during photosynthesis than do C_4 plants.
As a result C_3 plants have highly depleted $^{13}C/^{12}C$ ratios, which are expressed as delta values in parts per thousand relative to a Cretaceous fossil cephalopod which is highly enriched in ^{13}C, resulting in most terrestrial plants and animals having negative carbon isotope ratios [more correctly, negative delta ^{13}C values].
C_4 plants, on the other hand, are much less depleted in ^{13}C (Smith and Epstein, 1971; Vogel, Fuls and Ellis, 1978).
[C_3 forbs, minus 27; C_3 trees, minus 27; C_4 grasses minus 12 (delta^{13}C values), with virtually no overlap at all of statistical distributions]. ……

Although there is some variability within the C_3 and C_4 groups, it tends to be fairly small within any given savanna ecosystem (Codron et al., 2005);
However, delta ^{13}C of C_3 vegetation, in particular, can change significantly between sere [dry…], open environments (relatively enriched delta ^{13}C), and damp, closed canopy forests (relatively depleted delta ^{13}C; Vogel, 1978b; Ehleringer and Cooper, 1988; van der Merwe and Medina, 1991). ……

We also know that the stable carbon isotopes in plants are passed down into the tissues of animals that eat them. ……

The carbon isotopes in bone and enamel readily provide dietary information in modern animals, but only enamel mineral has proved reliable with fossils (e.g., Lee-Thorp and van der Merwe, 1987; Cerling, Harris and Leakey, 1999; Sponheimer, 1999).
Bone is highly organic and poorly crystalline, which leaves it very susceptible to diagenetic processes, while enamel has very little organic content and is highly crystalline, making it essentially "prefossilized" and much less susceptible to postmortem alteration (see Le Geros, 1991). ……

Thirty-seven australopith specimens [dental specimens, judging from the reference titles] from Swartkrans, Kromdraai, Sterkfontein, and Makapansgat have now been analysed, the data for which are summarized in figure 8.2
[60 x C_4 feeders, delta ^{13}C = minus 1;
19 x *Australopithecus*, delta ^{13}C = minus 7;
18 x *Paranthropus*, delta ^{13}C = minus 8;
61 x C_3 feeders, delta ^{13}C = minus 11].

Analysis of variance and post hoc tests of the combined datasets demonstrate that the fossil C_3 consumers (e.g., giraffes) and fossil C_4 consumers (e.g., zebra) have very different delta ^{13}C (Scheffe [test] $p < .00001$ [likelihood of the results being that difference by chance alone], just as is the case today.

This shows that diagenesis does not obscure the dietary carbon isotope signal.
[In these selected environments, not necessarily representative of all environments inhabited, nor even of those mostly inhabited…]

The data also show that both *Australopithecus africanus* and *Paranthropus robustus* are highly distinct from the C_3 browsers and the C_4 grazers with which they are associated (Scheffe test, $p < .00001$), although they are not different from one another.
This indicates that both australopiths consumed foods enriched in ^{13}C, such as C_4 grasses, C_4 sedges, or animals that ate these foods. ......

Still, it is important to acknowledge that the australopith delta ^{13}C is highly variable, far more so than most modern and fossil taxa in South Africa (e.g., Lee-Thorp, van der Merwe and Brain, 1994; Sponheimer, Robinson et al., 2003; van der Merwe et al., 2003). Two or three of the 37 australopiths analyzed had nearly pure C_3 diets, while several were extremely enriched in ^{13}C, with delta ^{13}C values within the range of the grass-eating baboon *Theropithecus oswaldi* (Lee-Thorp, van der Merwe and Brain, 1989; Sponheimer et al., 2005).

We also know that modern chimpanzees (*Pan troglodytes*) do not typically eat detectable quantities of C_4 foods. This is hardly surprising for chimps in heavily forested environments where few if any C_4 resources are available, but it also holds true for those in semiarid woodland environments where C_4 foods are extremely abundant (Schoeninger, Moore and Sept, 1999; Carter, 2001). ......

Moreover, because many australopiths probably inhabited environments similar to those of today's woodland chimpanzees (McGrew, Baldwin and Tutin, 1981; Reed, 1997), [though there may not be/ we may not have australopith remains in/ from these *environments*, even if there be/ we have australop. remains in/from the same *locations*.] this strongly suggests that different dietary adaptations, rather than environmental differences, are responsible for the disparity in their consumption of ^{13}C-enriched foods. In other words, it appears likely that, even in identical woodland savanna habitats, chimpanzees would continue to consume their favored forest foods to the degree that was possible, while australopiths would extensively supplement these with C_4 resources. ......

A recent study of baboon (*Papio ursinus*) feces showed that populations in marginal environments may eat up to 50% ^{13}C-enriched foods such as grasses and CAM [a third, but minor, type of photosynthetic pathway] plants, while those in more hospitable woodland environments eat only about 10% C_4 vegetation (Codron, 2003).

Furthermore, some of the *Papio* and *Parapapio* specimens from Swartkrans and Sterkfontein show significant C_4 consumption. ...... This is not to say that papionin [baboons and their ancestors] diets are, or were, like those of australopiths but only that both taxa exhibit a dietary flexibility and willingness to use C_4 resources not paralleled in chimpanzees. ......

Thus, australopith diets may have been quite dissimilar in character from those of all living primates, although there is reason to believe there was considerable dietary overlap with extant hominoids [present-day apes] (e.g., Grine, 1986; Grins and Kay, 1988; Ungar, 2004). ......

In addition, we cannot meaningfully distinguish between diets as different as frugivory, folivory, and carnivory in forest ecosystems, because all are ultimately based on C_3 plants. ......

... ruminants ... methane ... If this is the case, then it might mean that most modern primates and early hominins have a diet-enamel fractionation of about 12%, but that the ruminants have a fractionation of 14%. This would mean that we cannot directly compare the delta ^{13}C of ruminants and hominins, but would rather have to adjust the values of the hominins by about 2% to make them comparable.

This would mean that we have underestimated the amount of ^{13}C-enriched foods that early hominins consumed, and that 50% would be closer to correct than the 35% to 40% we currently accept (Sponheimer et al., 2005). ……

We now return to stable isotopes and australopith diets.

We possess overwhelming evidence that australopiths consumed foods that are enriched in ^{13}C such as C_4 grasses, C_4 sedges, or animals that ate these plants, but the stable isotope studies have created a new unknown, namely, which of these foods were actually utilized?
Answering this question is important, as the use of these different resources might have a variety of physiological, social and behavioral implications.
[By at least this point, a more detailed presentation is necessary of just what potential foods the ranges of C_4 and C_3 plant foods include, for properly productive consideration of the likelihood of their consumption…]
For instance, if australopiths had a grass-based diet similar to the modern gelada baboon (*Theropithecus gelada*), this would almost certainly indicate that their diets were less nutrient-dense than those of modern apes, possibly placing important limitations on burgeoning hominin brains and sociality (Aiello and Wheeler, 1995; Milton, 1999).
[This is quite wrong – if a lesser but substantial amount of nuts, oily seeds and starchy tubers were eaten alongside the grass (particularly seed and corm) base, particularly if the nutrient-denser parts of the grass (seeds and richer corms) were selected, but even if not, there would be no such limitation – the same basic mistake in thinking applies to much theorizing about the small evolutionary decrease (reasonably inferred from comparison with present-day apes) in human large intestine and increase in human small intestine…]
The converse, that australopiths ate diets rich in animal foods, would indicate a leap in dietary quality over modern apes, which would have been a crucial step toward hominin encephalization [developing larger brains], the development of stone tool industries, and increased social complexity (Milton, 1999).
[Maybe not, as macronutrient (protein, mostly, which is generally not required to be added, and has to be broken down to derive energy instead, which is generally a less desirable process for humans) density is not necessarily equatable to quality.
And even if it was, consider that raw meat is about 70% water; raw nuts and oily seeds have very little water, the large majority of the weight being taken up by about 20% protein on average, and another 70-odd% being fully fat or carbohydrate – so obviously raw nuts (600 kcal / 100 g) (or oily seeds (600 kcal / 100 g), and also cereal grains (380 kcal / 100 g) for that matter…) are more energy dense than raw meat (190 kcal / 100 g, lean; 230 kcal / 100 g, regular). And, that fat is more required than protein.]
Similarly, it has been suggested that consuming the underground storage organs of plants like C_4 sedges could represent an increase in dietary quality over that of extant great apes because, when they are edible, they are lower in dietary fiber than ape fallback foods (Conklin-Brittain, Wrangham and Smith, 2002).
[Yes, but why fail to consider the likely contribution of nuts, oily seeds and starchy tubers to the diet?...]

Yet, we are still uncertain as to which ^{13}C-enriched foods were consumed by australopiths. ……

While it is certainly true that the underground parts of some sedges (e.g., *Cyperus papyrus*, *Cyperus esculentis*) would make an attractive food for tool-wielding hominins, at least seasonally, it was found that there are few edible C_4 sedges in South African environments (Peters and Vogel, 2005; Sponheimer et al., 2005), although they are abundant in perennial wetlands like the Okavango Delta (Ellery et al., 1995). Indeed only

28% of the sedges in Kruger National Park use C_4 photosynthesis (Sponheimer et al, 2005).
Thus, while sedges could certainly have contributed, it is unlikely that their consumption alone could account for the observed ^{13}C-enrichment of australopiths.
However, other early hominin habitats, such as the wetlands of the Eastern Lacustrine Plain at Olduvai Gorge (Hay, 1976; Deocampo et al., 2002), might have been better sources of edible C_4 sedges. The extremely enriched delta ^{13}C of *Paranthropus boisei* [Considered *less* likely than australopithecines to have had much meat in their diet…] might indicate that C_4 sedges were locally abundant and heavily used (van der Merwe, Cushing and Blumenschine, 1999). Puech and colleagues have also suggested that the dental microwear of early East African hominins is consistent with the consumption of such foods. ……

Baboons are known to eat grass-eating grass-hoppers (*Acrididae*) almost exclusively during temporary gluts (Hamilton, 1987).

Grass-eating termites represent another intriguing possibility, particularly given recent studies suggesting that bone tools from Swartkrans were used to extract termites from mounds (Backwell and D'Errico, 2001).
[And therefore other larvae also are potential contributors to ^{13}C-enrichment via feeding on C_4 plants. Many present-day gatherer-hunters eat larvae and/or insects.] ……

However, most evidence also suggests that C_3 mammals were very common, especially at Makarpansgat and Sterkfontein (Luyt and Lee-Thorp, 2003; Sponheimer and Lee-Thorp, 2003).
Thus, as with our discussion of sedges and termites, unless the hominins were eating very large quantities of these animals, or concentrating on C_4 mammals alone, mammals are not a likely source of the ^{13}C-enrichment on their own.
[The likelihood of the consumption of C_4 mammals is not well explored here - if the early hominins were often present in the C_4 plant areas, eating C_4 plants, wouldn't they therefore be more likely to encounter C_4 mammals?...]
[Also, again, a more detailed presentation is necessary of just what potential foods the ranges of C_4 and C_3 plant foods include, for properly productive consideration of the likelihood of their consumption.]

In short, what we now know about isotopic compositions of ^{13}C-enriched foods suggests that none, save C_4 grasses themselves, appear to be able of producing a high degree of enrichment on their own.
[But there is not a high degree of enrichment in the *Australopithecus* and *Paranthropus* samples assayed – it is a little on the lower side of being intermediate between C_4 feeders and C_3 feeders, and if we apply the suggested correction for the fractionation it is intermediate – or 50% consumption of C_4 foods, as the authors note…]
Indeed from an isotopic point of view
[This is incorrect, in this case, as with only the *two* isotopes assayed, it is from a dietary, ecological and environmental point of view…],
it seems far more likely that several C_4 foods were consumed in tandem, but this conclusion is based on a lack of evidence for the consumption of one ^{13}C-enriched food, rather than direct isotopic evidence for the consumption of a variety of resources.

We also do not know about possible seasonal differences in C_4 consumption.
Our traditional sampling protocols provide us with stable isotope abundances for enamel that is formed over many months and, in some cases, years (see Sponheimer et al., 2005).
Consequently, the stable isotope data are generally telling us something about an individual's "average" diet.

It would be a mistake, therefore, to assume that ^{13}C-enriched resources were important foods for australopiths all year round.

Primate diets are known to change from season to season, sometimes profoundly so (e.g., Altmann and Altmann, 1970; Dunbar, 1983; Wrangham, Conklin-Brittain and Hunt, 1998).
As such, it might well be that C_4 resources were largely fallback foods that made up the bulk of australopiths diets during the dry season when preferred resources were no longer available or when their nutritional quality was no longer acceptable.
Then again, australopiths might have consumed ^{13}C-enriched resources all year, perhaps C_4 grass and sedge underground storage organs (USOs) during the dry season and grass seeds during the rains. We simply do not know … ……

Thus, in practice, it might be argued that we know very little about the diets of the early hominins (ironically, perhaps, both more and less than we knew fifty years ago).
What we do have is a great many reasonable hypotheses for which there are greater or lesser degrees of theoretical support. ……
For one, it would be impossible to ever know which C_4 resources australopiths used because of insurmountable equifinality problems.
In other words, there are many diets that can conceivably produce the same carbon isotope signature. …… It would be impossible, for instance, for stable carbon isotopes of tooth enamel carbonate to ever distinguish between a purely vegetarian hominid that ate 50% C_3 fruits and 50% C_4 grass and a highly predaceous hominin that ate nothing but C_3- and C_4-derived mammal flesh in roughly equal proportions. ……

1.14 Cooking

Wrangham R. Chapter 16 The cooking enigma. In, Ungar PS. (Ed) (2007) Evolution of the human diet The known, the unknown, and the unknowable. Oxford University Press.

"There are two contrasting views.

The first, which is conventional wisdom, sees cooking as merely one of many extra-oral food-processing techniques (such as pounding, grinding, or drying) that can raise food quality. …… For example, it would not be considered to have led to fundamental changes in the human digestive system. …… The essential implication is that human biological evolution was not influenced in any major ways by the adoption of cooking and that evolutionists can therefore ignore it.

The radical alternative is that cooking is a core human adaptation that has importantly directed our evolution, or as Coon (1954) wrote, that cooking was "the decisive factor in leading man from a primarily animal existence into one that was more fully human."
……
Thus, the cost of digestion is a high proportion of total energy expenditure in all animals. In humans it has been measured at around 5%-15% of energy expenditure (Westerterp, 2004). ……
For instance, high-protein diets increase the cost of digestion by about 30% compared to high-fat diets (Westerterp-Plantenga et al., 1999). ……

As expected, the glycemic index is indeed consistently increased by cooking (Brand et al., 1985; Bjorck, Liljeberg and Ostman, 2000). ……

In habitats with little or no fruit, gorillas can survive by eating fibrous foods for 100% of their feeding time (Doran et al., 2002).
Chimpanzees never do so (Doran et al., 2002). ......
Thus, gorilla molars have long shearing edges compared with those of chimpanzees, and gut passage rates in gorillas are longer than in chimpanzees, allowing more opportunity for fermentation of plant fiber (Milton, 1999; Remis et al., 2001; Remis and Dierenfeld, 2004).
[Wrangham should also have mentioned that chimps like fat in food, whereas gorillas dislike it] ......

For example, it [cooking] should increase the range of edible foods and therefore allow extension into new biogeographical zones.
Other things being equal, it should also provide a more predictable food supply during periods of scarcity because it enables a range of otherwise inedible items to be used.
It should have further effects by softening food.
For example, it should lead to a greater ability of adults to provision infants, whose dentition is too immature to allow hard chewing, other than by giving milk.
It should likewise cause a substantial drop in the amount of time that individuals spend chewing, with large consequences for the species activity budgets (Wrangham and Conklin-Brittain, 2003).
[But cooking also requires time; and the smoke may attract unwanted attention that then has to be guarded against; and the food being cooked is less immediately portable than it was before should exiting the area become necessary or strategic; and food can be processed for young children using naturally shaped stone and wood tools...]
Unlike the ordinary feeding pattern of any nonhuman ape, therefore, this means that a cooking population is exposed to intragroup [and intergroup...] competition over a valuable accumulated food pile.

For example, among chimpanzees by far the most valuable type of concentrated food supply is meat.
[Firstly it needs to be noted that the most valuable, and valued (sought after) foods for chimpanzees are not "concentrated" foods, they are fruits.
Secondly, opportunities to eat meat are most often passed up by chimpanzees in seasons when the fruits are available, in favor of the fruits, as meat is more of a fallback food that is eaten more in seasons of shortage of the more preferred fruits.
Thirdly, there is no basis in evidence for stating that meat is more valuable to or for chimpanzees than insects or nuts.
Fourthly, many individual members of a chimpanzee group will actually eat almost no meat, and the large majority eat only a very small amount (less than 5% of calories) of meat (due to their inability (i.e. in males) to catch it, or their disinclination (i.e. in females with offspring, due to the danger) to catch it, or to their inability to successfully beg for it from those who have caught it), whereas other foods are eaten in far more equal proportions among members of a group.]
[Furthermore, Wrangham fails to mention palatability and natural taste preferences, and their relationship to dietary evolution, at all... Likewise, parasites...]

...... In short, the adoption of cooking is expected to be accompanied by a series of large influences on various important biological systems, such as foraging behavior, digestive strategy, infant development, geographical range, and the regulation of social competition.
...... Cooking was once considered to be an important influence on human evolution [Wrangham cites Charles Darwin, 1871] ...... But in the second half of the twentieth century, such ideas largely disappeared. The main reason appears to have been the pattern that has crystallized in archaeological data.

Data shows that fire was controlled in several sites in southern Europe during the Middle Paleolithic back to at least 250,000 years ago, and probably as early as 300,000 – 500,000 years ago (James, 1989; Straus, 1989; Gamble, 1993; Monnier et al., 1994; Brace, 1995). This evidence has been widely regarded as so much stronger than any indications of the control of fire in earlier times that the Middle Paleolithic is now conveniently interpreted as the first time that humans used fire. There is admittedly scattered evidence for earlier control of fire, but none of it is sufficiently convincing to persuade the skeptics (table 16.1).

[There needs to be differentiation amongst using fire fortuitously or opportunistically obtained, controlling fire fortuitously obtained for relatively long periods of time, and *making* fire as required, which are three very different things…] ……

According to such evidence, therefore, Homo must have relied on raw food before the Middle Paleolithic.

In that case, if the adoption of cooking strongly influenced human biology, a suite of changes should be visible in our ancestors' evolutionary anatomy around 300,000 – 500,000 years ago.

In fact, however, the evolutionary changes in anatomy that are recorded around that time were trivial. ……

Three kinds of solution have been suggested. ……

The Late Solution.

The Late solution suggests that cooking has been adopted too recently to have had time to influence our evolutionary biology.

[Wrangham's reference to Gould's (2002) statement that speciation can occur in less than 25,000 years, is meaningless without reference to the generation time of the species referred to. Also the average time of speciation, or the distribution of speciation times, were they known, might be more relevant. Also differentiation of the time required for DNA-degenerative change, time required for removal of that, time required for selection-from-DNA-variety change, and time required for neogenesis of new function, is relevant. (see also section 3.2, p265 below)] ……

To be reconciled with the Middle Paleolithic evidence for the control of fire, this hypothesis would have to suggest that fire was controlled for a long period without leading to cooking. That solution is hard to imagine. Even wild chimpanzees take advantage of fires to eat foods that have been cooked by chance.

[Fires that occur more often in the dry season, obviously enough, that is the season of preferred food shortages for chimpanzees, when they are more hungry than usual (very hungry in fact, as juveniles cease to grow…) – and eating the cooked corpses remaining after the fire has been through hardly constitutes much of a "use" of fire, and certainly nothing resembling *making* fire….]

[Again, there needs to be differentiation amongst using fire fortuitously or opportunistically obtained, controlling fire fortuitously obtained for relatively long periods of time, and *making* fire as required, which are three very different things, and almost certainly there was a substantial period of time of evolution from the first, through the second, and to the third as a general practice…]

The Sneak Solution.

The Sneak solution accepts the idea that cooking was adopted during the Middle Paleolithic and therefore concludes that it did little to affect human evolution or biology beyond eventually causing a reduction in tooth size (beginning around 100,000 years ago, Brace, 1995). In other words, cooking "sneaked" into human culture with minimal effect.

[Whoever thought up this singularly undescriptive uninformative name should try again! … Why not just the Minimal Effect Solution?...]
In its favor the Sneak solution provides a logical interpretation of the Middle Paleolithic archaeological evidence.
Furthermore, it is compatible with the idea that cooking has only trivial nutritional effects, which was suggested by Levi-Strauss (1969) and has not been completely abandoned.
Against it lies the challenge of explaining why the apparently important results of adopting cooking, including a large improvement in the diet
[What evidence is there for this, other than in palatability of some things, and amount, neither of which is equivalent to an improvement in quality?…
Statements to the effect that cooking results in a large improvement in the diet are generally not well considered.
Firstly, improving the palatability of some foods might actually decrease the quality of the diet if more suitable alternative foods were then not sought.
Secondly, changes to fiber in foods, to break it down and render it more digestible, might not be beneficial – there has been quite a lot of association of dental caries with cooked carbohydrate rich foods in the literature, as well as extracted uncooked carbohydrate – such material is more likely to stick in tooth-tooth and tooth-gum crevices than the same foods uncooked and whole.
Related to this, the increase of the speed at which the carbohydrate in a food is able to be absorbed into the bloodstream, a measure of which is the familiar ‘glycemic index’, by cooking, is very probably not a good thing.
Making it possible to eat a meal much more quickly than if the food were raw might not be good for digestion (a quicker and more complete digestion of carbohydrate is not the same thing as a more complete digestion of protein in the same meal, in fact it is probably somewhat incompatible with it, regards the protein food spending time in the stomach in gastric juice which is being diluted by incoming food) (see section 13.3, p392), as well as psychological satisfaction and social aspects.
The formation of ‘Maillard products’ in which the essential amino acid lysine is bound to a sugar and therefore not available for absorption from the digestive tract is not a good thing.
The oxidative damage of unsaturated fat subjected to high temperatures in the presence of oxygen to form peroxides and epoxides, reactive biochemicals which play a role in damage to artery surfaces and stimulation of macrophages to enter artery tissue and progress atherosclerosis (hardening of the arteries) is not a good thing.
The conversion of nuts into more allergenic foods by roasting is not a good thing.
On the other hand there is the beneficial inactivation of some toxic legume lectins, but this needs to be considered in the context of raw legumes having been reliably reported to be one of the very main foods of the large majority of the wide range of primates, allowing their ecological development as a group – is it possible that only a lesser number of legumes have harmful lectins, and that these lectins are less harmful if eaten in the context of a whole raw meal, perhaps because there are other chemical moieties in the intestinal residue that they can bind to instead of the exposed carbohydrate chains of the glycoproteins in the human cell wall membranes?
These considerations need to be made before thrusting forward one’s theory as a definitive or even potentially plausible statement of the state of affairs…]
and a major change in the way in which it was distributed
[What evidence is there for this?...],
did little or nothing to influence the course of human evolution.
A possible solution is that previous food-processing techniques (such as pounding) closely mimicked the effects of cooking (Wrangham, 2006).
[Techniques such as pounding and grinding, would somewhat “mimic” (serve the same function, more accurately…) some effects of cooking, such as making eating the food

easier, but not others, for example the effect on taste - which is not a very competent thing to omit to mention, as taste very much drives food choices…]
[Also, selection for a supralaryngeal vocal tract able to make all of the sounds of modern humans, which was purportedly (and I have found no disagreement in the literature) not finalized in evolution until 'fully modern man' evolved 35,000-odd years ago (see section 1.11, p139 above), needs to be considered in the final reduction of human tooth size and final jaw shaping.]

The Basal Solution.

The Basal solution is the radical hypothesis that cooking was adopted around the origin of *Homo erectus* and was responsible for many of the features that characterize human evolutionary changes from australopithecines. It was proposed on the basis that many of the evolutionary changes that accompany hominization are easily explicable as responses to cooking, such as the reduced jaw and teeth, evidence of smaller gut, and yet higher energy expenditure (Wrangham et al., 1999).
[These evolutionary changes are much more likely to be due to early tool use of digging sticks in digging for tubers, pounding rams in breaking up tubers, naturally formed hammer and anvil stones in i.e. nut cracking, naturally formed mortar and pestle stones in i.e. seed grinding, and naturally formed sharp-edged stones in tuber peeling, along with soaking of various foods.]
[And again, selection for a supralaryngeal vocal tract able to make all of the sounds of modern humans, which was not finalized in evolution until 'fully modern man' evolved 35,000-odd years ago (see section 1.11, p139 above), needs to be considered in the final reduction of human tooth size and final jaw shaping.]
This solution faces the challenge of explaining why evidence of the control of fire is scarce before about 400,000 years ago. It must also be reconciled with the traditional idea that meat eating was the prime dietary mover of the evolution of the genus *Homo*.
[What evidence and logic is there is there in support of this cherished traditional idea?... And see the previous comments just above here]

I now consider evidence relevant to each of the three solutions. ……

It is also known that a diet of raw plant food creates substantial energetic problems for humans under even the best conditions. …… Second, even under the most favorable conditions people who attempt to restrict their diets to raw foods do not thrive. …… In sum, the sparse [Indeed, and it must be *very* sparse!...] current evidence suggests that raw-food diets produce inadequate energy for humans even under excellent conditions, at least when the diet is dominated by plant items. …… Third, even the most committed raw-foodists find it difficult to keep to their regime because they are consistently hungry even when they eat as much as possible.
[And later on:]
Evidence from raw-foodists indicate that under subsistence conditions humans would not survive long on the kinds of raw foods that are available in the wild.

[I have myself carried out a personal *experiment* (*not* a belief system, please note, other than maybe a little very early on…) for more than 20 years (from 1986 to 2008) of eating only a raw vegan (exclusively plant food) diet, eating only 2 meals per day, late brunch and dinner, without any supplementation of vitamin B12 (and this, the lack of any vitamin B12, and very probably this alone is the risky part of the experiment, and I do not advise being without some source of vitamin B12 as an ongoing thing, and I make explicit here that it would be unethical to subject a child to it).
To start taking vitamin B12 supplementation in 2008 was virtually the only change I have made to this, and this to reduce risk, because I had achieved informatively valuable

endpoints (good health, with performance measures of it) already by that stage of the experiment.

I stake all the credibility that my qualifications in science can muster on my claim that this is the absolute truth, with virtually no deviation or exception whatsoever, other than maybe 10 times or less a tiny scrap of cheese (to total a combined weight of certainly less than 10g, and very probably only a fraction of that), that was missed by my very thorough scrutiny of a salad made by someone else (rarely eaten), may have found its way into me.

Nor was there any oral sex other than less than half a dozen instances, to provide whatever small amount of vitamin B12 might come from that (And no, not due to lack of sex drive (The presence of females at university and elsewhere in fact inflamed my sex drive to an annoying and distracting height, with persistent erections and sexual thoughts intruding on study), rather due to focus on large workloads, restraint of the urge, and lack of fixed abode such as would better facilitate entering into an intimate life partnership).

In general vinegar flies were prevented from laying their eggs in my food supplies, the very large majority of which were not organically grown for budgetary reasons, and although some very very small amount of vitamin B12 probably was supplied by vinegar fly or other insect eggs and larvae, it would certainly have been orders of magnitude below that amount that apes get incidentally in nature from only their plant food, without even considering what they get from adult insects.

However, I have throughout eaten a large green salad with virtually every meal (only two per day, late brunch and dinner), and green leaves are the main source of cobalt atoms that are not already incorporated in vitamin B12 – therefore it is possible, indeed probable, that my large intestine bacteria did manufacture some vitamin B12 from that, which found its way down a concentration gradient in the moist raw-food-derived large intestinal contents, back up through the ileocecal valve and into the small intestine just on the other side of that valve, where vitamin B12 is absorbed by the human body.

My vitamin B12 levels for most of that 20-plus year experimental period were 40-100 picomolar (normal range 150-450 picomolar) (around 50 picomolar for at least the last 10 years of that period) and my homocysteine levels were 80-100uM (normal range less than 12uM).

Regards body muscle and nervous system health:

At the end of that period of 20-plus years of raw vegan diet unsupplemented by any vitamin B12, at the age of 45 years, I finished the 2007 Ironman Korea Triathlon (a 3.8 km swim followed by a 180 km bicycle ride followed by a 42km marathon run, the same distance as at the famous Hawaii Ironman Triathlon) at the 15th percentile, that is, with 15% of the field in front of me and 85% of the field behind me.
That is my whole of life best objectively measured athletic result, notwithstanding that it was the one most trained for (for three months of set regime coming off a substantial mix of add-lib running and cycling for a base) and tried hardest for. I have never manifested athletic champion genetic ability, and have carried since the age of 18 or so a hamstring tear with adhesions, and a partly rehabilitated prolapsed intervertebral disk. I was in employment, though on less than the normal 30 (about 24) teaching hours per week, as an ESL English teacher throughout the time of the Ironman training.

Working on farms harvesting fruit in between contiguous years of university study, I have consistently averaged 6 full workdays per week of work, for the 3 month periods, and my reputation as a hard and good worker on those many farms and amongst those

I've worked alongside is good, better than average, though I have no proof of this to offer (farmers, Mareeba, Queensland: Howe, Nastase, Rogato, Morgante, Alba, Wever, Bosnic, Di Stefano, Hartley, Del Ben, Mete, Iacutoni, Ravanello, Christensen, Cazzola, Nucifora). I always had nothing to eat and only water to drink till lunchtime at midday while working on these farms – lunchtime was usually one hour, but sometimes only half an hour – when lunchtime was only half an hour I cut my leafy green consumption to one mouthful at each of the start and finish of the meal, and spent the rest of the short time eating peanuts and rolled oats (chewed dry and properly), and fresh raw mangoes (eaten more quickly than I like to (which is slowly), but not bolted) – my work performance in the afternoons was, like that in the mornings, above average.
Sometimes (say twice a week) I ran for 10-odd km after work, with a 1-hr-+ bicycle ride thrown in maybe once a week after work. Yes, the evening meal included a good gutful of mangoes, but the oily nut and seed intake was no more than 100-odd grams – and, somewhat annoyingly, it was usual for me to have more body fat (though not that much) at the end of the harvest season than at the start of it – mangoes are not as concentrated in energy as bananas, but they are at the high end of the range for fresh raw fruits…

And the old saying "Hunger is the best sauce" was well borne out at those farm work-day lunch hours, the raw peanuts and dry rolled oats tasting *incredibly* good, to an extent that the average westerner with their sense of taste deranged by food preparation and flavouring techniques, and their lack of proper physiological hunger due to chronic overeating relative to requirements, would be far from able to comprehend – and this is/was not due to me talking myself into it – as a psychology graduate I am fairly well armed against that – it was very real, and so marked as to be actually fairly surprising.

Since the start of the experimental raw vegan diet 20-plus years ago my resting heart rate has almost never been above 48 beats per minute, with the minimum ever a couple of 39 bpm amongst the 42bpm of my fitter early focal triathlon period in 1986 to 1991 – by most standards that indicates, in the absence of some pathological process, a pretty good level of fitness.
Only now at age 48 and a half years does it appear that I must concede some barely noticeable ageing, with my failure to go under 40 minutes for the 10 km 2010 Bridge to Brisbane fun run (40:55) on training that I think would have sufficed in 2008 (when I did 39:49), notwithstanding a little extra upper body muscle (pushups, chin-ups, dips…) and fat (1 kg ?), this compared to my personal best of 38:10 in 2001 and comparable half marathon personal best of 1:27 in 2001 and wind(gale)-adjusted equivalent in 2006 of 1:28 (this would plausibly adjust back to 1:27 or less (for the howling gale), the personal best time I had trained fairly hard for), though at these times I had been doing more running and there was less (say 1-2kg) of upper body muscle and fat. Half marathons in 2010 of 1:34 for both of the only two.

There has throughout the experiment virtually always been mostly daily enthusiastic physical exercising, running and martial arts mostly, with a variable good serving of bicycling (road and MTB) thrown in when convenient.

Regards nervous system i.e., brain health:

In 1996, 10 years into the experimental raw vegan diet, I achieved approximately (the results distribution graph was very small and not lined so as to facilitate reading it very accurately) the 15th percentile (~15% of the others getting a better score) of the GAMSAT test that Australian university students must sit to gain entry to postgraduate medical schools to study medicine to be medical doctors – approximately 25% of those who sit the test are, after an interview assessment, admitted to the courses. Had I completed senior high school maths and physics before sitting that exam, which I hadn't due to being an almost singularly unapplied and rudderless student in high school (as a junk-food eating omnivore, notwithstanding the lack of almost any steering committee…), I would almost certainly have achieved a score higher again than the 15th percentile.

(High school physics and Pure Maths were completed in 2000, 14 years into the experimental raw vegan diet, at the Queensland TAFE's Open Learning Centre, with the highest possible grade category of the three passing categories being achieved, with a bare minimum of tutor assistance).

In my most recently completed university degree, Master of Public Health, at one of Australia's most highly ranked universities, the University of Queensland, in 2008, just after 22 years of the experimental diet, and some months into vitamin B12 supplementation (which did not result in any detectable difference in subjective feeling, notwithstanding me being of course non-blinded to my conditions), my grade point average was 6, the possible maximum being 7 (though I strongly objected to the marking of work in one subject (graded 5) for which I had probably better qualifications for the assessable items than did the course tutor and coordinator, and in which it seemed my rightfully destructive criticism of the poor work we were required to write about was poorly received due its antipathy to the networking practices often employed by the incompetent in academia in securing each other's little niches in the avoidance of the dreaded manual labour…) though that was done on only a part-time basis while I carried out another research project, a monograph on the biomedical significance of homocysteine (which is raised in vitamin B12 or folate deficiency, among other conditions), which will soon be finished and uploaded to the Multifactor Health website, sections of it being there already.

And the state of my brain function may be further assessed by examination of this writing here.

I have done most of my university degrees, commenced in 1991 after 5-odd years of raw vegan diet, at an overload of subjects (none repeated subjects), for example 50% overload in the third year of the BSc, 66% overload of PGDip PH&TM and second year psychology (this did have some, but not a full, content overlap in the statistics area), 75% overload (all subjects passed, at a D or 6/7 gpa, though I did have some prior knowledge) in one semester of Master of Medical Statistics and Master of Nutrition & Dietetics (which degree I was prevented from finishing, during the very *un*objectively assessed practical placements, the final semester of the course, by a most silly little coterie of liars and mental midgets (such as professor(!!...) _oger _ughes, very probably the stupidest PhD candidate that Prof Kerin O'Dea and Dr Shawn Somerset have ever nursed through that degree (I challenge them to deny it…), and the stupidest blockhead I've ever had as a Head of Department or lecturer anywhere in my university studies, who would not have relished my adversarial and much better qualified and competent presence in the field of public health nutrition that he has elected to network his easiest possible to arrange of paychecks in…) such as are not in short supply in the dietetic profession, but that's another story and I'm sure the intelligent reader could surmise readily enough the general theme of my idiot-inflicted problems, due to my vegetarian"ism", there, for example by fellow student _llison _piegel (previous 'leadership' course attendee somewhere…) who achieved (we corrected each other's exam paper) 6 out of 24 for a multiple choice clinical dietetics exam of 24 questions with 4 answers to choose from per question (The result you'd be most likely to get if you knew nothing *at all* about the examined material and selected answers totally randomly – my result was 22/24, the highest result in the class on that occasion…)).

Of course these academic challenges were far from easy for me – I submit that very few people would have found them easy… The regular daily physical exercise was continued alongside that.

Regards the eating/meal schedules:

Since 1986 or so I've virtually only ever eaten two meals a day, late brunch and dinner, with nothing other than water in between (and this water usually mostly in summer, the meals having their own generally nicely sufficient water content), excepting a couple of liters of fruit juice in the middle of the 6-odd hour Ironman training rides and the Ironman race (No-one else does these things without taking in some food along the way either, generally…) and a couple of marathon (42 km) runs and one 9 hour running traverse end-to-end of Korea's Baekunsan Provincial Park's mountain ridge system and a 6 hour run followed by 7 hour fastish walk to traverse the even more formidable Jirisan National Park end-to-end in one day (Yes, as a pure run, which was the intention, Jirisan beat me, but the one day full traverse just in itself is nevertheless quite a challenge – Hwaomsa Temple up to Nogodan along to Cheonwangbong peak (1,950m) down to Daewonsa Temple, if anyone's interested, and I'd advise not calculating your estimated time of completion solely on the basis of the map distance…).

I do spend some substantial time eating my food, say a total of 3 hours a day (but almost nil time in preparation and washing up…), but I am a slow eater, by choice, because I enjoy what I am eating, and do not like to rush my meals, and I have seen many others eat the foods I linger over in a mere fraction of the time I spend.

Even so, far from always feeling hungry, often enough I choose lower-calorie foods to prevent my tendency to overeat (which I have had since a child, though it has never lead to obesity as formally defined, only to, say, 5 kg of excess fat, mostly for some periods of childhood and adolescence) from adding unwanted and physiologically unnecessary fat to my carcass – for example, although I like bananas well enough, I usually reserve them for dinners before a race or long bicycle ride (more than two hours) or run (more than one hour).
I have only very, very infrequently eaten dried fruit in the 20-plus years of the experimental diet (not at all for most of the first 15 years – I did fall for the dried persimmons in Korea in 2000, and again in 2006-7 and 2009-10, but not heavily), mostly lately in an effort to be a bit more relaxed about life, but I have stopped this again now (for the present, I will have some more dried persimmons, hopefully, and Chinese dates, in the future…) as only a small amount of dried food at the end of a meal was adding unwanted fat to my carcass, and was reducing a little my appreciation of raw fresh fruit, which I usually find delicious, not merely tasty.

Furthermore I generally do not eat more than 100-200 grams of nuts and oily seeds per day, as to do so would be a gross excess of protein intake, and also bring my energy intake to an excess – and I can crack brazil nuts, almonds, macadamia nuts etc, with naturally shaped stones, faster than I can eat them – so obviously Professor Wrangham's (of Harvard University…) statements here and in his book Catching Fire How Cooking Made us Human (2009), that humans cannot meet their caloric requirements from raw plant foods, are so far wrong as to be laughable, if the possible reinforcement and spreading of this gross ignorance were not a very serious matter regarding the public health.

In short, If I was to spend more than approximately one hour per day eating the more concentrated raw natural plant foods available to me, at the speed that I've seen many people eat them, I would become more and more overweight by the week… if I was to spend two hours per day doing that, I would become morbidly obese, despite my regular exercise, soon enough!...

An old saying: "It is better to remain silent and be thought a fool, than to speak and remove all doubt." …]

"Humans are not known to be able to survive on raw food
[I must have died many years ago, if so – isn't it strange the way that feeling of extancy lingers on?… And see old saying immediately above…],
which suggests that during our evolution, we became physiologically committed to eating foods of such high quality that in most circumstances they had to be cooked.
[More arrant nonsense, see above.] ……

First, although raw plant food is evidently a poor diet for humans, a sufficient inclusion of raw meat could, in theory, create an energetically adequate diet.
[Consider the basic absurdity of this statement: raw meat is about 70% water; raw nuts and oily seeds have very little water, the large majority of the weight being taken up by about 20% protein on average, and the another 70-odd% being fully fat or carbohydrate – so obviously raw nuts (600 kcal / 100 g) (or oily seeds (600 kcal / 100 g), and also cereal grains (380 kcal / 100 g) for that matter…) are more energy dense than raw meat (190 kcal / 100 g, lean; 230 kcal / 100 g, regular) - the maths here is at about a primary/ elementary school level – just open up a standard food nutrient content tables and do the elementary school maths!...
Furthermore note that high-protein diets are reported to increase the cost of digestion by about 30% compared to high-fat diets.] ……

The *theoretical* reasons why a raw-food diet (at least when dominated by plant items even of the highest quality) is *expected* to be difficult for humans have not yet been elaborated.
[(My emphases) Consider the utterly circular reasoning in this statement!...] ……

Milton (1987, p.103) noted that human gut proportions were similar to those of Cebus and Papio, species [of monkeys] that eat many insects and process their food with their hands. She concluded that the similarity represented "similar adaptive trends in gut morphology in response to diets made up of unusually high-quality dietary items that are capable of being digested and absorbed primarily in the small intestine."
[Quite possibly – macronutrient-dense foods like nuts, oily seeds, cereal seeds, starchy tubers, fruits and insects being just as likely, if not more likely, to meet that criteria as meat…] ……

Likewise, although the human digestive system does not seem to have been compared systematically with those of carnivores
[and what an oversight that is, at that time (2006/7)!.. though apparently plenty of writers are prepared to make vaunted (assumed) similarities a crux of their ill-formed arguments, despite the general gross anatomical appearances of the guts of humans and apes and frugivorous monkeys being very similar…],

It is clear that gut kinetics are radically different. Important, for example, although gut passage rates are similar between humans and dogs, dogs retain food in their stomachs for much longer (around 4 to 12 hours) than humans (around 1 to 2 hours; Ragir, Rosenberg and Tierno, 2000).

In addition humans show little evidence of being able to survive purely on meat diets, even when cooked (Speth and Spielmann, 1983). ……

Current data give us little hope of knowing much about the detailed origins of cooking. For example, how was fire first controlled? How long was it used without being made? How long was it controlled without being used for cooking? And what methods of food preparation were used by humans before cooking? There are no obvious ways to answer these questions. …… we may still be uncertain whether a given population of fire users cooked their food. ……"

[Here I note the commonly reported association of cooked starchy foods (it is generally agreed, including by Wrangham, that cooking makes the starch more easy to break down, in fact breaking it down aforehand, into glucose, the free refined form of glucose being also associated with tooth decay) with dental caries (tooth decay) while the raw foods are much less so associated (Moynihan 2005).
Franck (1959) noted, while working in Africa at Albert Schweitzer's bush hospital, that African college students from either side of a sharp line of dietary demarcation differed greatly: All had massive decay of all molars and most front teeth from the area with manioc (starchy root eaten cooked as a dietary staple), peanuts, sugar cane and some meat as their diet, whereas there was hardly any decay at all in students from the area with fish as the staple food.
Similar observations were reported by dentist Weston Price (1939) notwithstanding his substantially faulty grasp of epidemiological research methodology (it was very early in the evolution of epidemiology as a science), and the multitude of imperfections in some of his reasoning on what diet might be best for humans. Unfortunately his present-day followers of the Price-Pottenger Nutrition Foundation have almost the same degree of misunderstanding as Price did back then…]

[I'll finish here by noting that Wrangham (2009) in his book Catching Fire How Cooking Made us Human, is just as fatally flawed in his arguments that we are evolutionarily genetically physiologically adapted to a diet of cooked food high in animal food.
Just briefly here I'll list the main flaws of the first chapters, leaving the fuller detail for a book review planned to be uploaded to the Multifactor Health website before the end of March 2011, to which (as for this my writing here) he will be offered the right to full and unabridged reply to what will be an extensively destructive review of his work, as he is basically a person, Harvard Professor notwithstanding, who has some cherished ideas and beliefs about his cherished lifetime dietary habits, that he attempts to support to others and rationalize to himself with a complicated structure of selected (leaving aside the contradictory) facts and faulty reasoning, in the manner of a theologian, rather than in the manner of a scientist:

In the Introduction, Wrangham fails to note that most "hunter-gatherers" are in fact better described as "gatherer-hunters" because most of their diet is made up of plant foods;
fails to note that present-day and historical gatherer-hunters live in more marginal areas because the more choice areas were taken over by agriculture;
fails to note the concept of 'fallback foods' eaten in seasons, and, very much the point here, places, where preferred foods are unavailable;
uses "butchering habilines" to describe our ancestors of two million years ago, which without any qualifier indicates that butchering was a main or the main activity of theirs, which is very unlikely from the evidence, evidence that he does not at all address in any case;
claims that fossil evidence, evidence that he does not at all address in any case, indicates the cooking of food genetically physiologically influenced our evolution from before the arrival of Homo erectus, which is sheer conjecture mostly disagreed with by others in the field of knowledge, and not at all able to be stated as a fact even in the least conservative reasonable assessment of the evidence.

In Chapter 1 Wrangham makes (the whole chapter is) the same utterly false claims that humans cannot readily meet their caloric (energy) requirements from raw food, a claim that I have comprehensively demolished above in this chapter section here.

In Chapters 2 and 3 Wrangham again fails to so much as even mention the alternative to cooking that is well known to all scientist/academics in the knowledge area – that food processing by the use of naturally shaped stone and wood tools, which wild chimpanzees and even some wild monkeys are well documented to do, provides the same enhanced breakdown and selective consumption of components of otherwise difficult foods.
He also fails to provide those simplest of relevant nutritional facts, that oily nuts and seeds are roughly 600 kcal / 100 gram, cereal grains roughly 380 kcal / 100 grams, and meat roughly 200 kcal / 100 grams, in talking about cooking and meat having been responsible for humans having evolved a smaller large intestine and larger small intestine, failing also to refer to the evolutionary triplication of the human salivary amylase (enzyme for breaking down starch) and starchy root vegetables and cereal grains, not to mention all the other such details of human metabolic biochemistry that I deal with in chapter 3 here.
And fails to address the proper chewing of food, and how most people in modern societies have become very much in the habit of not doing so, and how these relate to the digestibility of starch in raw foods.

In chapter 4 Wrangham fails to understand that humans still climb trees well enough to make nests in them as apes do;
fails to understand that with the appropriate use of sentries in a sufficiently large enough group sufficiently well armed with naturally shaped clubs and stone projectiles, fire is not necessary for survival while sleeping on the ground at a strategic location (i.e., under climbable trees, with weapons cached in them, with cover and clear ground in appropriate locations in the field of potential operations) instead of in the trees;
fails to differentiate amongst using fire fortuitously or opportunistically obtained, controlling fire fortuitously obtained for relatively long periods of time, and *making* fire as required;
fails to address what proportions of archaeological sites at the different archeological times had evidence of fires and did not have evidence of fires; ….

And the flaws go on and on, page by page, which I will detail in a separate book review planned to be uploaded to the Multifactor Health website before the end of March 2011.

Approximately the only useful information to be derived from (the first four chapters of) this dreadfully flawed book is that cooking probably does increase the digestibility of starchy foods, and that cooking has had a psychological, social and cultural effect on human evolution, but the lay reader will be put to much effort to disentangle these facts (available better dealt with elsewhere) from the nonsense, and the cognate reader either much annoyed at the faults or pleased to have themselves a better understanding of the matter than Wrangham, probably both at once…

And how amazing that all those in the long list of prestigiously placed people quoted in praise of the book in the first four pages inside its cover should have such little understanding of the topic, yet are so forthcoming in voicing their effusive (and probably reciprocated somewhere down the line or in the network to mutual personal financial if not intellectual benefit?...) praise for the book…

Again I quote, "It is better to remain silent and be thought a fool, than to speak and remove all doubt."…]

1.15 Energetic models

Leonard WR, Robertson ML, Snodgrass, JJ. Chapter 18 Energetic models of human nutritional evolution. In, Ungar PS. (Ed) (2007) Evolution of the human diet The known, the unknown, and the unknowable. Oxford University Press.

"Comparative analyses indicate that at walking speeds (between 2.4 and 6.0 km/hour), humans expend significantly less energy than most quadrupeds (Rodman and McHenry, 1980; Leonard and Robertson, 1995, 1997b).
At running speeds, however, bipedalism is less economical, as humans expend significantly more energy [per body weight per distance travelled] than quadrupeds (Taylor, Heglund and Maloiy, 1982). ……
Over the entire range of walking speeds, human males expend 13% less energy than their quadrupedal counterparts, whereas human females expend 25% less [as they carry less upper body musculature than males].

In contrast to humans, the available data on modern apes indicates that they have relatively high energy costs for locomotion. Taylor and Rowntree (1973) and Taylor, Heglund and Maloiy (1982) found that chimpanzees expend 35% to 40% more energy for quadrupedal locomotion than an "average" mammalian quadruped of the same size. ……

These differences in typical movement patterns between humans and apes are evident when we compare the day ranges of modern apes to those of human hunter-gatherers [/gatherer-hunters…]. Modern apes move only about 1.8 km/day, whereas human foragers move an average of 13.1 km/day (Leonard and Robertson, 1997b). ……

[This following is for "an animal the size of an early australopithecine"] For the smaller day ranges [2.0 and 5.0 km/day], the energetic savings accounts for about 1% to 3% of estimated resting metabolic rate [per day]; however, for day ranges similar to those of modern human foragers, the savings are 12% to 16% of RMR, a substantial portion of the daily energy budget.
[What *sort* of animal is very much the crucial point! It's not the size of the thing that matters anywhere as near as much as the shape/biomechanics! And in this regards early (or any!) australopithecines are not the equivalent of modern humans! They do note this later, but why mention the model at all without referring immediately to these things?...]
……
Selection for energetically efficient locomotion is therefore likely to be more intense among animals that must move over larger areas because they have the most to gain by being more efficient. ……

If day ranges for the earliest australopithecines were similar to those of modern chimpanzees, then the initial benefits of upright movement would have been more modest. If true, this would suggest that the full energetic benefits of bipedal locomotion were not gained during a limited period of time but were accrued over a larger span of evolutionary time.
The archaeological evidence suggests that with the drying of the African landscape between 4 and 1.5 million years ago, hominins were moving over larger areas and adapting to more open grassland environments (Reed, 1997; Potts, 1998). It is over this time that we also see the evolution of humanlike body size and proportions that become fully realized with the advent of Homo erectus at 1.8 million years ago (McHenry and Coffing, 2000). …..

In humans, brain metabolism accounts for 20% to 25% of RMR, as compared to 8% to 10% in other primate species, and 3% to 5% in nonprimate mammals (Leonard and Robertson, 1994). ……

Even among chimpanzees, the most predatory of the great apes, only about 5% of their calories are derived from animal foods, including insects (Teleki, 1981; Stanford, 1996). Meat and other animal foods are much more dense in calories and nutrients than most of the plant foods typically eaten by large-bodied primates. This higher-quality diet means that humans need to eat less volume of food to get the energy and nutrients that we require.
[Consider the basic flaws in their thinking here: raw meat is about 70% water; raw nuts and oily seeds have very little water, the large majority of the weight being taken up by about 20% protein on average, and another 70-odd% being fully fat or carbohydrate – so obviously raw nuts (600 kcal / 100 g) (or oily seeds (600 kcal / 100 g), and also cereal grains (380 kcal / 100 g) for that matter…) are more energy dense than raw meat (190 kcal / 100 g, lean; 230 kcal / 100 g, regular) - the maths here is at about a primary/ elementary school level – just open up a standard food nutrient content tables and do the elementary school maths!
Furthermore, the proportions of energy/protein in raw nuts are much closer to that required by apes or humans than is that of meat.
The evidence clearly indicates that raw meat is not a preferred food for chimpanzees, but a fallback food, and nuts are also used as a fallback food.
Now, a certain amount of water is needed every day – if it does not come in the food, then it must come from elsewhere – why wouldn't food be a good place to be getting most of the required water from?
And, without detailing of just which "nutrients" are better (which is not at all the same thing as more) provided by animal foods, isn't that part of their statements here just another motherhood saying about the diet they are habituated to? … the only nutrient of any real concern here on average is vitamin B12, and we need little (proportional to animal food contents) (2 to 3, probably < 1 *micro*grams, not *mille*grams) of that…
If one fails to consider these fairly elementary aspects, surely better that one not voice opinion on the matter?...] ……

Humans, however, have relatively enlarged small intestines and a reduced colon.
[This needs to be quantified!]
This morphology is more similar to a carnivore
[The morphology is in fact more (and apparently *much* more) similar to that of apes than it is to carnivores! It is changed a *small* amount in the direction of the carnivores, but this could easily be accounted for by an increase in more macronutrient (carbohydrate, fat, and protein, in order of optimal proportions, from most to least) dense plant foods such as nuts, oily seeds, cereal seeds, starchy tubers and legumes, with or without tool use for processing, without any theoretical need for the invoking of substantial meat eating, particularly in the absence of any evidence of it being generally the case other than in some areas in archaeologically very very recent areas, such that the relatively very very recent increases in meat eating would not have been the driver or facilitator of the gut changes, being mostly another fallback strategy, as well as availing escape of male gathering drudgery into hunting excitement; and displays of martial power to neighboring groups, males within one's own group, and females within one's own group, as detailed in section 7.3, p355, and elsewhere here.] ……

The comparative dietary and energetic data thus suggest that the dramatic expansion of brain size over the course of human evolution likely would have required the consumption of a diet that was more dense in energy and nutrients than is typically the case for most large primates.

This does not imply that dietary change was the driving force behind major brain expansion during human evolution. Rather, the available evidence indicates that a sufficiently high-quality diet was probably a necessary condition for supporting the metabolic demands associated with evolving larger hominin brains. [OK] ......
Yet, despite having smaller teeth and jaws, *H. erectus* was a much bigger animal than the australopithecines, being humanlike in its stature, body mass, and body proportions (McHenry, 1992, 1994b; Ruff and Walker, 1993; Ruff, Trinkhaus and Holliday, 1997; McHenry and Coffing, 2000).
Together these features indicate that early *H. erectus* was a consumer of a richer, more calorically dense diet with less low-quality fibrous plant material.
[OK, and likely, more calorically dense plant material, like nuts, oily seeds, cereal seeds, starchy tubers and legumes, with or without tool use for processing…] ......

Two of the most important unresolved issues are
(1) how [*If* it was!...] energy metabolism was altered during the course of human evolution to support the costs of larger brain size and
(2) what kinds of dietary/nutritional changes likely occurred to promote the initial increases in brain size that we find with the evolution of early *Homo*. ......

Relative to other primates and other mammals, humans have lower levels of muscle mass and higher levels of body fatness (Leonard et al., 2003).
[*Which* humans, and *when*?!... Surely this is a plainly silly thing to say without specifying these things?... Surely the reader, like myself, has seen pictures of various gatherer-hunters, primitive agriculturalists, and other humans that are nearly as lean as a mammal can get, and of other mammals fatter than this?!... Surely the availability and preferredness of foods needs to be specified for the statement to have any meaning at all?!...]

...... First, because fat has lower-energy requirements than muscle tissue, replacing muscle mass with fat mass results in energy "savings" that can be allocated to the brain.
[This is plainly nonsensical! Muscle is able to be used in food acquisition and combat, whereas fat is not, generally hampering food acquisition and combat, and requiring energy to be carried around – carriage of excess fat is only adaptive insofar as it can provide needed energy when suitable food to do so is unavailable, such as in seasons or other periods of hardship, as they note in their next sentence, and when it can beneficially insulate against body temperature loss in cold environments in the absence of dry clothes, which they do not note.]

Additionally, fat provides a ready source of stored energy that can be drawn on during periods of limited food availability. ......
The composition of all mammalian brain tissue is similar with respect to these two fatty acids [docosohexaenoic acid, DHA, and arachidonic acid, AA] (Crawford et al., 1999).
[And the elephant has the biggest brain on the planet, twice the size that of the blue whale and four times the size that of humans – and not generally known to eat much (or any?) meat. … And the apes being relatively big-brained, yet eating generally very little meat….]

In their raw form, the starch in tubers is not absorbed in the small intestine and is passed through the body (Tagliabue et al., 1995; Englyst and Englyst, 2005).
[This is very incorrect, if one actually chews the stuff properly first, instead of bolting it unchewed like a dog, then the large majority of it will in fact be digested and absorbed – an easy enough experiment to try by oneself (a radical suggestion, apparently, for some armchair theorists and parrot-like-vocalizers in nutrition and dietetics, but there we go…), even without the simplest of blood sugar assay equipment…] ......

Likewise, while we can estimate the increased energy costs associated with the larger brain and body sizes of *H. erectus* relative to the australopithecines, we will never know with accuracy how much energy our hominin ancestors were consuming or where those calories were derived from.
[If we can think a little more correctly, clearly, objectively and comprehensively, we might be able to suggest it with high enough probability…] ……

Obviously, the predictions of any model are only as good as the data and assumptions that go into that model.
[Indeed, but Leonard, Robertson and Snodgrass don't exemplify thinking according to this fact in their own theoretical modeling (even if not explicitly called modeling) here…]

1.16 The brain and intelligence

Foley RA, Lee PC (1991) Ecology and energetics of encephalization in hominid evolution. Phil Trans R Soc London B 334: 223-232.

"It is calculated here that modern human infants have energetic requirements approximately 9% greater than similar size apes due to their large brains.

…. In considering, for example, the development of human foraging strategies, increased returns for foraging effort and food processing may be an important pre-requisite for encephalization, and in turn a large brain is necessary to organize human foraging behavior. …..
This suggests that whatever selective pressures there may be driving the size of the brain up, these are satisfied only in the context of there being sufficient energy. ……

… a diet of 'imbedded foods' – foods that are neither visually obvious nor available without considerable time and energy costs associated with processing, and hence extracted from a secondary source – require complex cognitive skills (Gibson 1986).
……
… encephalization is primarily a characteristic of the genus *Homo* rather than of the hominids as a whole ….. .

…… it may be argued that tool use increases extractive efficiency or widens the range of resources available (Torrence 1983, Gibson 1986) and that meat represents a high quality resource, both of which would constitute increases in the net rate of energetic returns necessary for further encephalization.
[Not enough consideration has been given to the possibility of the use of spears as a defensive weapon for ground foraging, and the use of (accurately thrown) spears and stones to defend tree fruit from competitors such as monkeys, apes and birds, and the cognitive requirements of this as a strategy requiring a step up, but a plausibly achievable step up in planning for the near future.
Furthermore, that the full exploitation of nuts ripe on the ground, that only some chimpanzee groups have learned to use anvil and hammer rocks to start doing, would be facilitated by spears, as hominids, with their feet differentiated from the apes and monkeys, while still able to take shelter in trees as unencumbered adults with sufficient speed to escape predation, would not be as fast at doing so as apes and monkeys, particularly if encumbered with infants and very young children.
Also, that if the human gastrointestinal tract does in fact differ from that of the apes and more frugivorous monkeys such as to suggest a proportionately greater intake of 'higher quality' (really denser in energy and/or (less likely) protein, which is not the same thing as being of higher quality, much as most authors do not properly relate quality to the need

or benefit of any and all nutrients, not merely energy or protein) food, then an increase in intake of nuts would very certainly meet that criterion, as would an increase of smaller such things generally termed seeds, likewise grains. These are in order of decreasing ease of extraction, large nuts being very readily extracted with the correct use of hammer and anvil stones, seeds less so (though not all may require extraction prior to eating), and grains maybe less so again (though not all may require extraction prior to eating at all stages i.e. sprouting or soaked or ground up.)]

….. The energetic costs of brain maintenance have been assessed …… . This accounts for 20% of human basal oxygen uptake and energy expenditure. Growth costs are additional to this. ……

Chimpanzees are born with a neonate brain mass that is 47% of adult brain mass, and adult brain mass is reached by 4 years of age.
In contrast modern humans have neonate brain masses that are 25% of adult mass and by 4 years have reached only 84% (Passingham 1982). …… Overall the energetic costs of brain maintenance for modern humans are about three times those of a chimpanzee. Growth costs will also be commensurately larger.

….. The implications of the data presented here is that a small increase in the level of energy and protein available through meat eating may have major consequences for the nutritional status of the populations concerned. Given that other plant foods, especially underground resources (root vegetables), may also be incorporated into the diet, then a small increase in the amount of scavenging and hunting over the levels observed in chimpanzees may be sufficient. In considering the evolution of human carnivory it may be that a level of 10-20% of nutritional intake may be sufficient to have major evolutionary consequences, and that the conditions under which protein starvation (Speth, this symposium) may result will seldom occur.
[But as protein will not be the limiting factor, but energy will, the use of starchy root vegetables could potentially meet this requirement, without even mentioning increased nut, seed and grain consumption – taken together, that these four sources could make up the energy requirement seems highly plausible…..]

….. The observation of slower rates of development and an extended period of immaturity among *Homo* suggests that mothers were indeed under energetic and time-budget constraints which could not be compensated for entirely by a shift to a higher quality diet. …… .As a result there was selection for a shift in life-history parameters towards slower development.
[Note that the ratio of protein/energy in human breast milk is very low, only a fraction of what it is for all of the several other mammals I've found data for, indicating that we remain after evolution a naturally physiologically and biochemically low-protein diet consuming animal.]
…..Discussion

JHP Jonxis: In the first months after birth the energy metabolism of the brain is far higher than in a somewhat older baby. In those first months the brain covers about half of the total heat production."

Martin RD (1984) Chapter 3 Body size, brain size and feeding strategies. In, Chivers DJ, Wood BA, Bilsborough A (1984) Food acquisition and processing in primates. Plenum Press, New York.

"Regression analysis showed that in frugivorous bat species (N = 19) brain size, relative to body size, was typically almost twice as large as in aerial insectivore species, with no overlap between the two groups. This distinction was all the more convincing in that the frugivore category included two quite distinct groups of bats which have probably adopted fruit-eating habits independently (the micro-chiropteran Stenoderminae and the macrochiropteran Pteropidae).
Analysis of brain size in bats with other dietary habits, relative to the regressions for frugivores and insectivores, showed that nectarivorous bats had relatively large brains and generally overlapped with the frugivores, whereas the remaining bats (carnivores, piscivores, sanguivores and foliage gleaners) typically had brain sizes intermediate between frugivores and insectivores. ......

Clutton-Brock and Harvey (1980) also found some evidence of a positive correlation between comparative brain size and home range size in primates, after effects for body size had been taken into account."
[Note: folivores need a smaller range than frugivores to find their food in, generally.]

[Note: however, relative brain size might not be entirely the point, absolute brain size may be somewhat or even more relevant, as neurons are limited in the smallness they can achieve, and in the number of inter-neural connections they can make for a given size of membrane (although thicker membrane may be able to form a lesser number of inter-neural connections than thinner membrane) – ultimately the absolute number of inter-neural connections, their plasticity (ability to form new and different connections) and brain regional location may be the main factors in determining absolute intelligence. However the general association is accepted to be real and significant, notwithstanding variance within, and exceptions to, the rule.
But perhaps overriding this consideration in the animal world viewed overall, bear in mind that evolution selects for the lightest relative brain size in animals that fly, like birds; middling relative brain size in terrestrial animals; and in aquatic animals allows the largest relative brain size; considering only the factor of the ease of the brain being transported by the body – best not to assume that birds are unintelligent, as sufficient work on crows, pigeons and parrots has shown that these birds are quite intelligent animals…]

Steele J (1989) Hominid evolution and primate social cognition. Journal of Human Evolution 18, 421-432.

"The principal alternative [To feeding-niche complexity] hypothesis for the evolution of primate intelligence is focused not on spatio-temporal complexity of resource distributions, but on social complexity and the cognitive demands of primate sociality. … Whiten & Byrne (1988b) discuss primate cognitive operations of differing orders of complexity:
discovery behaviours (curiosity and exploration of physical objects, and social play for gathering information about social relationships);
innovation or tactical flexibility as responses to novel associations of events (spatio-temporal fluctuations in food availability, or in social interactions the abstraction of "conditional social rules" as a basis for social actions, in relation to specific actions by other con-sociates whose past social-interactive patterns can be stored in short-term and long-term memory);
cognitive representations of the intentionality and attentional states of others, perhaps involving attributions of several levels of intentions ("X thinks Y knows that X wants Y to do such-and-such", or in more complex tripartite interactions "X thinks Y knows that Z wants Y to do such-and-such", etc);

and (perhaps only in humans) *self-reflective monitoring* as a means of abstracting an explanatory model of human action.
[Note: these are in ordering of ascending difficulty, with apes able to do, and doing, more of the complex things than monkeys] ......

Modern human gut morphology, compared to that of the great apes, is characterized overall by a relatively small gut, and within the gut by a relatively small colon and enlarged small intestine. Variation in primate gut morphology is correlated with dietary preference, these human traits indicating a common ancestor with a relatively high-quality,
[Note: read energy- and protein-dense, which is not the same thing.],
relatively nutrient-dense and rapidly absorbed diet.
[Note: that the absorption would be more rapid is not obvious.]
(Milton, 1988; cf Martin et al, 1985).
This pattern may derive from new dietary patterns in *H. erectus*, for which paleopathological (Kennedy, 1985) and dental and masticatory evidence (Lucas et al., 1985) indicate a shift to greater use of higher grade plant foods,
[Note: read energy- and protein dense, which is not the same thing, really]
and to seasonally-varying, heavy consumption of meat.
[Note: the evidence for this claim is not at all good, nearly always this claim is pure assumption, and here it is very conjectural, based on hypothetical models of tooth structure, and the abstract indicates no consideration of nut consumption, which is essential for a sensible consideration.]

However, as Milton (1988) points out, this need not imply a causal relationship of new dietary strategy demanding greater cognitive capacity, but could equally well indicate (conversely) new dietary patterns answering the nutritional and metabolic demands of maintenance of a more Encephalized nervous system (cf. Harvey and Bennett, 1983), itself functionally adapted to some other suite of environmental problems. ......

It would be purely speculative to argue that predation was the major determinant of later hominid grouping patterns in the absence of any supporting evidence.
Certainly, however, an increase in body-size coupled with large social groups would demand greatly increased home range areas (Ripley, 1984: 40), and this appears to be confirmed by lithic transport over successively greater distances in the later hominid artifactual record (Toth, 1987).
If predation risk promoted later hominid grouping patterns entailing larger group sizes than were optimal simply for foraging, then increased levels of intragroup competition and decreased group stability would have occurred. In such contexts
[including the development of more potent weaponry and its use, even if initially for defense]
the capacity for social cognition, and that for complex language as "a delicate instrument for establishing and modulating ever more intricate relationships between human beings" (Burling, 1986: 14-15; cf. Bruner, 1984), would have come under strong selection pressure as a mechanism for learning, communicating and manipulating shared mental-model representations of the contexts and constraints on action. .... Tool use in non-human primates is relatively rare, although "primatologists repeatedly emphasize the ability of subjects to use other individuals as 'social tools' to achieve particular results" (Cheney et al., 1986: 1365).

Selection for corticalization and complex language in later hominid evolution (genus *Homo*) indicates an adaptive context favouring increased capacity for mental-model representation (implicated in intentional action), and for social exchange and modification of such representations. ......

Non-human primate grouping patterns and social systems are determined by three major factors: dietary characteristics, predation risk, and social competition.

If social systems are determined by grouping patterns, and the latter by factors other than those of optimal foraging
[grouping patterns seem likely to be at least partly determined by optimal foraging factors – nut-cracking, seed-grinding and cracking, and digging make sounds that may attract both predators and competitors (i.e. from rival hominid groups), requiring more defense than quieter foraging.]
then these two hypotheses for primate encephalization (foraging complexity and social cognition) refer to two distinct sources of selection pressures. ……

Evidence for scavenging and for increased predation risk for early genus *Homo* groups suggests that optimal foraging models are not sufficient to predict later hominid social strategies.
[On the basis of my comment immediately above, this therefore seems dubious.]
Increased social competition resulting from larger grouping patterns, the latter a response to predation risk [itself related to optimal foraging factors], is hypothesized to have presented the dominant selection pressure for human social cognitive and socially manipulative group-living skills, and thus for hominid brain and language evolution."
[It might have been a lesser selection pressure, and come later (even if as a product of the other), and still had a magnifying or interaction effect.]

1.17 Cracking and grinding tools, digging tools

[Modern chimpanzees have been well documented to use, in their 'natural' habitat, naturally shaped hammer and anvil stones, naturally shaped mortar-and-pestle stones, naturally shaped grinding stones, and sticks/twigs for termite fishing …(see chapter 2 below), and capuchin monkeys have been documented to use, in their 'natural' habitat, naturally shaped hammer and anvil stones, naturally shaped mortar-and-pestle stones, naturally shaped grinding stones (see section 2.16 of chapter 2 below).

Such naturally shaped tools are invisible in the archaeological record because the twigs, sticks and branches decay fully into dust, and the naturally shaped stone tools look, for all practical purposes in the present state of technology and method/practice, like just like stones among other stones.

However, it seems to me that if present-day apes and monkeys can use naturally shaped stones and sticks as tools then even our most distant hominin ancestors would almost certainly have done so too. And lack of ability to reason using language does not indicate the lack of ability to reason with mental pictures.

Wild animals are, perhaps above all in their behavior, observant – failure to be so is very likely to result in removal from the gene pool at a very early age – I find it very unlikely that even our earliest hominin ancestors were incapable of watching a pig dig up roots on serial occasions and realizing that they themselves could also do so with the use of a stick – in fact, such reasoning in animals has been pretty conclusively demonstrated and reported on by now in 2010, though I will not provide any references in this present version of this book.
Therefore, to reason in mental pictures that if nuts could be processed (cracked open) between two hard objects, that it might be worth a try using some variation of that as constitutes grinding i.e. of grains, seems well within the mental capabilities of even our

earliest hominin ancestors to me, notwithstanding the frustrating lack of the mental capabilities that one might think reasonable to expect, in many modern humans…

It seems very probable to me that even our very earliest hominin ancestors used naturally shaped hammer and anvil stones, naturally shaped mortar and pestle stones, and digging sticks. The invisibility of this use in the archaeological record due to decay of the sticks and undifferentiability of the stones from other stones clearly make such absence of evidence (which is not the same thing as evidence of absence) not at all a good argument against such use having occurred.]

1.18 Defense needs and weapons

[I doubt that with their legs, pelvis and spine not as well evolved as ours for moving on the ground, and their small stature (1.1 meters), that australopithecines would have had a snowball's chance in hell of surviving on the ground, which is where their feet (greatly resembling our own) strongly suggest they were spending a great deal of time, without weapons. In my opinion anyone thinking otherwise has a totally inadequate understanding of combat reality and the unarmed defense capabilities of such a creature in the environment it shared not only with large cats and hyenas but also with baboons (more grass-eating then as they purportedly were) the size of a modern gorilla and very probably other large animals outside my knowledge.
And I think much the same applies to the larger *Homo erectus* that came later and was more our size – without fangs and claws, or the speed of a gazelle, or the size of an elephant, or the hide and tusk of a rhinoceros, or the reproduction rate of a rabbit, or the taste of a stink-bug, or the camouflage of a stick insect, some other method of survival is required. Weapons were almost certainly it, obviously enough, but apparently not to some, as the subject seems generally very poorly dealt with in the literature, notwithstanding that I haven't read or canvassed all of it.]

1.19 Scavenging, power scavenging, and hunting

[There is a side debate which is somewhat peripheral to the main questions addressed by my book here – that debate is about whether our earlier hominin ancestors acquired what meat they did eat by scavenging, power scavenging, or hunting (Lewin 1984; Shipman 1986; Lyman 1987). Power scavenging involves driving off whatever animal is feeding on a carcass and taking it over for eating.

It seems to me that by the time our earliest ancestors had evolved feet like ours, they, although still very proficient climbers of trees for the purposes of getting fruit, would not have been good enough climbers to hunt monkeys in the way that chimpanzees do (though the use of thrown weapons may have improved their ability here a little).

However, they probably were able to hunt on the ground in a manner resembling that of chimpanzees observed hunting piglets and baby deer. However, in the more open country that it seems likely that our early hominin ancestors spent a lot of time in, it is uncertain that this would have been as feasible as it was for the chimpanzees in their forests.

Furthermore our early hominin ancestors already had lost the elongated canine teeth such as chimpanzees (though mostly in combat, not in predation…) and carnivores use in killing and tearing – there are other ways to kill, such as by battering or beating, but this lack of elongated canine teeth would reduce the size of animals able to be killed.

I doubt that with their legs, pelvis and spine not as well evolved as ours for moving on the ground, their shoulders probably not as well evolved for throwing things and maybe not even as well evolved for club use as ours, their smaller stature than ours, their at least earlier lack of quality spears, and their lack of bows and arrows, that our earlier hominin ancestors, the australopithecines, were able to hunt anything other than very small animals, and these probably were not taken by active hunting and chasing, but seized or cornered more by chance as baboons do with small animals and young animals who use freezing as a defence, that they come upon more by chance. Baboons have been recorded as chasing down hares, but it seems very unlikely to me that an australopithecine could run at the speed that that would require – I also doubt that a chimpanzee would be capable of it, and it seems unlikely to me that pure running down would be the method that chimpanzees use in their recorded occasional predation on young baboons.

It seems very likely to me that, probably used only in times of great hunger, both scavenging and power scavenging were the only access our earlier hominin ancestors had to the meat of large animals. They probably needed weapons for either method, as scavenging is likely to involve making more sound than the average made for gathering, and moving quietly was probably a central requirement of survival for early hominins other than armed and at the ready with clubs, stones and sharp sticks, and in groups – having such weapons, they would probably have been capable, as a group, of driving off at least sufficiently sated carnivores of some size larger than themselves.

Raw meat was probably not very palatable to our early (and maybe nearly all) hominin ancestors, who probably preferred raw fruits, nuts, and some vegetative plant parts when available, and probably even legumes and starchy tubers, to raw meat, and almost certainly to raw meat in various stages of rotting.

This is of course conjecture, but we are all limited here to trying to think sensibly about the capabilities and taste preferences of our hominin ancestors – conclusive evidence is unlikely to ever be available.]

1.20 Proposed implications of studies on early hominin diets: Cordain; Eaton; Ungar; in Ungar (2007); and Eaton et al's (1988) Paleolithic Prescription (includes sodium:potassium ratios (p192) and acid/base contributions (p191) of foods)

Cordain L. Chapter 19 Implications of Plio-Pleistocene hominin diets for modern humans. In, Ungar PS. (Ed) (2007) Evolution of the human diet The known, the unknown, and the unknowable. Oxford University Press.

"With regard to diet and health, food staples and food-processing procedures introduced during the Neolithic and Industrial era[s] have fundamentally altered seven crucial nutritional characteristics of human diets: (1) glycemic load, (2) fatty acid composition, (3) macronutrient composition, (4) micronutrient density, (5) acid/base loads, (6) sodium potassium ratio, and (7) fiber content. ……

Similar to historically studied hunter-gatherers [/gatherer-hunters] (Cordain et al., 2000, 2002b), there would have been no single, universal diet consumed by all [20 or so] extinct hominin species. Rather, diets would have varied by geographic locale, climate, and specific ecological niche.
[Far more the point is that within any one extinct hominin species, in particular the ones that are our ancestors, diets would have varied by geographic locale, climate, and specific ecological niche…] ……

Observational studies of wild chimpanzees demonstrate that during the dry season meat intake is about 65g per day for adults (Stanford, 1996).
[This is very incorrect, as many adult chimpanzees eat little or no meat because they are not proficient enough hunters, and lack the social connections to receive meat that they ask for (see various sections of chapter 2 below) – unlike the intake of the foods that are more evolutionarily physiologically naturally theirs.
Furthermore, the growing young, that have more need of food and need a higher protein/energy dietary ratio, will get much less than many adults, unlike the situation in almost any more natural carnivore.
Furthermore, the implication that the figure of 65g/day is generally applicable to all groups of chimpanzees in all places is incorrect, as meat intake is quite variable.]

Accordingly, it is likely that the very earliest hominins would have been capable of obtaining animal food through hunting and scavenging in a manner similar to chimpanzees.
[Firstly, even the earliest hominins would not have been agile enough climbers (hominins have by definition feet more like our own than the chimpanzees' feet with their opposable big toe) to prey on monkeys, which are the main prey of chimpanzees when they are hungry enough to do so.
The earliest hominins may have had the facility to prey on some terrestrial animals, like piglets, as some, and only some, groups of chimpanzees do, but lacking the enlarged canine teeth that the chimpanzees have and use in predation for both the killing and the eating, as well as the probably more important combat, it is by no means sure that they would have been successful, without weapons to kill, and tools to process, in preying on anything larger than animals small enough to be more 'gathered' than 'hunted'.
Secondly, chimpanzees very very rarely scavenge anything other than carcasses very freshly killed by the still feeding possessor, who is displaced by force from the carcass by 'power scavenging' – and this itself is a rare event.]

Carbon isotope data also support the notion that early hominins were omnivorous. By about 3 million years ago (Mya), *Australopithecus africanus* obtained a significant portion of food from C_4 sources (grasses, particularly seeds and rhizomes; sedges; invertebrates, including locusts and termites; grazing mammals; and [and/<u>*or*</u>!] perhaps even insectivores and carnivores; van der Merwe et al., 2003).
Other fossils of early African hominins, including *Australopithecus robustus* and *Homo ergaster*, maintain carbon isotope signatures characteristic of omnivores (Lee-Thorp, Thackeray, and van der Merwe, 2000; Sponheimer and Lee-Thorp, 2003). The finding of C_4 in *A. robustus* fossils refutes the earlier view that this hominin was vegetarian (Lee-Thorp, Thackeray, and van der Merwe, 2000).
[Firstly, one does not find C_4 in fossils! C_4 is a photosynthetic process in plants – one finds higher ^{13}C derived directly or indirectly from eating those plants… This is the sort of mistake that gives a strong indication that the person making it has not a sufficient competence in the area of science to be properly giving voice to opinion – an old saying: it is better to remain silent and be thought a fool, than to speak and remove all doubt…
Secondly, as Sponheimer, Lee-Thorp and De Ruiter (2007) (see section 1.11 above) state in their conclusions: "For one, it would be impossible to ever know which C_4 resources australopiths used because of insurmountable equifinality problems. In other words, there are many diets that can conceivably produce the same carbon isotope signature. …… It would be impossible, for instance, for stable carbon isotopes of tooth enamel carbonate to ever distinguish between a purely vegetarian hominid that ate 50% C_3 fruits and 50% C_4 grass and a highly predaceous hominin that ate nothing but C_3- and C_4-derived mammal flesh in roughly equal proportions."
Again, the old saying: it is better to remain silent and be thought a fool, than to speak and remove all doubt…] ……

The development of stone tools and the increased reliance on animal foods allowed early African hominins to colonize northern latitudes outside of Africa where plant foods would have been seasonally restricted.
Early *Homo* skeletal remains and Oldowan lithic technology appear at the Dmansi site in the Republic of Georgia (40 degrees north) by 1.75 Mya (Vekua et al., 2002), and more recently Oldowan tools dating to 1.66 Mya have been discovered at the Majuango site in North China (40 degrees north; Zhu et al., 2004). Both of these tool using hominins would likely have consumed considerably more animal food than prelithic hominins living in more temperate African climates.
[Firstly, one needs to differentiate the hunting of animals for hides to make warm clothes, from that for meat to eat – the clothes are an absolute necessity in cold climates, whereas the meat is not an absolute necessity but only a likely one – and there is every chance that raw meat was not found as palatable as say, raw almonds or hazelnuts, etc…
Secondly, to have colonized an area at one time, does not imply the occupation of that area ever after – early hominins may have moved into such areas in warmer periods and moved back out in colder periods.
Thirdly, it should be kept in mind when theorizing about related matters that there has to be some drive for living in the cold eating much raw meat instead of living in the warm eating raw nuts, seeds and grains, and more fruit and tubers, and more vegetables, with or without some little raw meat – it is unlikely hominins moved there purely for the novelty or to demonstrate their hardiness – it is more likely that they moved there (albeit gradually, generation by generation) because of population pressure and territorial control going against them (climate change would likely have been a factor in this, both of moisture and temperature) – what the ramifications of this are for human evolution (incl. via population generation and population sinkage in different areas) is not clear.] ……

In this regard, hominins, like felines, have experienced a reduction in gut size and metabolic activity along with a concurrent expansion of brain size and metabolic activity as they included more energetically dense animal food into their diets (Leonard and Robertson, 1994; Aiello and Wheeler, 1995; Cordain, Watkins and Mann, 2001).
[To repeat my comments of section 1.13 above:
Firstly, the human and ancestral hominin gastrointestinal morphology is in fact more (and apparently much more) similar to that of apes than it is to carnivores! It is changed a *small* amount in the direction of the carnivores, but this could easily be accounted for by an increase in more macronutrient (carbohydrate, fat, and protein, in order of optimal proportions, from most to least) dense plant foods such as nuts, oily seeds, cereal seeds, starchy tubers and legumes, with or without tool use for processing, without any theoretical need for the invoking of substantial meat eating, particularly in the absence of any evidence of it being generally the case other than in some areas in archaeologically very very recent areas, such that the relatively very very recent increases in meat eating would not have been the driver or facilitator of the gut changes, being mostly another fallback strategy, as well as availing escape of male gathering drudgery into hunting excitement, displays of martial power to neighboring groups, males within one's own group, and females within one's own group, as detailed in chapter 7 here below.
Secondly, consider the basic flaws in Cordain's nutritional knowledge and thinking here: raw meat is about 70% water; raw nuts and oily seeds have very little water, the large majority of the weight being taken up by about 20% protein on average, and another 70-odd% being fully fat or carbohydrate – so obviously raw nuts (600 kcal / 100 g) (or oily seeds (600 kcal / 100 g), and also cereal grains (380 kcal / 100 g) for that matter…) are more energy dense than raw meat (190 kcal / 100 g, lean; 230 kcal / 100 g, regular) - the maths here is at about a primary/ elementary school level – just open up a standard food nutrient content tables and do the elementary school maths!
Furthermore, the proportions of energy/protein in raw nuts are much closer to that required by apes or humans than is that of meat.

Thirdly, a certain amount of water is needed every day – if it does not come in the food, then it must come from elsewhere – why wouldn't food, such as fruits and tubers, be a good place to be getting most of the required water from?...]

Further, similar to obligate carnivores (Pawlosky, Barnes and Salem, 1994), humans maintain an inefficient ability to elongate and desaturate 18 carbon fatty acids to their product 20 and 22 carbon fatty acids (Emken et al., 1992). Since 20 and 22 carbon fatty acids are essential cellular lipids, then evolutionary reductions in desaturase and elongase activity in hominins indicate that preformed dietary 20 and 22 carbon fatty acids (found only in animal foods) were increasingly incorporated in lieu of their endogenously synthesized counterparts derived from 18 carbon fatty acids.
[Firstly, for this to have any useful evidence at all, the ability to elongate and desaturate 18 carbon fatty acids to their product 20 and 22 carbon fatty acids must be quantified – for carnivores, for humans, and for apes, at a minimum – without providing these absolutely necessary facts, why bother to attempt to discuss the matter at all?...
Secondly, the assumption (which may well be incorrect) of the biochemical necessity for the provision or synthesis of the same amount of those fatty acids per kg body weight per unit of time must at least be noted – preferably, the actual respective needs for these fatty acids per kg body weight per unit of time needs to be known for carnivores, humans, and apes (see p281: vegan women's milk and wild ape milk being similar in LCPUFA)]

Finally our species has a limited ability to synthesize the biologically important amino acid, taurine, from precursor amino acids [methionine and cysteine] (Sturman et al., 1975; Chesney et al., 1998), and vegetarian diets in humans result in lowered plasma and urinary concentrations of taurine (Laidlaw et al., 1988).
Like felines (Knopf et al., 1978; MacDonald, Rogers and Morris, 1984), the need to endogenously synthesize taurine may have been evolutionarily reduced in humans because exogenous dietary sources of preformed taurine (found only in animal food) had relaxed the selective pressure formerly requiring the need to synthesize this conditionally essential amino acid.
[analogously to my comment immediately above:
Firstly, for this sort of discussion to have any useful evidence at all, the ability to synthesize taurine must be quantified – for carnivores, for humans, and for apes, at a minimum – without providing these absolutely necessary facts, why bother to attempt to discuss the matter at all?...
Secondly, the assumption of the biochemical necessity for the provision or synthesis of the same amount of taurine per kg body weight per unit of time must at least be noted – preferably, the actual respective needs for taurine per kg body weight per unit of time needs to be known for carnivores, humans, and apes… Regarding this, I quote Stipanuk (2006) p549 of Shils et al: "The mean taurine content of milk of lactating vegan women is lower than that of lactating omnivores, but overlap of values is still considerable, and the taurine concentration of milk of vegan mothers is about 30 times the level in the cow's milk-based infant formulas used before the mid-1980s." (cites Rana and Sanders, 1986)
So, some animals (i.e. cows or calves) have less need for taurine than others (i.e. humans. And, the vegan mothers' breast milk taurine is close to (and overlapping with) that of the omnivore human mothers' breast milk taurine. And, taurine is derived from the amino acids methionine and cysteine, which can vary in the diet.
So, the reality is that the implication of the relevant facts is that human taurine requirements are, more probably than not, met by any diet, including a vegetarian one, that is sufficient in protein, within the context of legumes being a little low in methionine and cereal grains being a little high (and the converse for lysine); and, in the aforementioned context, that the limitation of our species to synthesize the taurine we need is probably not significant. (see also chapter 3)]

There is little evidence to the contrary that animal foods have always played a significant role in the diets of all hominin species.
[Whether significant mainly as a fallback food, or significant in affecting our evolution much, is very much the question. And part of the evidence to the contrary is our and our hominin ancestors' lack of elongated canine teeth; our very low breast milk protein/energy ratio (a mere fraction of all of the several other animals I was able to obtain data for); the ability of many people (see my own 20-plus year raw food vegan B12-unsupplemented experiment detailed in section 1.14 above and chapter 10 below) to go without any apparent dietary source of vitamin B12 for many years without any apparent harm, notwithstanding some cases where harm does occur; and our inability (in contrast to most other mammals) to synthesize vitamin C… In fact, all of the evidence from human metabolic biochemistry I present in chapter 3 here below strongly suggests only a very minor role for meat in hominin diets other than the very most recent…]

Increased reliance on animal foods not only allowed for enhanced encephalization and its concomitant behavioral sophistication (Leonard and Robertson, 1994; Aiello and Wheeler, 1995; Cordain, Watkins and Mann, 2001),
[This is not known to be matter of fact! Far from it, it is probably incorrect – It is not known whether enhanced encephalization was allowed by increased reliance on animal foods – it is at least as likely to be due to increased intake of macronutrient-dense plant foods such as nuts, oily seeds, cereal grains, legumes, and starchy tubers, with or without the use of tools such as naturally shaped hammer and anvil stones, mortar and pestle stones, pounding logs, and soaking arrangements. To state as fact what remains very obviously only conjecture is to illustrate one's great lack of understanding of the evidence. And again, the old saying: it is better to remain silent and be thought a fool, than to speak and remove all doubt…]
but this dietary practice also permitted colonization of the world outside of Africa.
[It may well have been other things that mostly or equally permitted colonization of the world outside of Africa, such as animal skin acquisition for warm clothing, and periodic warming, as well as changes in what types and species of plant foods were consumed, apart from the fact that regions outside of Africa require differentiation on the basis of climate, as they are far from equal in that respect!...]

An unresolved issue surrounding Plio-Pleistocene diets is the relative amounts of plant and animal foods that were typically consumed.
[Indeed, and more the point, *when* in the Plio-Pleistocene, because it must be differentiated on this, as well as on where!...]

Before the advent of Oldowan lithic technology about 2.6 Mya, quantitative estimates of hominin energy intakes from animal food sources are unclear, other than they were likely similar to [ok], or greater than [not necessarily, i.e. hominins don't have the chimpanzees' enlarged canine teeth], estimated values (4% - 8.5% total energy) for chimpanzees (Sussman, 1978; Stanford, 1996).

Although all available data point to increasing animal food consumption following the arrival of lithic technology, the precise contribution of either animal or plant food energy to Plio-Pleistocene hominin diets is not known.
[The data for the period of millions of years after the arrival of lithic technology is quite ambiguous about there being anything other than a very small increase from an already very small amount, to result in a still very small amount – only relatively very very late in archaeological and evolutionary timescales is there any good evidence of a major increase in the consumption of animal foods by hominins – such as in colder glacial Europe and when population pressure on plant foods forced consumption of more animal foods and cooking made consumption of meat palatable.]

Obviously, then as now, no single (animal/plant) subsistence ratio would have been necessarily representative of all populations or species of hominins.
[OK]

However, there are a number of lines of evidence which suggest more than half (>50%) of the average daily energy intake for most Paleolithic species and populations of species was obtained from animal foods.
[This is sheer nonsense! At risk of inflicting boredom on the reader I repeat yet again "Secondly, as Sponheimer, Lee-Thorp and De Ruiter (2007) (see section 1.11 above) state in their conclusions: "For one, it would be impossible to ever know which C_4 resources australopiths used because of insurmountable equifinality problems. In other words, there are many diets that can conceivably produce the same carbon isotope signature. …… It would be impossible, for instance, for stable carbon isotopes of tooth enamel carbonate to ever distinguish between a purely vegetarian hominid that ate 50% C_3 fruits and 50% C_4 grass and a highly predaceous hominin that ate nothing but C_3- and C_4-derived mammal flesh in roughly equal proportions."
Cordain gives the examples of the Neanderthals of ~28,000 - 29,000 years BP (before the present) and *Homo sapiens* from Britain of ~11,700 – 12,380 BP referring to ^{13}C and ^{15}N isotope studies – the Neanderthals are not of our evolutionary ancestry; and the *Homo sapiens* here are so close to the agricultural revolution, and in a period of increasing and maybe even climaxing pre-agricultural population stress, that their relevance to the question of what humans are evolutionarily suited to eat is approximately nil…]

[Cordain's discussion/reference here to historically studied hunter-gatherers/ gatherer-hunters, as collectively eating more animal food than plant food, is flawed by its failure to refer to the marginal nature of the areas occupied by those groups, the better areas having been taken over by agriculture – furthermore in Table 19.1 there are included two different proportions of animal/plant food for the !Kung, 33/67 by Lee (1968) and 68/32 by7 Yellen, 1977, that give us some indication of the great unreliability of such assessments, an unreliability that has been noted as being the case by many scholars in the area, and see chapter 4 below in this regard – furthermore, other such collations of multiple studies have concluded the converse, that more plant food was eaten than animal food, and those studies should have been at least dealt with here (see chapter 7 below)]

[Cordain's statement that cereals would have contributed little or no energy in the archetypal preagricultural hominin diet (cites Cordain et al., 2000) is very likely to be wrong, and likely to be very wrong – there is plenty of evidence in the literature of grain consumption by apes, monkeys (and not only gelada baboons), and Australian aborigines eating grains, and using tools (the apes and humans) to process them, see chapters 2 and 7 here below, and only a person very ignorant of some very important evidence would fail to observe that our and our ancestors' molar and pre-molar teeth are possibly the best-adapted of all animals for chewing grains (and see section 1.9 here above), whether as fallback foods or staples.]

Unrefined wild-plant foods, like those available to contemporary hunter-gatherers, typically exhibit low glycemic indices (Thorborn, Brand and Trusswell, 1987)
[As do most fruits commonly eaten in modern societies.]

Milk, yogurt, and ice cream, despite having relatively low glycemic loads are highly insulinotropic [causing insulin levels to rise], with insulin indices comparable to white bread (Ostman, Liljeberg Elmsthal and Bjorck, 2001).

In the current US diet, the ratio of n-6/n-3 PUFA [polyunsaturated fatty acids] has risen to 10:1 (Kris-Etherton et al., 2000), whereas in wild-animal-food dominated [which not

all, or not most, are…] hunter-gatherer diets it has been estimated between 2:1 and 3:1 (Cordain et al., 2002a).
[A statement of the food sources of n-3 and n-6 PUFA would have improved the informativeness of this writing.]

Refined sugars are essentially devoid of any vitamin or mineral. ……
A similar situation exists for refined vegetable oils, except that they contain two fat-soluble vitamins (vitamin E and vitamin K [and pre-vitamin A!...]). ……

Adequate dietary intake of both folate and vitamin B6 prevent the accumulation of homocysteine in the bloodstream.
[Vitamin B12 is more important than vitamin B6, as vitamin B6 is fairly ubiquitous in foods, whereas vitamin B12 is only present in animal foods and a few seaweeds… Furthermore, it is *not* an accepted fact that mildly elevated homocysteine (12-20uM) is an independent risk factor for cardiovascular disease – this was savagely debated and is now all but disproved (Clarke et al (2010), meta-analysis of randomized controlled trials…)]
……

[Cordain's discussion of the nutritional effect of the displacement of fruits, vegetables, lean meats and seafoods by whole grains and milk, and wild-plant foods by domestic varieties is fairly meaningless as it is without reference to the particular nutrients.]

After digestion, absorption and metabolism, nearly all foods release either acid or bicarbonate (base) into the systemic circulation (Frassetto et al., 1998).
[Table 19.6 shows that:]
fish [+14.6],
meat [+12.4],
poultry [+7.8],
eggs [+7.3],
cheese [+3.3],
milk [+1.3] and
cereal grains [+1.1]
are net acid producing [metabolized Met, Cys to H_2SO_4; Lys, Arg, His to H^+],
whereas
fresh fruits [-5.2],
vegetables [~-20],
tubers [-5.4],
roots [-17]
and nuts [-1.1] [units of meq/418kJ]
are net-base producing [(mainly) potassium salts; metabol'd Asp, Glu to OH- (?)].
Legumes yield near-zero mean acid values, reflecting an overlapping distribution from slightly net acid producing to slightly net base producing.
Not shown in table 19.6 are energy-dense, nutrient-poor foods, such as separated fats and refined sugars, that contribute neither to the acid nor base load.
Additionally, salt (NaCl) is net acid producing due to the chloride (Frassetto et al 1998).

The typical western diet yields a net acid load estimated to be + 50 meq/day (Lemann, 1999). As a result, normal adults consuming the standard US diet sustain a chronic, low-grade pathogenic metabolic acidosis that worsens with age as kidney function declines (Frassetto, Morris and Sebastian, 1996).
Virtually all preagricultural diets were net-base yielding because of the absence of cereals and energy-dense, nutrient-poor foods – foods that were introduced during the Neolithic and industrial eras and which displaced base-yielding fruits and vegetables (Sebastian et al., 2002).

[Any diet that was predominantly meat and/or fish wouldn't have been, as these are, according to the table (19.6) given by Cordain, the two highest acid producing per kilojoule foods in the list. Such as, the diets of the Neanderthal and very late preagricultural *Homo sapien* examples given by Cordain (see above)]
[Furthermore, the great difference in acid load between flesh (fish, meat and poultry) and cereals is not properly noted by Cordain – the cereals are relatively close to zero, and if they were eaten along with fruit and vegetables the net result would still be strongly net-base yielding…] ……

The known health benefits of a net-base yielding diet include preventing and treating osteoporosis (Sebastian et al., 1994), age-related muscle wasting (Frassetto, Morris and Sebastian, 1997), calcium kidney stones (Pak et al., 1985), hypertension (Morris et al., 1999), exercise induced asthma (Mickleborough et al., 2001), as well as slowing the progression of age- and disease-related chronic renal insufficiency (Alpern and Sakhaee, 1997). ……

Three dietary factors are primarily responsible for the dietary [typical US] ratio of Na/K, which is greater than 1.0.
First 90% of sodium in western diets comes from manufactured salt (NaCl); hence, the sodium content of naturally occurring foods in the average US diet (~330mg) is quite low.
Second, vegetable oils and refined sugars, which are essentially devoid of potassium, make up 36% of the total food energy. ……
Third, the displacement of vegetables and fruits by whole grains and milk products may further reduce the potassium intake because potassium concentrations in vegetables are 4 and 12 times higher than in milk and whole grains, respectively, whereas in fruits the potassium concentration is 2 and 5 times higher than in milk and whole grains (First Data Bank, 2000).
[It needs to be stated whether the potassium differences are by energy, by protein, or by wet/dry weight…]
[Cordain fails to explore the effect of the displacement by flesh foods of plant protein foods (and also plant energy foods, as humans need such little protein) (increased acidity), and what applies when grains are used only to contribute to protein requirements and as only a part of meeting energy requirements (negligible acidity), and these are not trivial to the question of what the generally optimal diet for humans might be… (see also Ch. 4)]

Taken together, the addition of manufactured salt to the food supply and the displacement of traditional potassium-rich foods by foods introduced during the Neolithic and industrial periods caused a 400% decline in the potassium intake
[This is not possible (which does not bode well regards competency in the appropriate or crucial quantitative comparisons…) – only a 100% decline is possible… perhaps what was meant was a decline to 1/4 of the original intake, a 75% decline…]
while simultaneously initiating a 400% increase in sodium ingestion (Frassetto et al., 2001).

Eaton SB. Chapter 20 Preagricultural diets and evolutionary health promotion. In, Ungar PS. (Ed) (2007) Evolution of the human diet The known, the unknown, and the unknowable. Oxford University Press.

[Eaton begins this chapter with a bemoaning that "epidemiology's record has been unimpressive" regarding the (lack of) reduction of essential hypertension, diabetes mellitus, age-related fractures, obesity, depression melanoma, breast cancer and asthma, these stated (and I don't dispute this) to be occurring as or more frequently now than thirty years ago.

If Eaton had a better grasp of the science of epidemiology, which is a science that helps (though certainly does not ensure…) people to make sensible assessments about what factors are contributing to various disease states, he might have addressed the other factors that have not been constant in the last 30 years – less manual labour; the massive and perverted use of advertising to persuade the populace that satisfaction lies in increased consumption of an increasing variety of physical- and mental-health-destroying consumables; and the increasing retention of bad gene mutations in the population due to medical care preventing diseases from removing them, to name just three.
To be a little more blunt, if many of the people proposing various theories about the evolution of the human diet had a better understanding of the science of epidemiology, which done properly is a quite comprehensively methodologically sophisticated science, they would give voice to a lot less nonsense – the analogous remark applies to economics, and it is just that lack of methodological understanding and rigour that gave rise to the term "the abysmal science" as a pejorative for economic theorizing. Although epidemiology does indeed have much better datasets (populations of many individual people(s)), to apply that methodology to, than do archeology, anthropology and economics, the understanding of the methodology involved in good epidemiology improves the understanding of the limitations of one's ability to draw conclusions from the evidence at hand in archeological, anthropological and economic studies …

However, I will certainly state, on the side of Eaton, though he does not mention it himself, that it seems highly probable to me that lack of knowledge of the science of psychology often greatly compromises the competence of both epidemiologists and health promoters, as well as the ability of the general public to understand why they do what they do and think what they think, this understanding being in my opinion a precondition to people replacing destructive behaviour (And there are some very large and pervasive vested interests that prevent the science of psychology being appropriately included in the school curriculums – I leave it to the reader to surmise what these might be, because it is not a matter I will directly address in this book).

Also, unfortunately, many epidemiologists, in particular those coming from a medical background, have a quite marginal understanding of nutrition…]

Eaton:
"Since the appearance of agriculture, about 10 kya [thousand years ago], there have been few generally recognized genetic changes (e.g., hemoglobinopathies, adult lactose intolerance, etc.).
To an overwhelming degree, our genome and its epigenetic regulatory mechanisms remain adapted for a Paleolithic lifestyle.
[But *which* period of the Paleolithic is the point, as the very late periods (say the last 40,000 years or so of it) are very probably not representative, particularly with regard to the proportion of meat in the diet, of the very large majority (the previous 2 million years) of the Paleolithic… And it also requires to be addressed the degree to which those first 2 million years of the Paleolithic differ (or not, more likely) from the 3 or 4 millions of years before that with regards to the evolution of our diet-related genes…]
While our genes have been relatively constant,
[Whose genes, in what ethnic-originating locations, is perhaps not a trivial question…]
our culture has changed to an almost indescribable extent.
This has produced discordance or mismatch between our genes and our lives, and this discordance promotes the chronic degenerative diseases that are responsible for most morbidity, mortality and health expenditure in developed, westernized nations. ……
The retrojections presented in the following section pertain to ancestors in East Africa, roughly 60 kya, about the time and place from which the modern human diaspora originated."

[Eaton then proceeds to address various nutrient groups that have changed in the human diet since 60 kya:
total fat intake;
saturated fats;
trans fatty acids;
polyunsaturated fats and their n-3/n-6 ratio;
cholesterol;
protein (5.9g/kg/day (about 7 times the recommended daily allowance for humans, which is about twice or thrice the experimental minimum achieved with purified amino acids), is given by Eaton as the high end of the dietary intake range for primates, which could hardly possibly apply to relevant primates (apes), and probably is that of the small predominantly insectivorous primates that are irrelevant to the human diet protein requirement/benefit question. Furthermore, he fails to note and address the significance of the fact that human breast milk protein/energy ratio is a mere fraction of that of approximately all several other mammals for which I have been able to find data (which does not include apes), a fact that *very* strongly suggests that humans are an animal adapted to a relatively very low-protein diet);
carbohydrate (Eaton states as a fact that our ancestors rarely used cereal grains, which is not at all able to be stated as a fact – the dentition and the ^{13}C isotope studies do not allow such a statement to be made (see also sections 1.8 and 1.13 here above)), and the triplication of salivary amylase gene (see also chapter 3, section 3.7) argues against this;
fiber (now less digestible) and phytic acid (which can bind some minerals and reduce their absorption into the body, now more present due to cereal grain consumption);
acid-base considerations (now more acid-yielding) (Eaton fails to note that flesh foods are about 10 times more acid-producing per unit of food energy content than cereals are… (see also my discussion of Cordain's work above in this section));
Na/K (sodium/potassium) (now increased 5 times)]

Ungar PS. Chapter 21 Limits to knowledge on the evolution of hominin diet. In, Ungar PS. (Ed) (2007) Evolution of the human diet The known, the unknown, and the unknowable. Oxford University Press.

"There is no clear consensus on what is "known" about early hominin diets. Indeed, there are as many differing opinions as there are researchers to express them. ……

More recent work has demonstrated considerable overlap in microwear patterns between the species, suggesting that diet differences were likely not in preferred foods but in occasional ones with challenging fracture properties (Scott et al., 2005).
This pattern also seems to hold when comparing microwear of early *Homo* species (Ungar, Grine, and Teaford, 2006).

In this case, a perceived limit of microwear analysis, the fact that microwear reflects only a few days diet, actually allows us to look at short-term variations when reasonably large sample sizes [numbers of individual teeth, preferably from different individual hominins] are available.

Microwear patterns are consistent with a diet dominated by soft fruits or other foods with less-challenging fracture properties but still occasional fall back on items that were more difficult to fracture. ……

… early hominins from South Africa tended to have ^{13}C enriched diets. This suggests that these hominins probably had fairly broad diets, including both C_3 foods, such as forest fruits, and C_4 plant resources, such as grasses, sedges, or perhaps animals that eat these."

[Ungar fails to properly address the loss of the elongation of the canine teeth, molarization of the premolar teeth, backward replacement of the jaw, and flattening of the face, as well as the final and very recent (finalized only 35,000-odd years ago with the evolution of fully modern humans) stages of the evolution of the supralaryngeal vocal tract (i.e., including the mouth cavity) for the enabling of the formation of the full range of vocal sounds possible in the modern human.]

Despite debates about methods of acquisition, processing, and transport of carcasses, it is clear that by 2.5 Mya [million years ago], hominins at least [and maybe only] occasionally consumed meat, marrow, and perhaps other tissues from a variety of animals (Bunn, chapter 11).
[Ungar should have noted the need for consideration of the differentiation of hunting or carcass involvement for meat acquisition and hunting for the acquisition of hides to make clothes from.
Also he does not note the general, almost total, lack of attempts at even approximate quantification of the proportion that meat might have constituted of the diet, from the archaeological data available (i.e., see my calculations in section 1.6 here above)] ......

Still, these simple flakes and cores [early manufactured stone tools] may well have been what Shea (chapter 12) refers to as general-purpose instant technology, providing access to a broad range of potential foods [not just large animal parts] ......

Reed and Rector (chapter 14) argue, for example, that ecomorphology and other lines of evidence indicate environmental changes during the Plio-Pleistocene that would have reduced fruit availability compared with earlier times. This implies possible selective pressures to increase leaf consumption. ...... Peters (chapter 13) elaborates on the potential plant food niche, suggesting that environmental reconstructions indicate a hominin demographic sink, with many resources, including fruits, leaves and shoots, seeds and pods, and roots, all readily available, depending on rainfall and other variables. He ranks potential foods by nutritional quality and suggests that hominin generalists would have done well on the East African plateau.
[This does not make sense – a demographic sink is a place that a population spills out into from more habitable places, where the occupants then do more dying there than reproducing, precisely because resources for it to be a place of demographic growth are *not* readily available. This unless predation differences or some other important factor was at play, but I have not come across any informative discussion of such in the literature I have reviewed...
The point is, it seems to me, that these less suitable places for hominid occupation are where the fossil remains are best preserved, due to the climate, so that the hominid fossil remains that are found, and their associated assemblages of other fossil remains, give an unrepresentative, biased, impression of what the life of most early hominids was like.
It is theoretically possible that evolution could proceed in a different direction in the demographic sink, even though there is more dying than reproducing there, but this aspect was not addressed in the book, nor elsewhere in the literature I have covered (which is not all of the literature...), and in any case that would imply evolution of different human species or at the very least strains (races), that would therefore be genetically different in ways probably including those related to diet, and again, this aspect was not addressed in the book, nor elsewhere in the literature I have covered (which is not all of the literature...).
Furthermore it seems very strange for Ungar to have mentioned only leaves in the first sentence above just here, instead of also mentioning nuts, oily seeds, cereal grains, and tubers, even though he mentions these a few sentences further on (see above)] ......

She [Sept, chapter 15] posits that, for example, if we consider meat to have been an important fall-back resource for early hominins, there would have been less competition for dwindling plant resources during crunch times.
[In this context, the word "important" does not necessarily imply a high percentage of diet, only that it facilitated survival through the lean times…] ……

Ethnographic evidence shows that meat, tubers, and other resources have been important food sources for many recent foraging peoples (Wrangham, chapter 16; Cordain, chapter 19; O'Connell, Hawkes and Jones, 1999), though researchers have yet to reach a consensus on ratios of meat to plant matter consumed (Milton, 2000).
[Ungar fails here to address the marginality and hence (un)representativeness of the areas occupied by these modern-day foraging peoples, and also the effects of their interactions with other peoples they have come/ are in contact with, for example infectious disease, emigration and trade…]

Ethnographic studies also demonstrate that cooking is a cultural universal, and so must have its origin in the distant past.
[Cooking requires the use of controlled fire, and daily cooking requires the *making* of fire, and this was probably an evolutionarily very very recent development (probably not before 200,000 – 300,000 years ago, and maybe later, for the making of fire on demand, but this is much debated), compared to the millions of years of evolutionary relevance…]

As with other tools, fire offers unprecedented access to nutrients in foods that would otherwise be inaccessible to hominins (Lucas, chapter 3; Wrangham, chapter 16).
[At least as much the point is the palatability of foods, for example of raw versus cooked meat, to hominins…]

While there remains debate as to when cooking was established, this tool would certainly have been available to hominins by the middle Pleistocene [0.8 million years ago].
[This is probably incorrect, if the ability to *make* fire on demand is the prerequisite, rather than the ability merely to use it infrequently when it was available by chance.]

Studies of nonhuman primate diets likewise provide potential insights into the diets of our ancestors (Stanford, 2001; Rodman, 2002). As Wrangham (chapter 16), and I (Ungar, chapter 4) point out, for example, African apes today have a penchant for easy to digest, sugar-rich fruits.
It is mostly at times of resource stress that gorillas "fall back" on tougher, lower-quality food resources.
Early hominins, too, then may have preferred higher-quality food sources such as fruits except at "crunch times."
Lambert (chapter 17) develops this idea further distinguishing different types of fallback foods. Her comparisons of chimpanzees [living in more seasonal habitats] and bonobos [living in less seasonal habitats] suggest possible links between biome variability, diet and tool use by hominins.
[Ungar should here have better dealt with the issues of carnivory (less than 5% of calories overall) as a fallback food in shortage seasons in chimpanzees, and its very unequal distribution amongst members of any chimpanzee group in contrast to other foods eaten.]

……One thing that should be obvious from this discussion is that we really do not know much at all about early hominin diets. ……

This "Paleolithic Prescription" has received a great deal of attention in the media and has lead to many popular diet books (e.g., Eaton, Shostak and Konner, 1989; Baudette and

Gilchrist, 1996; Cordain, 2002; among others). This approach has been criticized on both practical and theoretical grounds.
First, as is clear from the chapters in this volume, we have little knowledge of specific nutrient contents [of the diets] of our hominin ancestors.
Second, given the wide range of environments and broad spectrum of resources consumed by past peoples, it is clear that there was no single ancestral paleodiet to which we are adapted (Ungar and Teaford, 2002). Even if there were, because selection is so dependent on niche, we cannot expect to use early hominin models to infer an optimal or ideal diet for us.
[But the point is to, objectively, not so as to be rationalizing one's own dietary habits, and logically, combine and analyse what data is available from archaeology, anthropology, primatology, anatomy, physiology, biochemistry, food nutrient analyses, genetics, epidemiology, sociology, and psychology, to draw what generalizations are available from the collective informations, assessing roughly the likelihoods of the generalizations, and enumerating and elaborating on what type of, and if possible what particular, exceptions to the generalizations are likely to exist or known to exist (for example in certain geo-ethnic locations).... And, to the extent that the theorist does not have a sufficient grasp the relevance of all these knowledge areas to the question of the optimal diet for humans, they will usually fail, other than in rare cases by luck, to comprehensively answer the question to the extent possible...]

Nevertheless, there is practical value in drawing attention to early hominin diets. Some vegans and raw-food enthusiasts claim that eating animal products and cooking are somehow "unnatural" (M. Plavcan, personal communication).
[The point is, that what is *natural* is not necessarily what is *optimal* – in nature there will be distributions of variability in optimality of the sort that fit the 'normal' or Gaussian distribution (the 'bell-shaped curve'). Nature is what allows ongoing reproduction, not what gives the maximization of health and lifespan to all members of the species reproducing itself.
Virtually doubtless is that most early hominins ate some meat at some time in their life – insofar, it is therefore natural that hominins ate/eat some meat.
It is documented that Galapagos tortoises, normally grass-eaters, had to fall back on eating the remains of their fellows who had starved to death before them, in order to survive starvation themselves until enough grass grew back – insofar, it is therefore natural that Galapagos tortoises ate/eat some meat. It is well enough documented that humans have fallen back on cannibalism in times of food shortage (and a lot more often than most western 'civilization' people think...) – insofar, cannibalism is natural for humans. The proportion of the diet comprised, and its relative benefit is the point!!
Whether eating meat is a better nutritional practice than eating insects instead is doubtful. Whether there is any benefit from eating more animal products than necessary just to meet the requirement for vitamin B12 is debatable.
Whether the health would have been/ be better if the preferred foods were/are available all year round is also a question that needs to be asked. The answer to which is complex and involves the comparison of the immigration rate from population growth areas into specific population sinkage areas with the endemic birth and survival rates in those population sinkage areas, regards the possibility of divergent evolutions in these different areas.
The different nutritional environment of even different population growth areas (seaside versus mountain, regards iodine for example) probably have led to more divergent evolution than is known (population-differential adult lactase retention is known, for example) (see also section 3.11 here below).
Furthermore, it is likely enough that cooking, as probably an evolutionarily relatively very recent practice in the context of the prerequisite for continual daily use being the ability to *make* fire, is indeed *somehow* unnatural, and fairly obviously so – and not all of

the changes it brings to food are beneficial, for example the generation of mutagens and carcinogens (cancer) and oxidatively damaged unsaturated fatty acids (atherosclerosis) and Maillard products (reduction of Lysine availability) by cooking at high temperatures exposed to air, and the ability to consume food with less chewing and in a shorter time, is not necessarily a good thing (I once sat next to a fellow who devoured a lump of meatloaf the size of a red house-brick and a large pile of French fries in less than the time that it took me to properly eat and enjoy just one apple of the several that were only a part of my meal…) (see also section 1.14 above)]

Atkins writes [in Robert Atkins' New Diet Revolution] in this 10 million plus bestseller, "The human animal was able, for millions of years, to remain strong and healthy in conditions of often savage deprivation by eating the fish and animals that scampered and swam around him, and the fruits and vegetables and berries that grew nearby." (Atkins, 2002).
[To survive long enough to reproduce is not at all the same thing as to remain strong and healthy – this is an example of a sort of naturalistic utopian delusion… 'natural' is not the same thing as 'optimal'; though they are strongly related, 'optimal' is one point in the gradation or continuum of optimality that is 'natural' or of Nature regarded not just as a surviving entity but composed of individuals in different life situations]
Such statements clearly go way beyond the limits of our knowledge. As Peters (chapter 13) notes, "limitations in our paleoanthropological knowledge sometimes invite overly rich interpretations and speculations uninformed in ways that are critical."
[Indeed.]

How can we hope, for example, to compare ingestive behaviors in species of early Homo when the combined published sample of *H. habilis*, *H. rudolfensis*, and *H. erectus* includes only 5 maxillary central incisors (Ungar et al., 2006)? ……

Taphonomic [burial science-related] alteration of specimens further limits our samples, particularly for studies of epigenetic [environmental effects on gene expression] phenomena, such as dental microwear and stable isotope ratios. ……

The ultimate sample-related limitation rests with our efforts to infer dietary behaviors that leave no traces and therefore no samples for us to analyse. As Blumenschine and Pobiner (chapter 10) note, for example, "assigning aspects of hominin carnivory to 'the known, the unknown, and the unknowable' depends on how well these can be linked to the production of dietary trace fossils."

Methodological Limits.
Most authors use the present to interpret the past. Uniformitarian assumptions limit our retrojections to the way things work today (Leroi et al., 1994; Doughty, 1996). As Walker (chapter 1) writes, "paleontologists and historians both know the danger of the pull of the present (presentism for historians) whereby the past can be distorted by our knowledge and immersion in present day circumstance." In the worst-case scenario, this would lead us to reconstruct "a world of the past that is an exact replica of the present" (Kelley, 1993).
[And the best example of that in the present context of this writing might be the effect the high meat consumption of modern western societies, and in particular its belovedness as a personal habit of some dietary and nutritional theorists, on some of the theories propounded or even presented as fact…] ……
As Sponheimer and coauthors (chapter 8) aptly write, "there can be little doubt that some 50 years hence, many will look back at what we 'know' today and shake their heads with wry amusement."

[I hope that I have provided, with this book here, the opportunity to do that much earlier than 50 years hence…] ……
Hypotheses can be shown to be robust by colligating disparate lines of evidence.
[Indeed.] ……

… help us better understand the relative roles of preferred foods and fallback resources in early hominin paleoecology."
[Indeed!]

Eaton SB, Shostak M, Konner M (1988) The Paleolithic Prescription. A program of diet & exercise and a design for living. Harper and Row, New York.

[Chapter one includes a storytale description of Paleolithic life that is a rank idealization and as much the product of E, S & K's imaginations as of any proper assessment of the evidence, and also quite an overgeneralization.
Of use is the noting that fossil bones and teeth of the period are stronger than those of modern-day peoples despite the absence of dairy products – making a good point that in nature, as a rule, milk-drinking ceases after infancy, and usually without the ill-effects that the dairy industry would have us think are due to insufficient consumption of dairy products.]

[Chapter two fails to consider that weapons can be used, and might even have been more used, for defense as well as offense.]
"The Cro-Magnons, among the first true modern people – *Homo sapiens sapiens* – lived in Europe from around 35,000 years ago until between 15,000 and 10,000 years ago. When this time period – called the late Paleolithic – ended, a number of radical transformations had taken place.
The last of the four great Ice Ages which had visited the planet over the previous 600,000 years was ending and receding glaciers in Europe and North America were precipitating warmer and wetter climates. [More the reverse direction of causation?]
[Note the importance of these major climatic changes as potential influences in periodically changing the diet of our ancestors… which E, S & K do not properly address…]
But from the human viewpoint, the most dramatic development was the debut of an entirely new subsistence pattern: agriculture. …… Others reason that agriculture emerged in response to population pressure.
[And rightly so, very probably, but not enough attention is given to the effect of the increasing population pressure, in the millions, and in the hundreds of thousands of years up to this threshold point, on human diet – and the questions of what proportions of fruits, leaves, legumes, nuts, oily seeds, cereal grains, sedge grains and corms, tubers, and meat comprised the diet of our hominin ancestors at different times and places due to population pressure] ……

The practicality of agriculture also increased as the Ice Age ended and the climate became warmer and wetter. This change proved a boon to wild grasses – including the ancestors of today's cereal grains – which became abundant, widespread, and ever more attractive as food sources.
[Most cereal grains are temperate, not tropical plants, so in the absence of something else happening, such as preagricultural deliberate or even incidental sowing of them, I don't think it probable that climate alone did anything to change the distribution of cereal grains that it hadn't done before in similar climatic conditions – nor that the change would have been anything other than that of location rather than of volume…] ……

To look at it yet another way, before agriculture actually took hold around 10,000 years ago, 100 percent of the estimated 5 to 10 million people alive were subsisting as hunters and gatherers. By AD 1500, the world population had grown to over 350 million and only 1 percent were still maintaining themselves in this mode. ......
[Chapter two includes a description of hunter-gatherer (/gatherer-hunter) life that seems at least a little idealized – beautifully egalitarian groups with no dominance issues to resolve within or between groups or between the sexes – too good to be true? – (see also Chapter 7 here below.)] ......

Even in the heart of today's Kalahari Desert in southern Africa – an environment that early hunters and gatherers [gatherer and hunters?...] would probably have shunned – the /Gwi San (Bushmen) gather around 60 varieties of food plants while their more northern neighbors, the !Kung San, choose from among 105 edible species. ......

Although the earliest human hunters were probably limited in their technology and their hunting success, people living between 40,000 and 10,000 years ago were supreme big-game hunters, pursuing a variety of animals in diverse locales.
[E, S & K fail to address the issue of what role population pressure on preferred food sources resulted in hunting, and to what extent hunting was more or less a proxy display of martial power (via witnessing of the act itself or of carcass remains) to warn neighboring groups that engagement in combat with them could have dire consequences, particularly as weaponized combat was far more potentially lethal than the non-weaponized combat of our distant very ape-like ancestors. And also, what analogous role it might play within a group, as well as constituting a mating display to females (see also chapter 7, section 7.3, 7.6, 7.8, here below)]
[E, S & K state the scope of the hunting to have been "staggering" as exemplified by the Solutre remains (see section 1.6 here above) – if they had actually attempted some quantification of what percent of diet the meat acquired from the Solutre kills might have been, in the manner I have done in section 1.6, they would have found the actual extent of the carnivory indicated to have been far far less than "staggering" – much more like trivial or insignificant, as a percent of diet...] ......

But except in the most extreme environments, edible plants are abundant and available in sufficient variety to allow a wide range of choices. From fruits, nuts, beans, roots, tubers, stalks, bulbs, berries, melons, and gums, leaves, fungi and flowers a highly diversified, nutritionally rich, and often extremely tasty diet can be obtained.
[And I agree, whereas Wrangham (2007, see section 1.14 here above) does not...]

[In their consideration of why Paleolithic people seem to have been bigger and stronger than agricultural people, and even modern gatherer-hunters (they do rightly consider as a factor, the marginal nature of the areas occupied by hunter-gathers in the agricultural era), E, S & K fail to consider some of the social epidemiology – when a small band fights another small band, all males are likely to be involved in the fighting, and larger males are more likely to survive the fighting and go on to reproduce their genes than small males; agricultural males do not all take part in warfare – they are represented by armies of warriors – and those selecting the warriors choose on average larger rather than smaller males – so being a small agricultural male is more likely to lead to being passed over for selection for the army, and survival to go on and reproduce their genes.
As well as which, there is the factor of infectious disease and its increasing rate with increasing population density, which agriculturalism allowed.
The effect of infectious disease is also related to warfare, with its increased population density of warrior groups and those other people they are amongst, including but not limited to, in occupied territory, and the population shifts and intermingling of advancing, retreating, fleeing, occupying and occupied people.]

[In chapter three, E, S & K advance the Discordance Hypothesis – basically that our genes are adapted to our prehistoric ancestral lifestyles not to our modern lifestyles, the discordance leading to the proliferation of 'lifestyle' diseases in modern societies] ……

The transition from *Homo habilis* to *Homo erectus* about 1.7 million years ago was apparently marked by significant improvements in hunting technique, so that larger game could be pursued
[That it was a change in technique rather than an increase in body size, is not at all clear from the evidence available, nor is it clear that the increased hunting ability was not merely a byproduct, and maybe a little-used one of, of increased defensive ability…],
But thereafter there was little change for over a million years. Archeologists have been impressed with the almost complete lack of cultural change during the time of *Homo erectus* [this based on stone tool characteristics]. Living sites dated hundreds of thousands of years apart show essentially identical tools and subsistence patterns.
[The manufactured stone tools are clear enough, the subsistence patterns are not visible, they are assumed…]
…… But after anatomically modern humans became widespread – between 40,000 and 30,000 years ago – the rate of cultural innovation gradually increased.
[E, S & K fail to note that this period of time is that approximately just after the final evolution of the human supralaryngeal vocal tract (the mouth cavity) to be able to make the full range of vocal sounds that modern humans can, and to discuss how this might relate to language evolution and its role in culture.] ……

Genetically, our bodies are virtually the same as they were then [40,000 years ago]. Except that "then" was a world we had spent hundreds of thousands of years adapting to [some 6–odd million years, probably more the point…], …… The pace of biological evolution is glacially slow. …… Instead tens of thousands of years are often necessary for minor changes to occur, hundreds of thousands or even millions for more significant changes.
[This is an overgeneralization – for example infectious diseases can alter the gene pool for the Major Histocompatability Complex within a generation, as could warfare depleting genes for large size and strength.
Also it needs be noted that a genetic function can deteriorate through deleterious DNA-*degenerative* mutations being allowed to remain in the DNA coding for a function that is no longer required for or even conducive to survival, much more quickly on average than mutations building up a new function that did not exist before – the examples, of faster evolution, of the sickle red blood cell, and persistence of lactase into adulthood, are both examples of DNA degeneration – in the case of the lactase, it is the degeneration of the gene that regulates the lactase to be switched off at a certain age in childhood, curious that it is that the gene came to be in the first place (maybe as a weaning inducer?)…]

[Pleasing is that E, S & K note the confounding of the much-vaunted moderate alcohol consumption over total abstinence for better health 'finding' through faulty epidemiological methodology, for example, that most investigators grouped lifelong teetotalers together with reformed drinkers, some of whom had previously savagely imposed on their hapless bodies, we may assume to ill effect, in the "nondrinker" category. And they hint at the range of other, psychosocial and socioeconomic factor confounders that I think very likely to be in action here – and this in 1988, a good 10 years before large numbers of less competent epidemiologists had swallowed that nonsense, probably in a solution of the favorite alcoholic beverage of their cherished habit, in psychological rationalization of it, in the common enough absence of the psychological knowledge background that might have helped them to see what they were doing – my compliments to E, S & K on this, but unfortunately their application of epidemiological principles to other matters dealt with in their book is far from as good...]

[In chapter four E, S & K present a table (from Page et al. in Friedlander, ed., 1987) which I reproduce here (I have rounded off figures that could not possibly be statistically significantly accurate, for example 118.00 to 118, 387.50 to 390, and I have emphasized the calcium (the calcium difference is similar to that between crustaceans (which also have exoskeletons) and meat):

Nutrients in insects and wild game. Components of a 100-gram portion

Nutrient	Insects	Wild Game
Fat	4.3 gm	3.9 gm
Protein	22 gm	24 gm
Energy	118 kcal	141 kcal
Calcium	*83 mg*	*10 mg*
Sodium	97 mg	69 mg
Potassium	210 mg	390 mg

The first hominoids – great apes – living 30 million years ago, subsisted largely on fruit,
[This is not known, it is an assumption, even though I think it is a very plausible assumption…]
although fossil remains show that their teeth were also suitable for processing other vegetables and meat [as they had elongated canine teeth, which we humans do not…]. This pattern – fruits primary, with other vegetable foods and meat a distant second and third – seems to have prevailed for about 20 million years. [Assumedly…]

Around 7 million years ago – when the protohuman line diverged into two paths, one leading to chimpanzees, the other to humans – meat began to assume greater importance, [This is an assumption! There is no solid evidence to this effect – in fact the reduction of the canine teeth from being elongated to not being elongated at all, is solid evidence suggesting (not proving) that the opposite may well have happened; that meat may have begun to assume a lesser importance – it is at least as plausible that other foods like nuts, oily seeds, cereal grains, sedge corms and seeds, and/or starchy tubers were the foods that began to assume a greater importance!...]

Although whether it was obtained by hunting, scavenging, or both is not clear. ……

The role of vegetable foods has been harder to assess, since plant remains and even gathering tools – most are made of wood or plant fiber – are poorly preserved. But, in most cases, these foods are likely to have been the mainstay of the hunting and gathering diet, then as now.
[In which case, better to use the term gathering and hunting…]

A probable major exception was a relatively short period beginning about 35,000 years ago in Europe, not long after truly modern humans appeared. With well-developed equipment, sophisticated hunting techniques, and human population densities low in relation to numbers of game animals,
[But perhaps not low in relation to preferred plant foods, there perhaps being population pressure in that regard, which is probably a very important point…]
people such as the Cro-Magnons concentrated on big-game hunting. In some areas during this time meat probably provided over 50% of the diet.
[In some areas, at some times, it probably did…]
But this didn't last long in evolutionary terms;
[which is a very important point.]
even before agriculture appeared 10,000 years ago, overhunting, climatic changes, and population growth had caused a shift from big-game hunting toward a broader spectrum of subsistence activities. ……

… suggesting that widespread use of aquatic foods is relatively new. (Whether this was true along the seacoast as well is not known since those regions were permanently submerged when the glaciers melted.)
[It seems doubtful that coastal hominids would not consume some seafoods at some times, though whether they were preferred foods, therefore constituting a large proportion of diet, is another matter, particularly in the absence of cooking to improve palatability…])

Remains of small game animals and tools for processing plant foods – such as grindstones and mortar-and-pestles – also date from this period, which is known as the Mesolithic, a period between the Paleolithic and the Neolithic, or from around 20,000 to around 10,000 years ago.
[It is very important to bear in mind that these tools are *manufactured* stone tools, which are therefore visible in the archaeological record – almost certainly the prototypes of these tools were naturally shaped stones used for the same purpose, and probably for millions of years, as modern day chimpanzees are known to use naturally shaped stone tools for (processing food – cracking, grinding, pulping) – and being naturally shaped these stone tools are not visible in the archaeological record – which causes some theorists (those apparently unaware of the significance of the basic facts of chimpanzee stone tool use…) to suggest or even state that early hominins did not use stone tools…]

Within a few thousand years of agricultural life, vegetable foods came to make up as much as 90 percent of the diet. At the same time, the broad range of wild plant foods typical of hunting and gathering [gathering and hunting…] diets was replaced by a narrow range of cultivated crops. The result? People living after the advent of agriculture were 4 inches shorter, on average, than early European *Homo sapiens sapiens* living 30,000 years ago. Even *Homo erectus* of 1.6 million years ago was tall; a remarkably complete fossil skeleton recently found in Kenya by Richard Leakey and Alan Walker suggests that these remote humans were as tall as or taller than we are today.
[And I repeat: In their consideration of why Paleolithic people seem to have been bigger and stronger than agricultural people, and even modern gatherer-hunters (they do rightly consider as a factor, the marginal nature of the areas occupied by hunter-gathers in the agricultural era), E, S & K fail to consider some of the social epidemiology – when a small band fights another small band, all males are likely to be involved in the fighting, and larger males are more likely to survive the fighting and go on to reproduce their genes than small males; agricultural males do not all take part in warfare – they are represented by armies of warriors – and those selecting the warriors choose on average larger rather than smaller males – so being a small agricultural male is more likely to lead to being passed over for selection for the army, and survival to go on and reproduce their genes.
As well as which, there is the factor of infectious disease and its increasing rate with increasing population density, which agriculturalism allowed.
The effect of infectious disease is also related to warfare, with its increased population density of warrior groups and those other people they are amongst, including but not limited to, in occupied territory, and the population shifts and intermingling of advancing, retreating, fleeing, occupying and occupied people.]

Since the industrial revolution, the animal-protein content of western diet has again increased. This may help explain the increase in the average height of Americans and Europeans observed within the last 150 years. We are now nearly as tall as were the first biologically modern humans.
[Firstly, this increase in height is more likely to be due to the rectification of energy intake deficiency as to be due to the increase in animal protein intake – bear in mind that the apes eat very little animal protein and are massively strong compared to modern humans on a body weight basis and equal training basis.

Secondly, in the era of modern weaponization, body size matters much less both to recruitment to the armed services and to survival in warfare – there was a good reason why the Colt 45(?) pistol was affectionately nicknamed "the equalizer"...
Thirdly, overreliance on one staple is probably more the point than plant versus animal protein sources, see chapter 3, section 3.5, p269 regarding the metabolic toxicity resulting from protein constituting more than 50% of calories, and section 3.6, p276 regarding the nutrients that meat is in fact a deficiency food in – for example, calcium, a major component of bones...]

Nevertheless, hunters and gatherers [gatherers and hunters, mostly] of today are different from their Paleolithic counterparts in significant ways. Contemporary preagricultural people live in marginal environments – the desert, the tropical forest, or Arctic regions. The food quest in such settings is more difficult than it would have been 25,000 years ago, when the most fertile and productive areas were available and game was more abundant.
......
An overview of contemporary preagricultural diets shows great variability – from the Aborigines of central Australia, who obtain only 10 to 20 percent of their food from meat, to the Arctic Eskimos, for whom animal foods provide 90 to 95 percent of the subsistence [and whose cardiovascular health has been documented to not be as good as many have vaunted in the usual absence of evaluation of the actual evidence available...] – and suggests an average pattern of 35 percent meat and 65 percent plant foods. ......

Cereal grains – an agricultural staple for the past 10,000 years – play only a minor part in the diet of recent hunters and gatherers [gatherers and hunters, mostly...]
[Because the areas suited to growing the stuff have been taken over by agriculture, obviously, surely?...]

(The almost total absence of grindstones and mortars-and-pestles in archaeological material dating from before 20,000 years ago suggests that this was true in the Late Paleolithic as well.) Roots, beans, nuts, tubers, fruits, flowers, and gums provide hunters and gatherers [gatherers and hunters, mostly...] with a greater quantity and wider variety of vitamins and minerals than farming did for agriculturalists.
[And I repeat: It is very important to bear in mind that these tools are *manufactured* stone tools, which are therefore visible in the archaeological record – almost certainly the prototypes of these tools were *naturally shaped* stones used for the same purpose, and probably for millions of years, as modern day chimpanzees are known to use naturally shaped stone tools for (processing food: cracking, grinding, pulping,) – and being naturally shaped these stone tools are not visible in the archaeological record – which causes some theorists (those apparently unaware of the significance of the basic facts of chimpanzee stone tool use...) to suggest or even state that early hominins did not use stone tools...
Furthermore, such tools are not strictly necessary for the consumption, via chewing, of all seeds and cereal grains, which are moister and softer and more chewable at the time of harvest than weeks or months later when dehydration has occurred, and when soaked after such storage has dehydrated and hardened them...]

By compiling studies which detail the diets of recent hunters and gatherers [gatherer and hunters, mostly...] it is possible to reconstruct a prototype of what was probably the nutrient content of our ancestral diet. Since the estimates average the diets of many groups, they should be considered ballpark figures, encompassing local variations in climate, season, terrain, and other environmental conditions, as well as differences in the specific plants and animals available.

[Firstly, present-day gatherer-hunters live in marginal areas that therefore will not be representative of the ancestral preagricultural diet(s), for reasons that include the fact that these areas are precisely those where cereal grain does not grow well naturally. Secondly, our ancestors probably ate, on average, very little meat before the last 35,000 years or so, and evidence such as the singularly low protein:energy ratio of human breast milk in comparison to that of other animals, and the fact that many individual humans are able to live healthily for decades without any intake of dietary or supplemental vitamin B12, suggest that we are evolutionarily still much more the product of the millions of years of ape ancestral life followed by the 6 million years of much much less than the 35% of calories meat diet theorized, than of the last 30,000 years or so.]

[I deal with most of the nutrients E, S & K address in particular in chapters 3 and 4 below, but in relation to what modern western food groups and in what proportions relative to requirements they are found, but I think it worthwhile to include the following excerpt here:]
An analysis of over one hundred plant foods used by recent hunters and gatherers [gatherers and hunters, mostly…] supports the notion that high levels of calcium intake are desirable. Wild vegetable foods average about 130 mg of calcium per 100 g portion; since late Paleolithic people were eating about 1,460 g of such food each day, plant foods would have provided over 1,800 mg of calcium while the meat they ate would have given another 100 mg.
Chewing bones from fowl or small mammals might have substantially increased this last value.
[The main points I will make here are the greatly reduced intake of fruit of vegetables in modern western countries (and in many developing countries where the culture is to look down on green vegetables as things that only the unfortunate poorest eat in any quantity), not only by what people are actually eating, but by what the dietetic orthodoxy are recommending they eat – in particular, that dairy foods are the optimum source of calcium for humans, and that it would be grossly impractical, unreasonable and inconvenient to expect people to eat enough plant foods to meet their (proposed) calcium requirements. (The analogous situation applies for iron and folate regards red meat…). And western (and other…) culinary practices have been and are still usually such as to greatly derange the evolutionarily natural palatability of largish dietary proportions of green vegetables…
Also, however, that my work in chapter 4 here indicates that the fruit and vegetables and other plant foods generally available to us in the modern western world have substantially less than that calcium content, which causes me to want to know what foods they assayed to get their figure, and how representative they are, and to what extent soil Ca, farming etc has to do with it.]

[E, S & K's statement on p83 that before agriculture people ate no cereal grains is rank, and very probably incorrect, assumption! The human teeth are extremely well suited to the grinding of cereal grains – if one was to consider why the human teeth differ in the small ways they do from chimpanzee teeth, one would have to consider the possibility that the non-elongated canine teeth suggested even less consumption of meat than chimpanzees with their less than 5% of diet meat consumption, and that the flattened surfaces of the molars, and the premolar teeth becoming more like the molars, suggested an increase in grain consumption, including cereal grains… And cereal grains can be eaten (successfully ground up with the molar teeth) raw and whole if fresh, and soaked if stored and dehydrated over time – I myself commonly eat a few mouthfuls of raw soaked barley, wheat or rice on a daily basis, usually with soaked legumes – and that the soaking process initiates the sprouting process probably improves the vitamin content of them, while of course the mineral content remains the same, although the sprouting may beneficially affect the absorption of the minerals into the body.]

[E, S & K begin chapter five by asking if it is possible to eat as our ancestors did 25,000 years ago. It is probably much more appropriate to ask if it is possible to eat as our ancestors did 50,000, 100,000, 200,000, 500,000 or 1 million years ago - the greatest differences in diet are very probably to be found between diets of the two most recent dates above, 25,000 and 50,000 years ago, due to population pressure on preferred food resources, but also there is the effect of cooking in changing the palatability of foods to consider – probably the best ancestral diet(s) to refer to in asking such a question is that/those of the time period before increased population pressure increased the need for meat consumption and for hunting as a culturally-derived proxy display of martial power for the retention of territorial control, and before the advent of fire-*making* changed the palatability of meat and other probably naturally less palatable foods so as to substantially increase their consumption…]

Most vegetables and plant foods, in contrast [with animal foods], provide incomplete protein, which lacks one or more of the essential amino acids.
[This is gross ignorance of the most basic sort! The sort of statement made by people who have not taken the little time required to look at a tabulation of food amino acid contents – for example the qualified dietician who managed to be allowed to give me a failing grade in a public health nutrition course at JCU – whose error in making just that statement I had pointed out in a very appropriately unflattering tone, though without naming her exactly, in an assignment she marked,.
The fact of the matter is that the proteinaceous plant foods are only a *little* low, relative to the proportion to other amino acids required, of some amino acids – most notably cereal grains and nuts and seeds being generally a *little* low in lysine, and a *little* high in methionine, and legumes being generally a *little* low in methionine, and a *little* high in lysine. (see chapter 5 below)
Furthermore, it is not "very important" to carefully combine these two groups in exact ratios in a meal to optimize amino acid ingestion, it merely improves (and that only potentially, as other foods eaten in the meal also contribute their, lesser, amounts of various amino acids to the total complement) it a little. Almost invariably we eat more protein than we need, because we need so little of it – and because the limiting amino acid is only limiting to a relatively small extent, most often a meal with legumes but no grains, nuts or seeds, or a meal with grains, nuts or seeds but no legumes, will provide more than enough of each and all of the essential amino acids needed.
Furthermore, other than supposedly eggs (which by the way are reported to have only 1/6 the vitamin B12 absorbability of other common animal foods), all the animal foods have their own limiting amino acid too – and in the case of cooking at the usual temperatures other than boiling, Maillard Products (Lysine-sugar) will be formed, reducing the amount of lysine that can be absorbed by the gastrointestinal tract from ingested food, rendering lysine even more so the limiting amino acid that it commonly is in animal foods to start with…]

[In their comments on the generational height increase trends in immigration-originating USA families, E, S & K very likely get the epidemiological science wrong by failing to differentiate, via adjusting for confounding effects, the "effects" of the 3 factors they note: (1) increased caloric intake; (2) improved general health during childhood; (3) increased protein intake. Probably, for the first two, probably *not*, for the third, as we need so little protein… Furthermore, they fail to note that obesity also causes (or has been touted to) an increase in height, though this might not have been 'known' at that time.]

Yet meat is not an absolute necessity. Strict vegetarians eating a vegan diet (which except for breast milk avoids all animal products including eggs and dairy foods) can be basically healthy, although there is some evidence that children raised exclusively on such diets have slowed growth and development.

[The crucially important point here is whether there was vitamin B12 supplementation or not – without knowing this there's no point discussing the matter. Although I went for more than 20 years as a vegan without any vitamin B12 supplementation, apparently without any damage to my nervous system, it is a dangerous experiment, as those who carry deleterious mutations in the DNA coding for vitaminB12 gastrointestinal uptake, transport or/and storage due to the mutations not having affected the survival of their ancestors because they were (in the last 35,000-odd years) eating a lot of meat, would be likely enough to get brain (not only spinal cord, the only part noted by E, S & K on p129) or other nervous system damage from undertaking that experiment. I'm not sorry I did, as it has demonstrated that at least some of us need only an extremely small intake of vitamin B12 to allow a relatively high state of health, on some, in my case one, but not necessarily all types of vegan diets. (see also sections 1.14 above and chapter 10 below). Certainly no child should be forced to undertake the experiment – a vegan child should have vitamin B12 supplementation.]

[For interest, from p109, tabled data from Watt and Merrill, 1975:
Percentage of polyunsaturated fats in selected meats:
Domestic meat:
Beef, 2.0; pork, 9.6; lamb, 2.7; veal, 4.2; chicken 17.0
Wild game:
Cape buffalo 30.0; eland, 35.0; hartebeest, 32.0; giraffe, 39.0; kangaroo, 36.0; warthog, 43; caribou, 22.0; grouse, 60.0]

[Unfortunately, no mention of the oxidative damage that cooking at common high temperatures in the presence of air does to unsaturated fat, causing atherogenic fat…]

[Chapters six through ten of E, S & K's book The Paleolithic Prescription will not be addressed here as they are more peripheral to the focus of my work here – perhaps I will do a complete book review in the future, as many people have read the book and follow its advice, later revamping, rehashing or purported refining by E, S &/or K or others notwithstanding.]

References for chapter 1

Alexander RM. Section 2.9 Human locomotion. In, Chapter 2 The life of primates. In Jones S, Martin R, Pilbeam D, Bunney S (Eds) (1992) The Cambridge Encyclopedia of Human Evolution. Cambridge University Press.

Andrews P, Martin L (1991) Hominoid dietary evolution. Phil Trans R Soc London B 334: 199-209.

Blumenschine RJ, Pobiner, BL. Chapter 10 Zooarcheology and the ecology of Oldowan hominin carnivory. In, Ungar PS. (Ed) (2007) Evolution of the human diet The known, the unknown, and the unknowable. Oxford University Press.

Bunn HT. Chapter 11 Meat made us human. In, Ungar PS. (Ed) (2007) Evolution of the human diet The known, the unknown, and the unknowable. Oxford University Press.

Clarke R, Halsey J, Bennett D, Lewington S (2010) Homocysteine and vascular disease: review of published results of the homocysteine-lowering trials. Journal of Inherited Metabolic Disease, e-publication

Copeland SR (2007) Vegetation and plant food reconstruction of lowermost Bed II, Olduvai Gorge, using modern analogues. Journal of Human Evolution. 53: 146-175.

Cordain L. Chapter 19 Implications of Plio-Pleistocene hominin diets for modern humans. In, Ungar PS. (Ed) (2007) Evolution of the human diet The known, the unknown, and the unknowable. Oxford University Press.

Daegling DJ, Grine FE. Chapter 6 Mandibular biomechanics and the paleontological evidence for the evolution of human diet. In, Ungar PS. (Ed) (2007) Evolution of the human diet The known, the unknown, and the unknowable. Oxford University Press.

Dean C. Section 2.4. Jaws and teeth. In, Chapter 2 The life of primates. In Jones S, Martin R, Pilbeam D, Bunney S (Eds) (1992) The Cambridge Encyclopedia of Human Evolution. Cambridge University Press.

Delson E, Tattersall I, Van Couvering JA, Brooks AS. (Eds) (2000) Encyclopedia of human evolution and prehistory (2nd Ed). Garland Publishing Inc, New York.

Demes B (1988) Bite force, diet, and cranial morphology of fossil hominids. J Human Evolution 17: 657-70

Eaton SB. Chapter 20 Preagricultural diets and evolutionary health promotion. In, Ungar PS. (Ed) (2007) Evolution of the human diet The known, the unknown, and the unknowable. Oxford University Press.

Eaton SB, Shostak M, Konner M (1988) The Paleolithic Prescription. A program of diet & exercise and a design for living. Harper and Row, New York.

Eaton SB, Konner M, Shostak M (1988) Stone agers in the fast lane: chronic degenerative diseases in evolutionary perspective. American J Med 84: 739-749.

Egeland CP, Dominguez-Rodrigo M (2008) Taphonomic perspectives on hominid site use and foraging strategies during Bed II times at Olduvai Gorge, Tanzania. Journal of Human Evolution 55: 1031-1052.

Fabig A, Herrmann B (2002) Trace elements in buried human bones: intra-population variability of Sr/Ca and Ba/Ca ratios – diet or diagenesis? Naturwissenschaften (89): 115-9.

Fleagle GJ. Section 2.8 Primate locomotion and posture. In, Chapter 2 The life of primates. In Jones S, Martin R, Pilbeam D, Bunney S (Eds) (1992) The Cambridge Encyclopedia of Human Evolution. Cambridge University Press.

Foley RA, Lee PC (1991) Ecology and energetic of encephalization in hominid evolution. Phil Trans R Soc London B 334: 223-232.

Franck FS (1959) Days with Albert Schwietzer. Greenwood Press, Westport Connecticut p66

Freeman LG. Chapter 4 The fat of the land: notes on Paleolithic diet in Iberia. In, Harding RSO and Teleki G (Eds) (1981) Omnivorous Primates Gathering and hunting in human evolution. Columbia University Press.

Harding RSO, Teleki G. Chapter 1 Introduction. In, Harding RSO and Teleki G (Eds) (1981) Omnivorous Primates Gathering and hunting in human evolution. Columbia University Press.

Henry AG, Wood W. Chapter 2 Whose diet? An introduction to the hominin fossil record. In, Ungar PS. (Ed) (2007) Evolution of the human diet The known, the unknown, and the unknowable. Oxford University Press.

Isaac GLI, Crader, DC. Chapter 3 To what extent were early hominids carnivorous? An archaeological perspective. In, Harding RSO and Teleki G (Eds) (1981) Omnivorous Primates Gathering and hunting in human evolution. Columbia University Press.

Isaac GLI, Isaac B (1977) Olorgesailie Archeological studies of a middle Pleistocene lake basin in Kenya. The University of Chicago Press.

Jablonski NG, Chaplin G (2004) Becoming bipedal: how do theories of bipedalization stand up to anatomical scrutiny. Chapter 13 in Anapol F, German RZ, Jablonski NG Shaping primate evolution Form, function and behavior. Cambridge University Press.

Kelley J Section 6.4 Evolution of apes. In, Chapter 6 The primate fossil record. In Jones S, Martin R, Pilbeam D, Bunney S (Eds) (1992) The Cambridge Encyclopedia of Human Evolution. Cambridge University Press.

Klein RG. Chapter 5 Later stone age subsistence at Byeneskranskop Cave, South Africa. In, Harding RSO and Teleki G (Eds) (1981) Omnivorous Primates Gathering and hunting in human evolution. Columbia University Press.

Leonard WR, Robertson ML, Snodgrass, JJ. Chapter 18 Energetic models of human nutritional evolution. In, Ungar PS. (Ed) (2007) Evolution of the human diet The known, the unknown, and the unknowable. Oxford University Press.

Lewin R (1984) Man the scavenger. Science 224: 861-862.

Lieberman P Section 3.5 Human speech and language. In, Chapter 3 The brain and language. In Jones S, Martin R, Pilbeam D, Bunney S (Eds) (1992) The Cambridge Encyclopedia of Human Evolution. Cambridge University Press.

Lucas PW. Chapter 3 The evolution of the hominin diet from a dental functional perspective. In, Ungar PS. (Ed) (2007) Evolution of the human diet The known, the unknown, and the unknowable. Oxford University Press.

Lucas PW (2004) Plant mechanics and primate dental adaptations: an overview. Chapter 9 in Anapol F, German RZ, Jablonski NG Shaping primate evolution Form, function and behavior. Cambridge University Press.

Lyman RL (1987) Hunting for evidence of Plio-Pleistocene hominid scavengers. American Anthropologist 89: 710-716.

Macho GA, Shimizu D (2009) Dietary adaptations of South African australopiths: inference from enamel prism attitude. Journal of Human Evolution 57: 241-247.

Mann AE. Chapter 2 Diet and human evolution. In, Harding RSO and Teleki G (Eds) (1981) Omnivorous Primates Gathering and hunting in human evolution. Columbia University Press.

Maresca B, Schwartz JH (2006) Sudden origins: a general mechanism of evolution based on stress protein concentration and rapid environmental change. The Anatomical Record (Part B: New Anat.) 289B: 36-46.

Martin RD (1984) Chapter 3 Body size, brain size and feeding strategies. In, Chivers DJ, Wood BA, Bilsborough A (1984) Food acquisition and processing in primates. Plenum Press, New York.

Olsen SL (1989) Solutre: A theoretical approach to the reconstruction of Upper Paleolithic hunting strategies. Journal of Human Evolution 18, 295-327.

Peters CR. Chapter 13 Theoretical and actualistic ecobotanical perspectives on early hominin diets and paleoecology. In, Ungar PS. (Ed) (2007) Evolution of the human diet The known, the unknown, and the unknowable. Oxford University Press.

Potts R. Section 9.1 The hominid way of life. In, Chapter 9 Early human behaviour and ecology. In Jones S, Martin R, Pilbeam D, Bunney S (Eds) (1992) The Cambridge Encyclopedia of Human Evolution. Cambridge University Press.

Price, Weston A. (1939) Nutrition and Physical Degeneration: A Comparison of Primitive and Modern Diets and Their Effects. Paul B. Hoeber, Inc; Medical Book Department of Harper & Brothers.

Reed KE, Rector AL. Chapter 14 African Pliocene paleoecology Hominin habitats, resources and diets. In, Ungar PS. (Ed) (2007) Evolution of the human diet The known, the unknown, and the unknowable. Oxford University Press.

Richards MP (2002) Review A brief review of the archeological evidence for Paleolithic and Neolithic subsistence. Eur J Clin Nutr 56: 1-9.

Schoeninger MJ. Chapter 9 Reconstructing early hominin diets Evaluating tooth chemistry and macronutrient composition. In, Ungar PS. (Ed) (2007) Evolution of the human diet The known, the unknown, and the unknowable. Oxford University Press.

Sept J. Chapter 15 Modeling the significance of paleoenvironmental context for early hominin diets. In, Ungar PS. (Ed) (2007) Evolution of the human diet The known, the unknown, and the unknowable. Oxford University Press.

Shea JJ. Chapter 12 Lithic archaeology, or, what stone tools can (and can't) tell us about early hominin diets. In, Ungar PS. (Ed) (2007) Evolution of the human diet The known, the unknown, and the unknowable. Oxford University Press.

Shipman P (1986) Scavenging or hunting in early hominids: Theoretical framework and tests. American Anthropologist 88: 27-43.

Sponheimer M, Lee-Thorp J, De Ruiter D. Chapter 8 Icarus, isotopes, and Australopith diets. In, Ungar PS. (Ed) (2007) Evolution of the human diet The known, the unknown, and the unknowable. Oxford University Press.

Sponheimer M, Lee-Thorp J, de Ruiter D, Codron D, Codron J, Baugh AT, Thackeray F (2005) Hominins, sedges and termites: new carbon isotope data from the Sterkfontein valley and the Kruger National Park. Journal of Human Evolution 48: 301-312.

Sponheimer M, Lee-Thorp J (1999) Isotopic evidence for the diet of an early hominid, Australopithecus africanus. Science 283 15 Jan: 368-370.

Steele J (1989) Hominid evolution and primate social cognition. Journal of Human Evolution 18, 421-432.

Teaford MF. Chapter 5 What do we know and not know about diet and enamel structure? In, Ungar PS. (Ed) (2007) Evolution of the human diet The known, the unknown, and the unknowable. Oxford University Press.

Teaford MF. Chapter 7 What do we know and not know about dental microwear and diet? In, Ungar PS. (Ed) (2007) Evolution of the human diet The known, the unknown, and the unknowable. Oxford University Press.

Ungar PS. Chapter 4 Dental functional morphology The known, the unknown, and the unknowable. In, Ungar PS. (Ed) (2007) Evolution of the human diet The known, the unknown, and the unknowable. Oxford University Press.

Ungar PS. Chapter 21 Limits to knowledge on the evolution of hominin diet. In, Ungar PS. (Ed) (2007) Evolution of the human diet The known, the unknown, and the unknowable. Oxford University Press.

Wrangham R. Chapter 16 The cooking enigma. In, Ungar PS. (Ed) (2007) Evolution of the human diet The known, the unknown, and the unknowable. Oxford University Press.

Wrangham R (2009) Catching Fire How cooking made us human. Basic Books, New York.

Chapter 2. Our closest animal relatives: apes (chimpanzees, bonobos, gorillas, orangutans, and gibbons) and monkeys

2.1 Summary of chapter 2 p214

2.2 Different types of primates and their diets. Harding (1981) p218

2.3 Gombe chimpanzee detailed food list; dubious calculation of meat consumption (13 % of calories) without any detailing of denominator group, season (i.e. of preferred food plenty or scarcity), or time trend across the 13 year period covering the four periods sampled; human influence; cooperative hunting; only adult males hunt; strong lack of carrion-eating; power-scavenging of fresh kills; transportation of food and objects; 20-fold increase of severe attacks on other chimpanzees at predatory events; travel preferred across more open ground on trails; terrestrial travel with arboreal feeding; stone thrown as weapon. Teleki (1981) p221

2.4 Omnivory not a very informative term regards position in that continuum; digestibility of food types. Chivers, Hladik (1984) p226

2.5 Legumes as important diet mainstays in many primate species in varied climates; faulty statement of broadness of omnivory due to non-differentiation of apes from other larger primates; failure to properly address frugivore diet seeds, nuts, legumes and grains; brain size and sociality; failure to address need for large brain in memorizing seasonal timing and spatial location of fruits, leaves, seeds, nuts, legumes and grains, and in tool-using processing of some of these; failure to note that nuts, oily seeds, legumes and grains are *more* energy-dense than flesh. Ripley (1984) p227

2.6 Lowland gorillas' and chimpanzees' diet of preference is mostly fruit; in seasons of fruit scarcity gorillas eat more non-fruit plant parts whereas chimpanzees eat more meat; highland gorillas eat 80% non-fruit plant parts, highland chimpanzees eat 40% non-fruit foods. Yamagiwa, Maruhashi, Yumoto, Mwanza (1996) p229

2.7 Borneo orangutans' diet complex, and mostly fruit other than in some months; no meat-eating recorded. Galdikas (1988) p229

2.8 Chimpanzee meat-eating mostly a fallback in seasons of scarcity of preferred plant foods in more seasonal habitats; gorilla and orangutan meat-eating approximately nil; initial disgust and eventual habituation of captured apes to meat as a food; recalcitrance and illogicality of paleoanthropologists in failing to properly extrapolate these facts into theories of ancestral hominin diet. Kortlandt (1984) p230

2.9 Exaggeration and sensationalization of Gombe chimpanzee carnivory. Stanford (1999) p233

2.10 Better detailing of the Gombe chimpanzee situation and less than 5% of diet calories as meat. Stanford (1998) p236

2.11 Failures: to quantify Kibale Forest chimpanzee carnivory and to note its seasonality (My quantification suggests less than 0.5 – 1 gram monkey per chimp per day), to detect and quantify insectivory, and to mention nut-cracking and grain-eating. Wrangham, Conklin, Chapman, Hunt (1991) p238

2.12 Gabon chimpanzees, diet fruits 68%, leaves 28%, animals 4%, great hunger in fruit shortage season (growth stops), uneven distribution of vertebrate meat. Hladik (1981) p239

2.13 Gabon chimpanzees and gorillas: similar teeth, guts and gut transit times; no nut-cracking with tools here at Lope Reserve; both prefer fruit-dominated diet with leaves, stems, piths, seeds and insects; gut entodiniomorph ciliates more present in gorillas; gorillas always eat some foliage; in seasons of fruit scarcity gorillas fall back on leaves, stems and piths, whereas chimpanzees split up to continue seeking fruit. Tutin, Fernandez, Rogers, Williamson, McGrew (1991) p240

2.14 Gombe chimpanzees: infanticidal cannibalism due to poor mothering; eating of human babies. Montgomery (2009) p241

2.15 Kenya Pumphouse Gang baboons, predator (lions, leopards) reduction by humans, baboon and ungulate population growth, long-distance hunting by baboons catalyzed by ordinarily dangerous behavior of unorthodox individual male evolves into cooperative hunting which disappears with immigration and emigration of particular males. Strum (1981) p243

2.16 Monkeys and apes; tooth shape & enamel thickness; digestive systems; energy-density of foods; chimpanzee hammer, anvil, and mortar and pestle use; capuchin monkey hammer, anvil, mortar and pestle, and digging tool use; seasonality and climate shift, fallback foods versus preferred foods. Lambert (2007) p248

2.17 Tooth structure and function have variable relationships; bilophodont molars associated with high-fiber diet; high, pointed cusps for puncture-crushing hard and brittle foods, e.g. bone and cartilage of vertebrates. Chivers, Andrews, Preuschoft, Bilsborough, Wood (1984) p251

2.18 Primate elongated canine teeth, dominance hierarchies, defense, prey killing and eating. Mace (1992) p252

2.19 Bearded Saki monkey canine teeth used to crack hard-shelled fruits and nuts. Van Roosmalen (1984) p252

2.20 Gelada baboons: grass seeds most of diet; enlarged premolar and molar teeth for grinding; high manual dexterity; erect sitting position; fatty deposits on buttocks; buttock-shuffling feeding locomotion. Foley (1992) p253

2.21 Tooth enamel ^{13}C study - extinct baboons ate more grass than modern South African savanna baboons; faulty statement of broad-spectrum omnivory due to non-differentiation of apes from other larger primates. Codron, Luyt, Lee-Thorp, Sponheimer, de Ruiter, Codron (2005) p253

2.22 Pleistocene African savanna baboons were nearly as large as modern gorillas; became extinct with the appearance of man. Coe (1984) p254

2.23 Protein content (dry weight) of legumes and leaves is high, of cereals, oily nuts and seeds and (underground) storage organs is substantial but generally lower, of fruit flesh is 20% that of leaves; diet of primates normally ensures an adequate supply of water; B vitamins; vitamin C; gut flora of all mammals capable of some cellulose digestion; plant defenses, tannins, lignins, proteinase inhibitors, anti-microbials, toxins. Waterman (1984) p254

2.24 Gibbons defend territory, baboon groups avoid each other in extensive areas of territorial overlap, langurs have intergroup dominance relationships Harvey (1992); Apes savagely defend territory, other than bonobos; hunting versus combat violence; population stress on food supplies; infanticide; generalisability of the psychology of killing p258

2.25 Chimpanzees dominance, leverage, cooperation, friendship rules and outrage from their breakage. De Waal (1996) p259

2.26 Corralled chimpanzees: reciprocity with food sharing for grooming, and sharing for sharing. De Waal (1989) p259

References for chapter 2 p260

2.1 Summary of chapter 2

[The distant ancestors of all primates (humans, apes, and monkeys; and lemurs, lorises, galagos and tarsiers) were insectivores – most of their diet was insects.

Our genetically closest animal relatives are the apes, animals that look like monkeys and are closely related to them but have no tails. The apes are the chimpanzees, bonobos (sometimes termed pygmy chimpanzees as they are very similar to chimpanzees in appearance), gorillas, orangutans, and gibbons (gibbons are less similar, and they are an ape, being tailless, but not a 'great ape'). Chimpanzees are generally agreed to be genetically the closest to humans, having roughly 97% exactly the same DNA, and it is generally agreed that the chimpanzee (and bonobo) evolutionary line split from the human evolutionary line about 6 million years ago. The gorilla evolutionary line split a little earlier, and the orangutan line a little earlier still.

All of the four present-day great apes, in their 'natural' habitats, are predominantly frugivorous very predominantly vegetarian omnivores – they all eat some animal foods, but only a little (mostly insects), usually and on average, and among the plant foods they prefer to eat mostly sweet fruits.
They also eat green leaves, legumes, nuts, piths, and seeds including oily seeds and cereal grains (cereal grains not very often).
Chimpanzees like fat-containing foods, whereas gorillas reportedly do not, although how this fits with their insect eating would be interesting, because often the most favored insects for humans and presumably chimpanzees have a high fat content.

Of the apes, chimpanzees eat the most meat, but animal foods are generally agreed to total less than 5 % of total calories, and of this they eat more insects than meat.

The meat hunted by chimpanzees is mostly colobus monkeys (mostly young/infants), but also piglets, small deer called dikdik, and baboons (mostly young/infants).

All chimpanzees eat a significant amount of insects (eggs, larvae/grubs, and adults) (such as would meet their vitamin B12 requirements), but only some adult males eat most of the hunted meat. This is in sharp contrast to naturally more carnivorous (including piscivorous, fish-eating) or insectivorous animals where the growing young receive proportionately more of the animal foods, as they have a higher protein:energy (the ratio) requirement than adults.

However, chimpanzees generally do not set out to hunt for meat, but do so, only sometimes (usually not), only in the course of looking for the plant foods that are their preference (sweet fruits), if they happen upon an opportunity.
Furthermore, hunting by chimpanzees increases in the seasons when their preferred foods are in short supply – and it should be noted that at these times they are generally *very* hungry, as it is reported that the growth of young chimpanzees stops during these seasons.

Therefore, it makes more sense to consider meat to be very much only a fallback food of chimpanzees, resorted to mostly in seasons of scarcity of their preferred plant foods.

Bonobos live in a much less seasonal environment than chimpanzees, and they eat even less meat again (nearly none) than chimpanzees.
This reinforces the conclusion that meat is very much only a fallback food of chimpanzees, resorted to mostly in seasons of scarcity of their preferred plant foods.
Bonobos eat enough insects to supply their vitamin B12 requirements.

Orangutans have only very rarely been observed to eat meat (squirrels), and it should be noted that orangutans also experience seasons when they are unable to eat even close to the amount of sweet fruits that they would like to, such that it is no longer the mainstay of their diet that it is in the seasons of plenty. Orangutans eat enough insects to supply their vitamin B12 requirements.

Gorillas have apparently not been observed to eat any meat. They favor sweet fruits, but in some highland regions where such fruits are always in short supply eat mostly other plant matter all year round. Gorillas eat enough insects to supply their vitamin B12 requirements.

There are many more species of monkeys than there are of apes, and some among them, for example capuchin monkeys, are very substantially carnivorous, much more so than chimpanzees. However, taken as one group, the monkeys are predominantly frugivorous or folivorous predominantly vegetarian omnivores.

Apes and humans all have color vision, which was regained in evolution by our earlier common ancestors after being previously lost earlier in evolution (some crustaceans see in 5 or more colors, compared to our three colours). This is probably related to the identification of ripe fruit.

The human gastrointestinal tract differs from that of the apes in having less large intestine, and more small intestine, such that we are less suited to breaking down cellulose in the large intestine (but keep in mind that we are still suited to digesting some dietary fiber, probably including some amount of cellulose, in the large intestine, because the descending third of the large intestine prefers the short chain fatty acids that result from that digestion as its energy source, compared to most other body tissues that prefer glucose, fats, or ketones).
The suggested physiological difference that humans are suited to a more energy-dense diet, contrary to some authors' theories would very probably be better accounted for by increased dietary energy-dense plant foods like nuts, oily seeds, cereal grains and starchy tubers, than by meat.

The human teeth differ from those of the apes in having canine teeth that are not elongated, which suggests, the reduced need as a combat weapon notwithstanding, that humans are, insofar, less suited for carnivory than any of the apes.
Also, in having flatter molar tooth horizontal surfaces, suggesting that humans are better adapted for grinding things like seeds (including grass seeds i.e., cereals).

Also in having premolar teeth that are more like the molar teeth ('molarized premolars'), which again suggests that humans are better adapted for grinding things like seeds (including grass seeds i.e., cereals).
Also, in having thicker enamel, which again suggests that humans are better adapted for grinding things like seeds (including grass seeds i.e., cereals, which may have abrasive dirt particles in them, being near the ground).
(Humans have thicker enamel than orangutans, which have thicker enamel than chimpanzees, bonobos and gorillas.)
The grinding of tubers, and the resistance to abrasion from dirt remaining on them, would also be at least accommodated, and facilitated, respectively, by the premolar and molar surface shapes and the thicker enamel.

In many of the places where chimpanzees have been observed, there is a strong possibility that human activity has increased the amount of chimpanzee carnivory.
For example, at Gombe National Park, feeding chimpanzees bananas, which would have reduced their dietary protein:energy ratio, and which also put them in closer contact with the baboons that were also being fed bananas, the chimpanzees then greatly increasing predation on the baboon young and infants (which would increase protein:energy ratio).
At Gombe, which is not a large park, and is surrounded by human habitation, poaching had probably depleted the population of leopards (leopard skins sell for far more than antelope meat…), such that chimpanzee population density probably increased such as to increase population pressure on food sources.
Note that the predation of chimpanzees on colobus at Gombe had not been quantified before the banana feeding and the greatly increased predation on baboons that it facilitated, but only after that time.
Also, note that the chimpanzees, which were allowed to, and frequently did, wander at will through much of the Gombe Primate Research Station, almost certainly would have seen meat being handled as a food and consumed by carnivorous primatologists and their henchpeople there.
(Note however that chimpanzee predation on colobus monkeys has been found in effectively all places in which chimpanzees in their 'natural' environment have been studied – such that some small amount of colobus hunting must reasonably be said to be an evolutionarily natural behavior, even if mostly a fallback one for times of preferred food scarcity. Also, in contrast to true carnivores, most females and young get little to nearly none of the meat. Also in contrast, they eat the most fat-rich part, the brain, first)
Also, chimpanzees are able to feed on oil palm nuts (seasonally) in plantations, which may (if the proteinaceous kernel were not what is accessed for absorption) also cause a reduced dietary protein:energy ratio, which might increase the tendency to predation to provide a balancing higher protein:energy ratio).
(Chimpanzees are very probably, like humans very probably are (see section 3.5, p271), an evolutionarily naturally low-protein diet adapted animal, but bananas and palm oil are at the very extremes of low food protein:energy ratio.).
Furthermore, over many years, with a probably exponential increase over the last couple of hundred years, chimpanzees have had the opportunity to observe an animal that they probably recognize (humans) (at least to the extent that would lead to a greater tendency to imitation of them) as being more similar to them than, i.e. leopards and lions, hunting and killing and eating other animals.

Following on from the discussion of factors influencing chimpanzee hunting activity, the study of the 'Pumphouse Gang' baboon troop, which will surely remain for all time a classic of environmentally and socially modified hunting behavior in a higher primate, shows the following:
Baboons (which are not as closely related to humans as apes are) do naturally eat a significant amount of meat.

When human activity (turning the area into a ranch/range farm) reduced the numbers of the animals that naturally prey on them (lions, in particular), the baboon population increased.
So did the population of ungulates, notably, Thompson's gazelles.
Probably there was population pressure on the baboon's plant foods, and the other (ungulate) species' increased populations probably increased that pressure to some extent, so that the baboons may have been hungrier than they ordinarily would have been.
Active hunting at a long distance from the baboon troop by solitary males, in the absence of the lions, was no longer selected against in natural selection, and so it increased from experimental isolated occurrences by one particularly predisposed male to become fairly prevalent by males.
Then, after experiencing the success of a small number of chance arrangements of baboon hunters to form a cooperatively functional group, cooperative long-distance hunting became prevalent.
Baboons learned to select Thomson's gazelle herds of females and young, not of males, as hunting targets.
The Thomson's gazelles then learnt to regard baboons as dangerous predators and to increase vigilance behavior and to set much greater alarm and flight distances between themselves and the baboons.
Then, following immigration of more powerful males that prevented the successful baboon hunters from actually eating their prey, but who themselves preferred to stay close to the troop as baboons usually do, and the subsequent emigration of a couple of the successful hunter males, the practice of cooperative long-distance hunting ceased.
Focus on hunting in general declined, but increased solitary hunting by immature males, closer to the troop, persevered.
Baboon dismemberment of the prey, in the process of eating it, was often clumsy enough to indicate that it was not as strongly natural an activity as it would be for the dedicated carnivores, who carry out that process in a fairly standard and proficient manner, learning from infancy from the parents notwithstanding.

This is a classic example of how human influence, itself not at all obvious even to the long-term observer, even with a primate less intelligent than a present-day great ape and presumably even our most distant bipedal hominin ancestors, can modify the apparently natural behavior of a primate in a present-day apparently natural environment.

Tool use by chimpanzees in their 'natural' habitat includes the use of naturally-shaped hammer and anvil stones to crack open nuts, naturally-shaped mortar-and-pestle stones for grinding seeds and tubers, and naturally-shaped pounding logs, as well as the use of twigs to extract termites from termite mounds. Also the use of thrown rocks as weapons in hunting has been reported.
Tool use by capuchin monkeys in their 'natural' habitat includes digging for tubers with stones, cracking open seeds and branches with stones, breaking tubers with stones, and using stones as hammers in combination with wooden anvils to crack seeds.
It is therefore probable that all of these fairly conceptually simple tool uses were carried out also, by even our most distant bipedal hominin ancestors.

Among the great apes, our closest animal relatives, chimpanzees (males mostly) have been reported to violently and lethally defend (and acquire or expand) their territories against other groups, gorillas perhaps even more so but maybe less lethally, the savagely solitary orangutans likewise, while the bonobos are reputed to be less violent in general.

It is possible that human encroachment on habitat, impact on large cat predators on apes (i.e., leopards, which I surmise are more populous in the bonobo's habitats (mostly closed forest), which I gather are less impacted by humans), and enactment of imitatable territorial and hunting violence, may have increased the level of territorial violence to some extent, but it does seem likely to be a natural inclination – in the absence of sufficient ape population losses to predators, disease, cannibalism and/or infanticide such behaviour is almost certainly adaptive to survival – in the chimpanzee case combat is most often pursued by a more powerful group, and avoided by a weaker group.

Also, the assumption that chimpanzees habituated to human observation have no reaction *at all* to that observation, for example aggression or violence (e.g. as a display to third as well as to second parties), however manifest, may well be a false assumption…

The basic relevance of this to the human situation should be obvious enough. And it seems rather foolish, to me, and wishful thinking, mostly by those enamoured of human (in particular their own) carnivory, to posit that hunting violence and violence to others of one's own species are not at all related – in the human case, surely killing large mammals whose emotional similarity (i.e., in distress) to oneself is obvious enough makes it easier for one to inflict violence on other members of one's own species? And conversely, if humans (in the absence of enough population losses to predation and disease to prevent population stress on i.e. plant food supplies) must kill their own infants or other adults to secure enough territory to get (nearer to) enough plant food, surely this killing makes humans more likely to kill animals? The former seems likely to be and to have been more in force than the latter, though.]

2.2 Different types of primates and their diets. Harding, in Harding and Teleki (1981)

Harding RSO. Chapter 6 An order of omnivores: Nonhuman primates in the wild. In, Harding RSO, Teleki G (Eds) (1981) Omnivorous primates Gathering and hunting in human evolution. Columbia University Press. New York

"… quantitative comparisons are not necessary to establish the points at issue: that the various primate species are not exclusively vegetarian, particularly when an entire annual cycle is considered; that the order Primates as a whole is omnivorous; and that the family Hominidae is no exception to this overall primate pattern.
[I accept this as correct, but the very important question that remains unanswered here, and for which quantitative comparisons are essential, is to what different *extents* the different apes, and humans, are by evolution naturally *predominantly* vegetarian animals.]

[In an apparently fairly thorough tabled collation of data for 129 primate species Harding reports:]

Eclecticism is the rule among primate diets.
Only two species, the tree shrews Dendrogale melanura and Tupaia tana, have been reported to subsist exclusively on one category of food (insects, in this case). ……
All the other primate species have much more varied diets, however; an average primate diet contains between 4 and 5 of the 9 possible categories.
At the opposite end of the spectrum from the tree shrews are those species which use food from all 9 categories: *Cercopithecus aethiops* [an African monkey], *Pan* [chimpanzees and bonobos genus], and the *Papio* [baboons genus] species.

It should also be noted that, despite the occasional dietary specialization of individual species, the same 4 food categories – fruits, soft plant parts, mature leaves, and insects – are the items found most frequently in the diets of each of the major taxa [zoological groupings]: the prosimians, Old World [African] monkeys, New World [American] monkeys, and apes.

When primate diets are ranked by food category the following conclusions emerge. [Here I have left the letters (A, B, C...) that Harding used in his ordering of the list, but I have rearranged the food groups they are attached to in order of the percent of species that includes the foodstuff in its diet, which is a more sensible ordering and better in accord with his own use of the word "ranked". Keep in mind that the %s refer to number of species that eat the food, and not to the % of the diet that is made up by the food.]

C. Fruit. The most widely used type of food, fruit is found in 90% of all primate diets. Even among prosimians, whose diets contain substantial proportions of insects, more eat fruit than any other food category (81% of all prosimian diets).

B. Invertebrates. Prosimians are the heaviest consumers of invertebrates (72% of all prosimian diets surveyed include invertebrates), but it should be stressed that invertebrates are eaten by every major primate group, even the "folivorous" columbines. The unavoidable implication is that insect-eating is not only an ancient but a consistent trait throughout primate evolution.
[It seems to me that possibly this category should have been the first in my list here, because it seems almost impossible for an animal to not eat some insect (be it adult, larval, pupal or egg) in the course of eating whatever other food type it may focus on]

E. Soft plant parts. This category is also heavily utilized by Old World monkeys (59%), with cercopithecine and colubine monkeys participating almost equally (92% and 94% of diets [of monkey species' diets]). Immature leaves, buds, shoots, and flowers are second only to fruit, being found in 79% of all primate diets.

F. Mature leaves. This food category is of special interest to the Old World monkeys, a conclusion that would seem inevitable given the leaf-eating specializations of the colubine monkeys. However, other Old World monkeys also consume mature leaves (they are found in 79% of all cercopithecine diets), while the comparable figures for other major taxonomic groups vary between 58% and 70%.

D. Seeds and nuts. These items are important components of the diets of Old World monkeys (59%), but are less significant to New World monkeys, prosimians and apes. Seed-eating may thus be a relatively recent primate adaptation, associated with colonization of habitats that are either relatively rich in this resource or poor in some of the other primate staples such as fruits, soft plant parts, and insects.

I. Human crops. The exploitation of this resource (where it is available), is much more widespread than expected, but is still limited to the Old World monkeys (52% of all diets [of monkey species' diets]), the great apes (75%), and to one prosimian family, the Lorisidae (11%).

A. Animal foods [other than insects]. More than one-third (37%) of all primate diets contain vertebrate or crustacean flesh or eggs.

H. Tree parts. Consumption of barks, gums and rotted wood varies between 50% (prosimians) and 7% (New World monkeys) of all diets.

G. Grasses and roots. Only Old World monkeys and apes appear to make any use of this kind of food, and less than one third of the diets in each of these two groups include grass or roots. ......

Accidentally eaten invertebrates, such as the larvae that often infest the fruit of Ficus [figs], are never included in estimates of insect consumption, and yet under some circumstances they may contribute substantially [particularly vitamin B12] to the diet. Primates, then, can only be described as omnivores; they are definitely not vegetarian animals.
[An important point missed here is that the primates evolutionarily closest to humans definitely can be described as *very predominantly vegetarian* omnivores, which is much the point, because without that additional information, the term 'omnivore' is almost meaningless as it conveys very little information about just where in the wide spectrum of omnivory an animal is located.]
I have not attempted to assess quantitatively the relative importance of various items in each primate's diet.
[And therefore, as just noted, that important information is lacking here]

[When Harding says:
"… the diet of almost every living wild primate is characterized by great diversity. There is no reason to suppose that the diet of the common ancestor of humans and apes differed in any major way. Substantial amounts of animal food must already have been present in the diet of our prehominid ancestors."
He fails to note that the primates most closely evolutionarily related to humans, the great apes, are much more accurately characterized as being *very predominantly vegetarian* – chimpanzees, even those in environments of dubious closeness to those they were in before human contact and influence, eat less than 5% of calories as meat, and many individuals of any one group eat much less than even that; whereas bonobos, gorillas and orangutans all eat much less than 5% of calories as meat.
The real biological significance of animal food consumption is that of insects, and it relates much more to vitamin B12 than it does to protein or probably any other nutrient, which Harding apparently fails to understand here.]

[When Harding says:
"There is every reason to suppose that this latter behavioral trait [predation on small vertebrates i.e. lizards] was widespread within the primate order well before the Pliocene, when the first of the hominids developed. If meat-eating is not a radical innovation of our earliest ancestors, then hunting cannot be the major adaptation which first set them on the pathway to modern humanity. Thus, the extreme theories that attribute the many forms of present-day human aggression, including war and murder, to this hypothesized innovation (Ardrey, 1961; Lorenz, 1966) can no longer be sustained."
He fails to note that large mammals look a lot more like humans than any size of reptiles, let alone small ones, and that the emotions of large mammals when they are being killed and otherwise distressed by predation can be much more obvious in their similarities to the emotions of humans – and that the closer the similarity of what a human is killing to a human, the smaller is the psychological step to be taken to arrive at killing a human.
The role of hunting large mammals in human intraspecific violence is probably only partial, though, as another part of the problem is weaponization making a natural level of aggression that previously was much less potentially lethal, much more potentially lethal, and the psychological effect of killing from a distance (enabled by weaponization) greatly facilitating the willingness to kill.]

2.3 Gombe chimpanzee detailed food list; dubious calculation of meat consumption (13 % of calories) without any detailing of denominator group, season (i.e. of preferred food plenty or scarcity), or time trend across the 13 year period covering the four periods sampled; human influence; cooperative hunting; only adult males hunt; strong lack of carrion-eating; power-scavenging of fresh kills; transportation of food and objects; 20-fold increase of severe attacks on other chimpanzees at predatory events; travel preferred across more open ground on trails; terrestrial travel with arboreal feeding; stone thrown as weapon. Teleki (1981)

Teleki G. Chapter 9 The omnivorous diet and eclectic feeding habits of chimpanzees in Gombe National Park, Tanzania. In, Harding RSO, Teleki G (Eds) (1981) Omnivorous primates Gathering and hunting in human evolution. Columbia University Press. New York

"According to reports from several field studies conducted since 1959 in Tanzania, Uganda, Rio Muni, and Senegal, and from a study of captive chimpanzees reintroduced to a natural habitat in Gabon, the omnivorous diet of these apes includes the following broad categories of food: leaves and shoots, buds and blossoms, fruits and berries, grains and seeds, husks and pods, nuts, seeds and grasses, vine stems, barks and resins, lichens, galls and larvae, ants and termites, caterpillars and cocoons, birds and eggs, honey, various mammals, minerals, and water. ……
If we narrow the focus to vertebrate foods, the list still includes 10, possibly 11 other species of primates; 4 species of ungulates; and at least 3, but possibly 4, species of small mammals.
In addition to these, 9 avian species are consumed, but aside from guinea fowl most of these are fledglings collected from nests rather than actively hunted.
Moreover predatory behavior involving vertebrate prey has now been recorded at all major study sites in equatorial Africa, from Uganda and Tanzania to Sierra Leone and Senegal.
I expect that the known geographical distribution of predatory behavior will continue to expand as new chimpanzee projects are launched, though it is probable that some populations practice this behavior little or not at all while others do so regularly and systematically (Teleki, 1975). ……

Performance variations across populations, and probably also across social communities, may result partly from differences in environmental conditions, but behavioral, social, and psychological factors may be equally or even more influential in the differential formation and dissemination of activity patterns of social customs (McGrew and Tutin, 1978; Teleki, 1974, 1975) ……

My guess is that even the most meticulous estimates of food intake among gatherer/hunters have in the past inflated the faunal component of the diet because plant foods, particularly small items such as berries, are less likely to be brought to the camp for distribution. ……

… which indicates that fauna are the more attractive foods for all omnivorous primates. [Very wrong, on the evidence it seems much more probable that hunted animals are in fact mostly a fallback food for apes – chimpanzees, the most predaceous of the apes, usually do not take the opportunity of hunting animals when their more preferred plant foods i.e., fruits are available, and animal food seems generally agreed to comprise less than 5 % of diet by calories, including insects. Also, it is fatally faulty to the addressing of the human diet question not to differentiate apes from other primates…]
There is some indication that predation on mammals decreases during the insect-collecting season (Teleki, 1973a; Wrangham 1975), and so hunting may be partly a

response to shortage of invertebrate foods [and maybe partly also of plant foods] between February and August. ……

Suffice it to note here that when banana provisioning was dramatically reduced at Gombe in the middle of 1969, causing numerous changes in behavior, social structure, and population dynamics (Teleki, Hunt and Pfifferling, 1976; Wrangham, 1974), baboons and chimpanzees were no longer strongly attracted to the provisioning area, and as a result, chimpanzee predation on baboons declined sharply.

This shift was predictable some years ago (Teleki, 1973a), although it was not known that colobus monkeys would become the primary targets in subsequent years, from 1970 to 1974 (Wrangham, 1975; Busse, 1978).
[And therefore it was not known if colobus monkeys were *much* or *as much* eaten *before* banana (a low protein:energy ratio food) provisioning, notwithstanding that data from elsewhere suggests that it was very likely that some colobus eating took place before – *how much* is very much the point.
Note also that human influence on chimpanzee diet was probably substantial before the Gombe studies even started, by way of increased consumption of low protein:energy oil palm nuts from very prevalent oil palms grown in various sizes of plots, farms and plantations by humans, and reduction of population densities of species like leopards and other big cats that naturally prey on chimpanzees.
They have also had the opportunity to witness animals similar to themselves (humans) hunting the animals they now hunt, for a long time.]

One interesting distinction in the predatory habits of Gombe chimpanzees is that they have not been observed to exploit any of the numerous fish, amphibian, or reptile species available in the park (Wrangham, 1977). The fish (Stolothrissa sp.) being dried on the beaches by humans are frequently bypassed, as are skinks (Mabuya sp.) and monitor lizards (Varanus sp.). These resources are not ignored by collector/predator baboons at Gombe and elsewhere (e.g. Ransom, 1971)
[Which is suggestive of predation not being naturally a strong primary drive.]

Gombe chimpanzees do not habitually travel in areas of canopy forest; rather, they are most likely to travel across open grassland or sparse woodland, moving over level or gradually sloping terrain on worn trails.
Also, they are more arboreal than terrestrial in only one activity: feeding on plant resources, especially the fruit and nuts that grow mostly on trees and clinging vines. [sleeping?] Accordingly these apes should not be categorized as arboreal, forest-dwelling, brachiating primates (see Napier and Napier, 1967). ……

Sketchy as this picture of chimpanzee life still is, it lends considerable support to the proposition of Mann (1972) and others that chimpanzees are ecologically, behaviorally, socially, and psychologically best suited to serve as a model for reconstructing hominid origins. No other extant nonhuman primate species has equivalent tendencies and talents to exploit an array of resources (leaves, fruits, vertebrates, etc) in mosaic habitat conditions with complete multidimensional mobility
[Maybe, if it is kept in mind that “best” does not equate to “perfectly”, and can only be taken to be a statement of average suitability overall that does not imply reference to all aspects of importance.]

Wrangham’s more detailed picture (1975) based on 82 of the most reliable predation records in the Gombe Stream Research Centre files, covering four different periods [we need to know the total time to consider the correctness of the calculation!...We also need to be able to consider any trend across the four different periods!...]

of investigation from 1960 to 1973 and excluding all kills made at the provisioning station, implies that there is a kill rate of 33 colobus monkeys, 23 bush pigs, 14 bushbucks, 2 redtail monkeys, 2 blue monkeys, and 1 baboon per year by the 48 chimpanzees exploiting a home range area of about $8km^2$ in the central study area. ……
Basing his calculations on an estimated predation rate of 75 kills per year by a chimpanzee community, and on rough estimates of 3-5kg per prey individual
[This seems an overestimate as Stanford has noted that the most common kill was a baby colobus of 1kg],
Wrangham (1975:4.34) proposes that 321 kg of mammalian meat are consumed by 31 adult chimpanzees
[Note that in species where meat is a truly natural part of the diet regards protein, the young will be fed it preferentially to the adults, who eat a less proteinaceous diet, because they are not growing and need less protein:energy in their food than growing young]
for an average intake of 10kg per chimpanzee per year. ……
[This is 27 grams per chimp per day, or roughly 150-odd kcalories (for lean meat) compared to very roughly 2,000 calories as a daily energy requirement is 13% which is a lot more than the less than 5% of calories as animal food including insects that seems to be commonly agreed upon and given by Stanford, who worked at Gombe…Is the discrepancy due to the use of the number of chimpanzees actually eating the meat rather than the number of chimpanzees in the whole group? This aspect should have been clarified, as the meaning is obscure without it…]

With the exception of a few baboon cases observed at Gilgil in recent years (Strum, this volume), nonhuman primates that actively prey on mammals do not seem highly motivated to eat carrion (Teleki, 1975:159). All experiments conducted with collector/predator baboons and chimpanzees at sites where predatory behavior is frequent – experiments in which various kinds of meat, including carcasses of species normally preyed upon, were offered for examination – have thus far failed to stimulate a meat-eating response (van Lawick-Goodall, 1968). ……
The only behavior in chimpanzees that resembles scavenging is the appropriation of bushbuck carcasses from Gombe baboons (Morris and Goodall, 1977). The meat is always quite fresh, however, because such pirating occurs within seconds or minutes of capture by the baboons.

No less interesting is the fact that 75% of the bushbucks eaten by Gombe chimpanzees were appropriated from baboons (McGrew, 1979). [Power scavenging] ……

The fact remains, however, that Gombe chimpanzees do pursue various prey species, including baboons and colobus monkeys (Teleki, 1973a; van Lawick-Goodall, 1968, 1971), in a coordinated manner, with 2-9 adults positioning and repositioning themselves to maintain an enclosure, sometimes for an hour or more, to effectively anticipate and cut off all potential escape routes of the prey. ……

Moreover, bioenergetic models of predatory behavior, such as the one outlined by Busse (1977, 1978), may fail to uncover solid explanations because they exclude social motivations for acting cooperatively. ……
During 43 hours of meat-eating activity observed during 1968-1969, segments of carcasses, ranging in size from small slivers to entire haunches, were widely distributed among 4-15 participants in predatory episodes, with consumption being directly proportional to attendance numbers, regardless of prey weight (Teleki, 1973a).
Meat is distributed in several ways: by removing pieces (without asking permission) from an owner's portion, by overtly requesting shares from owners, or by simply recovering dropped fragments.

Excluding the third method, which by definition is always successful, Gombe chimpanzees share meat at an average rate of once every 8.6 minutes.
Requests, which involve active solicitation with postures, gestures and vocalizations, occur every 6.5 minutes and are successful every 22.5 minutes (Teleki, 1975).
Voluntary offers of small and sometimes large portions have also been observed.
Sharing of plant foods is perhaps less common and less intense but does occur with some regularity in Gombe national Park (McGrew, 1975, 1979; Wrangham, 1975).
The fact that a 3-kg carcass can be distributed among as many as 15 chimpanzees who spend up to 9 hours sharing and consuming the meat contributed greatly to the premise that predatory behavior among these apes has a social as well as a nutritional basis (Teleki, 1973a:171-73).
[There seems to be much more sharing here than in Stanford's descriptions. Question: hunger/season? Question: sharing as function of taste preference – inversely related?]

Other amicable interactions, ranging from copulation to grooming, are also common at predatory episodes involving many chimpanzees, so the overall [including sharing, apparently among select groups] rate of interaction is even higher than these figures indicate.

… the frequency of aggressive attacks is generally low among Gombe chimpanzees, averaging 0.2 attacks per day, and the number of attacks does not rise with increase of group size.
In this light, competition is certainly greater at predatory events, where 4-5 severe attacks occur per observation day (12 hours) according to my estimates.
[Increased hunger?]

Predation on mammals is, in effect, an adult male domain, with males performing most pursuing, capturing and consuming activities, and with control males presiding over meat distribution (Teleki 1973a). ……

Thus seniority (i.e. experience?) as well as sexual attractiveness and kinship ties are important prerequisites in obtaining meat at predatory episodes…….

In complete agreement with McGrew (1979), I suggest that these sex differences are an incipient form of labor division that evolved, in the hominid line (perhaps in tandem with a shift towards hunting larger game) towards the more rigidly patterned forms of labor division observable today among gatherer/hunter band societies (Galdikas and Teleki, 1981).

And on 26 October 1972, Frans Plooij (1978) observed an adult male chimpanzee throw a large stone, measuring 25 x 10 x 10 cm, at an adult bush pig during a cooperative predation on a group of pigs. The stone hit but did not visibly injure the pig, yet it retreated sufficiently from its protective stance over a small piglet for another chimpanzee to make the capture.

… yet it is clear that chimpanzees can prey efficiently upon mammals without recourse to tools; this in turn implies that early hominids may not have applied technological skills in the hunting context until big-game hunting tactics emerged over [only] a million years ago

… technological skills may well have been applied initially in plant- and invertebrate-collecting activities.
The grasses, vines, and sticks used by chimpanzees for so many purposes are excellent prototypes for the digging sticks carried by many gatherer/hunters. These simplest of

implements are useful in exploiting various subsurface fauna and flora (see also Mann, 1972; Coursey, 1973)

Wild chimpanzees carry many kinds of objects, especially prized food items, in a variety of situations (Teleki, 1975; McGrew 1979), and can do so while travelling quadrupedally, tripedally, or bipedally (Teleki, 1977a, 1978). That chimpanzee carrying behavior is less efficient than human methods of transporting objects while striding bipedally (Hewes, 1961) cannot be denied. But that discrepancy is not enough to justify the claim that nonhuman primates "neither share nor carry" (Lancaster, 1978:86), for chimpanzees, baboons, and other non-human primates are known to carry objects from one place to another (e.g. Bauer, 1977; Rose, 1976, 1977).
[However, this behavior may be/ have been copied from humans, as there have very probably been long periods of observation by apes of humans carrying things – and the analogous possibility applies to hunting....] ......

Transportation of food and other objects may therefore have developed largely in conjunction with the advent of temporary group settlement at favorable sites (see Speth and Davis, 1976).
A strong impetus toward labor division may also have appeared in connection with these developments.
But neither object transportation nor labor division was developed by early hominids; both must have been inherited, in rudimentary if not complex forms, from common ancestors of apes and humans.
[This cannot properly be stated to be an absolute fact - notwithstanding that chimpanzees may be more closely related to humans than the other three apes, the behavior in this regard of the other three great apes must be examined.]

But all that seems certain about human origins is that they always remain enigmatic, no matter how much we pressure and posture in favor of one scenario or another. ......

Human behaviors are often attributed to early hominid innovation simply because we lack or ignore first-hand knowledge of what other primates actually do when left to their own devices in their native habitats. ......

By way of conclusion, I submit the following general points for consideration.
1. Despite the availability of an array of niches within a mosaic habitat that encompasses grassland, woodland, and forest zones, Gombe chimpanzees travel mainly through the flatter, more open areas on a network of permanent trails. Cross-country travel is rare, and dense vegetation is often avoided during movement through the home range. Gombe chimpanzees are clearly adapted, both morphologically and behaviorally, to utilize different vegetation and terrain types; I find it difficult to conceive how such a generalized primate would be deterred from quickly colonizing a transitional, fringe woodland habitat or a savannah-woodland home comparable to that which supposedly launched early hominids in a unique evolutionary direction. ......
4. Gombe chimpanzees possess a variety of behavioral, social, and psychological attributes that fulfill many of the conditions used in previous years to define human uniqueness. Cooperative hunting, food sharing, and incipient labor division are merely three of the more obvious instances."

2.4 Omnivory not a very informative term regards position in that continuum; digestibility of food types. Chivers, Hladik (1984)

Chivers DJ, Hladik CM (1984) Chapter 9 Diet and gut morphology in primates. In, Chivers DJ, Wood BA, Bilsborough A (1984) Food acquisition and processing in primates. Plenum Press, New York.

"Although these two dietary extremes ["faunivory", eating animals, and "folivory", eating leaves] have a special significance, we are clearly dealing with a biological continuum, which is not clarified by lumping all intermediate forms in the category of "omnivore".
The very nature of the specializations for faunivory mean that no species can feed on a balanced mixture of animal and leaf matter.
There is in fact a third type of food – fruit – which provides a focus for intermediate dietary adaptations, of which primates provide an excellent example.
These reproductive parts of the primary production, including flowers and immature fruit, seeds, and even tubers, contrast structurally and biochemically with the vegetative parts of the primary production, including barks and gums as well as leaves (mature and young), shoots and stems, and even more with invertebrate or vertebrate matter from the secondary or tertiary level of the production pyramid.

As pointed out by Ripley (1979), animal foods are the easiest to digest but the most difficult to collect, whereas "leaves" are the easiest to collect but the most difficult to digest, with fruit intermediate in both respects.
[This is probably incorrect insofar as at least most fruits consumed by modern man, and many wild fruits, are quite easy to digest, in particular probably requiring nowhere near as much stomach digestive juice acidity as animal foods do].

[Chivers and Hladik note that it is relatively easy with wild primates to record accurately the details of feeding bouts in terms of time, but much more difficult to estimate the amounts ingested in terms of fresh (or dry) weight over a reasonable length of time, and to analyse food composition.]

Regressing data for mature wild individuals [of various types, not only primates] separately according to diet reveals that only frugivores follow the expectation of absorptive area being proportional to metabolic body size; faunivores seem to have relatively less small intestine when larger, whereas large folivores, especially those with large stomachs, have relatively larger small intestines ……

No mammal mixes large quantities of both animal matter and leaves in its diet without including fruit.
Since faunivory and folivory seem to represent contrasting and incompatible adaptations, the quantity of fruit in such a mixed diet is always considerable. ……

Thus
[by regressing log {surface area of small intestine + one half that of (stomach + caecum + colon)} on log {body length}, adjusted (differentially for folivores, frugivores and faunivores) for body size]
it is possible to devise a scale of negative and positive values to reflect the direction and degree of adaptation to faunivory and folivory respectively.
Most primates fall within these limits, but the concentration of frugivorous species closer to the upper line reflects the greater morphological similarity to folivores than to faunivores."
[Unfortunately Chivers and Hladik made no attempt to relate these results to humans.]

2.5 Legumes as important diet mainstays in many primate species in varied climates; faulty statement of broadness of omnivory due to non-differentiation of apes from other larger primates; failure to properly address frugivore diet seeds, nuts, legumes and grains; brain size and sociality; failure to address need for large brain in memorizing seasonal timing and spatial location of fruits, leaves, seeds, nuts, legumes and grains, and in tool-using processing of some of these; failure to note that nuts, oily seeds, legumes and grains are *more* energy-dense than flesh.

Ripley S (1984) Chapter 2 Environmental grain, niche diversification and feeding behavior in primates. In, Chivers DJ, Wood BA, Bilsborough A (1984) Food acquisition and processing in primates. Plenum Press, New York.

"Insofar as frugivores must look beyond fruits (their energy source) to animals or leaves for protein, they are generalized feeders, sometimes called "omnivores", albeit inaccurately (Chivers and Hladik, 1980). ......
[Ripley fails to properly address the role in the various frugivore diets of seeds, nuts, legumes and grains.]

... primates are the broadest of generalists ... .
[As per my comment for Codron just above: This statement is only correct in the context of a very (somewhat over-) generalized statement made for *all* primates as a group – it would not be correct, and would become obvious as an overgeneralization, if only the *apes* are considered rather than looking at *all* primates – because there is much more difference among the species of primates as a whole group than there is among just the species of the subset of apes (the subset most relevant to humans, on average) – *none* of the *apes* are "the broadest of generalists"; they are all predominantly frugivorous very predominantly vegetarian omnivores...]

Most significant for our concerns is Hladik and Chivers' (1978) demonstration that intermediate-density "frugivores" must select their protein either as animals or leaves, ...
[Again Ripley fails to properly address the role in the various frugivore diets of seeds, nuts, legumes and grains.] ......

In mammals, successful reproductions of these social organizations are certain to involve large brains capable of assimilating diverse programs (MacLean, 1978). Such a brain seems likely to have evolved in a species whose diet is very broad (Ripley, 1970), including, but not specializing on colonial invertebrates [insects that live in colonies, i.e. ants and termites] and vertebrate prey. Both such resources are known to be made more accessible to primates by the use of tools or weapons and by cooperation (Goffman, 1969), including polyspecific associations [i.e. use of hunting dogs], capabilities facilitated by the possession of large brains.
[Ripley fails to address the need for a large brain in the keeping in memory of the seasonal timing and spatial location of not only fruits and leaves, but also of seeds, nuts, legumes and grains (exemplified by the *extremely* solitary orangutan, which is savagely antisocial other than when mating), and in the processing of some of the latter by the use of hammer and anvil stones and (probably) primitive natural grinding stones.]

Complexity and flexibility in social organization can compensate frugivores for the competitive disadvantages of moderate biomass density and intermediate body size. This is especially so if the aggression and protection of small groups can be enhanced by "weapons."

Faunivores are prevented [from evolving social enhancement of competition in the maximizing of population numbers] by spending much time searching for food, solitary

foraging and low biomass densities, with the exception of the special social carnivores in particularly productive habitats. ……
In contrast, long-lived vertebrate generalists must individually acquire learned traditions appropriate to local conditions and seasons. These traditions are stored in, retrieved from and transferred intergenerationally between large, complex brains housed in genetic and social competitors.

Dentally-specialized sciurid rodents have captured the rich but hard seed-nut niches, … [Note: again no mention of ape and human use of these probably very significant resources.] ……

Generalist primate frugivores can further increase their body size by broadening the diet to include sources of animal food in compact large units: middle- and top-level carnivory. [Ripley fails to properly address the role in the various frugivore diets of nuts, oily seeds, legumes and grains – most of which are *more* dense sources of energy than flesh, because all flesh is approximately 70-odd% water, and also sources of more than enough protein alongside the carbohydrate and fat.
It should also be noted that gorillas eat less animal-food than other apes and are far and away the biggest and strongest, notwithstanding that they are also the best adapted to digesting cellulose.
Also, it should be noted that as excess protein cannot be stored as such for later use, and that as (in humans, and probably in other large primates) consumption of protein in excess of 50% of energy intake is metabolically toxic from un-metabolizable ammonia waste from metabolism of amino acids surplus to tissue rebuilding/ enzyme replacement requirements, and that as human breast milk is maybe the lowest known among mammals in protein:energy ratio (at only a fraction of the ratio of other mammals for which data is available), indicating fairly clearly that at least humans (and probably other large primates) are by evolutionary nature an animal suited to a low-protein diet, smaller prey closer in series would be better than more occasional large prey.]

Chimpanzees achieved denser biomass, and thus larger body size and increased sociality by adding hard-shelled nuts to this [termites] rich food source (Hladik, 1975: 29; Rahm, 1971), an invasion of the squirrel niche that rudimentary tool use also renders accessible to a frugivore …
Larger body size in turn permitted a reduction of metabolic rate, and thus increased longevity, longer gestation and larger brains; which in turn reduced reproductive output and intensified parenting, play and learning. These trends together made it possible to evade predators better, outwit competitors, recall seasonally-appropriate foraging tactics, extend and remember kin relationships, and communicate with occasional acquaintances; the denser biomass facilitated the expression of complex sociality and the other unique talents of chimpanzees.
[So why fail to consider the use of hard-shelled nuts (and oily seeds, other seeds, cereal grains, and starchy root vegetables), in the case of humans?]

Numerous recent primate studies (Clutton-Brock, 1977) show again and again the predominance of one plant family, the Leguminosae, in diets. … This co-evolutionary pact with fertilizing (nitrogen-fixing) micro-organisms enables members of this extremely successful family (Stebbins, 1974) to do as well in the humid tropics (Klinge et al., 1975) as in arid seasonal environments (Givnish, 1978). Thus, a dietary focus on the Leguminosae by many primate lineages aided the order's exploitation of both extremes. This plant family has also provided humans with their major source of plant protein, their abundance under cultivation lifting human biomass densities well above the restrictions of their former hunter-gatherer [gatherer-hunter, more often, really] base."

2.6 Lowland gorillas' and chimpanzees' diet of preference is mostly fruit; in seasons of fruit scarcity gorillas eat more non-fruit plant parts whereas chimpanzees eat more meat; highland gorillas eat 80% non-fruit plant parts, highland chimpanzees eat 40% non-fruit foods. Yamagiwa, Maruhashi, Yumoto, Mwanza (1996)

Yamagiwa J, Maruhashi T, Yumoto T, Mwanza N. Chapter 7 Dietary and ranging overlap in sympatric gorillas and chimpanzees in Kahuzi-Biega National Park, Zaire. In, McGrew WC, Marchant LF, Nishida T (1996) (Eds) Great Ape Societies. Cambridge University Press

"However, the sympatric populations of gorillas and chimpanzees in lowland forests show extensive dietary overlap. ...... Fruit parts accounted for 55% of gorilla and 76% of chimpanzee plant food items, and the overlap of fruit species between them was greater than that of other food classes (79% for gorillas and 82% for chimpanzees). These results suggest that both species of apes strongly prefer fruits in general and tend to have common dietary features. ......

In the montane [mountain] forests of Kahuzi, ... Fruit parts accounted for only 16% of gorilla and 38% of chimpanzee plant food items. ......

Tutin and Fernandez (1993) reported that the diets of gorillas and chimpanzees showed greatest divergence at Lope Reserve (lowland Gabon) during the dry season when fruit was scarce.
In contrast, both species of ape consumed the most kinds [not the same thing as amounts, note] of fruit during the long dry season in the Kahuzi region.

This study showed that the chimpanzees enlarge their daily food breadths and consumed various animal foods to complement their nutritional requirements under the conditions of fruit scarcity.

Chimpanzees remained in the wooded areas, but their frugivorous and omnivorous strategy has enabled them to enlarge their range into arid areas (Nishida 1976)."
[This reference has a title referring only to bark eating, which is something gorillas are known to do also...]

2.7 Borneo orangutans' diet complex, and mostly fruit other than in some months; no meat-eating recorded. Galdikas (1988).

Galdikas (1988) Orangutan diet, range, and activity at Tanjung Puting, Central Borneo. Int J Primatology 9 (1) 1-35.

"Previously, wild orangutan feeding and ranging behaviors have been described only from populations in hilly or mountainous regions.
The Tanjung Puting study focuses on an orangutan population in a swampy lowland area near sea level.
Tanjung Puting also differs from other areas in the virtual absence of large figs, which are significant orangutan food sources elsewhere.
During a 4-year period and 6804 hr of observation, focal orangutans were recorded in 11,338 foraging bouts accounting for 3805 hr.
The orangutans were predominantly frugivorous, with fruit-eating accounting for 61% of the foraging time.
However, the overall variety in their diet was remarkable; 317 different food types have been identified, including fungus, insects, and honey.

Orang-utans were strongly opportunistic foragers, with the composition of their diet varying markedly from month to month.
During most months orangutans fed on a complex mix of fruit, leaves, bark, insects, and small vines.
During some months fruit was not the major component of the diet."

2.8 Chimpanzee meat-eating mostly a fallback in seasons of scarcity of preferred plant foods in more seasonal habitats; gorilla and orangutan meat-eating approximately nil; initial disgust and eventual habituation of captured apes to meat as a food; recalcitrance and illogicality of paleoanthropologists in failing to properly extrapolate these facts into theories of ancestral hominin diet. Kortlandt (1984)

Kortlandt A (1984) Chapter 5 Habitat richness, foraging range and diet in chimpanzees and some other primates. In, Chivers DJ, Wood BA, Bilsborough A (1984) Food acquisition and processing in primates. Plenum Press, New York.

"... chimpanzees ... Others are too small, e.g. most grass seeds, or too time-consuming to collect." ......
[This may well be incorrect, as gelada baboons' dietary focus is very much on grasses including seeds. The reality may well be that the time spent on the ground collecting or eating them is *dangerously* long, and that unlike nuts they are less readily carried to a safer processing or eating place (e.g. at the foot of a rapidly-climbable tree) without a container of some sort – the geladas are in a fairly treeless environment, but their safety is found over the edge of the cliffs kept near to their foraging and eating area.]

However, chimpanzees [unlike the !Kung] do not feed on roots, bulbs, truffles, gums, palm hearts and mushrooms as a rule. ......

Gorillas have never been observed to eat honey, eggs, insects or meat, not even when they were sitting or nesting almost on top of a honeycomb or a bird's nest, except for one single case of honey-eating reported by Sabater-Pi (1960), and except for the "slugs, larvae and worms" which constituted 1% of the food item observations recorded by Fossey (1974).
No animal remains have been found in gorilla dung, except for one case presumably indicating cannibalism (Fossey, 1981).
[This is incorrect regarding insects – their remains have been found in gorilla dung.]

Orang-utans eat honey, insects and, occasionally, bird's eggs, but no vertebrates (ref. above).
[This seems to be incorrect - they have been reported as rarely eating squirrels on the basis of just one or a very few observations.]

The Old World monkeys, except perhaps for some of the specialized leaf-eaters, commonly eat insects and probably also bird's eggs, but both mostly in small quantities. Predation on vertebrates (other than their eggs) has been reported for all well-studied genera of Cercopithecoidea, except for the gelada baboon and the specialized leaf-eaters (e.g. Clutton-Brock, 1977; Gaulin and Konner, 1977; Harding 1981; Kortlandt and Kooij, 1963; Montgomery, 1978; Napier and Napier, 1967).
However, such predation is almost always a rare phenomenon among the Old World monkeys. Many experienced field workers have never seen it, others only a few times. Still others report that it is a seasonal and/or local phenomenon, or is caused by prolonged drought and food scarcity, etc. (Kortlandt and Kooij, 1963).

Under special ecological conditions, however, large-scale predation on relatively large prey has been observed to develop among vervet monkeys and Papio baboons (Galat and Galat-Luong, 1977; Strum, 1981; see also Skinner and Skinner, 1974).
This behaviour suggests that such behaviour is, so to speak, primarily an emergency measure for special needs.

Chimpanzees, on the other hand, spend a remarkable amount of time, mental effort and tool use on searching out insects and feeding on them in every place in which they have been extensively studied. Hladik and Viroben (1974) have shown that this insect food is nutritionally important in order to compensate for a deficiency of certain amino acids in the plant foods
[This is almost certainly not the case – relative to needs, amino acids are in abundance.], even in the rich environment of the Gabon rain-forest. (For comparison: the Aka pygmies eat more than 40 insect species, even though they avail of plenty of meat food [so obviously amino acids are not in shortage here…]; Bahuchet, 1978.)
Another indication of this issue is found in the different frequencies of egg and meat eating in different habitats.

In the rain-forest habitat of Beni, for instance, I set up a dummy bird's nest containing hen's eggs along the chimpanzee trail. In nine tests, when groups comprising on average 9.4 apes passed by a total of 16 times and noticed the eggs, only six times did an individual take away and (probably) eat the eggs, and these eggs were presumably eaten by only three individuals altogether (Kortlandt, 1964b).
In the poorer environment of Gombe, on the contrary, an estimated 60% of the chimpanzees ate the hen's eggs at the camp site (J. Goodall, pers. comm.).
[Note: ideally the experiment should have also offered plant foods *in the same way*, although perhaps this has already been done to an extent considered exhaustive and so well known as to require no further addressing – it did need to be addressed by Kortlandt here in some way though.]

Similarly, predation by chimpanzees on vertebrates (other than their eggs) is undoubtedly a rather rare phenomenon among rain-forest-dwelling populations, as is indicated by:
(1) the absence (or virtual absence) of hair, feathers or scales in the stomachs and intestines of hundreds of chimpanzees which, in past times, have been shot, dissected, investigated and described by dozens of museum collectors, anatomists and parasitologists (e.g. Reichenow, 1920; Vandebroek, pers. comm..),
(2) the rarity of parasites indicating carnivorous habit (Kuntz and Myers, 1969; Myers and Kuntz, 1972; Rahm, 1966),
(3) the rarity of pertinent field observations (Badrian et al., 1981; Hladik, 1973; Sugiyama, 1981; Suzuki, 1971), and
(4) the responses when I placed live as well as dead potential prey animals along the chimpanzee paths at Beni (Kortlandt, 1967b; Kortlandt and Bresser, n.d., Kortlandt and Trevor, n.d.).

In the poorer mosaic and savanna landscapes of Tanzania and Senegal, on the other hand, predation on vertebrates appears to be a much more common phenomenon, even considering the easier observation conditions (for reviews and viewpoints, see Busse, 1977; Kawanaka, 1982; Kortlandt, 1972; van Lawick-Goodall, 1968; McGrew et al., 1978; Nishida et al., 1979; Plooij, 1978; Ransom, 1981; Rijksen, 1978b; Suzuki, 1975; Teleki, 1973, 1975).

In conclusion, the wealth of flora and insect fauna in the rain-forest provides both chimpanzees and orangutans [Note: why not mention also gorillas?] with a dietary spectrum that is apparently wide enough to satisfy their normal requirements of various

proteins, vitamins and minerals, so that they do not, as a rule, need to have recourse to the hunting, killing and eating of vertebrates.
In the poorer environments of mosaic landscapes, woodlands and savannas on the other hand, even a flora of several hundred species and the accompanying insect fauna appear to be nutritionally insufficient.
[Note: it would be more accurate to have said that the total energy, and maybe but probably not the protein, content was insufficient.]
Carnivorous behaviour then becomes inevitable, even when the quantities of meat obtained are minimal and perhaps insufficient to meet the basic needs.

The same conclusion applies, of course, to the hominids, i.e. not only our contemporary conspecifics, but also all australopithecines, i.e. including Paranthropus and Zinjanthropus.
[Note: this may be very plausible, but the same amount of evidence is not available in the case of the australopithecines – of course, neither is it available to support the extensive carnivory claimed for them by many…..]
It is strange that most paleoanthropologists have never been willing to accept the elementary facts on this matter that have emerged from both nutritional science and primate research.
[Note: not so strange in the light of understanding the psychological tendency of humans to rationalize their own behaviour, in this case slaughter, butchery and carnivory.]

[Note: It also needs to be considered if (close, as at some camps where chimpanzees are living together or interacting closely with) contact with humans has increased the amount of carnivorous behaviour in chimpanzees, who probably at least see and smell raw meat that they are able to identify as such, and observe it to be consumed, even if after cooking.] ……

… Another example of lack of interest is, perhaps, even more striking. In 1920 the then renowned German zoologist Reichenow published an interesting paper on two ciliates that live mainly attached to the mucous membrane of the caecum and large intestine of, respectively, the chimpanzee and the gorilla: Troglodytella abrassarti and T. gorilla (Reichenow, 1920 [Note: in German.]). He described that these infusoria ingest and digest chiefly cellulose and some protozoa, reproduce quite rapidly and are largely digested by the host while they are still in the intestine.
[Note: digestion seems unlikely in the *large* intestine, unless it occurs associated with fermentation.]
He also noticed that Troglodytella occurred abundantly in the intestines of both shot and the just-captured wild chimpanzees and gorillas, as well as in their faeces, but disappeared soon after capture. This disappearance was followed by intestinal disorder [Note: some later authors state the Troglodytella to be the cause of intestinal disorder, though the detail in the reports would seem to have the error in attribution with these latter reporters, unless Reichenow just invented the detail of temporal sequence abundance > decrease > disorder…].

Furthermore, newly captured apes showed "disgust" (Abscheu) when offered meat
[Note: need to know what the response was to plant foods offered in the same way at the same time.],
but after habituation they learned to enjoy large quantities of it.

From all this Reichenow inferred that the symbiotic function of these ciliates probably is to convert vegetable food into what he called "animal food for a vegetarian that requires nutrients of animal origin.". This function would be similar to the digestive processes performed by other ciliates in the stomachs and intestines of many hoofed animals.

[Note: cobalt is found in green leaf matter more than in other plant parts, and so the synthesis of vitamin B12 is a possibility, notwithstanding that generally in ungulate herbivores this function is generally attributed to bacteria not eukaryote microbes.] ……

… rain-forest-dwelling chimpanzees form larger communities than their woodland-, savanna- and mosaic-dwelling conspecifics. Again, this is contrary to the general rule when we compare forest-dwelling with open-land-dwelling mammal communities … chimpanzee communities normally split up into small searching parties that wander around all over their range, survey the available sources of food, and call loudly when they have found something highly rewarding. On the one hand, they have little to fear from their only natural enemy, the leopard,
[Note: *not* the only enemy when on the ground in that environment.]
and consequently they can afford to travel in quite small parties.
[Note: maybe small parties can get to and up the scarcer trees in response to approach or attack by predators more successfully]
On the other hand, the vastness of the area of their resources favours cooperative exploitation [Note: and predator location?] by means of communication.

Broadly speaking, one may distinguish two types of scientists: the data-collectors and the causalists. The first know a lot and systematize their knowledge very well as a rule, but tend to explain very little, while the second always try to explain, but occasionally [only occasionally?...] appear to [actually do?...] forget those facts that do not fit their theories."

2.9 Exaggeration and sensationalization of Gombe chimpanzee carnivory. Stanford (1999)

Stanford CB (1999) The hunting apes. Meat eating and the origins of human behavior. 1999 Princeton University Press.

Chapter 1 The indelible stamp. (p3-14)

[Stanford begins with a description of one day's foraging by some chimpanzees, including the eating of colobus monkeys, by calling them (p3) "The hunters …" despite his later noting, consistent with other writers, that chimpanzees do not set out to hunt, but rather set out to gather fruits and leaves (and nuts and seeds, which foods he never addresses in the book), and only pursue prey species that they come across incidentally, and then only sometimes, and then mostly in seasons of scarcity of the foods they actively seek. In so doing he has started his book with a misrepresentation.

Questions about the chimpanzee monkey-eating incident that need to be answered to provide the correct context to determine the probable meaning of it include:
In what season did the incident occur (Gombe Park, in which the incident occurred, has more- and less-plentiful seasons regards chimpanzee plant foods)?
How hungry were the chimpanzees (in poor seasons juvenile growth stops…)?
How often do such incidents occur (and of course relate this to season)?
What percentage of dietary intake does the eating of vertebrates make up?
Had the chimpanzees been induced to increase their meat-eating through interactions with the Gombe research station's primatologists and support staff in which they would have been able to see and smell that meat was at least being handled, and probably able to know that it was being eaten (There was a banana feeding station at the research station, which the chimpanzees made very regular use of to get and eat bananas, as well as moving at will through the research camp, other than being physically barred from entering some buildings at some times)?

Had observation of the increasingly present human hunters' and poachers (who look more like chimpanzees than other perhaps more natural predators like leopards do) hunting activities induced any increase in chimpanzee hunting behavior at any time up to the present (Gombe is a relatively small park with humans all around it)?
Had the eating of freely or plentifully available bananas at the banana feeding station at the research station caused the chimpanzees' diet to have become lower in protein derived from plant sources than it had been before (almost certainly it did, because bananas are lower in protein:energy ratio than the average of the chimpanzee plant food diet)?

Chapter 2 Man the hunter and other stories. (p15-51)

Stanford repeatedly exaggerates, usually using sensationalizing hyperbole like "voraciously", "crave", "relish", "love", "avidly", "highly desired", "devote hours", without actually quantifying, the amount of meat that chimpanzees eat (< 5% of calories, Leonard et al 2007)
Given that, as he says himself numerous times in his book here, they don't actively seek it, and quite often don't take it when they find it, even when placed trailside experimentally, and that other primatologists are in very consistent agreement that they don't, this is quite incorrect, or, if I were to match hyperbole with hyperbole, sheer nonsense.
Stanford fails here to understand that, accordingly, their taste preferences have probably evolved for those things that they are able to access in an energetically efficient manner, the end result likely being that chimpanzees probably do not find (raw, note…) meat as palatable as plant foods, such that in the present, it is taste preference that directs the behaviour, not abstract considerations of energetics – and of course, the hungrier the animal, such as in seasons of shortage of preferred food, the more acceptable almost anything edible will eventually become.

Stanford's correlation of meat eating, tool use, and large relative brain size in *selected* primates is plainly *very* faulty because he has not considered first the group of primates that are most closely related to humans, the apes – among which, chimpanzees eat some (only sometimes, nutritionally speaking) substantial amount of meat, whereas gorillas, orangutans, bonobos, and gibbons do not. His correlation of brain size and cooperative hunting in nonprimate animals is likewise faulty due to selection of only those examples that are consistent with his theorizing – examples that are not consistent with it are manifold and likewise relatively obvious, for example, piranhas, ants, elephants, sharks and blue whales.

Stanford's dealing with stone tools by our early hominin ancestors addresses no possibilities for their use outside of butchering carcasses (though he notes correctly that the implements of gathering made and used by women might not remain archeologically).
He fails to note that the tools referred to are by no means known to be used only or even primarily for dealing with carcasses; that the dealing with carcasses is by no means known to be more for the acquisition of meat to eat than for the acquisition of hides and sinews for making clothing and bags from; and that natural stone tools such as hammer and anvil stones used in nut cracking, and mortar (shallower depression) and pestle stones used in breaking and grinding of seeds and grains, and any rock depressions used for soaking seeds are not visible in the fossil record.
And his statement that our stone tool-using capabilities evolved at the late date of 2.5 million years ago is very probably very wrong, given that modern chimpanzees are known to use naturally formed hammer (and anvil) stones to crack open nuts, as well as naturally formed mortar and pestle stones/logs for grinding seeds, and that capuchin monkeys have been reported to use mortar and pestles to break up tubers.

Stanford, in stating that "… in the course of primate evolution, the size of the brain has coevolved in inverse relation to the size of the digestive tract.", fails to differentiate the large from the small intestine – it is in fact said that humans have more small intestine, and less large intestine, than other apes, in the direction of true carnivores. And then, in theorizing that "This would partially explain why fruit-eating and meat-eating primates have shorter guts; these foods are more easily digested, allowing more energy to be devoted to brain size.", he fails to consider the role of legumes, nuts, seeds, and starchy root vegetables (tubers), enough of which are macronutrient-dense foods requiring less digestion than leaves (when leaf cellulose is to be digested to provide energy, which is a different requirement than that of extracting protein, minerals and vitamins from leaves), and that an increasing dietary proportion of any of these could very plausibly bring the same alterations in gut proportions as that brought by increasing dietary proportions of fruit or meat.

Chapter 3 Ape Nature. (p52-102)

Stanford notes that "Chimpanzees are alone among the four species of great apes in being frequent predators. We have no reason to believe that the digestive systems of the other apes could not process meat protein. Yet it is not consumed by gorillas at all, and by orangutans and bonobos only rarely."
Furthermore, that there is a great diversity of chimpanzee behavior in different regions of Africa regards various styles of tool use, hunting, and social behavior, and that these appear to be learned traditions similar to those passed on in traditional human societies. And that among higher primates, in which there are cultural traditions and a strong learned component to being a successful hunter, prey may not even be recognized as food items unless the hunters have been taught to do so.

Stanford notes that the majority of hunting by chimpanzees is by males.
Also, that within a chimpanzee community, males are dominant to females and at times behave quite brutally in their subordination of them. However, females actively choose their mates by inciting male competition when they are sexually receptive (advertised by the presence of fluid-filled swellings on their genital regions), and with their infants form the nuclear units of the social system. The females forage independently from males to minimize competition for food and socialize with the males mainly when it suits their own reproductive interests.
He notes that at Gombe National Park, male chimpanzees who have meat will sometimes offer bits of meat to swollen females, in exchange for which they receive matings. (However a recent work re-analyzing the data suggested that no immediate increase in matings received by particular males from the particular females was evident – this does not preclude later receipt of sexual favors though, which seems plausible enough…)

He notes that Manson and Wrangham (1991) consider this issue in comparison to the kinds of intergroup relations that most traditional human societies have. They considered females, whether human or ape, to be valuable alienable resources for which males should and do compete intensely. This competition places male cooperation at a premium, since each male benefits by being part of a strong alliance. They conclude that this set of benefits leads to males (and men) tending to remain in their natal area and obtaining females from other neighboring groups. Since males who remain in the home group tend to be related to one another, the incentive to compete rather than cooperate may be lessened. As Manson and Wrangham's paper was titled Intergroup aggression in chimpanzees and humans, it seems doubly unfortunate that Stanford does not here, nor elsewhere in the book, address defense issues.]

2.10 Better detailing of the Gombe chimpanzee situation and less than 5% of diet calories as meat. Stanford (1998)

Stanford CB (1998) Chimpanzee and red colobus The ecology of predator and prey. Harvard University Press.

"(p26) Gombe, like the rest of western Tanzania, has a highly seasonal climate, with two main rainy periods between November and May and a prolonged dry season the rest of the year. The forest's appearance changes dramatically between the wet and dry months. …… The behavior and ecology of both chimpanzees and red colobus also change from the wet to the dry months, as we shall see. ……

(p27) For a variety of reasons, Gombe National Park is not rich in large-animal species. The forested area is small and surrounded on the north, south and east by cultivated land, and on the west by Lake Tanganyika. About 240 bird species are present (Stanford 1994; Stanford and Msuya 1995), but many large animals are not. Ungulates such as buffalo, waterbuck, and roan antelope occurred in the recent past but owing to poaching are now locally extinct (Goodall 1986).
There are, however, numerous predators other than the chimpanzees themselves [Table 2.1 lists leopard, crowned eagle, martial eagle, olive baboon and African python]; leopards are seen occasionally, and at least one chimpanzee death in recent years was attributed to a leopard kill.
The leopard population must be very small given the size of the park, though large cat tracks and dung that I have seen on the rift escarpment suggest that the cats may hunt in both the forest and the farm area outside the park.

(p29) The question of human influence.
In the early days of research at Gombe, bananas were used to provision the then-unapproachable chimpanzees and hasten the process of habituation to human observers. Most of the observations were made in a cleared feeding station, where the chimpanzees gathered each day to wait for – and sometimes fight over – the banana ration. Artificially heightened levels of intragroup aggression resulted (Goodall 1986), and in the late 1960s the provisioning was curtailed.
Some scholars have questioned whether the behavior of chimpanzees at Gombe and also in Mahale National Park, which was also provisioned in the early years of research, may be different from naturalistic patterns of chimpanzee behavior. Power (1991) has even suggested that our view of chimpanzees living in societies driven by dominance among males, intense territoriality, and lethal intercommunity aggression may be an artifact of the provisioning in Gombe and Mahale. That claim was refuted by the behavior of other, nonprovisioned chimpanzee populations that exhibit the same pattern of intercommunity aggression seen at Gombe (Brewer 1978; Wrangham and Peterson 1996).
[However, the question of whether, at all of these places, reduction in leopard numbers by human poaching (leopard skins could be sold at a much higher price than ungulate flesh or skins, and farmers of animals nearby would be keen to curtail stock losses to the leopards) had led to increased chimpanzee population growth (and also the growth of populations of other leopard prey species) and pressure on food resources, causing not only recourse to perhaps less favored foods (maybe including red colobus) but also increased aggression (which might find some outlet in hunting behaviour), remains unmentioned, unaddressed, and unresolved].

(p30) Chimpanzee hunting patterns may have been colored by the provisioning of the chimpanzees during the 1960s (Goodall 1986). Baboons, drawn to the same banana supply, came in close contact with chimpanzees day after day, and the easy availability of baby baboons led to high levels of predation on them. A few scholars have suggested that

the hunting was aberrant behavior induced by human disturbance. This claim incited fierce debate (de Pelham and Burton 1976; Teleki 1977); at stake was whether hunting itself was natural chimpanzee behavior.
Observation of chimpanzees hunting for a wide variety of mammalian species, including an occasional Gombe baboon, in all ecological contexts in the ensuing decade has silenced the argument.
[This does not address the question I raise above on the effect of reduced leopard and other big cat predation on prey species numbers and food pressure.]
It is noteworthy that chimpanzees still encounter baboons frequently in the forest and camp in Gombe, but rarely hunt them.
Ransom (1972) and Goodall (1986) hypothesize that baboons gradually learned that chimpanzees were dangerous predators and consequently became more difficult prey.

(p30) The problem is more complex than eliminating direct interference.

At Gombe the most important year-round plant-food species for chimpanzees is the oil palm (Elaeis guineensis). This tree is native to coastal West Africa but is cultivated widely [including in many plantations] in the East (Hartley 1977).
It is used by some chimpanzee populations but not by others (McGrew 1992), and its highly nutritious fruits provide Gombe chimpanzees with nutrients that might not otherwise be available to them. It thereby exerts an influence on survival and reproduction that may be a much stronger, though less obvious, human influence than the effects of provisioning.
[Stanford fails to note that, if chimpanzees utilize the pericarp and not the kernel (elsewhere mention has been made of chimpanzees cracking open the nuts – the matter needs to be clarified…) this is likely to give the chimpanzees more fat and less protein in their diet than they would naturally get on average, which might induce an unnatural desire for protein from other foods, such as may be obtained from flesh, and though this seems contradicted by the fact that the fat-rich brain of the prey is eaten first or preferentially in contradistinction to large carnivore (i.e. cats) practice, the seasonality of the oil palm, and the nutritional composition of the trees they displace, needs further investigation with regard to how the chimpanzees nutrient supplies and habits are affected – it is even possible that during the oil palm season they acquire an unnaturally high taste preference for fat as a calorie source, a taste preference that is unable to be met from that source outside the oil palm season and contributes to increased hunting.]

(p189) With less than 5 percent of their diet composed of mammalian meat, chimpanzees are not obligate carnivores and do not appear to rely on red colobus as a source of essential calories or nutrients.

Of the four living species of great apes, chimpanzees, bonobos, gorillas and orangutans – only chimpanzees hunt and consume meat frequently.

(p195) My work on carcass consumption (Stanford 1996) indicated that Gombe chimpanzees seek fat from the brain and from limb-bone marrow over other body parts after a kill. ……

Even for slower harvesters, enough [(oil)] palm nuts could be obtained in one hour of foraging (30 palm nuts = approximately 1500 calories) to equal the fat and protein that all hunters except the captor and the most-favored meat recipients would get from an adult colobus carcass.
In a study of the nutrients obtained from figs by chimpanzees in the Kibale forest, Wrangham and his colleagues (1993) found that some fig species provide as many as 40 calories per fruit [and some insect protein and vitamin B12, especially in figs].

It appears that chimpanzees are able to obtain both calories and nutrients from their consumption of plant foods without resorting to hunting for animal prey."

(p202) I have argued that if Gombe chimpanzees hunt purely for nutritional gain, they are investing far more time and energy than their nutrient payoff is likely to justify.
(p233-4) It appears that the uses of meat and the motivation for hunting differ among the several chimpanzee study sites, suggesting that cultural variation exists not only in the pattern of hunting but also in the causes of hunting behavior."

2.11 Failures: to quantify Kibale Forest chimpanzee carnivory and to note its seasonality (My carnivory quantification suggests less than 0.5 - 1 gram monkey per chimp per day), to detect and quantify insectivory, and to mention nut-cracking and grain-eating.
Wrangham, Conklin, Chapman, Hunt (1991)

Wrangham RW, Conklin NL, Chapman CA, Hunt KD (1991) The significance of fibrous foods for Kibale Forest chimpanzees. Phil Trans R Soc London B 334: 171-178

"Chimpanzees are primarily frugivores. When fruit is rare, chimpanzees, like other frugivores, must migrate to more productive areas, reduce energy expenditure, or broaden their diet. ……

….poor food conditions favor a generalized increase in diet diversity involving stems, barks, underground storage organs (root vegetables) and insects.

[They observed 50 chimpanzees, for 3 years in the Kanyawara area of the Kibale Forest Reserve, an area at 1500m elevation and having mosaic vegetation of mid-altitude moist forest, secondary forest, tall grassland, swamps and softwood plantations (human activity nearby).
They state that "Vertebrates are eaten regularly" but fail to do the very simple calculation from their data to quantify this including determining the proportion of total diet that vertebrates comprise.
They observed the chimps to eat 14 monkeys (mostly colobus species) – this is approximately 0.1 monkey per chimp per year – at a very roughly approximate weight of 5kg per (large adult) monkey or 2.5kg edible parts per monkey this would be 250g of nutritionally useful monkey per chimp per year at the most, and if the monkeys were younger and/or smaller maybe only half of that or less, 125g monkey per chimp per year, or less than 0.5 gram per day – obviously this is only a *very* small fraction of food or protein intake.
Furthermore, not all chimps in the group would share equally (some not at all) in the eating of those monkeys.
Now, what if not all monkey eating had been observed, so that the real amount of monkey eating might be greater? And what of eaten animals other than monkeys?
In 1250 chimpanzee dungs examined, only 2.9% had any vertebrate remains detected in them – that is *not* 2.9% of dung weight/volume being vertebrate remains, but only 2.9% of the number of dungs having *any* amount of vertebrate remains detected in them – so that by weight/volume the whole dung sample must have been *much less* than 2.9% comprised of detectable vertebrate remains.
Figures given for other reports are 1.9% of 6,000 dungs from Senegal, 1.8% of 800 dungs from Mahale Mountains, Tanzania, and 5.8% of 2,000 dungs from Gombe, Tanzania.
However, vertebrate food might leave less remains to be detected than plant foods do – and this is one more thing that should have been, needs to be, and is easily enough able to be quantified, but was not.

Furthermore, seasonal variation or bias in collection of dung was not considered.
In any case it is quite evident that the proportion of chimp diet, at this location, constituted by vertebrates has not been at all properly quantified – and what quantification that can be readily done from the data, as here, suggests that the proportion of chimp diet (in this location) constituted by vertebrates of any size taken together is very low, and vary variable according to environmental conditions.
Of additional concern is that they report finding no predation on insects or other invertebrates, nor having found invertebrate remains in dung, despite the abundance of a variety of potential prey species, which seems highly implausible.]

Among hominoids thin-enameled teeth are restricted to the African apes, the only species that rely heavily on terrestrial piths.
Orangutans, by contrast, have thick-enameled teeth and only occasionally eat piths, although they spend as much time eating leaves as do chimpanzees.
This suggests that the divergence of African apes and hominoids, which was probably also a division into thin-enameled and thick-enameled groups marked a divergence into pith eaters and non-pith eaters.
The thick enameled Miocene Apes can be expected to have had a different vital food resource from the thin enameled apes of today, with consequent differences in many ecological and social variables. ……
This issue is of interest partly because the occurrence of an essential food resource in the ground layer may have influenced the evolution of terrestriality."
[They fail to consider here that ripe nuts and the stones to crack them open (as chimpanzees are known to eat and use) and grains (as baboons are known to eat) are all on the ground layer, and that it is a very small step from using stones to crack nuts to using stones (maybe even the very same stones…) to roughly grind grains.
Nor do they attempt to explain why humans and orangutans both have thick enamel while the chimpanzees and gorillas, both more closely related to humans have thin enamel.
Note that some characteristics of an animal are related to the conditions of their present existence that have continued from their ancestors who evolved the characteristic in response to the same conditions, while others are a holdover from conditions that are no longer extant, the characteristic having not yet randomly mutated away, or having been turned to some other use for which it was fitted and so been retained in natural selection.]

2.12 Gabon chimpanzees, diet fruits 68%, leaves 28%, animals 4%, great hunger in fruit shortage season (growth stops), uneven distribution of vertebrate meat. Hladik (1981)

Hladik C-M. Chapter 7 Diet and the evolution of feeding strategies among forest primates. In, Harding RSO, Teleki G (Eds) (1981) Omnivorous primates Gathering and hunting in human evolution. Columbia University Press. New York

[For a small group of relocated chimpanzees in Gabon, proportions of major food types ingested: fruits 68%, leaves 28%, animals 4%.]

"Kay and Hylander (1978) present evidence that the anterior part of the mandible, and especially the incisors, are well developed in frugivorous primates.
Folivores, by contrast, have less need to prepare food with the incisors but perform more mastication. Accordingly in these animals the posterior part of the mandible is predominant and the molars include different features for shearing and crushing fibrous materials. ……

There are important seasonal variations in the quantity and composition of food resources. This seasonality is very marked in tropical deciduous forests, and it is also apparent to a lesser extent in rain forests. ……

In the chimpanzees observed in Gabon (Hladik 1973), the growth rate is the highest when the production of fruit is at a seasonal peak between November and January, whereas body weight remains approximately constant during the months of low productivity, from February to June.
[For immature chimpanzees to stop growing, (other than by total weight, at very particular ages) hunger must be great, such that foods ordinarily not eaten due to taste preferences will become acceptable, and eaten…]
Because of the important annual variations in food production in every type of habitat, the growth curves of all primate species normally are irregular. ……
[I assume this refers to individuals still growing to full adult size?
Note, this implies great hunger, which might be a prerequisite for much hunting behavior, which occurs more in seasons of shortage of the preferred fruit…]

Predation on larger vertebrates, which has been documented by Teleki (1973a,b 1975) and others, does not serve as a major source of animal protein for a chimpanzee population as a whole (Hladik 1977; Hladik and Viroben, 1974) because females and young do not benefit significantly from these prey. However, there is some evidence that females consume more insects than males do (McGrew, 1974, 1979), which may compensate for the uneven distribution of vertebrate meat."
[Both the hunting and the termite gathering are more likely compensates purely for caloric hunger in the lean season, not for protein shortage, nor for vitamin B12, both of these being already met by the other foods eaten.]

Their crops, in turn, may have attracted and locally increased small-mammal populations that could be captured and eaten (see Linares, 1976). In thus adding small prey to fruits and seeds obtained from their own crops, human populations continued the primitive primate trait of dietary flexibility."

[Comment in passing: consider the relative cultural disgust to consuming a novel animal, to that to consuming a new fruit – the former much greater than the latter? Why? What might this imply?…]

2.13 Gabon chimpanzees and gorillas: similar teeth, guts and gut transit times; no nut-cracking with tools here at Lope Reserve; both prefer fruit-dominated diet with leaves, stems, piths, seeds and insects; gut entodiniomorph ciliates more present in gorillas; gorillas always eat some foliage; in seasons of fruit scarcity gorillas fall back on leaves, stems and piths, whereas chimpanzees split up to continue seeking fruit. Tutin, Fernandez, Rogers, Williamson, McGrew (1991)

Tutin CEG, Fernandez M, Rogers ME, Williamson EA, McGrew W (1991) Foraging profiles of sympatric lowland gorillas and chimpanzees in the Lope Reserve, Gabon. Phil Trans R Soc London B 334: 179-186.

"Gorillas and chimpanzees living in the same area in the Lope Reserve, Gabon, 1983-1990, had similar diets dominated by an equally diverse array of fruit species complemented with vegetative plant parts, seeds and insects.
Gorillas eat a higher proportion of vegetative plant parts than chimpanzees do.

In times (i.e. the characteristic 3-month dry seasons) of fruit scarcity gorillas stayed together and were able to subsist entirely on vegetative foods, while the chimpanzees split up into smaller foraging groups to maximize fruit acquisition.

Gorillas and chimpanzees have similar tooth and skull anatomy, guts and gut transit times.

Both gorillas and chimpanzees regularly consume insects. Remains of at least one species of insect were found in 30% of gorilla faeces and 31% of chimpanzee faeces [This would mean insect remains were some part of 30/31% of dungs, not 30/31% by weight or volume] The species eaten most often by both apes was the weaver ant, but there was little overlap with other species eaten.

The entodiniomorph ciliates, ubiquitous in faeces of lowland gorillas in Gabon (Goussard et al 1983), are symbiotic cellulose digesters (Collet et al 1984).
Chimpanzees also have ciliates, but they are less frequent in faecal samples (75% of samples compared with 100% for gorillas), and less diverse (two species for chimpanzees, at least four for gorillas) (Imai et al 1991, File et al 1976).
Gorillas can survive on a totally folivorous diet but it appears that chimpanzees cannot ……. .
It is not clear why gorillas avoid high-lipid foods, or how they cope with foods with high tannin content.

Apes have large brains, and the challenge of finding and exploiting keystone food resources, as well as keeping track of widely dispersed fruit foods, may have been an important selective pressure in the evolution of intelligence.

The kernels of large, well-protected nuts provide a high-quality food, but the essential tool-use technology exists in only some of the chimpanzee populations with access to the resource.
Chimpanzees at Lope do not crack nuts with tools, but if they did, intra-specific feeding competition in the dry season might be reduced ……

Gorillas will still eat some foliage when fruit is abundant, whereas chimpanzees might not, and perhaps this has something to do with 25% of chimpanzees not having entodiniomorph ciliates found in their faeces in the report noted above."

2.14 Gombe chimpanzees: infanticidal cannibalism due to poor mothering; eating of human babies. Montgomery (2009)

Montgomery S (2009) Walking with the great apes Jane Goodall, Dian Fossey, Birute Galdikas. Chelsea Green Publishing.

"On two occasions in the 1940s African babies had been seized by chimps in the area [Gombe] One of the babies had been killed and partially eaten. The other baby was rescued by his six-year-old brother. The chimps had eaten part of the baby's face. (p27) ……

Jane [Goodall] observed that good chimpanzee mothers raised confident, socially adept, competent offspring.
She often compared Flo's mothering styles with those of other Gombe mothers: look at Flo, Jane told her audiences, who would distract her offspring with play and grooming instead of cuffing them when they misbehaved, as Olly sometimes did with Gilka. Flo

would always run to her sons or daughters at the slightest distress call, unlike Passion, who ignored daughter Pom's whimpering. And look at Flo's family: her older son Figan eventually became the alpha male of the community, and Fifi became an exceptionally efficient and loving mother. Olly's Gilka became a timid, sickly adult; Passion's Pom became a wildly aberrant female, an infant-killing cannibal. (p29) ......

In the fifth month of her study, Jane made two discoveries that caused worldwide excitement. ...... the chimps were eating a piglet. (p84) ......

...; by 1972 [Goodall started the Gombe study in 1960] more than a hundred people ... were supporting a study of fifty-four chimpanzees and the troops of baboons who shared their land. By this time most of the chimps were regularly visiting Jane's banana feeding station. .....
[Bananas are higher in carbohydrate and lower in protein than the average chimpanzee natural diet, and this could have caused greater hunting activity in the chimps, as also might have observation and smelling of camp people handling and eating meat – also it might have given them an unnatural amount of spare time on their hands...]

[100 pound chimp as strong as two men – 17-year old male chimp grabbed a grown man by the ankle and dragged him several dozen yards through the foliage. (p101)]

... during the first decade of research no chimpanzee ever seriously injured a human observer at Gombe; this seemed a powerful testament to the animals' inherent gentleness. It bore out the portrait Jane had been painting of them: excitable at times, but intrinsically peaceful beings, an attractive ancestor of the potentially peaceful human species.
But with other eyes watching them, during the second decade of Gombe research, "My Friends the Wild Chimpanzees" turned out to be cannibals, infant killers, and warmongers. ......

"The single most important reason they didn't see these things" [cannibalism and warfare], John quickly adds, "is the fact that they're tough animals to study, and these are rare events; but these women also had ingrained biases. And this isn't a criticism," he emphasizes. "It points to the need for perseverance." (p102) ......

Passion [chimp], with the aid of her adolescent daughter, Pom, had seized Olly's [chimp] three-week-old granddaughter. The mother-daughter team had then spent five hours [which is a far longer time than Stanford (1999) gives for the time taken ("minutes") for chimpanzees to eat an infant monkey] eating the body of the infant. ......
But she [Goodall] vividly remembers that Passion, when Pom was born in 1965, had proved "extraordinarily inefficient and indifferent" as a mother. ......
Over the next two years Jane's African staff saw Passion and Pom kill and eat three infants [chimps] of the Kasakela community. Between 1974 and 1977 two other infants vanished during their first month of life. (p105) ......

Yet for all its horror, the violence at Gombe proved beyond doubt the centerpiece of Jane's thesis: that the animals' individual temperaments, family backgrounds, and decisions were the basis of their history. (p107) ......

... they [gorillas] ate mainly leaves, stems, bark, fungus, earth, snails and grubs that they dug from rotting bark with their great hands. (p122) ......

... gorillas will sometimes eat their own dung to recycle nutrients. (p128) ......

For the solitary orangutans Birute was studying, hierarchy has little meaning.

For the large community of chimpanzees at Gombe, who travel in constantly changing smaller groups, social rank is of limited importance.
But for a close-knit family of gorillas, one's place in the group is all-important. (p129)
……
She [Galdikas] carefully catalogued the hundreds of plants and insects that orangutans ate. (p152)"

2.15 Kenya Pumphouse Gang baboons, predator (lions, leopards) reduction by humans, baboon and ungulate population growth, long-distance hunting by baboons catalyzed by ordinarily dangerous behavior of unorthodox individual male evolves into cooperative hunting which disappears with immigration and emigration of particular males and increased ungulate vigilance. Strum (1981)

Strum SC. Chapter 8 Processes and products of change: baboon predatory behavior at Gilgil, Kenya. In, Harding RSO, Teleki G (Eds) (1981) Omnivorous primates Gathering and hunting in human evolution. Columbia University Press. New York

"The data presented here were collected on Kekopey Ranch near Gilgil, Kenya, roughly 115 km northwest of Nairobi in the Great Rift Valley, an area of high-altitude savanna with volcanic soils. ……

During the past 20 years the area has been altered for cattle raising by bush clearing and construction of a system of cattle troughs that bring water into previously dry areas. ……

Large carnivores have been shot or trapped since the early part of this century, but despite their reduced numbers, hyena, lion and cheetah have occasionally preyed on the cattle. A sizeable leopard population remains, along with three species of jackals, bat-eared fox, and feral domestic dogs that hunt in packs.

Baboon and ungulate populations have increased rapidly, probably as a result of ecological modifications at Kekopey (additional grassland and water plus lowered predator density.
[the latter leading to eventual population pressure on preferred food sources in any case?]
Although a rough estimate of 50 Thomson's gazelle in the central area in 1956 (A. and T. Cole, personal communication) is probably imprecise, the order of magnitude seems correct.
By the middle of 1970 the population had expanded to 2,000 (Blankenship and Qvortrup, 1974).
At this time, in the central area there also were approximately 1,400 impala, 1,000 reedbuck, 200 zebra, 100 buffalo, and 150 eland, in addition to the smaller antelope species.
[This marked increase in prey animal populations in the presence of reduced numbers of natural predators is probably important regards behavior such as baboon food choices and Thomson's gazelle vigilance]

In 1970, 7 baboon troops were identified in this area, with group sizes ranging from 35 to 125 and a population density of 10.3 baboons/km^2 (Harding, 1973b), at the upper range for "savannah" baboons.
Since 1970 most baboon troops have doubled in size, and many have split. In 1977 there were a minimum of 12 troops within the same area, yielding a density of roughly 19 baboons/km^2. ……

With minor exceptions, the species preyed on by PHG ["Pump House Gang"] baboons have remained the same over the years, but the proportional representation (the prey profile) differs from year to year (Table 8.2).
Not all available prey species are captured, and baboon troops in the same population differ in their selection of prey.
The PHG baboons capture the young of several antelope species (Thomson's gazelle, impala, dikdik, steenbok, reedbuck, klipspringer) plus individuals of all ages of the smaller antelopes, such as dikdik and steenbok. Cape hare and button quail, several smaller ground-nesting birds, and some tree-nesting bird species are also frequent prey.
......
It is difficult to distinguish between types of hunting that lie along a continuum of behavior, but the contrasts between types of hunting at either end of the continuum are clear.
I suggest that one baboon moving 3m to pursue a small animal for a few minutes and five male baboons moving synchronously over 1,600m to contact a prey animal, then spending 2 1/2 hours in following, scattering, searching, and chasing represent different hunting patterns. ......

Chances for opportunistic capture arose during the day, as the troop searched for vegetable foods.

Because hares and infant antelopes attempted to avoid detection by remaining concealed and immobile, a baboon often fed as close as 10cm to a potential prey before it ran away. However, concealment may not be as effective a defense against baboon predators as it is against predators that rely on movement cues.

Not all successful kills were the result of accidental discoveries.
In simple hunting two methods were employed.
In the first, a baboon left the troop to approach a nearby herd of Thomson's gazelle, then moved through the herd and scrutinized the surrounding area. These searches sometimes resulted in the location and successful capture of infant gazelles. Since dikdik and steenbok prefer denser cover and are not clearly visible to the baboons during troop movement, searching also included detouring through thickets, with the result that prey was flushed, chased, and sometimes captured.
[now a description of a baboon running down a hare from 150m away at the start to close that distance to nothing by 450m, which seems to involve a running speed that no human could achieve, notwithstanding reports of human ability to run down ungulates over very long distances, though this latter is not perfectly described so as to differentiate running from one serial stalk through cover to another, from a purely non-stop run out in the open with both parties fully aware of the location of the other at all times, which are two very different things...] ......

RD's [a baboon] behavior at the end of this example verges on what I have termed complex hunting, for it was systematic hunting, sophisticated in its searching and chasing techniques, and pursuit was persistent and prolonged. ......

Holding the antelope to the ground with both hands and feet, the consumer took the first bite from the underbelly. After the viscera were eaten, the meat on the ribs, the vertebral column, and the limbs was consumed. In the process of eating the limbs, the baboons often turned the skin inside out. The consume alternated between these various parts of the body until only the limbs and head remained. Once the fleshy parts were gone, the consumer had to exert more and more effort to free the meat from the bones and skin.

Force was applied using the incisors to pull in opposition to the hands and feet.
[Does this have relevance for the evolution of human incisor evolution or retention? Probably not in evolution in most cases, but rather in retention, if at all? What about the Neanderthals' relatively larger incisors (front teeth)? But almost certainly by the time hominins were eating more than only a very small amount of meat, they had manufactured sharp stone tools to use instead. So, probably there's not strong evidence here that the use in meat eating was anything other than an adaptation of a function used for ridding plant foods of undesired coverings…]

The head was consumed last, although marrow eating often followed brain consumption and the carcass might be reexamined for further edible portions. Before the brain was attempted, the ears and eyes were eaten. The brains of prey were not always eaten. Baboons consumed the brains of smaller species, such as Cape hares, as a matter of course, but they often left the brains of the larger antelopes intact. ……
[Whereas chimpanzees are reported to eat the head/brain of red colobus monkeys first… chimpanzees are reported to favor fatty foods, whereas gorillas are reported to dislike them – do baboons dislike fatty foods?...]

As was often the case with Thomson's gazelle, the prey had outrun the first baboon that pursued it and probably would have escaped if additional baboons had not intervened.

…… baboons consistently and frequently searched for and obtained eggs. ……

Lions have been eliminated and leopard numbers have been reduced at Kekopey. ……

Only certain adult females showed definite interest in predation. ……

Although the amount of meat that each individual obtained may have been quite small, meat-eating was widespread, particularly during 1973. In that year, for example, 58 members of the troop (76%) ate meat; participation ranged from individuals who ate in 63% of the predatory episodes to those who ate in only 1%. ……

Resolution of such conflicts ranged from consort [accompanying a female in heat] males entirely ignoring meat-eating opportunities to males allowing their consort females to have, keep and eat the carcass. ……

… males gave predatory behavior a lower priority than either sexual behavior with receptive females or social behavior with other males. ……

The main type of predation during 1970-1971 was opportunistic capture of concealed prey, but techniques of active searching (simple hunting) were also employed, though infrequently.
Harding also observed an unsuccessful sequential chase of an infant Thomson's gazelle by male baboons.
The basically opportunistic nature of predatory behavior was still evident at the beginning of the 1973 study, but simple hunting increased in sophistication and persistence during 1973, culminating in the development of elaborate searching, stalking, and chasing, or what I have termed complex hunting. ……
The sequential chase of Thomson's gazelle observed by Harding was both unsuccessful and probably fortuitous; there was no evidence that the baboons were consciously acting together.
Midway in the 1973 study, I observed another instance that succeeded. From that point onward, male baboons changed the orientation of their chases when other hunting baboons were near. Prior to this incident, the prey individual was chased in any direction,

often out onto the open plain. Thereafter, baboon males chased Thomson's gazelles toward other hunters, and several kills resulted from sequential chases. Coordination of hunting activities between several males increased, as evidenced by the rise in the number of simultaneous hunters, from two in April to five in August. ……

Thomson's gazelle herds take three forms: all-male herds; mixed herds containing adult males, females and young; and female herds with young. Initially the baboons did not distinguish between different types of herds during their complex hunting, and spent much time following, scattering, and searching through herds containing only adult males. As complex hunting developed, the baboons began to look briefly at all-male herds and then ignore them in favor of mixed herds or female herds. ……
Gazelle herds did not flee from or avoid PHG baboons in 1970-1971 or for the first 6 months of the 1973 study. But in May 1973, the month with the highest complex-hunting rate, herds that had been hunted most recently or most frequently changed their behavior. They began to show vigilance, and a flight distance of 20-25m was established. This distance increased during the next few months to 45-275m the normal range distance from predators such as lions (Walther 1969). The change in response of Thomson's gazelle towards baboons was a process of adjustment in which at least certain herds began to recognize baboons as potentially effective predators.
[All of this suggests that hunting is not naturally much of a baboon activity, but became so when population pressure on more naturally preferred foods increased due to human intervention. Note also that the behavior of the baboon who catalyzed the development of hunting behavior in the PHG, in ranging far to the periphery of the group's feeding area, would have been less conducive to survival in the presence of lions etc, being behavior that is ordinarily selected against for survival.] ……

Elsewhere in Africa, baboons have not been observed to share food of any kind, either between a mother and her offspring or among adults (e.g. Altmann and Altmann, 1970; De Vore and Hall, 1965; Kummer, 1968; Ransom, 1971; Rowell, 1966; Stolz and Saayman, 1970), although a single piece of food, such as meat, is often consumed in serial fashion by several individuals.
Changes in the pattern of attendance and consumption affected the distribution of meat within the troop.
The amount that some individuals obtained during the year was relatively small, but the widespread nature of meat-eating within the troop in 1973 contrasted with the 1970-1971 period.
Two additional changes in behaviour were seen midway through the 1973 study:
first, individual baboons moved to one side, leaving a significant portion of meat and allowing other individuals (of the other sex) to approach and feed in proximity;
second, individuals fed from the same carcass simultaneously.
Because the distribution of meat between individuals during these episodes differed from the serial distribution of meat seen at other times and was more active in its form, I classify these interactions as "sharing". ……

Overall, the increase in hunting by one male appeared to change the frequency of hunting in other males. …… A different social milieu could have prevented or hindered the flow of information and the facilitation of behavior crucial to the elaboration of the baboons' pursuit behavior. ……

The primate group, and especially the baboon troop, has been viewed by some as a basic adaptation to problems of defense against predation. …… The possibility of attack by large predators is irrelevant to opportunistic baboon hunting that occurs within or near the troop, but is important in considering activities that take an individual away from and out of contact with the group. ……

Perhaps the most obvious change since 1975 is the cessation of complex hunting and the reduction in amount of simple hunting. ……
Although predatory behavior had already changed by the time of the 1975 study, major ecological factors can be excluded as explanations for the change. The drought continued through this period, and there were no changes in rainfall pattern or in prey availability
[This may well be incorrect, as the Thomson's gazelle, as noted, had significantly changed their vigilance and flight behavior with regard to the baboons.]
Only social and demographic conditions changed, and since these factors were implicated in the previous modification of PHG predatory behavior, it seems reasonable that they influenced the second set of changes as well.
[RD, the baboon who had been the leader in the evolution of the hunting behavior, had become unable to keep his kills, immigrating males tended to stay close to the troop, and two other male baboons very forward in the hunting had emigrated]
Although prey availability may have been different between the study periods because of the complex interaction between prey movements and predator ranging patterns
[this seems to be only the aforementioned increased vigilance and flight behavior of the Thomson's gazelles],
the baboons ignored too many potential opportunities each month to make that a convincing explanation.
[It seems to me to have very probably been one major factor, and the aforementioned baboon social factors another major factor] ……

Adult male baboons cooperate with each other in response to threats from within the group or outside of it, and this cooperative experience may have facilitated the coordination of male predatory behavior seen in 1973. ……

Variability in primate behavior and personality can be an important source of innovations (Frisch, 1968; Green 1975; Itani and Nishimura 1973).
Once an innovation occurs, individuals acquire the new behavior through learning.
How and from whom they learn depends on the type of behavior and the nature of communication and affiliative networks within the group (for other primate examples, see Beck, 1975; De Vore, 1963; Hall, 1962, 1963, 1968; Jay, 1963; Kawai, 1965; Kawamura, 1959; Rogers, 1973). ……
As the more recent PHG baboon data illustrate, behaviors can become extinct as rapidly as they are elaborated, and reversals or rapid alternations may have been common in human evolution. ……

But if elaborate hunting behavior has important bioenergetic determinants, then some habitats would be especially suitable for the hominid development of large-game hunting.
For example, a forest habitat has few species of large animals and few communal species; prey exists there in low density and wide distribution, and visibility is poor. These conditions would make the development of predation on large animals unlikely.
By contrast, the savannah, with its great numbers of large, group-living prey, would provide frequent opportunities for this type of hunting to develop.
Thus, if the evolution of predation within the primate order was a graded process, the adaptive shift that included large animals as prey probably occurred in open savannah or open woodland. ……
The predatory behavior of the earliest hominids may have been similar to that of today's baboons or chimpanzees, and could have originated many times, among groups in different populations living under differing ecological conditions."
[Note that neither chimpanzees are, nor early hominids were, close to as fast as baboons on the run in more open spaces – chimpanzee hunting is more by encirclement entrapment in more closed forest…]

2.16 Monkeys and apes; tooth shape and enamel thickness; digestive systems; energy-density of foods; chimpanzee hammer, anvil, and mortar and pestle use; capuchin monkey hammer, anvil, mortar and pestle, and digging tool use; seasonality and climate shift, fallback foods versus preferred foods. Lambert (2007)

Lambert JE. Chapter 17 Seasonality, fallback strategies, and natural selection. In, Ungar PS. (Ed) (2007) Evolution of the human diet The known, the unknown, and the unknowable. Oxford University Press.

"From the perspective of a plant eating animal, climate shifts and their impact on food resource availability and abundance vary in predictability. For example, longer-term (e.g., glaciation) or more extreme (e.g., monsoons [I think Lambert means a shift between monsoonal and non-monsoonal patterns, not the shift from one season to the next inside one monsoonal year]) shifts are relatively unpredictable and can have important evolutionary consequences, particularly with regard to speciation and extinction. ......

That is, a trait's function is critical to the harvesting or ingestion of a particular resource (fallback foods); the crucial or "critical" aspect of this argument relates not to the overall utility of this feature for consuming a food type in high quantity or in high frequency, but that it instead has extreme utility under particular – limiting – environmental circumstances, during which time an animal must fall back on foods that are uncommonly consumed (Robinson and Wilson, 1998). ......

As such "fallback" is used variously as an indicator of food preference, as an index of fallback food availability, either in absolute abundance or relative to preferred foods, or to simply refer to the absolute inherent quality and nutrient density of the food ......
In this discussion, I place fallback *foods* into two broad categories that ultimately relate to two different *strategies*, including (i) fallback foods of lower nutritional density (lower inherent energy yield). ...... anatomical adaptation ..... (ii) fallback foods can be of higher nutritional density (higher inherent energy yield). ...... behavioral adaptation. ...... It should be noted that neither strategy precludes the other, so that, for example, species that use tools to access a high energy, but mechanically protected, food may also have thick dental enamel and a robust masticatory apparatus (e.g., Cebus spp.). ......

[Lambert's goal 2 in her chapter here] To argue that the evolution of such strategies is influenced by the likelihood of an environment undergoing periods when food availability shifts (critical periods), and that environments differ in this regard as a function of latitude, altitude, rainfall and area – with some habitats being more inherently vulnerable. ......

As such [predators on plants that have evolved various defensive mechanisms], there is ample evidence that primates are extremely selective feeders and consume only a fraction of the plant species available to them in a habitat (Oates, 1977, 1987; Glander, 1978; Milton, 1984). ......

During the study period, ripe fruit and other preferred, softer, and more commonly consumed foods were unavailable to these two species [*Lophocebus albigena* (with thick dental enamel), *Cercopithecus ascanius*, sharing the same area ("sympatric")]. The study animals instead consumed fallback foods, which were not only different in form and type, but also hardness.

We found no difference in dietary hardness when the puncture resistance of all fruit (over 1991-1997) consumed by the two species was compared. However, when the hardness of only fallback foods was evaluated, *L. albigena* exploited a diet with greater resistance to puncture and crushing than *C. ascanius*. This difference was largely explained by the higher percentage of bark and seeds consumed by the *L. albigena* during this period. The diet of the two species can have high overlap, and *L. albigena* can be highly frugivorous; however, *L. albigena* fall back on very hard foods, while the guenon [*C. ascanius*] increases leaf intake. Bark is a highly lignified and particularly hard resource, and these results corroborate the hypothesis that it is not so much what is consumed most commonly (i.e., soft, fleshy fruit) that selects for enamel thickness but the hardness of foods that are consumed during critical periods, when other more preferred foods are not available.

Like many primates, including chimpanzees, lowland gorillas show a preference for ripe, soft fruit (Remis, 1997; Remis et al., 2001). Nishihara (1995), for example, has found that western lowland gorillas (*Gorilla gorilla gorilla*) spend 63% of their feeding time on fruit. And, in a recent interview that included data from all long-term western lowland gorilla sites, Rogers et al. (2004) documented that these apes will maintain fruit in their diet throughout the year; the authors indeed call *G. gorilla gorilla* "fruit pursuers, with strong preferences for particular and often rare species, for which they will incur significant foraging costs" (p175).

However, as noted by Ungar (2004), the dietary differences between gorillas and chimpanzees become most apparent in analyses of fallback foods, such that during periods of preferred fruit scarcity, gorillas will tend to fall back on more fibrous foods than those consumed by chimpanzees. ……

Ungar (2004) argues that it is the key differences in the mechanical properties of the fallback foods that can account for the longer shearing crests and steeper cusps in gorillas and flatter molar surfaces in chimpanzees. Gorillas are argued to be able to consume their lower-quality fallback foods as a consequence of the aforementioned adaptations, as well as a high number of cellulose digesting ciliates and overall body size, which facilitates the slower digestive processing required for fermenting fiber (Demment and van Soest, 1983; Collet et al., 1984; Remis et al., 2001). ……

Yet, as it turns out, both cercopithecoid [monkeys] subfamilies have relatively longer digestive retention times than the much larger African apes (Lambert, 2002a). This is to be expected among the Colobinae, who have specialized, sacculated, and alkaline [relatively, not absolutely?...] stomachs for consuming a diet high in fiber. On average, their mean digestive retention times range from roughly 40 to 60 h (refs in Lambert, 1998; Caton, 1999). ……

Although their colon is proportionately larger [=?!...] than what is seen in *Homo sapiens*, chimpanzees have digestive retention times that are comparable to those of humans and, contrary to what might be predicted for their body size, consume an unexpectedly high-quality diet (Milton, 1999).
[Lambert notes that chimpanzee behavioral strategies for seasons of preferred food scarcity include the breaking up of feeding parties into smaller groups, and the use of tools to overcome mechanical protection of foods like nuts and termites]

Chimpanzees will use a variety of tools regardless of overall food availability (Whiten et al., 1999). ……
Chimpanzees in this forest [Bossou, Guinea] employ a number of feeding-related tools, including hammer and anvil, pestle pounding, and ant dipping [with a twig]. ……
During periods of extreme scarcity, nut [oil palm nuts, which may and often do in some locations, come from human plantations, not from natural wild trees…] cracking and

pestle pounding for pith increase dramatically, providing critically needed calories from lipids and carbohydrates. ......

Overall the bonobo is argued to live in habitats where food resources are spatially and temporally more abundant than those of common chimpanzees (Malenky and Wrangham, 1994; Thompson, 2003). ...... Indeed, relative to the forests in the Democratic Republic of Congo [where the bonobo live], savanna woodland is drier with more seasonal rainfall patterns, has xeric[to dryness]-adapted plant species with hard-shelled seed or nuts and underground storage organs, and has more patchily distributed preferred resources as a function of greater seasonality (Moore, 1996). The common chimpanzee is also well known for its diverse toolkit, while feeding-related tool use in wild populations of *P. paniscus* [bonobos] has not been reported (Boesch, Hohmann and Marchant, 2002; McGrew, 1992). Are these facts related? [Very probably…] ......

... although, given the inherently more labile nature of behavior over, say, evolution of dental traits, it might be expected that on those occasions where food is limiting, behavioral solutions might arise. ......

At the same time, capuchins [monkeys] have been demonstrated to use tools to facilitate access to resources during periods when preferred foods were not available. ...... In the Caatinga dry forest of northeastern Brazil, capuchins have been observed to commonly use tools and to do so during the extended dry season of this region. ...... Several tools and tool-facilitated behaviors have been found in four habituated [to tolerate human observation] capuchin groups foraging in these areas, including digging for tubers with stones, cracking open seeds and branches with stones, breaking tubers with stones, and using stones as hammers in combination with wooden anvils to crack seeds. ......
... in another dry region of Brazil, where capuchins commonly use hammer-and-anvil technology to crack nuts during the dry season when preferred resources are scarce (Fragaszy et al., 2004). ......

This southern region [the southernmost of the bonobo's range, "transitional grassland habitat"] is characterized by different soils and a marked seasonal distribution of rainfall. *Pan paniscus* [bonobos] are known to move through and feed in these areas. At least three grassland species [grasses, I assume…] have been consumed by *P. paniscus*, and there are behavioral correlates of being in a different habitat type, too, including no vocalizations (Thompson, 2003). ......

Foley (1993) argued that hominin origins are closely linked to the development of increasingly seasonal conditions; he also argues that differences among African ape species were driven by the degree to which their ancestral and extant environments are seasonal and differences in the selection for traits for dealing with food scarcity in more seasonal habitats. ......
I suggest that the *P. troglodytes* [chimpanzee] – *P. paniscus* [bonobo] comparison, along with knowledge of cercopithecoid [monkeys] and gorilla feeding, can lend insight into the evolution of behavioral solutions to facilitate a fallback strategy on relatively high-quality foods. ...... *Pan troglodytes* occupy habitat that can have upward of five to seven dry (rainfall <100mm) months/annum;
Australopithecus afarensis has been found in habitats that probably had even more dry months (Foley, 1993; see Reed and Rector, chapter 14, in this volume). ......

With regard to hominin diet and adaptations, Scott et al. (2005) have recently argued that there is greater dietary overlap and variability in *Paranthropus* [larger-jawed, not of our line] and *Australopithecus* [of our line] than previously thought. They indeed suggest "early hominin diet differences might relate more to microhabitat, seasonality, or fallback

food choice than to oversimplified, dichotomous food preferences" (Scott et al., 2005, p. 694).
I agree, but I would add another degree of complexity; there are differences among both fallback *foods* and, more importantly, overall fallback *strategies*. On the basis of the above framework, we can envisage a continuum of fallback strategies: at one end, we see the evolution of *anatomy* to exploit *lower-quality* (less nutritionally dense) foods; at the other end, we see evolution of *behavioral* mechanisms for maintaining *higher-quality* foods, with various degrees of mixed strategy in between (fig 17.4).
Only in relatively aseasonal habitats like that of *P. paniscus* [bonobos] would we see relaxation of fallback strategies.
We know that the evolution of hominin species is closely related to increasing seasonality, so some fallback strategy at a point along this continuum is expected. In the case of a comparative *Paranthropus*, *Australopithecus*, early *Homo* evaluation, we might expect, based on anatomy alone, for *Paranthropus* to be at one end of the continuum, early *Homo* at the other, and *Australopithecus* somewhere between the two."

2.17 Tooth structure and function have variable relationships; bilophodont molars associated with high-fiber diet; high, pointed cusps for puncture-crushing hard and brittle foods, e.g. bone and cartilage of vertebrates. Chivers, Andrews, Preuschoft, Bilsborough, Wood (1984)

Chivers DJ, Andrews P, Preuschoft H, Bilsborough A, Wood BA (1984) Chapter 25 Food acquisition and processing in primates: concluding discussion. In, Chivers DJ, Wood BA, Bilsborough A (1984) Food acquisition and processing in primates. Plenum Press, New York.

"Lucas, Kay and others then developed this correspondence between these dietary categories and different features of the anterior and posterior dentition.
For example, incisors are small and piercing in faunivores, broader in frugivores and narrow in folivores, whereas the cheek teeth vary from puncture-crushing (and shearing) in faunivores, through crushing in frugivores, to shearing and grinding in folivores. ……

It was possible, therefore, to arrange each variable (and its range of variation) in accordance with the spectrum from 100% faunivory, through maximal frugivory (about 80%) to 100% folivory (fig 25.3).

For example, apart from being small and scarce, [(primate)] faunivores live in very small groups (often being solitary) and are the most nocturnal of primates; they progress by leaping or quadrupedal walk-run, often clinging [in trees] when stationary.
They have light jaws, large gape, little lateral jaw movement, thin enamel on the teeth and low shearing blades; the anterior dentition is small, but often modified, and the cheek teeth possess high, pointed cusps for puncture-crushing hard and brittle foods, e.g. bone and cartilage of vertebrates, the cuticle of invertebrates, and, sometimes, nuts and seeds if there is a fruit component to their diets.
Chemically their foods are high in protein and lipids, moderate in free carbohydrate but low in structural carbohydrate, moderate in toxins and low in digestion inhibitors; this includes any plant component in their diet since nuts are high in lipids and may contain toxins (which can be avoided by manipulation, i.e. by eating the seed but not the 'fruit' testa).
The guts are small, relative to body size, and dominated by small intestine. …..

Structure, function and biological role have variable relationships, so that similar structures may have different biological roles, and different structures may have similar biological roles. …
The biological systems of extant species include a substantial element of preadaptation and/or retention of ancestral features in their structure. … For example, the complex molar structure of monkeys (designated as bilophodonty [having molariform teeth with two ridges] for simplicity) may have biological roles in either leaf- or fruit-eating; likewise, some monkeys and apes have similar biological roles in their diets but these are achieved by different dental structures.
Such discrepancies may be more apparent than real, however, especially if the actual function of the structure can be determined.
For example, some squirrels, fruit bats, cebids and hominoids [apes] subsist mainly on fruit, yet their teeth are different in structure. They all share, however, large occlusal basins (on the principle of the pestle and mortar, Lucas and Luke, chapter 14), which are the key similarity of structure and, hence, of function. ……
… bilophodont molars seem to be associated with high-fibre diets, and so they are preadapted to all diets, whether fruit or leaves, that are high in fibre.
Similarly, a capacity to cope with plant secondary compounds (associated with eating more leaves) would have preadapted monkeys to eating unripe fruit, thereby giving them a competitive edge over the apes.
The morphological pattern of a species includes structures that are peculiar to that species, as well as those that are inherited unchanged from ancestral species, but it is the former that are likely to show a closer correlation between structure and biological role. Thus, these two sets of phenotypic characters require separate consideration before inferring function from structure."

2.18 Primate elongated canine teeth, dominance hierarchies, defense, prey killing and eating. Mace (1992)

Mace G. Section 2.3 Differences between the sexes. In, Chapter 2 The life of primates. In Jones S, Martin R, Pilbeam D, Bunney S (1992) The Cambridge Encyclopedia of Human Evolution. Cambridge University Press.

"In general, amongst primates, longer canine teeth are found in polygynous species (where they are used in establishing dominance hierarchies amongst males) than in monogamous species, and in terrestrial species (where they are used in defense against predators) than in arboreal species."
[In all instances of meat eating by monkeys or apes that I have read about, unless the prey is so small that it may be chewed to bits crunching bones and all whole or continuously, the elongated canine teeth are necessary, for use in the same way that more predominantly or exclusively carnivorous animals use them in separating meat from bone, though not always used in administering a killing bite before commencing the eating.]

2.19 Bearded Saki monkey canine teeth used to crack hard-shelled fruits and nuts. Van Roosmalen (1984)

Van Roosmalen (1984) Chapter 7 Subcategorizing foods in primates. In, Chivers DJ, Wood BA, Bilsborough A (1984) Food acquisition and processing in primates. Plenum Press, New York.

[Van Roosmalen states that the large wedge-like canines of the bearded Saki monkey are used to crack hard-shelled fruits and nuts – does this imply that these food casings are opened at natural cracks and seams, molars being possibly preferable to use otherwise as in crushing?]

2.20 Gelada baboons: grass seeds most of diet; enlarged premolar and molar teeth for grinding; high manual dexterity; erect sitting position; fatty deposits on buttocks; buttock-shuffling feeding locomotion. Foley (1992)

Foley R. Section 9.2 Studying human evolution by analogy. In, Chapter 9 Early human behaviour and ecology. In Jones S, Martin R, Pilbeam D, Bunney S (1992) The Cambridge Encyclopedia of Human Evolution. Cambridge University Press.

"Perhaps the best known analogue model based on baboons is the *seed-eating hypothesis*.

The baboons in question are a particularly specialized species, the gelada baboon (*Theropithecus gelada*). These live exclusively in the highlands of Ethiopia, where they are adapted to highly seasonal, open and relatively nutrient-poor environments. They survive by specialized harvesting of grass (though they also eat other things), the only resource that occurs abundantly.
To this end, gelada baboons have undergone a series of evolutionary modifications. Their hands are particularly dexterous, to enable them to manipulate the fine grass stems. As they spend much of their day feeding, they have also evolved a bottom-shuffling locomotion that enables them to move along while grazing without standing up. To facilitate this movement, geladas, unlike other baboons, have fatty deposits on their buttocks and maintain an erect trunk while in this sitting position.
Their teeth also reflect their specialized diet: they have reduced front teeth (incisors and canines) and enlarged back teeth (premolars and molars) because the grass seeds that constitute most of their food are hard and small, requiring little or no preparation with the front teeth but considerable grinding.
These characteristics parallel many found in hominids – a move to erectness, high manual dexterity, and small front teeth compared to the back ones.
[And the enlarged (molarized?) human/hominid premolars, which might be particularly significant regards eating grass seeds.]
On the basis of this analogy, early hominids could have gone through a seed-eating phase while they were developing upright stance and bipedal locomotion."
[It's quite possible they didn't go "through" a seed-eating phase so much as come into it and maintain it as a part of the diet…]

2.21 Tooth enamel ^{13}C study - extinct baboons ate more grass than modern South African savanna baboons; faulty statement of broad-spectrum omnivory due to non-differentiation of apes from other larger primates. Codron, Luyt, Lee-Thorp, Sponheimer, de Ruiter, Codron (2005)

Codron D, Luyt J, Lee-Thorp JA, Sponheimer M, de Ruiter D, Codron J (2005) Utilization of savanna-based resources by Plio-Pleistocene baboons. South African journal of Science 101, May/June: 245-248.
[Enamel carbon-13 (^{13}C) plant C_4 use study]

"Utilization of grass by some of the extinct baboon specimens was much greater than is the case with modern South-African savanna dwelling individuals. ……"

[Their statement that larger non-human primates are broad-spectrum omnivores is only correct in the context of a very (somewhat over-) generalized comparison amongst *all* primates – it would not be correct, and would become obvious as an overgeneralization, if only the individual *ape* species were compared to each other rather than looking at *all* primates – because there is much more difference among the species of primates as a whole group than there is among just the species of the subset of apes (the subset most relevant to humans, on average) – therefore one may not logically suggest that the indication from *ape* diets is that humans are evolutionarily broad spectrum omnivores, because none of the *apes* are "broad spectrum" omnivores; they are all predominantly frugivorous very predominantly vegetarian omnivores…]

2.22 Pleistocene African savanna baboons were nearly as large as modern gorillas; became extinct with the appearance of man. Coe (1984)

Coe M (1984) Chapter 1 Primates: their niche structure and habitats. In, Chivers DJ, Wood BA, Bilsborough A (1984) Food acquisition and processing in primates. Plenum Press, New York.

"We must remember, however, that an element of habitat choice on the part of an individual may be ultimately determined by inter- and intra-specific competition for space (Partridge, op. cit.). ……

The greatest range of forms is found amongst the baboons, whose larger body size allows them to range over larger areas of the savanna in search of their broad omnivorous-opportunist diet.
[Note: grass eating Gelada baboons are one exception to this]

Baboon genera that lived in the savannas of Africa during the late Pleistocene were nearly as large as a modern gorilla
[Note: and would therefore probably have been an extremely formidable and dangerous opponent in combat with hominins of their times, such that combat with them was probably more avoided than sought…],
but seem to have become extinct with the appearance of man, with whom they may well have competed."

2.23 Protein content (dry weight) of legumes and leaves is high, of cereals, oily nuts and seeds and (underground) storage organs is substantial but generally lower, of fruit flesh is 20% that of leaves; diet of primates normally ensures an adequate supply of water; B vitamins; vitamin C; gut flora of all mammals capable of some cellulose digestion; plant defenses, tannins, lignins, proteinase inhibitors, anti-microbials, toxins. Waterman (1984).

Waterman PG (1984) Chapter 8 Food acquisition and processing as a function of plant chemistry. In, Chivers DJ, Wood BA, Bilsborough A (1984) Food acquisition and processing in primates. Plenum Press, New York.

"For the major foodstuffs [Note: he is dealing with plant foods only here] of man protein levels are highest in some seeds, notably legumes, and in leaves, when measured on a dry weight basis (Brouk, 1975).
Protein content of cereals, oily nuts (oily seeds) and storage organs is appreciable, but generally lower, while that of fruit flesh averages about 20% of that in leaves. ……

Protein concentration appears normally to be highest in the immature leaf (Hladik, 1977a; Milton, 1979; Coley 1980; Oates et al., 1980; Waterman et al., 1980). ......
According to Heath (1978) leaf protein is able to supply sufficient amounts of all amino acids, although there is a relative shortfall in S[Sulfur]-containing components. ......

The B-complex components thiamine, riboflavin, pyridoxine and niacin are found in highest concentrations in cereals and nuts but dietary requirements can usually be met from foliage or other seeds.

Fruit flesh appears notably deficient in thiamine and riboflavin
[Note: in human nutrition, with commercially grown varieties, on average, this would be quite incorrect – as, per unit of energy, by dry weight, fruits on average have one half or more of these as cereals do, and 25% more than what would be required to meet the recommended daily requirements (see chapter 4, section 4.6, p335 here below)].

Vitamin C, if required, is plentiful in fleshy fruits, foliage, storage organs and some seeds.

...... The water content of immature leaves and fruit flesh generally exceeds 80% of the wet weight and in most other food classes exceeds 50%. Thus the diet of primates will normally ensure an adequate supply of water (Nagy and Milton, 1979b). ......

... and it now appears probable that all mammals harbor populations of a gut microflora capable of some degree of cellulose digestion.
In man the contribution of products from this microbial fermentation to the energy budget is small but in some monogastric primates it is very substantial (Clemens and Phillips, 1980; Milton et al., 1980)
[Note: In man, one needs to note the difference between physiological normality (function being as it should) and statistical normality (the average actual function of the population) – the human descending colon prefers as fuel the short chain fatty acids that are the product of microbial fermentation of cellulose, and it may be able to uptake and provide these for other tissues to a greater extent than is the average in many (especially Western) populations that consume a low-fiber diet.] ......

In most herbivorous primates there is an obvious preference for fleshy fruits and nuts, the obvious sources of readily available energy, and/or for leaves, flowers and seeds, the obvious sources of protein, minerals and most vitamins. ......

Plants must seek to minimize predation by either becoming invisible to potential predators or by actively protecting their products. This may be achieved by a number of 'passive' strategies such as the irregular production of a super-abundance of a potential food source interspersed with periods of low or nil productivity, which may prohibit the buildup of a guild of predators (Janzen, 1974), or the mechanical protection of the item, usually a seed, by a thick, tough coat, or by maintaining nutrient levels in a plant part that are insufficient for cost-effective processing.
It is now widely held that a major response of plants to predation has been the development of a vast array of chemicals (secondary compounds) that are able to confer some protection from herbivory and there is a substantial body of data to confirm the defensive potential of these compounds (Freeland and Janzen, 1974; Rosenthal and Janzen, 1979).
In view of the enormous pressure insect herbivory places on plants and the long period of co-evolution between the two, it seems probable that selection of anti-feedants has been primarily concerned with protection from insects and resulting benefits against other herbivores are best considered as incidental. ......

Tannins. … Tannins [Note: a group of polyphenolic compounds] act by forming insoluble complexes with several types of compounds, including proteins, nucleic acids, polysaccharides and alkaloids. The chemistry of the tannin-protein complex is not well understood but appears to involve the interaction of ortho-quinoid forms, generated under acid conditions, with amino acids, notably proline and lysine, and further hydrogen bonding between phenolic hydroxyls of the tannin and amide carbonyls of the protein. The complexes formed have generally been considered to be strongly associated at acidic pH but to associate at alkaline pH. However, recent work by Hageman and Butler (1981) suggests that the strength of the complex … while normally stronger at low pH, in some cases greatest complex stability occurs under alkaline conditions. Furthermore the relative affinity of a condensed tannin for a range of proteins varied over four orders of magnitude, suggesting a specific structure-activity relationship. ……
While the above represents a considerable body of data supporting the anti-feedant properties of tannins, several important field studies on insect herbivory have failed to observe the anticipated effects (Bernays, 1981).
Contrary to most findings regarding the effects of tannins on nitrogen retention by ruminants, Jones and Mangan (1977) report that tannins enhance retention by complexing with proteins in the rumen and protecting them from microbial metabolism; the complex then dissociating in the duodenum, liberating the protein for assimilation.
[Note: the action of tannins in humans thus warrants further investigation, as the human stomach is highly acidic, whereas the duodenum is much closer to neutral. Bidlack and Wang Ch 114 p1801 in Shils et al (2006) do not address the above aspect, and although some association has been observed with a reduction of some cancers, confounding by associated chemical is likely, and the general effect seems to be a negative one – if plants produce tannins as protection, for example against bacteria, then a negative overall effect on humans is more likely than a positive one…]
The unpalatable truth may well be that general hypotheses are untenable because the interactions that occur between different tannins and proteins produce complexes possessing widely differing properties. ……

Proteinase inhibitors … small proteins or polypeptides … [lectins???] ……

Simple anti-microbial compounds ……

Lignin … phenolic polymers … intimately linked with polysaccharide cell wall components, the two together forming cell wall fibre … reduces the susceptibility of the cell wall polysaccharides to enzyme degradation [digestion] ……

Toxins … Once it enters the body a toxin must be neutralized. Normal processes involve oxidation and/or conjugation to produce a less active compound that may be excreted; this being something of a 'hit and miss' process capable of yielding more toxic compounds. In mammals oxidations are mainly performed by the liver microsomes, the specificity of which varies between species and they are one source of the differential toxicity observed in the response of different animals to a toxin. ……

If this is so then a slow growth rate [of a plant] with energy and nutrients channeled into defense chemistry may be the optimum strategy in a nutrient-deficient site whereas rapid growth without great expenditure on defense chemistry, and the acceptance of the loss of a certain amount of photosynthate, may be more appropriate on a nutrient-rich site. ……

The life-span of a leaf, which varies from a few weeks in some ephemerals to several years in some evergreen trees, should influence the effort made to protect them. Both Stanton (1975) and McKey et al. (1981) have reported generalist herbivores feeding preferentially on the presumably shorter-lived leaves of deciduous rain-forest trees. …

Data concerning the relative distribution of tannins and other phenolics is less clear. A number of investigations of non-tropical species (Feeny, 1970; Dement and Mooney, 1974; Lawton, 1971) indicate highest concentrations in mature leaves.
In contrast our own studies on rain-forest trees (Gartlan et al., 1980; Oates et al., 1980; Choo et al., 1981) have regularly shown higher levels in immature leaves and this has been confirmed by other investigations of tropical and sub-tropical material (Milton, 1979; Coley, 1980; Macauley and Fox, 1980). ...... it is clear that, on average, levels of condensed tannins found in seeds are significantly lower [than in leaves, 4.0% vs. 7.8% $p<.001$]. Where tannins do occur in the embryo then the seeds are comparatively large. By contrast several types of toxin occur most widely in seeds where they may have the added advantage of acting as a reserve food supply after germination. Among the most well documented of these are the non-protein amino acids of the Luguminosae, modified fatty acids and phytohaemaglutinins (Rosenthal and Janzen, 1979). The testa covering the embryo is often thin and constitutes only a small fraction of the weight of the seed. ... the latter is often rich in tannins, even when the embryo is tannin free. ...... For fruits from a number of families Gartlan et al. (1980) found levels of tannins to be generally higher in the pod or capsule. ......

The loss of astringency that occurs during ripening may be due to increased polymerization of the tannin, impairing its ability to precipitate protein, or its complexation with polysaccharides or proteins in the fruit (McKey, 1979; Swain, 1979). ......
... The seeds of fleshy fruits are often characterized as being highly fibrous and lignified, possibly to protect them from mechanical damage during fruit processing by the frugivore.
Others, such as Strychnos, contain toxins, possibly to encourage rejection [or non-mastication before swallowing] of the seed. ....

... levels of tannin and fibre in flowers to be comparable to those in mature leaves of the same species. ......

Characteristics of bark structure vary considerably, both within and between plant families, but they are generally highly fibrous and heavily lignified. As a consequence they tend to be poorly digested, usually below 10% except where thin, green and newly formed. ......

In both cases [Nilgiri Langur and Black Colobus], and also in a recently completed study of vegetation available to the mountain gorilla (A. Vedder, D. Watts and P.G.Waterman – unpublished), correlation of selection with digestibility was greater than correlation of selection with protein content ... the outcome of this selectivity is that the averaged digestibility of all leaf intake by these primates is, in each case, 30-40% higher than the average digestibility of those leaves available at the site. ...

The general preference shown by most if not all primate folivores for immature leaves is almost certainly due to the lower fibre levels [Lignin, not cellulose per se? And what about water levels as correlated with lower lignin levels?] (Choo et al., 1981) as much as to their enhanced protein content. ...... The above seems to suggest that the role of tannins as anti-feedants may have been over-emphasised.
However, it must be remembered that the most detailed investigations have primarily concerned primates which are adapted to a folivorous diet. Observations on primates without gut adaptations and on the influence of tannins in fruits may suggest greater relevance. ......

The implication that tannins are feeding deterrents in the unripe fruit is strong and is sustained by the chimpanzee's method of processing the unripe fruit of the apocynaceous species *Saba florida* and *Landolphia lucida*. When ripe these fruits are consumed completely, with the seeds being either swallowed or spat out. When unripe fruit are processed the seeds are removed from the fruit pulp and consumed but the tannin-rich pulp is discarded. ……

SUMMARY
Most primates are wholly or largely vegetarian,
[Note: this is incorrect insofar as the smallest primates are very predominantly insectivorous, and also insofar as the larger primates probably all consume, both deliberately and incidentally, a substantial amount of insect (eggs, larvae and adult) foods]
the majority being predominantly frugivorous, a minority folivorous.

The distribution of nutrients in different plant parts is reviewed and points to foliage being the major source of protein and fruits the normal source of readily metabolized energy-rich carbohydrates and lipids.
Seeds and flowers also appear to offer sufficient protein but are chemically more variable."

2.24 Gibbons defend territory, baboon groups avoid each other in extensive areas of territorial overlap, langurs have intergroup dominance relationships Harvey (1992); Apes savagely defend territory, other than bonobos; hunting versus combat violence; population stress on food supplies; infanticide; generalisability of the psychology of killing.

Harvey PH. Section 4.3 Home range and territory. In, Chapter 4 Primate social organisation. In Jones S, Martin R, Pilbeam D, Bunney S (1992) The Cambridge Encyclopedia of Human Evolution. Cambridge University Press.

"Higher primates differ in the degree to which they will defend their territories from other conspecific groups – gibbons defend the territory, baboon groups avoid each other in extensive areas of territorial overlap, langurs have intergroup dominance relationships. [Among the great apes, our closest animal relatives, chimpanzees (males mostly) have been reported to violently and lethally defend (and acquire or expand) their territories against other groups, gorillas perhaps even more so but maybe less lethally, the savagely solitary orangutans likewise, while the bonobos are reputed to be less violent in general.

It is possible that human encroachment on habitat, impact on large cat predators on apes (i.e., leopards, which I surmise are more populous in the bonobo's habitats (mostly closed forest), which I gather are less impacted by humans), and enactment of imitatable territorial and hunting violence, may have increased the level of territorial violence to some extent, but it does seem likely to be a natural inclination – in the absence of sufficient ape population losses to predators, disease, cannibalism and/or infanticide such behaviour is almost certainly adaptive to survival – in the chimpanzee case combat is most often pursued by a more powerful group, and avoided by a weaker group.

The basic relevance of this to the human situation should be obvious enough. And it seems rather foolish, to me, and wishful thinking, mostly by those enamoured of human (in particular their own) carnivory, to posit that hunting violence and violence to others of one's own species are not at all related – in the human case, surely killing large mammals whose emotional similarity (i.e., in distress) to oneself is obvious enough makes it easier for one to inflict violence on other members of one's own species? And conversely, if humans (in the absence of enough population losses to predation and disease to prevent population stress on i.e. plant food supplies) must kill their own infants or other adults to secure enough territory to get (nearer to) enough plant food, surely this killing makes it more likely to kill animals? The former seems likely to be and to have been more in force than the latter, though.]

2.25 Chimpanzees dominance, leverage, cooperation, friendship rules and outrage from their breakage. De Waal (1996)

De Waal FBM Chapter 12 Conflict as negotiation. In, McGrew WC, Marchant LF, Nishida T (1996) (Eds) Great Ape Societies. Cambridge University Press.

[Regarding chimpanzees:]
"... leverage of subordinate or weaker parties increases with the need of the dominant or stronger parties for cooperation.
Negotiation takes place whenever it is more costly to disregard than to accommodate the partner's interests. ......
The dominant who terrifies subordinates may reduce cooperative tendencies upon which his own survival and that of his offspring depend. ......
So, animals settle for less than their initial objectives in order to secure the cooperation, tolerance, or survival of those on whom they depend. ......
Whenever a violation of expectations leads to protest, scientists should pay special attention. Protest may derive from an image of how the other ought to behave. Anecdotal evidence for 'outraged' reactions in chimpanzees are the sudden attack by Yeroen on Nikkie described in the previous section, a female's protest against a male's lack of reciprocation described by de Waal (1982, p207), and Nishida's (1984, pp390-1) observations of subordinates who single-handedly put to flight dominants who apparently broke rules of friendship. Such unusual sequences suggest that chimpanzee transactions may be governed by a crude moral sense."

2.26 Corralled chimpanzees: reciprocity with food sharing for grooming, and sharing for sharing. De Waal (1989)

De Waal FBM (1989) Food sharing and reciprocal obligations among chimpanzees. Journal of Human Evolution 18, 433-459.

"Food sharing has been proposed as the propeller of hominid evolution, particularly of the development of systems of mutual social obligation. ......

... chimpanzees in an outdoor corral ...... after adjustment of the data for proximity and dominance relationships, a significant degree of reciprocity remained ....
A was more likely to share with B after B had groomed A, but less likely after A had groomed B. This suggests a turn-taking rule in the exchange of social favors, which prevents one-sided accumulation of benefits.
Individuals who were recalcitrant to share (i.e. showed a low rate of food distribution) had a higher probability of encountering aggression when they themselves approached food possessors."

References for chapter 2

Chivers DJ, Andrews P, Preuschoft H, Bilsborough A, Wood BA (1984) Chapter 25 Food acquisition and processing in primates: concluding discussion. In, Chivers DJ, Wood BA, Bilsborough A (1984) Food acquisition and processing in primates. Plenum Press, New York.

Chivers DJ, Hladik CM (1984) Chapter 9 Diet and gut morphology in primates. In, Chivers DJ, Wood BA, Bilsborough A (1984) Food acquisition and processing in primates. Plenum Press, New York.

Codron D, Luyt J, Lee-Thorp JA, Sponheimer M, de Ruiter D, Codron J (2005) Utilization of savanna-based resources by Plio-Pleistocene baboons. South African journal of Science 101, May/June: 245-248.

Coe M (1984) Chapter 1 Primates: their niche structure and habitats. In, Chivers DJ, Wood BA, Bilsborough A (1984) Food acquisition and processing in primates. Plenum Press, New York.

De Waal FBM Chapter 12 Conflict as negotiation. In, McGrew WC, Marchant LF, Nishida T (1996) (Eds) Great Ape Societies. Cambridge University Press.

De Waal FBM (1989) Food sharing and reciprocal obligations among chimpanzees. Journal of Human Evolution 18, 433-459.

Foley R. Section 9.2 Studying human evolution by analogy. In, Chapter 9 Early human behaviour and ecology. In Jones S, Martin R, Pilbeam D, Bunney S (1992) The Cambridge Encyclopedia of Human Evolution. Cambridge University Press.

Galdikas (1988) Orangutan diet, range, and activity at Tanjung Puting, Central Borneo. Int J Primatology 9 (1) 1-35.

Harding RSO. Chapter 6 An order of omnivores: Nonhuman primates in the wild. In, Harding RSO, Teleki G (Eds) (1981) Omnivorous primates Gathering and hunting in human evolution. Columbia University Press. New York.

Harvey PH. Section 4.3 Home range and territory. In, Chapter 4 Primate social organisation. In Jones S, Martin R, Pilbeam D, Bunney S (1992) The Cambridge Encyclopedia of Human Evolution. Cambridge University Press.

Hladik C-M. Chapter 7 Diet and the evolution of feeding strategies among forest primates. In, Harding RSO, Teleki G (Eds) (1981) Omnivorous primates Gathering and hunting in human evolution. Columbia University Press. New York.

Kortlandt A (1984) Chapter 5 Habitat richness, foraging range and diet in chimpanzees and some other primates. In, Chivers DJ, Wood BA, Bilsborough A (1984) Food acquisition and processing in primates. Plenum Press, New York.

Lambert JE. Chapter 17 Seasonality, fallback strategies, and natural selection. In, Ungar PS. (Ed) (2007) Evolution of the human diet The known, the unknown, and the unknowable. Oxford University Press.

Mace G. Section 2.3 Differences between the sexes. In, Chapter 2 The life of primates. In Jones S, Martin R, Pilbeam D, Bunney S (1992) The Cambridge Encyclopedia of Human Evolution. Cambridge University Press.

Montgomery S (2009) Walking with the great apes Jane Goodall, Dian Fossey, Birute Galdikas. Chelsea Green Publishing.

Ripley S (1984) Chapter 2 Environmental grain, niche diversification and feeding behavior in primates. In, Chivers DJ, Wood BA, Bilsborough A (1984) Food acquisition and processing in primates. Plenum Press, New York.

Stanford CB (1999) The hunting apes. Meat eating and the origins of human behavior. 1999 Princeton University Press.

Stanford CB (1998) Chimpanzee and red colobus The ecology of predator and prey. Harvard University Press.

Strum SC. Chapter 8 Processes and products of change: baboon predatory behavior at Gilgil, Kenya. In, Harding RSO, Teleki G (Eds) (1981) Omnivorous primates Gathering and hunting in human evolution. Columbia University Press. New York.

Teleki G. Chapter 9 The omnivorous diet and eclectic feeding habits of chimpanzees in Gombe National Park, Tanzania. In, Harding RSO, Teleki G (Eds) (1981) Omnivorous primates Gathering and hunting in human evolution. Columbia University Press. New York.

Tutin CEG, Fernandez M, Rogers ME, Williamson EA, McGrew W (1991) Foraging profiles of sympatric lowland gorillas and chimpanzees in the Lope Reserve, Gabon. Phil Trans R Soc London B 334: 179-186.

Van Roosmalen (1984) Chapter 7 Subcategorizing foods in primates. In, Chivers DJ, Wood BA, Bilsborough A (1984) Food acquisition and processing in primates. Plenum Press, New York.

Waterman PG (1984) Chapter 8 Food acquisition and processing as a function of plant chemistry. In, Chivers DJ, Wood BA, Bilsborough A (1984) Food acquisition and processing in primates. Plenum Press, New York.

Wrangham RW, Conklin NL, Chapman CA, Hunt KD (1991) The significance of fibrous foods for Kibale Forest chimpanzees. Phil Trans R Soc London B 334: 171-178.

Yamagiwa J, Maruhashi T, Yumoto T, Mwanza N. Chapter 7 Dietary and ranging overlap in sympatric gorillas and chimpanzees in Kahuzi-Biega National Park, Zaire. In, McGrew WC, Marchant LF, Nishida T (1996) (Eds) Great Ape Societies. Cambridge University Press.

Chapter 3 Diet-related biochemistry and mutations – indicators of evolutionary adaptations to and for diet:

3.1 Summary of chapter 3 p264

3.2 Introduct.: evolutionary loss of function vs. development of a new function p267

3.3 Vitamin C biosynthesis evolutionarily allowed to deteriorate to nil function p268

3.4 Vitamin B12 uptake, transport and storage, mostly very well conserved; saturation of intestinal B12 uptake at a low 1.5-2.0 micrograms/meal in the context of the body's ability to store 3000-5000 micrograms; low breast milk B12 and low newborn liver B12 p268

3.5 Human dietary protein requirements related to food protein contents; human dietary protein requirements very low (20g purified amino acids/ 40g food protein / day/ person); food type protein/energy ratios; human breast milk protein/energy ratio lower than that of all other food types (less than 1/10 that of meat), and much lower (only a fraction) than milk of other mammals; human water requirements related to food water and food protein contents considered together p271

3.6 Protein metabolism for energy – limited to 50% of calories before overt toxicity; the metabolism of protein itself uses more energy than that of CHO and fats. Nitrogenous waste byproducts are diuretic p278

3.7 Starch digestion and evolutionary triplication of salivary amylase; starchy tubers and cereal grains p279

3.8 Supposed human inefficiency in biosynthesis of taurine, as in cats (obligate carnivores); vegan breast milk 30 times the taurine level of cow's milk-based infant formula indicates otherwise p280

3.9 Supposed human inefficiency in elongating and desaturating 18 carbon fatty acids to their product 20 and 22 carbon fatty acids, docosohexaenoic acid (DHA) and arachidonic acid (AA), similarly to obligate carnivores; comparisons of vegan breast milk with captive and wild apes and monkeys (similar levels found) indicates otherwise p281

3.10 Preformed Vitamin A from animal food is far in excess of requirements in western diets; ability to synthesize vitamin A from pre-vitamin beta-carotene from plant food sources retained through evolution p282

3.11 Lactose intolerance / adult lactase absence in most of the world's population; unsuitability of milk outside of infancy p282

3.12 Inborn errors of fructose metabolism are very low in frequency (1/20,000 or less); very common fructose malabsorption seems mostly due to unnatural modern context of rapid ingestion of some high fructose:glucose ratio foods in the absence of other meal-balancing protein and fat foods, and to be otherwise beneficial p283

3.13 Celiac disease 1% prevalence is only a very weak evidence, or none at all, for humans not being evolutionarily suited to eating cereal grains p286

References for chapter 3 p287

3.1 Summary of chapter 3

[The following facts and their derived meanings are arranged in order of the chapter sections, by number:

3.2 Firstly, most genetic mutations of significance involve a loss of function, rather than the development of a new function, for the fairly obvious reason that a single deleterious mutation may suffice to destroy the function of some protein coded for by the (now mutated) DNA, whereas to evolve a new or substantially different function usually requires at least several mutations, and maybe requires these to happen together – though they might happen singly and, if not harmful, stay around in the population for long enough to be inherited together in individuals and then *interact* to cause a beneficial effect that provided an advantage for evolutionary selection of those individuals and those genes together. Degenerative loss of function, even if this is loss of repression of a function that therefore allows the function, is therefore more likely to occur in evolution, and to require much less time to do so, than the generation of new functions in evolution.

3.3 The large majority of animals make their own vitamin C; humans no longer do, and so must get their vitamin C from their foods, indicating persistence of a genetic adaptation to a diet higher in vitamin C than that of most other animals.

3.4 Vitamin B12 intestinal uptake, transport and storage is mostly very well conserved. Saturation of intestinal B12 uptake occurs at a low 1.5-2.0 micrograms/meal in the context of the body's ability to store 3000-5000 micrograms.
Even western diet-eating omnivorous humans have a low breast milk B12 (one tenth that of cow milk), and their newborn babies have a low liver B12 concentration relative to adults (from 1/3 to 1/20).
A substantial amount of vegan experience with little or no B12 in the diet for years and no clinically detectable problems (imperfect, patchy and dubious as some of this evidence is) including this author's (my own) experience confirms generally low B12 requirements.
These facts are consistent with humans being genetically adapted to a low-vitamin B12 diet with a somehow irregular intake.
Most probably the diet to which humans are evolutionarily suited is one with a regular (seasonally variable) intake of incidentally and deliberately eaten insect matter, larvae and adults, with some meat (incidentally) able to be accommodated in lieu of insects.

3.5 Human dietary protein requirements are very low (20-25g purified amino acids/ 40g food protein / day / person).
Human breast milk protein/energy ratio is lower than that of all other food types (less than 1/10 that of meat), and much lower (only a fraction) than that of other mammals
Consideration of human water requirements clearly indicates that if enough fruit or root vegetables, and leafy vegetables, are eaten to meet water requirements, then most of the human protein requirements will be met from those sources, leaving only a small requirement for more concentrated protein sources like legumes, nuts, oily seeds, cereal grains, or animal foods, the first four of which would round out the energy requirement also in the amounts needed to round out the protein requirement, as they are comprised mostly of carbohydrate or fat, compared to meat (and especially lean meat, i.e., wild game other than of very cold climates), which is comprised mostly of water.
These facts clearly indicate that humans are genetically adapted to a diet very low in protein, which strongly argues against the possibility that we have become evolutionarily adapted to a diet with a high or even a moderate proportion of meat.

3.6 Generally, for humans more than 50% of dietary calories as protein is overtly toxic (from the nitrogenous waste resulting), and illness is felt within days.

The excess nitrogenous waste resulting from the breakdown of dietary protein for energy is diuretic – that is, it causes an increased excretion of water as urine. This is undesirable in habitats where water is in short supply, and maybe even in any case, as predators lurk preferentially near groundwater supplies.
High-protein diets for humans are reported to increase the cost of digestion by about 30% compared to high-fat diets.
Taken together, this information is evidence further suggesting that humans are not genetically adapted to a high meat diet, but to a predominantly fruit and vegetable diet.

Claims of meat being more energy-dense than plant foods are very obviously absurd: raw meat is about 70% water; raw legumes, nuts and oily seeds have very little water, the large majority of the weight being taken up by about 20% protein on average, and the another 70-odd% being fully fat or carbohydrate, so obviously raw nuts (600 kcal / 100 g) (or oily seeds (600 kcal / 100 g), and even also cereal grains (380 kcal / 100 g dry)) are more energy dense than raw meat (190 kcal / 100 g, lean; 230 kcal / 100 g, regular).

3.7 Human starch digestion has been enhanced in evolution by the triplication of salivary amylase after the last common ancestor of chimps and humans. The foods most probably involved in this were starchy tubers and cereal grains.
This information is evidence suggesting that we are evolutionarily genetically adapted to a diet with higher complex carbohydrate (starch) content than that of our common ancestor with the chimpanzee.
Legumes also have a substantial amount of starch, but consumption of these has probably been relatively constant throughout relevant periods of evolution, as legumes have been reported to be a dietary mainstay for virtually all relevant non-human primates (see also chapter 2, section 2.5).

3.8 Regarding the supposed human inefficiency in biosynthesis of taurine, as in cats (obligate carnivores), vegan breast milk taurine was reported to be 30 times the level of cow's milk-based infant formula, and to considerably overlap in distribution omnivorous human breast milk taurine, suggesting that humans are just as likely to be evolutionarily genetically adapted to be able to meet their taurine requirements on a very predominantly vegetarian diet as to not be able to.

3.9 Regarding the supposed human inefficiency in elongating and desaturating 18 carbon fatty acids to their product 20 and 22 carbon fatty acids, docosohexaenoic acid (DHA) and arachidonic acid (AA) (similarly to obligate carnivores), comparisons of vegan breast milk with captive and wild apes and monkeys, with similar levels found in vegan breast milk and the milk of wild apes and monkeys, indicates otherwise – that our relatively low fatty acid elongation and desaturation ability is evolutionarily genetically adapted to provide a sufficiency of the DHA and AA products on a very predominantly plant food/ vegetarian diet.

3.10 Preformed Vitamin A from animal food is far in excess of requirements in western diets. The human ability to synthesize vitamin A from pre-vitamin beta-carotene from plant food sources has been retained through evolution. That humans have not lost the ability to synthesize sufficient vitamin A from the pre-vitamin A (beta carotene), commonly available from many different types of plant foods, despite high-meat diets leading to liver stores (relatively low at birth) of an average 450milligrams, massively in excess of the 1milligram Recommended Daily Allowance (McLaren 1981), is clearly consistent with human high-meat (assuming the liver would have been eaten) diets not having been part of our evolutionary history for long enough to have caused us to evolve to be suited for high-meat diets, and is consistent with humans being evolutionarily a predominantly vegetarian animal.

3.11 Adult lactase persistence in humans is almost certainly the result of only a few or several thousand years of milk consumption after weaning. So, the persistent activation of the genes for lactase has become part of the genome in some small and generally geographically prescribed populations (i.e., in Europe) and their emigrants.
Clearly all other mammals on earth drink milk only as infants, then go on to eat the natural (if not always the optimal) foods of their species. The fact that a small number of human groups have evolved by a genetically degenerative (DNA *de*generation) process to be able to digest the milk sugar lactose is not at all evidence that dairy products are the best source of calcium for humans of any age, as other nutrients, and the combinations of foods in meals and diets, and even other factors, need to be addressed in deciding what is the best source of calcium.

People should be skeptical of assumptions that our genome is "adapted for a Paleolithic lifestyle" (Cordain, chapter 19; Eaton, chapter 20), for not only were Paleolithic diets quite variable in time and place, but natural selection may occur more quickly than commonly assumed.

3.12 Fructose is a sugar found in fruit, and some vegetables (i.e., sugar beets) where it most often is present in an equal or lesser concentration than glucose.
Inborn errors of fructose metabolism (in Fructokinase, (Fruct)Aldolase B, Aldolase A and Fructose-1,6-diphosphatase) are very low in frequency (1/20,000 or less).
The very common 'fructose malabsorption' seems mostly due to the unnatural modern context of rapid ingestion of some high fructose:glucose ratio foods in the absence of other meal-balancing protein and fat foods, and to be otherwise beneficial to the digestive tract.
Overall it seems that the human diet must have contained a large proportion of fruit throughout the periods of time that have produced the current rareness of inborn errors of fructose metabolism – the evidence here is for humans being by genetic evolutionary nature very much a fruit-eating animal. However, the evidence agrees with a moderate swing towards suitedness for starchy root vegetables, such that we might be suited for a diet of less fruit and more starchy root vegetables and cereal grains than apes are.

3.13 Celiac disease is a condition in which an allergic response to the cereal grain (wheat, barley, rye, oats, millet…) protein gluten causes structural and functional damage to the surface of the small intestine, causing reduced absorption of nutrients from digested food.
The prevalence of celiac disease, in those countries where reporting can be relied on, is considered to be about 1%.
A 1% prevalence of celiac disease in the context of the predisposing infectious diseases being more prevalent in the higher population densities of the agricultural (and modern urban) era is hardly a very strong argument for humans not being evolutionarily suited to eating cereal grains – particularly as other foodstuffs are also the subject of allergies: peanuts and tree nuts around 1% each in a variety of countries, milk 0.4% in the USA, egg 0.2% in the USA, egg 2% in UK children, 0.1% in adults, shrimp 1% in the USA and shellfish 5% in an Asian country. These figures are only very roughly comparable indeed, as they need to be related to exposure frequency and schedule, and meal context, and corrected for socioeconomic confounders such as affordability, weaning, and related infectious and nutritional deficiency diseases – obviously cereal grains are almost certainly *the* most affordable of these and all foodstuffs, and are therefore very likely confounded in their association with allergy by a more unfavourable collection of socioeconomic confounders such as weaning, and relevant diseases.
Celiac disease's 1% prevalence is only a very weak evidence, or none at all, for humans not being evolutionarily suited to eating cereal grains, nor for cereal grains not having been a substantial part of the diet in various places in the human evolutionary past, as we can safely assume peanuts, tree nuts, eggs and shellfish were also.]

3.2 Introduction: evolutionary loss of function vs. development of a new function

[I'll begin by introducing some basic concepts that are needed for sensibly considering what genetic nutrition-related changes might have occurred in human evolution.
Firstly, most genetic mutations involve a loss of function, rather than the development of a new function, for the fairly obvious reason that a single deleterious mutation may suffice to destroy the function of some protein coded for by the (now mutated) DNA, whereas to evolve a new or substantially different function usually requires at least several mutations, and maybe requires these to happen together – though they might happen singly and, if not harmful, stay around in the population for long enough to be inherited together in individuals and then *interact* to cause a beneficial effect that provided an advantage for evolutionary selection of those individuals and those genes together.
Single deleterious mutations happen more often, and are removed more often by natural selection.

So, in the sections that follow in this chapter, the general theme is to address what sort of mutation might have occurred (or has apparently not occurred) to make humans a more naturally or optimally predominantly or substantially meat-eating animal than they were at a previous point in our evolutionary history, in particular our very predominantly vegetarian evolutionary origins of 6 million years ago when our ancestors diverged from the ancestors of the chimpanzees.

Here below I'll start by providing an example of the sort of gene-damaging mutation that can be allowed to remain in the gene pool because the function the gene coded for was no longer required; the loss of ability to biosynthesize vitamin C.

Then, I'll address genes and their functions involving two nutrients that animal-food has in greater abundance than plant-food (vitamin B12 and protein), and that an evolutionary adaptation to eating a large proportion of meat should therefore have damaged in the same way that eating foods that supplied a continuously sufficient amount of vitamin C allowed the eventually total destruction of an ancestor of our's ability to biosynthesize vitamin C.

Other such nutrients and the genes involved in them addressed here include amylase (starch digesting enzyme), lactose (the particular sugar of milk), the long chain polyunsaturated fatty acids (LCPUFA) docosohexaenoic acid (DHA) and arachidonic acid (AA), the amino acid taurine, vitamin A, the fruit sugar fructose, and the cereal grain protein gluten.]

3.3 Vitamin C biosynthesis evolutionarily allowed to deteriorate to nil function

[The large majority of animals make their own vitamin C; humans no longer do, and so must get their vitamin C from their foods.
How did this come about? Far and away the most obvious and likely answer is that the common ancestor of humans and the apes had been eating fruits and green leaf vegetables as a major part of its diet for so long that although the animal(s) it had evolved from had a different diet and needed to make vitamin C at least sometimes, that common ancestor of humans and the apes no longer needed to make any vitamin C at all, because those fruits and green leaf vegetables had all the vitamin C that was required for the health of that common ancestor, and so one after another mutation that damaged the DNA coding for the vitamin C-synthesizing enzyme(s), and therefore resulted in changed and less-functional vitamin C-synthesizing enzyme(s), was allowed to stay in the DNA because it caused no disadvantage to the individuals carrying them, such that eventually that DNA sequence was mutated fully away, and the ability totally lost.
This is strong evidence of our very predominantly plant-food eating evolutionary origins]

3.4 Vitamin B12 uptake, transport and storage, very well conserved; saturation of intestinal B12 uptake at 1.5-2.0 micrograms/meal in the context of the body's ability to store 3000-5000 micrograms; low breast milk B12 and low newborn liver B12

[Now, there is a nutrient that is maybe more sharply differentiated between animal-food and plant-food than probably any other – Vitamin B12 is said to be found only in animal-food, some fermented foods, and some seaweeds. (Though most such seaweed molecules are in fact non-active B12 analogues that are worse than useless because they occupy the molecular places of active B12 molecules – so care is required here (although there are some physiological mechanisms to screen out the non-active analogues from absorption in the digestive system) – apparently those seaweeds with active B12 are green and purple lavers, and those sources with inactive pseudo-vitamin B12 include the blue-green algae at least recently used in supplements for humans (Watanabe 2007)).
It seems that some plant foods grown in soils containing feces (most natural soils to some extent) might absorb some B12 from such soils and might therefore contribute some useful part, even most in some cases, of human vitamin B12 needs (Mozafar 1994), but this is not fully established/ documented, and there is an analogues question here too.

Cobalt, the mineral uniquely a part of vitamin B12, is found in green leaves including grass, but only if the soil does not happen to be deficient in cobalt, which it sometimes is, such that ruminant animals like cows that get their vitamin B12 from gut bacterial production, get much the same nerve-damaging vitamin B12 deficiency disease that some humans get in the absence of sufficient dietary intake of vitamin B12. Note also that channel catfish have been reported to have B12-synthesizing gut bacteria providing useful quantities of B12, and golden hamsters likewise. Humans very probably have some small ability to do this also, certainly dependant on dietary cobalt levels and probably also on other dietary factors, and note that in humans B12 is absorbed in the very last part of the small intestine, next to the large intestine where the B12-synthesis is.

So, if humans had eaten such a preponderance of animal-foods during the course of their evolution that they had significantly evolved away from plant-food predominant eating towards predominantly carnivorous eating, it should be expected that, due to the relatively large amount of vitamin B12 being taken in, similar to what almost certainly happened in the case of vitamin C as noted above, that all, most, or a substantial proportion of (some of) the human population(s) should have lost some of their ability to uptake vitamin B12 from the gut, to transport it to the liver, and to store it in the liver, due to damaging mutations being allowed to remain in the population's(') gene pool(s), because these functions were no longer needed to be so good as they were. And those abilities were very good because the human liver can store enough vitamin B12 to last 4 to 5 years without any more coming in from the diet, a circumstance that is not known to apply to any other nutrient, especially water-soluble nutrients – which in itself suggests that our evolutionary origins were of very low and irregular vitamin B12 intake.

An important fact is that the breast milk of humans, even those eating much animal-food, is remarkably low in B12, at 0.4 microgram / litre vs. cow milk at 4 microgram / litre (Hay et al 2008), this in the context that there is roughly the same energy/calories per litre in human breast milk and cow's milk, but 4 times the protein/calories in cow's milk.
The newborn human is reported to have only roughly 1/3 the liver B12 concentration of children 0.5-2.0 years old and 20-40 year old adults (I cannot readily determine what population these were drawn from, but the context suggests a ~ 1966 western population), but the adult levels in other studies have been as much as roughly 20 times that of the newborn value here (i.e. in Texas, where meat intake is greater than for example the UK).
These breast milk and newborn liver B12 values are consistent with a natural requirement much lower than the amount supplied by common western diets, and suggest that the common western intake of vitamin B12 is far in excess of what we have evolved to require (see also McLaren 1981).

Vitamin B12 is stored mainly in the liver, less in the kidneys, and less still in the plasma and the rest of the body tissues – I cannot yet provide the quantification of this but in cows, the liver has a concentration some tens of times that of other tissues.
Just exactly (molecularly) where this storage of B12 takes place seems in doubt, as does its regulation; I quote Quadros (2010): "The precise mechanism by which this storage occurs is not clear. As very little Cbl [(cobalamin, B12)] is distributed as free Cbl in the tissue, the assumption is that it is mostly bound to the two enzymes [(that require it for function)], methionine synthase (MS) and methylmalonylCoA mutase (MMU). However, both enzymes are only partially saturated with Cbl even under replete conditions (Kolhouse et al, 1980) and no direct correlation between the activity of these enzymes and Cbl-replete status exists. The mechanism of Cbl release from its storage site under deficient conditions is also not known. Typically, the fall in serum Cbl precedes the fall in liver stores (Booth and Spray 1960). ... ... The accumulation of Cbl in some tissues like liver and kidney appears to be the exception that cannot be explained by the dynamic flux [(amongst the known uptake and transport proteins and the two enzymes named above)] and raises a number of questions. For example, what factors regulate the accumulation of Cbl in liver and kidney? How does the Cbl-replete or Cbl-deficient status affect this process and the two enzymes? Does the stored Cbl compartmentalize to a location other than the two enzymes? Does the Cbl status affect synthesis, turnover and saturation of Cbl-dependent enzymes?" Nevertheless we may proceed with analysis...

It should further be borne in mind that in eating a very predominantly plant-food diet in nature an animal (in particular our hominin ancestors) certainly incidentally and often unknowingly ingested small amounts of insect eggs, larvae and dung, all of which would add to whatever vitamin B12 would almost certainly have been consumed from eating bird eggs found in the course of getting fruit and green leaves out of trees, and deliberately eating insects, possibly including termites seasonally as do chimpanzees.
It has been reported that absorption of B12 from eggs is reported to be only roughly 10% compared to roughly 50% from meats (Watanabe 2007).
It has been reported that insects vary widely in their B12 content, from none at all (e.g. the common housefly), through intermediate values, up to termites having 940pg/ mg dry weight (Wakayama et al (1984) (note that termites rely on a large amount of gut digestion of cellulose by bacteria – and many other insects, or at least their larvae, have a similar

type of requirement for digesting dellulose – and note how much higher cow (have gut bacterial digestion of cellulose) milk is than human milk in B12 (and protein)(re growth). This is very roughly 10ug/ 100g termite wet weight (factor of 10 for dry-wet; 10^6pg in a ug) and may be roughly compared to 100ug/ 100g beef liver and 2.5ug/ 100g lean beef.

These figures then need to be related to the ability of humans to take up B12 from the diet – if 1250-odd kcal of lean (cow) meat (the maximum (1/2) of calories without toxicity) was eaten over 3 meals in a day, the human body could just (at the maximum uptake usually given, which might be doubled in deficiency) absorb its B12 content – If the termites were quite fatty, roughly close to the same might be said – for most insects other than termites the B12 is all absorbable - not even close to the same can be said about cow liver - and it seems that insect B12 levels have a wide variety of B12 levels ranging from much less than lean cow meat to much more than cow meat, but less than cow liver.

There are yearly seasons for insects, and marked seasons (dry, low-fruit) of high termite consumption have been noted for wild chimpanzees (see chapter 2).
Note that with modern agricultural/ horticultural use of insecticides, incidentally eaten insects is one source of vitamin B12 that would have largely gone out of the human diet.

Crustaceans were probably eaten at times also, though being much less ubiquitous in the (human) environment than insects. The same applies to shellfish (e.g. mollusks).

That saturation of intestinal B12 uptake occurs at a relatively low 1.5-2.0 micrograms/ meal (Shils et al 2006, Watanabe 2007), compared to the body's ability to store 3000-5000 micrograms of B12 (~ half in the liver, targeted, not merely by mass action and micro character…) (roughly 5 years' worth of minimum requirement (at roughly half the Recommended Daily Allowance) in the ostensible absence of any intake or absorption, e.g. after total gastrectomy (Weir & Scott 1999)), alongside the above facts, suggests that:

1. The RDA of 2-3micrograms/day is well in excess of our needs, because roughly 4-6 micrograms per day is enough to eventually deposit 3000-5000 micrograms in the body stores, and, it is unlikely that an adult human needs more B12/kcal than a growing infant.

2. We are probably adapted for going long, more or less continuous periods with hardly any external B12 intake; otherwise the ability to store roughly 5 years' worth of B12 would not exist, as this is far in excess of the body's ability to store any other nutrient, especially water-soluble vitamins. My own (p383) and others' experience confirms this.

3. In the periods of time (in seasonal or longer cycles) that we as individuals are taking in dietary B12, we *might* be adapted for taking in intermittent large doses of vitamin B12 such as would occur in the case of intermittent intake of large amounts of meat in single meals from large animal hunting (in which case half of the daily calories (which is the maximum that can come from protein without experienced metabolic toxicity) 1,250 kcal, would contain 10micrograms or 3micrograms per meal which is only a little over the maximum intestinal uptake ability), but probably this occurs very infrequently otherwise the relatively massive B12 storage capacity of the liver probably would not exist.
Also the adaptation is *at least as plausibly* due to an (initial) evolutionary adaptation to sporadic (i.e. seasonal) consumption of larger amounts of insects and/or their larvae.

Some populations of modern man may eat a large proportion of meat in the diet(s), but the evidence is that this practice has not been going on for long enough to change our genes to suit it – the evidence here is clearly most closely in accord with humans being (still, as ever) evolutionarily adapted to a very predominantly vegetarian diet with a relatively small intake of animal matter (probably mostly insect, seasonally variable).]

3.5 Human dietary protein requirements related to food protein contents; human dietary protein requirements very low (20g purified amino acids/ 40g food protein / day/ person); food type protein/energy ratios; human breast milk protein/energy ratio lower than that of all other food types (less than 1/10 that of meat), and much lower (only a fraction) than that of other mammals; human water requirements related to food water and food protein contents considered together

[Regarding protein, although many nuts, seeds and legumes have the same roughly 20% protein content as flesh foods (and some have more than 20%), if the ratio of protein to energy is compared, flesh foods have a much higher protein/energy ratio because more than half of all flesh is water, whereas in the grains, nuts, seeds and legumes there is much less than half, the balance being fat and carbohydrate.
It should be borne in mind that humans need far more energy-providing fuels than we need protein.
Raw nuts (600 kcal / 100 g) (or oily seeds (600 kcal / 100 g), and also cereal grains (380 kcal / 100 g dry) for that matter…) are more energy dense than raw meat (190 kcal / 100 g, lean; 230 kcal / 100 g, regular).

There is the question of whether we accept the fashionable theory that the meat that primitive humans ate was (very) lean compared to meat that modern humans eat, or whether we allow that if a taste habit for cooked fat (particularly after protein requirements have been met in the meal) was commonly acquired (and my personal experience and the large amount of cooked fatty foods that modern humans eat tells me that taste habit can, is and was acquired), that some of a prey's lean meat might have been discarded in favor of killing more prey to get more fat instead.
And indeed there is ample evidence that when chimpanzees prey on monkeys, which they do mostly when they are very hungry in the season of fruit scarcity, they favor the most fat-rich portions of the carcass (see section 2.3 and others in chapter 2), and when gatherer-hunters are similarly very hungry in seasons of scarcity, they may discard an entire carcass of an animal they have just hunted, if it does not have enough fat.
And see also section 3.6 below on more than 50% of dietary calories as protein being tangibly metabolically toxic for humans.
Populations of humans feeding (for tens of thousands of years, if there were any such……) on marine mammals in very cold water (or, very likely, even allowing for metabolic rate differences in response to environmental temperature, land mammals in very cold air…) would of course have had access to plenty of animal fat (the well documented whale and seal blubber eaten in large amounts by Eskimos – and it seems that their previously much vaunted cardiovascular health is actually not so good on that diet after all (Bjerregaard, Young, Hegele 2003), and would have to be considered as particular populations not very representative of other populations or of their own populations in different points in evolutionary time of different environmental temperature (and this could be very informative if investigated genetically regards the aspects discussed here, but this has not been done yet, nor is likely to be for some years if ever…).

However, as a starting point for some quantitative comparisons, I'll provide below two different fat contents (5%, 'lean', and 15%, 'fatty') of ground beef for inclusion amongst the other flesh foods dealt with here.

So, the protein/energy ratio of various *individual* representative foods are given below in Table 1
(See table 2 for the protein/energy ratios of the food type *groups,* which are best considered for the purposes of my investigation here)

(Notes for Tables 1 and 2:

Overall, the different groups of foods are arranged in decreasing order of protein/energy ratio, with the exception of the green leaf and flower vegetables, because this last group is substantially more limited in meeting human energy requirements than the other groups due to the large amount required to do that.

Within the different groups of food, the meats and milks are arranged zoologically; the legumes, seeds, nuts, grains, fruit and root vegetables in decreasing order of protein/energy ratio, and the green leaf and flower vegetables in decreasing order of protein content per 500g, which is a reasonable amount to expect a human to eat in the course of one day.

The amount of energy chosen for the ratio presented here, 2500kcal, is the amount of energy required by a fairly active 65kg person.
For such a person, not pregnant or lactating or ill, the recommended intake of protein is about 50g (Insel, Turner and Ross, 2007), and keep in mind that this already has a safety factor of about 2, or 100% extra, to allow for incomplete digestion of protein and absorption of amino acids from the intestines, over the only about 20-24g that would be required for an average adult male in the case of digestion and absorption being complete and the amino acid proportions with each other being close to perfect, and there being metabolic adaptation to a minimal rather than an excess protein intake (Milward 2001, and earlier experimental report by different author not readily to hand).

Values over 100 have been rounded to the nearest 5 so as not to obscure the relative magnitudes of the different food groups for readers less versed in numerical comparisons, and because even so the implication of even that lesser degree of accuracy is almost certainly false because food constituents are more variable than that degree of implied accuracy.

All of the values are calculated from the Appendix A Food Composition Tables in Insel, Turner and Ross (2007), other than those for human milk, which were calculated from Shils, Shike, Ross et al (2006), and those for Brazil nuts, Macadamia nuts, cassava, kale, broccoli, some beans (black, red kidney, white, navy and adzuki) and plums, pineapple, watermelon, jackfruit, mango, apricot, persimmon and undried apple, and Red, Savoy and Chinese cabbage(s), and squash and zucchini, which were calculated from Whitney and Rolfes (2008).

For convenience of comparison, the average values of the different food groups collated in Table 1 below are listed immediately below Table 1, in Table 2):

Table 1: Protein content and weight per 2500kcal of foods

Animal Flesh

Meat (5% fat ground beef, broiled):	390gProtein/ 2500kcal (1.5kg)
Meat (15% fat ground beef, broiled):	260gProtein/ 2500kcal (1.0kg)
Ham (lean, roasted):	400gProtein/ 2500kcal (1.6kg)
Lamb (lean, roasted):	370gProtein/ 2500kcal (1.3kg)
Chicken Meat (all, fried):	385gProtein/ 2500kcal (1.1kg)
Tuna (baked/broiled):	400gProtein/ 2500kcal (1.4kg)
Salmon (baked/broiled):	350gProtein/ 2500kcal (1.4kg)
Squid (baked):	335gProtein/ 2500kcal (1.8kg)
Octopus (raw):	460gProtein/ 2500kcal (3.0kg)
Crab (leg, boiled):	500gProtein/ 2500kcal (2.6kg)
Prawns (boiled):	570gProtein/ 2500kcal (2.5kg)
Clams (canned):	435gProtein/ 2500kcal (1.7kg)
Oysters (raw):	250gProtein/ 2500kcal (3.7kg)
Mussels (steamed/boiled):	340gProtein/ 2500kcal (1.5kg)
Average Animal Flesh:	**390gprotein/ 2500kcal (1.9kg)**

Egg (whole, raw):	**230gProtein/ 2500kcal (1.7kg)**

Milk

Milk (cow, whole):	135gProtein/ 2500kcal (4.2kg = 4.2L)
Milk (goat):	135gProtein/ 2500kcal (3.6kg = 3.6L)
Average Cow/Goat Milk:	**135gProtein/ 2500kcal (3.9kg = 3.9L)**

Milk (human, mature (after day10)): 33gProtein/ 2500kcal (3.7kg = 3.7L)

Legumes

Soy beans (cooked):	240gProtein/ 2500kcal (1.4kg)
Lentils (cooked):	195gProtein/ 2500kcal (2.2kg)
Black beans (boiled):	170gProtein/ 2500kcal (1.9kg)
Red kidney beans (canned):	155gProtein/ 2500kcal (3.0kg)
White beans (boiled):	155gProtein/ 2500kcal (1.8kg)
Navy beans (boiled):	150gProtein/ 2500kcal (1.8kg)
Green Peas (boiled):	150gProtein/ 2500kcal (3.0kg)
Adzuki beans (boiled):	145gProtein/ 2500kcal (2.0kg)
Chickpeas (cooked):	140gProtein/ 2500kcal (1.5kg)
Average Legume (cooked/soaked):	**165gProtein/ 2500kcal (2.1kg)**
Peanuts (P. butter = raw):	115gProtein/2500kcal (0.43kg)

Seeds

Sunflower seeds (dry = raw):	100gProtein/ 2500kcal (0.42kg)
Pumpkin seeds (roasted):	105gProtein/ 2500kcal (0.56kg)
Sesame seeds/ Tahini:	80gProtein/ 2500kcal (0.44kg)
Average Seeds (dry/raw):	**95gProtein/ 2500kcal (0.47kg)**

Nuts		
Pistachios:	95gProtein/ 2500kcal	(0.45kg)
Almonds:	92gProtein/ 2500kcal	(0.43kg)
Cashews (roasted):	68gProtein/ 2500kcal	(0.43kg)
Hazelnuts:	60gProtein/ 2500kcal	(0.40kg)
Walnuts:	58gProtein/ 2500kcal	(0.40kg)
Brazil nuts (dried = raw?):	54gProtein/ 2500kcal	(0.38kg)
Pecan nuts:	32gProtein/ 2500kcal	(0.36kg)
Macadamia nuts (raw):	31gProtein/ 2500kcal	(0.35kg)
Average Nuts (raw/dry):	**61gProtein/ 2500kcal**	**(0.40kg)**

Grains		
Oats (rolled):	105gProtein/ 2500kcal	(0.65kg)
Wheat (flour, whole):	98gProtein/ 2500kcal	(0.74kg)
Millet (cooked, 70% water):	72gProtein/ 2500kcal	(2.1kg wet, 0.63kg dry)
Barley (whole, cooked, wet):	65gProtein/ 2500kcal	(1.8kg wet, 0.63kg dry)
Rice (brown, cooked, wet):	58gProtein/ 2500kcal	(2.2kg wet, 0.68kg dry)
Rice (white, cooked, wet):	50gProtein/ 2500kcal	(1.9kg wet, 0.55kg dry)
Average Grains (4 wet, 6 dry):	**75gProtein/ 2500kcal**	**(2.0kg wet, 0.65kg dry)**

Starchy Root Vegetables		
Potatoes (boiled):	62gProtein/ 2500kcal	(2.9kg)
Sweet potatoes (baked):	48gProtein/ 2500kcal	(2.8kg)
Yams (boiled):	40gProtein/ 2500kcal	(3.3kg)
Cassava:	15gProtein/ 2500kcal	(1.6kg)
Average Starchy Root Vegetables (undried):	**41gProtein/ 2500kcal**	**(2.6kg)**

Carrots (raw):	**82gProtein/ 2500kcal**	**(6.0kg)**

Fruit		
Apricot:	75gProtein/ 2500kcal	(5.2kg)
Cantaloupe/Rockmelon:	66gProtein/ 2500kcal	(7.3kg)
Peach:	65gProtein/ 2500kcal	(6.4kg)
Watermelon:	54gProtein/ 2500kcal	(8.3kg)
Oranges:	52gProtein/ 2500kcal	(5.3kg)
Pawpaw:	42gProtein/ 2500kcal	(6.4kg)
Honeydew melon (cubes)	42gProtein/ 2500kcal	(7.0kg)
Plums:	41gProtein/ 2500kcal	(5.5kg)
Jackfruit:	38gProtein/ 2500kcal	(2.7kg)
Pear (fresh):	26gProtein/ 2500kcal	(4.3kg)
Pineapple (raw):	26gProtein/ 2500kcal	(5.0kg)
Bananas:	25gProtein/ 2500kcal	(2.8kg)
Dates:	22gProtein/ 2500kcal	(0.89kg)
Grapes (seedless):	22gProtein/ 2500kcal	(3.7kg)
Mango:	18gProtein/ 2500kcal	(3.8kg)
Persimmon:	11-16gProtein/ 2500kcal	(1.9kg)
Apples (average (un)dried):	12gProtein/ 2500kcal	(4.7kg, undried)
Average Fruit (undried):	**38gProtein/ 2500kcal**	**(4.8kg)**

Avocado (fatty fruit):	30gProtein/2500kcal	(1.6kg)

Green Leaf, Flower and Shoot Vegetables

Kale (raw):	164gProtein/2500kcal	(16gProtein, 250kcal)/500g
Broccoli (raw):	220gProtein/2500kcal	(14gProtein, 160kcal)/500g
Broccoflower:	235gProtein/2500kcal	(15gProtein, 160kcal)/500g
Cauliflower:	210gProtein/2500kcal	(10gProtein, 120kcal)/500g
Lettuce:	310gProtein/2500kcal	(10gProtein, 70kcal)/500g
Cabbage (Savoy):	185gProtein/2500kcal	(10gProtein, 140kcal)/500g
Chinese Cabbage:	325gProtein/2500kcal	(7gProtein, 60kcal)/500g
Bamboo shoots (boiled):	320gProtein/2500kcal	(7gProtein, 60kcal)/500g
Cabbage (Red):	115gProtein/2500kcal	(7gProtein, 160kcal)/500g
Cabbage (standard):	140gProtein/2500kcal	(7gProtein, 120kcal)/500g
Av. Leaf, Flower, Shoot Veg's:	**220gProtein/2500kcal**	**(10gProtein,130kcal)/ 500g**

Squash and Zucchini

Squash, sum., 5 var's (boiled):	110gProt/ 2500kcal	(4.5gProt, 100kcal)/500g
Squash, win., 5 var's (baked):	59gProt/ 2500kcal	(3.5gProt, 155kcal)/500g
Zucchini, summer (boiled):	110gProt/ 2500kcal	(3.5gProt, 80kcal)/500g
Average Squash and Zucch:	**93gProt/ 2500kcal**	**(3.8gProt, 110kcal)/500g**

For convenience, the average values of the different food groups collated above are listed here below in Table 2, with a few common oddments relocated underneath:

Table 2: Average Protein content and weight per 2500kcal of food groups

Average Animal Flesh:	**390gprotein/ 2500kcal (1.9kg)**
Egg (whole, raw):	**230gProtein/ 2500kcal (1.7kg)**
Average Cow/Goat Milk:	**135gProtein/ 2500kcal (3.9kg = 3.9L)**
Milk (human, mature (after day10)):	**33gProtein/ 2500kcal (3.7kg = 3.7L)**
Average Legume (cooked/soaked):	**165gProtein/ 2500kcal (2.1kg)**
Average Seeds (dry/raw):	**95gProtein/ 2500kcal (0.47kg)**
Average Nuts (raw/dry):	**61gProtein/ 2500kcal (0.40kg)**
Average Grains (4 wet, 6 dry):	**75gProtein/ 2500kcal (2.0kg wet, 0.65kg dry)**
Average Starchy Root Veg (undried):	**41gProtein/ 2500kcal (2.6kg)**
Average Fruit (undried):	**38gProtein/ 2500kcal (4.8kg)**
Av. Leaf, Flower, Shoot Veg:	**230gProtein/ 2500kcal (10gProtein,130kcal)/ 500g**
Average Squash and Zucchini:	**93gProtein/ 2500kcal (12.2kg)**

Carrots (raw):	82gProtein/ 2500kcal (6.0kg)
Peanuts (P. butter = raw):	115gProtein/ 2500kcal (0.43kg)
Avocado (fatty fruit):	30gProtein/ 2500kcal (1.6kg)

Now, I shall begin here by noting that "human milk total protein content is the lowest among species" (Shils, Shike, Ross et al, 2006), and one can see in the data tabled above that, at 33g protein per 2500kcal compared to cow and goat (herbivores, not carnivores, note) milk at 135g protein per 2500kcal, and other animals (horse, sheep, dog, striped skunk, mink and Guinea pig) milks at 90 – 145g protein per 2500kcal (Oftedal 1981), human milk is not only relatively low in protein, it is very low in protein in the absolute sense. And I proceed by asking, does this fact suggest that we are a species that requires a relatively high protein diet, or a relatively low protein diet? More specifically, does this fact indicate that those periods in our evolutionary history when our human/hominin ancestors did eat significantly more meat (for example very recently, say in the last 40,000 years, in colder parts of Europe) than previously (the rest of the 6 million years since our evolutionary line diverged from that of the chimpanzees) have caused us to evolve into an animal that requires anything more than just the very small proportion of meat or other animal foods (i.e., insects) necessary to meet our vitamin B12 requirements?

But before formulating an answer, I ask a related question, is it likely that a human infant, engaged in relatively more growing (for which proportionately more protein is required) than activity (for which proportionately more energy is required) compared to older humans, is likely to require proportionately more dietary protein than energy when it does grow older? The answer to this latter question is that almost certainly, like the hummingbird that feeds itself largely on energy-rich nectar while it feeds its growing young on protein-rich insects, as a human infant grows older and into adulthood it is likely to require proportionately less dietary protein than energy.

The answer to the first question here above, then, is that very probably the fact of the very low protein content (low protein/energy ratio) of human milk suggests that humans are an animal that by biological and evolutionary nature eats, and is suited to, a very low-protein diet. In other words, a diet that requires no more of meat or other animal foods (i.e., insects) than just the very small amount necessary to meet our vitamin B12 needs, alongside an also relatively small proportion of other (plant) protein-rich foods.

This is clearly in accord with our evolutionary closeness with the great apes, and the low-protein diet that they eat. Now, given the massive difference in protein/energy ratio between both the 33g protein per 2500kcal of human milk and the experimental minimum of 20-24g per day (per approximately 2000kcal) required for adult human males, and the average animal flesh content of 390g protein per 2500kcal or the high(15%)-fat herbivore (cow) meat content of 260g protein per 2500kcal, if the time-spans in humans' recent historical past during which humans ate proportionately more animal-food had been long enough, and the proportion of animal-food eaten been great enough, wouldn't it be reasonable to expect that, in the same way that through evolution humans lost entirely the ability to synthesize our own vitamin C, we would have lost some of that obvious tendency to use absorbed dietary protein (amino acids) for tissue building and repair rather than for energy?

The answer seems to be that if that time-span had been long enough, and the proportion of animal-food eaten great enough, then this should have allowed enough individual DNA-*degenerative* mutations to have occurred and not been removed from the gene pool due to them not reducing the likelihood of survival (the probability of the *set* of mutations required to confer some sort of DNA-*neogenerative* advantage is much smaller, note).

However, while *some few* individuals of *some of* the world's populations may indeed have inherited such mutations, it seems (from orthodox nutritional data, as detailed above) that the very large proportion of the world's population(s) has not inherited them, because the periods of time in question have not been long enough to allow that evolution to occur, probably in conjunction with the proportion of animal-food eaten not having been great enough, and because those periods have been followed by significantly long periods within which the proportion of animal-food eaten was much lower (For example, in much of Asia for at least the last several thousand years, and most of the European peasant populations for most of history).

Regarding the question of from which foods humans might be evolved to best or mostly obtain their protein requirements, it is necessary to consider the contexts of human water requirement (Adequate Intakes are given as 3.7 litres for males and 2.7 liters for females (pregnant 3.0 liters, lactating 3.8 liters) (Insel, Turner and Ross, 2007), and how this relates to the protein and water content of foods. Consider the following:

1. Grains have roughly 10% protein content;

2. Fruit and vegetables are low in protein in their wet weight, but considered in the amounts that would be eaten to round out the dietary water and energy requirements after only enough (substantially less than, actually, being the point) protein from the more protein-rich plant-foods noted just above to meet the dietary protein requirements (which are quite little, see immediately below), they could actually contribute a substantial fraction of the daily protein requirement (for example, 2kg of fruit and vegetables eaten in a day, at 1% of wet weight protein content, provides 20g of protein);

3. If amino acids are present in exactly the right proportions and are perfectly absorbed then only 20-25 grams per day of protein (that would be only roughly 100-200 grams of the protein-rich foods noted above: flesh, legumes, nuts, seeds, grains) is required per day for an adult human (figures between 40 and 60 grams of protein per day have been put forward to include a safety factor to allow for imperfect absorption), and this amount is very easily obtained from diets containing even only quite small quantities and proportions of flesh, legumes, nuts, seeds and/or grains alongside larger quantities and proportions of green leaf vegetables and fruits;

4. For humans, unlike sharks and some other very naturally and optimally carnivorous animals, the nitrogenous waste product (ammonia) resulting from the use of the breaking down of protein (amino acids) to meet energy requirements, is very toxic, and the molecule that humans condense this into (urea) is still a toxic waste product that has to be removed. Humans are probably best using protein (amino acids) to replace human proteins, and fat and carbohydrate to meet energy requirements.

Regarding food protein, the amino acid proportions are a less sensible place to look for human genetic changes relating to them, other than possibly in the cases of some unusual and unusually genetically homogenous (i.e. isolated) populations, because:

1. While grains are lower than flesh foods in the amino acid lysine (which is the limiting amino acid, that is the amino acid present in the lowest proportion relevant to human needs, in grains) they are high in methionine (which is the limiting amino acid of most flesh foods, in conjunction with the fact that legumes have the reverse amino acid situation to grains insofar as they are high in lysine but lower than flesh foods in methionine.

2. Cooking (other than maybe boiling or other use of heat less than 100 degrees C) any foods (including meat, because it has some blood and tissue glucose and other carbohydrates) causes lysine to be joined to a sugar (i.e. glucose) in a Maillard Product and thus be rendered un-absorbable from the gastrointestinal tract.

3. Some fruits are high in lysine, and most fruits and many green leaf and other vegetable amino acid proportions are unknown (see some values for modern varieties in chapter 5 below), but must be taken into consideration, as it is quite likely that throughout our evolutionary history eating patterns were of several different major types of food being eaten close enough together in a day to combine in supplying amino acids to body tissues needing them.]

3.6 Protein metabolism for energy – limited to 50% of calories before overt toxicity; the metabolism of protein itself uses more energy than that of CHO and fats. Nitrogenous waste byproducts are diuretic.

[Generally, for humans more than 50% of calories as protein is overtly toxic (from the nitrogenous waste resulting), and illness is felt (Bilsborough & Mann, 2006) – and accordingly, in some conditions of great shortage of plant foods in studied gatherer-hunter/ hunter-gatherers, prey that does not have enough fat is discarded even if the hunters are in a state of hunger.
This may be contrasted with the state in sharks, where the nitrogenous byproduct of metabolizing protein for energy is not a waste product, but serves a physiological function. Unfortunately I am unable to provide the relevant information for cats and dogs.

Claims of meat being more energy-dense than plant foods are palpably absurd: raw meat is about 70% water; raw nuts and oily seeds have very little water, the large majority of the weight being taken up by about 20% protein on average, and the another 70-odd% being fully fat or carbohydrate – so obviously raw nuts (600 kcal / 100 g) (or oily seeds (600 kcal / 100 g), and also cereal grains (380 kcal / 100 g) for that matter…) are more energy dense than raw meat (190 kcal / 100 g, lean; 230 kcal / 100 g, regular) - the maths here is at about a primary/ elementary school level – just open up a standard food nutrient content tables and do the primary/ elementary school maths...

Furthermore, note that high-protein diets are reported to increase the cost of digestion by about 30% compared to high-fat diets (Veldhorst et al 2009) – it may be more accurate to state that digestion, absorption, transport and metabolism of protein that is in excess of minimum requirements for body proteins, so as to derive energy from it, has an energetic cost 30% more than the equivalent calories derived from fat and carbohydrate, but I will not clarify this further, and the general point of there being some substantial excess metabolic cost for acquiring energy from protein rather than from fat or carbohydrate is probably enough for the purposes of the large majority of the readers of this.

Also, the excess nitrogenous waste resulting from the breakdown of dietary protein for energy is diuretic – that is, it causes an increased excretion of water as urine (Silberbauer 1981). This is undesirable in habitats where water is in short supply.

And, bear in mind that seasonal variation in foodstuff supplies generally applies in most geographical areas, but there is no seasonal variation in the human toxicity of more than 50% of calories being from protein – so that for energy requirements other than can be met by animal fat (i.e. by killing many more lean animals than would be required to meet protein requirements, eating the fat and discarding most of the meat), other sources of energy must be available at all times, and it is likely that some plant foods have seasons when plenty exceeds immediate requirements and may be collected, stored and eaten later.

Taken together, this (section 3.6) information is evidence further suggesting that humans are not genetically adapted to a high meat diet, but rather to a predominantly fruit and vegetable diet, including also legumes, nuts, seeds and cereal grains.]

3.7 Starch digestion and evolutionary triplication of salivary amylase; starchy tubers and cereal grains

Walker A. Chapter 1 Early hominin diets. In, Ungar PS. (Ed) (2007) Evolution of the human diet The known, the unknown, and the unknowable. Oxford University Press.

"For instance, humans have five copies of an amylase gene, three of which are expressed in saliva. Amylase is needed to break down amylin, the otherwise indigestible envelope of starch grains. Samuelson and colleagues (1990) in their discussion of amylase gene evolution show that a triplication of the salivary gene in the human lineage occurred after the last common ancestor of chimps and humans and suggest that this triplication took place around 1 million years ago. But this might be a relatively recent event contingent on the domestication of cereals or tubers and the ingestion of greater amounts of starch. [It is just as likely to have been contingent on *early* hominids eating more cereals and/or tubers than apes, anywhere up to 6 million years ago…] ……

Care must be taken with conclusions from genomic studies because single genes can have several functions. For instance, the amylase gene is also involved in bacterial resistance (H. A. Larson, personal communication). ……

We should be cautious, then, in assuming that our genome is "adapted for a Paleolithic lifestyle" (Cordain, chapter 19; Eaton, chapter 20), for not only was the Paleolithic lifestyle most probably very variable in time and place, but we can easily underestimate the power and speed of natural selection."
[And it must also be realized that evolution involving the making of a new gene for a new function will, on average, take much, much longer to occur than evolution involving the degeneration or changing of a gene that already exists, even though a multiplication, as here, probably less so.

This information is evidence suggesting that we are evolutionarily genetically adapted to a diet with a higher complex carbohydrate (starch) content than that of our common ancestor with the chimpanzee – the obvious candidates for the foods involved in this evolutionary change are starchy root vegetables (tubers and corms) and cereal grains.

Legumes also have a substantial amount of starch, but consumption of these has probably been relatively constant throughout relevant periods of evolution – legumes have been reported to be a dietary mainstay for virtually all relevant non-human primates (see also chapter 2, section 2.5).]

3.8 Supposed human inefficiency in biosynthesis of taurine, as in cats (obligate carnivores); vegan breast milk taurine 30 times the level of cow's milk-based infant formula indicates otherwise

Cordain L. Chapter 19 Implications of Plio-Pleistocene hominin diets for modern humans. In, Ungar PS. (Ed) (2007) Evolution of the human diet The known, the unknown, and the unknowable. Oxford University Press.

"Finally our species has a limited ability to synthesize the biologically important amino acid, taurine, from precursor amino acids [methionine and cysteine] (Sturman et al., 1975; Chesney et al., 1998), and vegetarian diets in humans result in lowered plasma and urinary concentrations of taurine (Laidlaw et al., 1988).

Like felines (Knopf et al., 1978; MacDonald, Rogers and Morris, 1984), the need to endogenously synthesize taurine may have been evolutionarily reduced in humans because exogenous dietary sources of preformed taurine (found only in animal food) had relaxed the selective pressure formerly requiring the need to synthesize this conditionally essential amino acid."

[Firstly, for this sort of discussion to provide or identify much useful evidence at all, the ability to synthesize taurine must be quantified – for carnivores, for humans, and for apes, at a minimum – without providing these absolutely necessary facts, why bother to attempt to discuss the matter at all?...

Secondly, the assumption of the biochemical necessity for the provision or synthesis of the same amount of taurine per kg body weight per unit of time must at least be noted – preferably, the actual respective needs for taurine per kg body weight per unit of time needs to be known for carnivores, humans, and apes…

Regarding this, I quote Stipanuk (2006) p549 of Shils et al: "The mean taurine content of milk of lactating vegan women is lower than that of lactating omnivores, but overlap of values is still considerable, and the taurine concentration of milk of vegan mothers is about 30 times the level in the cow's milk-based infant formulas used before the mid-1980s." (cites Rana and Sanders, 1986)

So, some animals (i.e. cows or calves) have less need for taurine than others (i.e. humans). And, the vegan mothers' breast milk taurine is close to (and overlapping with) that of the omnivore human mothers' breast milk taurine.
And, taurine is derived from the amino acids methionine and cysteine, which can vary in the diet.
So, the reality is that the implication of the relevant facts is that human taurine requirements are, more probably than not, met by any diet, including a vegetarian one, that is sufficient in protein, within the context of legumes being a little low in methionine and cereal grains being a little high (and the converse for lysine); and, in the aforementioned context, that the limitation of our species to synthesize the taurine we need is probably not significant to health.]

3.9 Supposed human inefficiency in elongating and desaturating 18 carbon fatty acids to their product 20 and 22 carbon fatty acids, docosohexaenoic acid (DHA) and arachidonic acid (AA), similarly to obligate carnivores; comparisons of vegan breast milk with captive and wild apes and monkeys (similar levels found) indicates otherwise

Cordain L. Chapter 19 Implications of Plio-Pleistocene hominin diets for modern humans. In, Ungar PS. (Ed) (2007) Evolution of the human diet The known, the unknown, and the unknowable. Oxford University Press.

"Further, similar to obligate carnivores (Pawlosky, Barnes and Salem, 1994), humans maintain an inefficient ability to elongate and desaturate 18 carbon fatty acids to their product 20 and 22 carbon fatty acids (Emken et al., 1992). Since 20 and 22 carbon fatty acids are essential cellular lipids, then evolutionary reductions in desaturase and elongase activity in hominins indicate that preformed dietary 20 and 22 carbon fatty acids (found only in animal foods) were increasingly incorporated in lieu of their endogenously synthesized counterparts derived from 18 carbon fatty acids."

[analogously to my comment in section 3.8 immediately above:

Firstly, for this to provide much useful evidence at all, the ability to elongate and desaturate 18 carbon fatty acids to their product 20 and 22 carbon fatty acids must be quantified – for carnivores, for humans, and for apes, at a minimum – without providing these absolutely necessary facts, why bother to attempt to discuss the matter at all?... Secondly, the assumption (which may well be incorrect) of the biochemical necessity for the provision or synthesis of the same amount of those fatty acids per kg body weight per unit of time must at least be noted – preferably, the actual respective needs for these fatty acids per kg body weight per unit of time needs to be known for carnivores, humans, and apes…]

[And following, a study of the sort necessary to provide the context necessary for understanding the meaning of the data Cordain misconstrues:]

Milligan LA, Bazinet RP (2008) Evolutionary modifications of human milk composition: evidence from long-chain polyunsaturated fatty acid composition of anthropoid milks. Journal of Human Evolution 55: 1086-1095

"Brain growth in mammals is associated with increased accretion of long-chain polyunsaturated fatty acids (LCPUFA) in brain phospholipids. The period of maximum accumulation is during the brain growth spurt. Humans have a perinatal brain growth spurt, selectively accumulating docosohexaenoic acid (DHA) and other LCPUFA from the third trimester through the second year of life. ……

Milk LCPUFA profiles of captive anthropoids [apes and monkeys] (consuming diets with a preformed source of DHA) are similar to milk from women on a western diet, and those of wild anthropoids are similar to milk from vegan women. ……

Results from this project suggest that low conversion of ALA [alpha-linolenic acid] to DHA by human mothers may be a shared physiological trait of all anthropoids."

[So, the reality is that the implication of the relevant facts is that human DHA and AA requirements are, more probably than not, well met by human enzyme action on substrates in appropriately-composed vegetarian diets, such that additional dietary DHA and AA from animal food sources is probably not necessary for human health.]

3.10 Preformed Vitamin A from animal food is far in excess of requirements in western diets; ability to synthesize vitamin A from pre-vitamin beta-carotene from plant food sources retained through evolution

[That humans have not lost the ability to synthesize sufficient vitamin A from the pre-vitamin A (beta carotene), commonly available from many different types of plant foods, despite high-meat diets leading to liver stores (relatively low at birth) of an average 450milligrams, massively in excess of the 1milligram Recommended Daily Allowance (McLaren 1981), is clearly consistent with human high-meat diets (assuming the liver was eaten) not having been part of our evolutionary history for long enough to have caused us to evolve to be suited for high-meat diets, and is consistent with humans being evolutionarily a predominantly vegetarian animal.]

3.11 Lactose intolerance / adult lactase absence in most of the world's population; unsuitability of milk outside of infancy

Walker A. Chapter 1 Early hominin diets. In, Ungar PS. (Ed) (2007) Evolution of the human diet The known, the unknown, and the unknowable. Oxford University Press.

"Adult lactase persistence segregates as a dominant trait in families, and the prevalence of persistence is correlated with a long history of milk consumption after weaning (Weiss, 2005). Populations that have practiced this have only had domesticated stock for a few thousand years, so it follows that the genes that remain activated for lactase persistence have reached fixation in a very short time. ......
[Actually it is DNA *degeneration* through retention of destructive mutations in a repressor gene that has occurred in this case, which generally requires much less time than the evolution of a new function.]

We should be cautious, then, in assuming that our genome is "adapted for a Paleolithic lifestyle" (Cordain, chapter 19; Eaton, chapter 20), for not only was the Paleolithic lifestyle most probably very variable in time and place, but we can easily underestimate the power and speed of natural selection."

[Clearly all other mammals on earth drink milk only as infants, then go on to eat the natural (if not always the optimal) foods of their species.
A small number of human groups have evolved by a genetically degenerative (DNA degeneration) process to be able to digest the milk sugar lactose.
This is not at all evidence that dairy products are the best source of calcium for humans of any age, as other nutrients, and the combinations of foods in meals and diets, and even other factors, need to be addressed in deciding what the best source of calcium is.
What other genetic changes may have come from drinking milk are unknown to me.]

3.12 Inborn errors of fructose metabolism are very low in frequency (1/20,000 or less); very common fructose malabsorption seems mostly due to unnatural modern context of rapid ingestion of some high fructose:glucose ratio foods in the absence of other meal-balancing protein and fat foods, and to be otherwise beneficial.

[Fructose is a sugar found in fruit, and some vegetables (, i.e., sugar beets), where it most often is present in an equal or lesser concentration than glucose.

Inborn errors of fructose metabolism:

Steinman B, Gitzelmann R, Van den Berghe G (2001) Chapter 70 Disorders of fructose metabolism. In, Scriver CR, Beaudet AL, Sly WS, Valle D, Childs B, Kinzler KW, Vogelstein B (Eds) The Metabolic & Molecular Bases of Inherited Disease (8th Ed). McGraw-Hill, New York.

Homes FA (1993) Inborn errors of fructose metabolism. Am J Clin Nutr 58(Suppl): 788-795S.

Genetic defects in fructose metabolism (occurring after absorption and therefore not due to malfunction of the lining of the small intestine, which may be due to some number of different gut diseases, some of which are relatively common) are very rare:

Fructokinase (for fructose incoming from the diet, specific for fructose metabolism, but an alternative pathway is available) deficiency (autosomal recessive: both genes must be defective to manifest the condition), even though it is considered to be a benign condition, is estimated to have a prevalence of only 1/100,000.

(Fruct)Aldolase B (in liver, kidney and intestine; for fructose resynthesized from breakdown products of dietary fructose?) deficiency (autosomal recessive: both genes must be defective to manifest the condition), resulting in a broad spectrum of severity of: poor feeding on introduction to dietary fructose, failure to thrive, vomiting, liver and kidney damage, and possibly eventual death, is highly variable in its frequency among ethnic groups, but even the highest prevalences of approximately 1/20,000 in Switzerland and Great Britain (and maybe France, Italy and Yugoslavia similarly) are rare, and Japan had only 5 cases identified.

Fructose-1,6-diphosphatase (in liver, kidney, jejunum and leukocytes; for fructose resynthesized from breakdown products of dietary fructose?) deficiency (autosomal recessive: both genes must be defective to manifest the condition), resulting in acidosis, hyperventilation, trembling, apneic spells, dyspnea and tachycardia, irritability, somnolence, and coma, is very rare with only 100 or so cases known in the year 2000, from mainly Europe, the USA, Japan and Lebanon.

Aldolase A (in muscle; conversion of fructose-1,6-diphosphate (resynthesized from breakdown products of dietary fructose?) into glyceraldehydes-3-phosphate and dihydroxyacetone phosphate;) deficiency, resulting in mental retardation, short stature, and hemolytic anemia, is known in only several patients, all having partial activity of the enzyme.

Obviously assessment of the possible significance of the global epidemiology of the inborn errors of fructose malabsorption is hampered by the difference in diagnosis rates among countries, in particular as a function of their level of development.
However, overall it seems that the human diet must have contained a large proportion of fruit throughout the periods of time that have produced the current rareness of inborn

errors of fructose metabolism – the evidence here is for humans being by genetic evolutionary nature very much a fruit-eating animal.

As breast milk's disaccharide lactose is made up of one molecule each of glucose and galactose, and as blood sugar is glucose and the body stores glucose in glycogen, inborn errors of glucose metabolism are much less likely to survive than inborn errors of fructose metabolism.

Fructose 'malabsorption' or incomplete absorption:

Steinman B, Gitzelmann R, Van den Berghe G (2001) Chapter 70 Disorders of fructose metabolism. In, Scriver CR, Beaudet AL, Sly WS, Valle D, Childs B, Kinzler KW, Vogelstein B (Eds) The Metabolic & Molecular Bases of Inherited Disease (8th Ed). McGraw-Hill, New York.

1/3 'healthy' adults and 2/3 'healthy' children incompletely absorb fructose when administered orally (in a pure solution with no other food) at a dose of 0.7 – 2 g/kg. (This is 42-120g; 160-480 kcal for a 60kg person, or 1/15-1/5 of a generous daily energy requirement of 2,500 kcal). However when the same amount of fructose is taken together with glucose (or galactose, another dietary sugar, or glycine, an amino acid of dietary protein) (giving 1/7 – 2/5 of a daily energy requirement of 2,500 kcal), it seems that less than 1/10 of the 1/3 and 2/3 experience any symptoms, that is, 1/30 and 1/15 of adults and children respectively.
These figures must be considered in the context of the recognition of the high prevalence in populations of some spectrum of celiac disease (1%) and 'irritable bowel syndrome' (5-10%), not all or possibly most of which have fructose malabsorption as more a cause than a result.
Again, if the fructose source is eaten as part of a meal in which glucose (Moukarzel & Sabri 1996), protein or fat are present, the rate of stomach emptying into the small intestine is slowed, therefore allowing more of the fructose to be absorbed.
Some processed foods containing an unnaturally high amount of refined sugar, even if the fructose:glucose ratio is approximately 1:1 i.e., high fructose corn syrup, seem to be major culprits.

As human evolution probably involved the eating of fruit such that the rate at which fructose was delivered to the small intestine was considerably less than that resulting from drinking a pure solution into an empty stomach with no other food in the meal, and that usually the fructose was accompanied by glucose, galactose, protein and fat, and as it is a very macro nutrient, it does not seem correct to suggest that our evolution has been *much* away from the suitedness for fruit that apes have and probably had back at the common ancestor of chimpanzees and humans, but the evidence is consistent with a moderate swing towards suitedness for starchy root vegetables, such that we might be suited for a diet of less fruit and more starchy root vegetables and grains than apes are. However this needs to be considered in the context of our modern fruits not being exactly representative of the wild fruits that our pre-agricultural evolutionary ancestors evolved with, nor purportedly those that chimpanzees eat – our present day western (and doubtless many Asian, if goldfish are any evidence to go by) cultivars have been developed for exceptional size and sweetness, and for easy chewability. Apples and pears are exceptional in having a 2:1 fructose:glucose ratio. And fruit juice can be ingested in a fraction of the time it takes to eat the equivalent amount of fruit, and the lack of fiber might have some relevant effect.

An interesting aside with unknown meaning for the central questions of this book is that black Africans had twice the rates (79% vs. 42%) of incomplete fructose absorption (Segal 2002).
Segal (and Roberfroid 2000) has suggested that in fact it is healthy to have some undigested carbohydrate reaching the large intestine for fermentation there, and that the common advanced western country practice of eating foods having carbohydrates that are totally absorbed in the small intestine is not healthy. Note that the preferred energy substrate of the last third of the large intestine is the short chain fatty acids produced by such fermentation.
Why white Europeans should have more fructose absorption than black Africans, given what is known about African gatherer-hunters is a mystery to me – perhaps in the African evolution there was a higher ratio of more starchy root vegetable eating to fruit eating, than in the European evolution.]

3.13 Celiac disease 1% prevalence is only a very weak evidence, or none at all, for humans not being evolutionarily suited to eating cereal grains.

[Celiac disease is a condition in which an allergic response to the cereal grain (wheat, barley, rye, oats, millet…) protein gluten causes structural and functional damage to the surface of the small intestine, causing reduced absorption of nutrients from digested food. The prevalence of celiac disease, in those countries where reporting can be relied on, is considered to be about 1%.

90% of celiac disease individuals carry the HLA-DQ2 immunologic protein and almost all of the rest carry the HLA-DQ8 immunologic protein. These proteins are associated with approximately 40% of the heritable risk of developing celiac disease. The other 60% is possibly due to shared environmental factors as mentioned below.

Infectious disease which damages the integrity of the surface of the digestive tract, such as rotavirus and potentially other viruses, bacteria, unicellular parasites, and parasitic worms, allow the increased penetration into the bloodstream of gluten protein fragments (i.e., gliadin and its fragments) that have not been fully broken down into the mono-, di- or tri-peptides that the digestive tract has specific absorption molecules for.

Nutritional deficiencies that reduce the integrity of the skin, such as of vitamin A and C, are expected to increase the risk of developing celiac disease.

Weaning (i.e., onto cereal grain-containing foods) that is too early and too abrupt is also very plausibly considered to increase the risk of developing celiac disease.
(Tack et al 2010, Boguerra et al 1998).

A 1% prevalence of celiac disease in the context of the predisposing infectious diseases being more prevalent in the higher population densities of the agricultural (and modern urban) era is hardly a very strong argument for humans not being evolutionarily suited to eating cereal grains – particularly as other foodstuffs are also the subject of allergies: peanuts and tree nuts around 1% each in a variety of countries, milk 0.4% in the USA, egg 0.2% in the USA, egg 2% in UK children, 0.1% in adults, shrimp 1% in the USA and shellfish 5% in an Asian country.
These figures are only very roughly comparable indeed, as they need to be related to exposure frequency and schedule, and meal context, and corrected for socioeconomic confounders such as affordability, weaning, and related diseases – obviously cereal grains are almost certainly *the* most affordable of these and all foodstuffs, and are therefore very likely confounded in their association with allergy by a more unfavourable collection of socioeconomic confounders such as weaning, and relevant diseases.

Celiac disease's 1% prevalence is only a very weak evidence, or none at all, for humans not being evolutionarily suited to eating cereal grains, nor for cereal grains not having been a substantial part of the diet in various places in the human evolutionary past, as we can safely assume peanuts, tree nuts, eggs and shellfish were also.]

References for chapter 3:

Bilsborough S, Mann N (2006) A review of issues of dietary protein intake in humans. Int J Sports Nutr Exerc Metab Apr;16(2):129-52 (abstract only read).

Bjerregaard P, Young TK, Hegele RA (2003) Low incidence of cardiovascular disease among the Inuit – what is the evidence? Atherosclerosis 166: 351-357.

Bouguerra F, Hajjem S, Guilloud-Bataille M, Khlat M, Khaldi F, Bennaceur B (1998) Breast feeding effect relative to age of onset of celiac disease. Arch Pediatr 5(6): 621-621.

Cordain L. Chapter 19 Implications of Plio-Pleistocene hominin diets for modern humans. In, Ungar PS. (Ed) (2007) Evolution of the human diet The known, the unknown, and the unknowable. Oxford University Press.

Hay G, Johnston C, Whitelaw A, Trygg K, Refsum H (2008) Folate and cobalamin status in relation to breastfeeding and weaning in healthy adults. Am J Clin Nutr 88: 105-114.

Homes FA (1993) Inborn errors of fructose metabolism. Am J Clin Nutr 58(Suppl): 788-795S.

Insel P, Turner RE, Ross D (2007) Nutrition 3rd Ed; American Dietetic Association; Jones and Bartlett, Boston, London.

McLaren DS (1981) The luxus vitamins – A and B12. Am J Clin Nutr 34: August p1611-1616.

Milligan LA, Bazinet RP (2008) Evolutionary modifications of human milk composition: evidence from long-chain polyunsaturated fatty acid composition of anthropoid milks. Journal of Human Evolution 55: 1086-1095.

Millward DJ (2001) Protein and amino acid requirements of adults: current controversies. Can J Appl Physiol 26 Suppl: S130-40. (Abstract only read).

Moukarzel AA, Sabri MT (1996) Gastric physiology and function: effects of fruit juices. J Am Coll Nutr 15(5 Suppl): 18S-25S. (Abstract only read).

Mozafar A (1994) Enrichment of some B-vitamins in plants with application of organic fertilizers. Plant and Soil 167 (2): 305-311 (Abstract only read).

Oftedal OT (1981) Milk protein and energy intakes of suckling mammalian young. A comparative study. PhD Diss. Cornell U, Ithaca, New York.

Quadros EV (2010) Advances in the understanding of cobalamin assimilation and metabolism. British Journal of haematology. 148(2): 195-204.

Robinfroid MB (2000) Fructo-oligosaccharide malabsorption: benefit for gastrointestinal functions. Curr Opin Gastroenterol 16: 173-177.

Segal I (2002) Physiological small bowel malabsorption of carbohydrates protects against large bowel diseases in Africans. J Gestroent Hepat 17: 249-252.

Shils ME, Shike M, Ross AC, Caballero B, Cousins RJ (2006) (Eds) Modern Nutrition in Health and Disease (10th Ed), Lippincott Williams and Wilkins, New York, London.

Silberbauer G. Chapter 12 Hunter/gatherers of the Central Kalahari. In, Harding RSO, Teleki G (Eds) (1981) Omnivorous primates. Gathering and hunting in human evolution. Columbia University Press. New York

Steinman B, Gitzelmann R, Van den Berghe G (2001) Chapter 70 Disorders of fructose metabolism. In, Scriver CR, Beaudet AL, Sly WS, Valle D, Childs B, Kinzler KW,

Tack GJ, Verbeek WHM, Schreurs MWJ, Mulder CJJ (2010) The spectrum of celiac disease: epidemiology, clinical aspects and treatment. Nat Rev Gatroenterol Hepatol 7: 201-213.

Veldhorst MA, Westerterp-Plantanga MS, Westerterp KP (2009) Gluconeogenesis and energy expenditure after a high-protein, carbohydrate-free diet. Am J Clin Nutr Sep; 90(3):519-26.

Vogelstein B (Eds) The Metabolic & Molecular Bases of Inherited Disease (8th Ed). McGraw-Hill, New York.

Walker A. Chapter 1 Early hominin diets. In, Ungar PS. (Ed) (2007) Evolution of the human diet The known, the unknown, and the unknowable. Oxford University Press.

Wakayama EJ, Dillwith JW, Howard RW, Blomquist GJ (1984) Vitamin B12 levels in selected insects. Insect Biochemistry 14 (2): 175-179 (Abstract only read).

Watanabe, F (2007) Vitamin B12 sources and bioavailability. Exp Biol Med 232: 1266-1274.

Weir DG, Scott JM (1999) Cobalamins. In, Encyclopedia of Human Nutrition. Academic Press.

Whitney E, Rolfes SR (2008) Understanding Nutrition (11th Ed) Thomson Wadsworth, Australia, UK, USA.

Chapter 4 Food and food group nutrient supply sufficiencies and deficiencies according to Recommended Daily Allowances (RDAs) – indicators of evolutionary adaptations to and for diet food group proportions

4.1 Introduction p289

4.2 Summary of chapter 4 p292

4.3 Calcium (Ca), Phosphorus (P) & Magnesium (Mg), and Fiber Content, and Weight, per 2500kcal of Foods, and their sufficiencies or deficiencies related to the Recommended Daily Allowances (RDAs) / Adequate Intakes (AIs) p306

4.4 Vitamin B12 (VitB12), Folate, Iron (Fe), and Zinc (Zn) Content, and Weight, per 2500kcal of Foods, and their sufficiencies or deficiencies related to the Recommended Daily Allowances (RDAs) / Adequate Intakes (AIs) p315

4.5 Vitamin C, Vitamin A, Vitamin E, and Selenium (Se) Content, and Weight, per 2500kcal of Foods, and their sufficiencies or deficiencies related to the Recommended Daily Allowances (RDAs) / Adequate Intakes (AIs) p325

4.6 Thiamin (VitB1), Riboflavin (VitB2), Niacin (VitB3) & Pyridox(al/ine/amine) (VitB6) Content, and Weight, per 2500kcal of Foods, and their sufficiencies or deficiencies related to the Recommended Daily Allowances (RDAs) / Adequate Intakes (AIs) p335

References for chapter 4 p341

4.1 Introduction

In this chapter I have selected representative foods, on the bases of common usage and variety, otherwise fairly randomly before looking at their nutrient data, from the following food groups:
animal flesh (Land animal; marine fish; marine mollusk and crustacean);
eggs (chicken);
milk (cow & goat, human breast);
legumes;
seeds;
nuts;
grains;
corn;
starchy root vegetables;
carrots;
fruits;
avocado;
green leaf, flower and shoot vegetables;
squash and zucchini

The nutrient values of the foods are given as the amount in enough of the food to give 2,500 kcal ('calories'), which is the amount of energy commonly given as a rough average human daily requirement.
The weight of the food required to give 2,500 kcal is also given.
When it is not possible or clearly inappropriate to be eating a foodstuff as a major source of energy, for example green leaf, flower or shoot vegetables, the nutrient values of the

foods are given as the amount in 500 grams of the food, very roughly the minimum amount of green leaf, flower or shoot vegetables that a sensibly healthy diet for an average sized person should include, and the protein content is given also.

The Recommended Daily Allowance (RDA) of the nutrient is provided underneath for ready comparison, and when a food or food group's level of a nutrient is less than the RDA, it is highlighted in the table for ready visibility.

It should be borne in mind that the RDAs are usually formulated on the basis of 'expert opinion', and not established by controlled experiment, though of course we have 'natural experiments' or 'quasi-experiments' also to draw (with appropriate caution) relevant information from.

It should be borne in mind that RDAs are formulated with the intent to quantify the amount of the nutrient estimated to provide a sufficiency to the very large majority (97-98%) of the individuals making up some population (perhaps of all of the world), and are therefore not estimated average requirements. For example the calcium RDA has been set to accommodate an absorption into the body of only 30% of calcium from food. Note that people in old age often require higher amounts of nutrients due to reduced absorption – this raises the RDA above what levels are most relevant to the title questions of my work here, because the nutritional requirements of the elderly amongst our evolutionary ancestors (including because most would not have survived to old age…) of the last 6 million years, are probably a weaker force in driving evolution than the nutritional requirements of those amongst them who were reproducing and therefore more a focus of natural selection regards genetic evolution.

The body tends to absorb more when it is depleted of some nutrient by upregulating the amount of the body's molecules and atoms involved in absorption, and to absorb less when it is fully or excessively loaded with that nutrient. Some such mechanisms are common to the handling of different nutrients (i.e., metal ions of a similar size and charge) so cross-effects from deficiency, sufficiency or excess of one nutrient to the absorption of another are possible.
There are very probably enough such facts now known and available in the scientific literature to improve the quality of my work here by incorporating them, but many of the facts are used unscientifically by vested interests to bolster sales of their food products by taking them out of context or overgeneralizing from one particular context without properly stating what is unknown regards extrapolating from one context to a general statement. This lack of detail will have to remain for the moment (maybe to be improved on in future versions) a small very non-fatal weakness of my work, that I note here, and the reader will have to consider this in the context of the overall strength of my analysis of the data that I have collected here to answering the central questions in the title of this work. In fact there is no such information in even very recent nutritional textbooks written by orthodox nutritional authorities (i.e. Whitney and Rolfes 2008; Shils et al 2006) that I do not at least mention somewhere in my work here, and that information seems not to potentially destroy or much weaken my overall analysis in addressing the title questions of this my work here. (Note though, that textbooks often do not include relevant subject matter that has appeared in the scientific journals in the year before publishing, and obviously none appearing after the publication of the textbook).

It should be borne in mind that the estimation of the RDAs may have been biased towards the 'advanced' western nations that are usually the home countries of the organizations that are carrying out the estimations of the RDAs. People whose prehistoric or historic ancestral evolution has taken place in different geographical locations from each other (i.e., rich vs. poor countries; mountains vs. plains vs. coasts/islands; near vs. far from the equator; dry versus wet climate) may well have evolved different nutrient requirements from each other (even if mostly through DNA *degeneration* rather than evolution of new DNA coding for proteins with new metabolic, uptake, transport or storage functions, as discussed above in section 3.2).

The availability of nutrients varies according to what other chemicals are part of the food (i.e., vitamin C in some plant foods helping the uptake of iron; and phytin/ phytic acid in some plant foods i.e. unsprouted legumes binding to and hindering the uptake of zinc; oxalate in some plant foods i.e. spinach binding to and hindering the uptake of iron; how the food is treated in preparation or manufacture (which are not much the subject of this work as they are less relevant to the central questions of my work here, or finer details of the answers), and what other foods the particular food is eaten together with.

The comparisons I make of the nutrient sufficiencies and deficiencies of the various food groups are therefore, and for other reasons, only one source of information that is useful to answering the central questions of my work here: What is the natural diet of humans, and what is the optimal diet of humans?

The nutrient values of the *foods* are given individually (in Tables 7 (Ca, P, Mg, fiber), 9 (Vit B12, folate, Fe, Zn), 11 (VitC, VitE, VitA, Se) and 13 (VitB1, VitB2, VitB3, VitB6)), and as the averages for each of the food groups listed above (in Tables 8, 10, 12 and 14). Food group subgroup averages are given also when they are substantially different (i.e., land animal flesh; marine fish flesh; marine mollusk and crustacean flesh). Protein is dealt with in section 3.5 of chapter 3 above.

Under each of Tables 8, 10, 12 and 14, which contain the average *food group* nutrient data, I will list each food groups deficiency or sufficiency, on a per 2,500 kcal basis (on a 500 g basis for green leaf, flower or shoot vegetables), of the four vitamins or minerals dealt with in the respective pairs of tables.

In short the average nutrient's deficiency or sufficiency of the various food groups will be identified.

Then the meaning of these facts for the central questions addressed in my work here (What is the natural diet of humans, and what is the optimal diet of humans?) will be addressed.

My writing underneath each set of two tables further below is reproduced immediately below, in section 4.2 Summary of chapter 4, without the tables – the duplication may seem excessive to some readers, but others may like the opportunity, particularly in the context of reading the work as a computer document, of being able to more readily verify what my writing claims the data shows, and also to view interesting detail. And others may like being able to read just the summary.
The **bolded paragraphs** (my key analytic conclusions) are a further summary imbedded in the summary (relating of quantitative facts and their proportional contexts, along with some reasoning and other conclusions thereon).

4.2 Summary of chapter 4

4.2.1 Calcium (Ca) (RDA 1000 mg / day):

That human breast milk would be deemed a calcium deficient food, supplying 830 mg of the 1000 mg RDA, given that an infant must be forming some times over the amount of bone per body weight than an adult, even though absorption from breast milk by an infant is probably greater than absorption from foods by adults, seems to me a fairly strong indication that the adult RDA is in fact excessive – probably due to the vested interests of the dairy industry and the research it funds, and due to faulty science that does not relate calcium intake to other important factors having a bearing on i.e. osteoporosis, such as weight-bearing exercise, Ca/P ratio, protein intake, sodium intake, and coffee.
The older World Health Organisation (WHO) levels of around 400 mg per day seem to me to be more correct in the context of those other factors being as they ought – and bear in mind that tissue calcification has roles in disease pathologies that are not yet fully understood (calcification of atherosclerotic arterial lesions is becoming used as a marker of risk in atherosclerotic disease imaging), and it is probable that excessive calcium has bad effects, as calcium is by no means a very micro nutrient, and the more macro a nutrient is, the greater the likelihood that a proportionate excess of it will have a bad effect (think about carbohydrate, fat, sugars and sodium, for example).

That cow and goat milk have 5 times the amount of calcium per unit of energy of human breast milk seems to me that for the purposes of the human diet they are unnecessarily high and perhaps undesirably excessive in calcium, depending on the sufficiency of the rest of the diet.
This seems likely to me to be related to the vested interests of the dairy companies being involved in falsely inflating the RDA for calcium, in order to maintain a market for the product, which might otherwise be rightly seen as unnecessary or even harmful in the diet.

Lean land animal flesh is markedly deficient in calcium, and considering that only 50% of calories may be supplied as protein without metabolic toxicity being experienced as illness, so is marine fish flesh, supplying, respectively, at 50% of calories, 70 and 180 mg of the 1000 mg RDA.

A noteworthy curiosity that I've never seen or heard remarked on is that land animal fat has much more calcium than lean meat (which is not chemically surprising given the known ability of cationic calcium (Ca^{2+}) to join to the anionic end of fatty acids (FA^-)), such that 1kg of fat contains roughly 1000 mg of calcium along with 9000 kcal of energy. So Eskimos, if marine mammal fat calcium content is the same, which is not at all certain (it's probably greater, in accord with marine fish flesh), or arctic land animal eating hunter-gatherer peoples, from 1,500 kcal (roughly half their daily requirement) of fat might get only very roughly 175 mg of Ca and maybe another 25mg of Ca if they ate lean meat for the other half of their caloric requirement – the maximum Ca available even if only enough meat was eaten to supply protein needs (which does not address what *other* nutrients might require more than that small amount (roughly 200 grams) of lean meat to be eaten in a context of there being little or no plant foods in the diet) and all the energy was to come from fat might be roughly 300 mg – that these people are not reported (the reporting is probably very lacking/deficient in any case) to suffer from osteoporosis is suggestive that the RDA of 1000 mg Ca per day is too high.

As marine fish flesh is higher in calcium than land animal flesh, it may well be that marine fish fat is similarly higher, which would put fatty (more likely to be from cold waters) marine fish, consumed such that fat was a large fraction of the intake of them, close to being sufficient in calcium – and the same might apply to marine mammals, as the seawater environment (Seawater's high calcium content) might have something to do with the other land-marine difference.

On average marine mollusk and crustacean flesh is sufficient in calcium on a per 2500 kcal basis.

Whole egg is marginally deficient in calcium on a per 2500 kcal basis according to the RDA, but is very probably sufficient in reality, the RDA very probably being excessive as discussed below.

Legumes at 840 mg Ca / 2500 kcal are deemed deficient in Ca by the RDA, but I strongly suspect that in reality they are very adequate, due to the RDA being artificially inflated as discussed above.

Nuts and seeds at 470 mg Ca / 2500 kcal are deemed deficient in Ca by the RDA, but I suspect that in reality they are adequate, due to the RDA being artificially inflated as discussed above.

(Cereal) Grains at 200 mg Ca / 2500 kcal are deficient in Ca, but not as lacking as lean meat, and their addition to the diet as a caloric replacement for some fraction of fruit would still leave the overall meal/diet very probably quite sufficient in calcium.
Corn is the exception at 55 mg Ca / 2500 kcal, raising the possibility that this low calcium content is part of the association of dental caries with cooked corn as a predominant staple of diet, along with the deposition of carbohydrate along the gum line and on or between teeth.

Starchy root vegetables at 630 mg Ca / 2500 kcal / 2.6 kg are deemed deficient in Ca by the RDA, but I strongly suspect that in reality they are adequate, due to the RDA being artificially inflated as discussed above.

Carrots are root vegetables but not starchy enough to be considered as starchy root vegetables – 1.5 kg would supply 500 mg Ca (which would probably in reality be adequate, due to the RDA being artificially inflated as discussed above), but only 625 kcal, and would be too difficult for even a carrot-liking committed raw food eater to enjoy chewing without grating or some other processing – however, their inclusion in the diet would not negatively impact the diet's calcium sufficiency.

Fruits at 610 mg Ca / 2500 kcal are deemed deficient in Ca by the RDA, but I strongly suspect that in reality they are very adequate, due to the RDA being artificially inflated as discussed above. The weight to be eaten for this is 4.8 kg but with other more energy-dense foods, such as the protein-plus-fat/carbohydrate sources this would reduce. Note that for humans using 2500 kcal of energy an average water intake of roughly 3 litres / day is considered desirable, such as might come from roughly 3.5 kg of fruit.

Green leaf, flower and shoot vegetables at 240 mg Ca / 500 gram / 10 gram protein / negligible kcal are quite a good source of Ca in a balanced meal/diet.

Squash and zucchini are deficient in calcium.

The meaning of the food calcium (Ca) data I interpret to be that it is significant that 500 grams of green leaf, flower and shoot vegetables eaten along with enough fruit to meet caloric requirements would give 850 mg Ca per day, which I strongly suspect in reality would be very adequate for the very large majority of the human population(s), due to the RDA being artificially inflated as discussed above.

This could provide a little too much water for comfort, but replacement of some fraction of the fruit with tubers, legumes, nuts and seeds, and even grains would still very probably provide a sufficient intake of calcium.

4.2.2 Phosphorus (P) (RDA 700 mg / day):

All of the animal foods, with the very dubious exception of human breast milk, are sufficient in phosphorus.
Legumes, seeds, nuts, grains, starchy root vegetables and fruit are, on group average, sufficient in phosphorus.
Green leaf, flower and shoot vegetables are a good source of P in a balanced meal/diet.

Squash and zucchini are deficient in phosphorus.

4.2.3 Calcium/Phosphorus ratio (Ca/P):

Mature (at 10 days of suckling) human breast milk is roughly 2:1 in Ca/P.
Land animals and marine fish meat on average are roughly 1:9
Marine mollusks and crustaceans on average are roughly 1:5
Eggs 1:3.5
Cow&Goat milk 1.25:1
Legumes 1:1.3
Seeds 1:5
Nuts 1:3.5
Grains 1:10
Fruits 1:1
Green leaf, flower and shoot vegetables 1.3:1

The meaning of the food Ca/P ratio data I interpret to be that perhaps the most meaningful comment that could be made on this data is that with the exception of dairy milk (which has 4 times the protein/energy of human breast milk and therefore probably should not make up a large proportion of the diet) the dietary proportionation of food groups that would make the dietary Ca/P ratio most consistent with both the RDA Ca/P and the human breast milk Ca/P would be to obtain most of the calories through fruit, to have also at least 500 g of green leaf, flower or shoot vegetables, and to have a relatively small proportion (all together) of the other foods.

4.2.4 Magnesium (Mg) (RDA 360 mg / day):

Human breast milk is very dubiously deficient in Mg at 120 mg/ 2500 kcal on the basis of the adult RDA / kcal.

Meat is marginally deficient at 310 mg / 2500 kcal and eggs deficient at 200 mg / 2500 kcal, compared to the RDA of 360 mg, but on consideration of the lower level in human breast milk, although there is not at all any guarantee that the ratios of nutrients required for an infant are the same as those required for adults (consider bone hardness differences, in which Mg is involved, for example), it might be possible to consider both these foods tentatively actually sufficient in Mg. All other food groups are on average sufficient in Mg, some to the extent of being possibly excessively so.

The meaning of the food magnesium data I interpret to be that perhaps the most meaningful comment that could be made on this data is that if we take the actual optimal requirements as being probably on average no higher than the RDA (as the

breast milk level at 120 mg / 2500 kcal is much lower), then seafoods, cow/goat milk, legumes and seeds (very roughly around 1000 mg / 2500 kcal) should be (all together) no more than a lesser proportion of the diet.

4.2.5 Fiber (RDA 32 g / day):

All of the animal foods; flesh, eggs and milks; are absolutely deficient in fiber, having none.

It is the opinion of many orthodox dieticians and nutritionists that 80 g of fiber / day would be a better target than the RDA here of 32 g, and eating fruit and green leaf, flower and shoot vegetables seems the best way of meeting or (perhaps to even greater benefit) exceeding that amount.

4.2.6 Vitamin B12 (RDA 2.4 micrograms (ug) / day):

Human breast milk has roughly 3 micrograms VitB12 / 2500 kcal. The human RDA is 2.4 micrograms / day.

All other animal foods far surpass this level:
Cow/goat milk has 12 micrograms VitB12 / 2500 kcal,
Land animal flesh on average has 20 micrograms VitB12 / 2500 kcal,
Marine fish flesh on average has 55 micrograms VitB12 / 2500 kcal,
Marine mollusks and crustaceans on average have 530 micrograms VitB12 / 2500 kcal,
Eggs have 25 micrograms VitB12 / 2500 kcal (though reportedly only 10% available vs. 50% for flesh foods, which would bring the amount available per kcal back to something probably very close to that from human breast milk),
Insects need to be considered, but I haven't the data – they are probably sufficient in B12.

All terrestrial plant foods are reportedly absolutely deficient in intrinsic vitamin B12, having none. Probably variable and small amounts of B12 adhere from the environment.

Some seaweeds, green and purple lavers, have B12 useful to humans, others, like blue-green algae ("Spirulina" for example, I think) have inactive B12 analogues that may be worse than useless as they occupy B12 sites in the body or digestive tract (although the digestive tract reportedly has a mechanism that prevents some of these being absorbed).

The meaning of the food vitamin B12 data I interpret to be that as the human infant is making more neural tissue (absolutely, not only relative to its bodyweight) and other things than adult humans are, but needs absolutely less energy (1000 kcal / day for a 10 kg infant, for example), and therefore receives only roughly 1 microgram per day from the breast milk it drinks in meeting its lower absolute energy needs, it seems logical that adult human requirements, at similar absorption plus intestinal production (?), should be (much?) less than 1 microgram of B12 / day. This would be met by only a very small amount of animal matter, and could probably all be met by the amount of insect matter, larvae and adults, that apes eat incidentally in plant matter eaten, with or without deliberately focused insect eating; or alternatively by occasional egg and small animal eating.
Note that RDAs are set to accommodate the very large majority of the population including those with reduced absorption, transport, storage and use of a particular nutrient.

4.2.7 Folate (RDA 400 micrograms / day):

Human breast milk has about 400 micrograms folate / 2500 kcal.

Cow and goat milk have much less at about 200 and 30 micrograms folate / 2500 kcal respectively (though the difference seems dubious), despite the fact that they naturally eat food that would supply much more folate than would the diet of carnivores.
Folate supplementation for neural tube defect (NTD) prevention is one of the few instances of supplemented isolated nutrients being substantially effective in improving a serious human health condition at the population (-representative) level.
These facts, together with those immediately below, suggest that humans are a very high-folate requiring animal compared to both carnivores and other mammals.

Land animal flesh and marine fish flesh are both quite deficient in folate at 130 and 170 micrograms folate / 2500 kcal respectively, particularly when considering that for humans only 50% or less of the calories may be eaten as protein without metabolic toxicity being experienced as illness – for 1250 kcal (half the daily energy requirement) the intakes would be 65 and 85 micrograms folate / 2500 kcal respectively, compared to the RDA of 400 micrograms, and that folate being a water-soluble, and therefore not a fat-soluble, vitamin, the fat of any carcass will contain less folate than the lean meat (consistent with the nutrient data), so fat eating would not make up for the deficiency.
Note that there is no mechanism for storing folate in fat as there is for ionic calcium (which is very water soluble, though perhaps it is taken up into fatty tissue already joined to fatty acid (the 'calcium salt' of the fatty acid), nor for the storage of folate in a similar fashion as vitamin B12 (also water soluble) is stored (in liver).

On average marine mollusk and crustacean flesh is sufficient in folate on a per 2500 kcal basis.

Eggs are 200% sufficient on a per 2500 kcal basis in folate.

Legumes are 600% sufficient on average on a per 2500 kcal basis in folate. Therefore consuming roughly enough legumes to meet protein requirements would meet the RDA for folate, and provide roughly 1/4 of the daily energy requirement.

Seeds are sufficient on a per 2500 kcal basis in folate on average for the only three types considered, but pumpkin seeds are very deficient (unless the data is wrong). Consuming roughly enough seeds to meet protein requirements would provide roughly 1/2 (only 1/30 for pumpkin sds.) of the RDA for folate, and roughly ¼ of the daily energy requirement.

Nuts are deficient on a per 2500 kcal basis in folate on average, at 210 micrograms folate / 2500 kcal. Consuming roughly enough nuts to meet protein requirements would provide roughly 1/8 of the RDA for folate, and roughly 1/4 of the daily energy requirement.

Grains (without corn) are marginally deficient on a per 2500 kcal basis in folate on average, at 360 micrograms folate / 2500 kcal. Consuming roughly enough grain to meet protein requirements would provide marginally less than the RDA for folate (which would very probably be sufficient for the very large majority of the population(s), and roughly all of the daily energy requirement. Corn has a roughly 300% sufficiency in folate on a per 2500 kcal basis.

Starchy root vegetables are marginally deficient on a per 2500 kcal basis in folate on average, at 350 micrograms folate / 2500 kcal. Consuming roughly enough starchy root vegetables to meet protein requirements would provide marginally less than the RDA for folate (which would very probably be sufficient for the very large majority of the population(s), and roughly all of the daily energy requirement.
Cassava is markedly deficient in many nutrients, including folate, unless the data or assays are faulty.

Fruit is 170% sufficient on a per 2500 kcal basis in folate on average at 670 micrograms folate / 2500 kcal. Consuming roughly enough mixed fruit to meet the average daily water requirement of 3 litres would provide roughly all of the folate RDA, roughly half of the protein requirements, and roughly 2/3 of the daily energy requirement.

Eating 500 grams of green leaf, flower and shoot vegetables per day will provide 1/2 of the folate RDA on average, roughly 1/7 of the average daily water requirements, and roughly 1/6 of the daily protein requirements.

Eating 500 grams of squash or zucchini per day will provide roughly ¼ of the folate RDA on average.

In considering calcium, calcium/phosphorus ratio, fiber, folate, protein, water and energy together, it seems that any more (as a replacement for any other of the food groups noted here) than the very small amount of animal food required for meeting vitamin B12 requirements would have a detrimental effect (with reference to the RDAs) on a diet composed of mostly (roughly 3kg, such as to roughly meet the average daily water requirements) fruit and starchy root vegetables, having at least 500 grams of green leaf, flower or shoot vegetables, and roughly 200 grams dry weight of combined legumes, seeds, nuts and cereal grains. The same seems to apply even if these proportions are varied quite a bit.

4.2.8 Zinc (Zn) (RDA 9.5 mg / day):

Human breast milk has roughly 7.4 micrograms Zn / 2500 kcal, which is (very dubiously, though absorption will be higher) deficient on a kcal basis according to the RDA. The human RDA is 9.5 millegrams / day.

As the growing human infant is making many times more proteinaceous tissue (relative to its bodyweight) than the adult human is, it seems logical that adult human requirements, in the absence of some malfunction, should need only a fraction of 7.4 millegrams of Zn per day. It seems probable to me that the vested interests of the meat companies, via the research that they fund, in conjunction with the cherished dietary habits of sundry authorities, has been responsible for the artificial inflation of the actual human requirements for Zn.

Land animal flesh has an average 600% sufficiency of Zn, marine fish has a sufficiency, and marine mollusks and crustaceans have a 5000% sufficiency (raising questions of possible long-term toxicity) of Zn on a per 2500 kcal basis.

Cow and goat milk are sufficient in Zn on a per 2500 kcal basis.

Legumes have a 250% sufficiency of Zn, seeds have a 400% sufficiency of Zn, nuts a sufficiency of Zn, and grains a 150% sufficiency, on average on a per 2500 kcal basis. (Phytic acid, sprouting, vitamin C, cooking, and probably other factors potentially significantly modify the availability of Zn from food.)

Starchy root vegetables are marginally deficient in Zn, at 8 mg / 2500 kcal, according to the RDA of 9.5 mg / 2500 kcal, but it is very probable that in reality this amount is sufficient for the very large majority of the human population(s).
Cassava is markedly deficient in many nutrients, including Zn, unless the data or assays are faulty.

Fruits are marginally deficient in Zn, at 7 mg / 2500 kcal, according to the RDA of 9.5 mg / 2500 kcal, but it is very probable that in reality this amount is sufficient for the large majority of the human population(s).

500 g of green leaf, flower and shoot vegetables on average provide only a surprisingly (given the 10 g of protein provided) low 1.5 mg Zn. Nevertheless, this is still a useful lesser fraction of the Zn requirement from a food group that ought be only a fraction of diet and is not considered as contributing to the meeting of caloric requirements.
500 g of squash and zucchinis likewise, but less surprisingly, provide 1.5 mg Zn.

4.2.9 Iron (Fe) (RDA 8 mg / day for males; 18 mg / day for menstruating females):

Human breast milk has roughly 2.2 millegrams Fe / 2500 kcal, which is (very dubiously, though absorption will be higher) deficient on a per 2500 kcal basis according to the RDA. The human RDA is 8(male)/18(female) millegrams / day.

As the human infant is making many times more blood and other tissue (relative to its bodyweight) than adult humans are, it seems logical that adult human requirements, in the absence of some malfunction, leaving aside menstrual losses of Fe in blood, should need only a fraction of 2.2 millegrams of Fe per day. It seems probable to me that the vested interests of the meat companies, via the research that they fund, in conjunction with the cherished dietary habits of sundry authorities, has been responsible for the artificial inflation of the actual human requirements for Fe.

Land animal flesh on average is sufficient in Fe on a per 2500 kcal basis.
Marine fish flesh on average is sufficient in Fe for males but slightly deficient in Fe for females, on a per 2500 kcal basis.
Marine mollusk and crustacean flesh on average is 900% sufficient in Fe (raising questions of possible long-term toxicity), on a per 2500 kcal basis.

Eggs have a 160% sufficiency of Fe on a per 2500 kcal basis.

Cow and goat milk are greatly deficient in Fe on a per 2500 kcal basis, providing less than 1/10 of pre-menopausal female RDA. (Roughly similar to human breast milk)

Legumes on average have a 250% sufficiency of Fe on a per 2500 kcal basis.

Seeds on average are sufficient in Fe on a per 2500 kcal basis.

Nuts on average are sufficient in Fe for males but marginally deficient for females, according to the RDA, but in reality are probably sufficient for the large majority of the population(s), on a per 2500 kcal basis.

Grains on average are sufficient in Fe on a per 2500 kcal basis. (Phytic acid, sprouting, vitamin C, cooking, and probably other factors potentially significantly modify the availability of Fe from food.)

Starchy root vegetables on average are sufficient in Fe on a per 2500 kcal basis.
Cassava is markedly deficient in many nutrients, including Fe, unless the data or assays are faulty.

Fruits on average are sufficient for males in Fe, but slightly deficient for pre-menopausal females in Fe, on a per 2500 kcal basis, but the reality is probably that fruits on average

are actually sufficient for the large majority of the human population(s), particularly given the lack of phytic acid and the presence of vitamin C.
500 g of green leaf, flower and shoot vegetables on average provide only 3.5 mg Fe. Nevertheless, this is still a useful lesser fraction of the Fe requirement from a food group that ought be only a fraction of diet and is not considered as contributing to the meeting of caloric requirements.
500 g of squash and zucchinis provide 2 mg Fe.

The meaning of the food data (for calcium (Ca), phosphorus (P), their Ca/P ratio, magnesium (Mg), fiber, iron (Fe), zinc (Zn), folate and vitamin B12) considered so far in this chapter I interpret to be that in considering calcium, calcium/phosphorus ratio, fiber, folate, protein, water, and energy, to which now are added zinc and iron, all together, it still seems that any more (as a replacement for any other of the food groups noted here) than the very small amount of animal food (and certainly of milk with its great deficiency of iron) required for meeting vitamin B12 requirements would have a detrimental effect (with reference to the RDAs) on a diet composed of mostly (roughly 3 kg, such as to roughly meet the average daily water requirements) fruit and starchy root vegetables, having at least 500 grams of green leaf, flower or shoot vegetables, and roughly 200 grams dry weight of combined legumes, seeds, nuts and cereal grains. The same seems to apply even if these proportions are varied quite a bit.

4.2.10 Vitamin C (VitC) (RDA 82 mg / day):

Human breast milk has a sufficiency of 450% of vitamin C on a per 2500 kcal basis. This relatively high level may relate to the need for more vitamin C for more connective tissue formation.
Cow and goat milk are both deficient in VitC at 0 and 45 mg / 2500 kcal respectively, though the 0 value seems dubious. It seems odd that such relatively quickly growing animals with their milk protein/energy ratio some times over that of humans would not need more VitC for the infant – maybe cows make their own VitC, and maybe calves eat grass very early and during suckling…
In any case, it seems to me that the figures suggest that the RDA of 82 mg VitC / day may be set lower than is really required for optimal health, including regards antioxidant function, and other fairly credible authorities share this opinion.

Land animal flesh is grossly deficient in VitC.
Marine fish flesh on average is very deficient in VitC.
Marine mollusk and crustacean flesh on average is sufficient in VitC per 2500 kcal.

Eggs are absolutely deficient in VitC, having none.

Legumes, seeds, nuts and grains (other than corn), all in the unsprouted condition, are very deficient in VitC. Sprouting increases the Vit C levels of such foods, for example supposedly to an average of 17 mg / 100 g (wet weight) for legumes, i.e. 500 g sprouts to meet the RDA.

Starchy root vegetables on average have a 400% sufficiency of VitC per 2500 kcal.

Fruits on average have a 1,300% sufficiency of VitC on a per 2500 kcal basis.

Eating 500 grams of green leaf, flower and shoot vegetables per day will provide a 300% sufficiency of VitC.
Squash and zucchini are deficient in VitC.

The meaning of the food vitamin C data I interpret to be that:
Taken together with the data named above, it seems safe to say that much more green leaf, flower and shoot vegetables should be eaten than squash and zucchini. Also it seems safe to say that on the basis of vitamin C content alone, the human diet generally should derive at least half of its calories from fruits and/or starchy root vegetables, and that this does not contradict the optimum proportion of diet of these foods suggested by consideration of the other nutrients done above.

4.2.11 Vitamin A (VitA) (RDA 800 micrograms(IU) / day):

Human breast milk has a sufficiency of 200% of vitamin A on a per 2500 kcal basis. This relatively high level may relate to the increased need in infancy for more vitamin A for tissue formation and protection, and growth.

Cow and goat milk also have a sufficiency of 200% of VitA on a per 2500 kcal basis.

Land animal flesh (other than liver) is grossly deficient in VitA.
Marine fish flesh on average is sufficient in VitA, but there is great variability, from 1/4 to 10 times the RDA per 2,500 kcal.
Marine mollusk and crustacean flesh on average is sufficient in VitA, but there is great variability, from 1/4 to 4 times the RDA per 2,500 kcal.

Eggs have a 300% sufficiency of VitA on a per 2500 kcal basis.

Legumes, seeds, nuts and grains (other than corn), all in the unsprouted condition, are very deficient in VitA. Sprouting increases the VitA levels of legumes, but the result is still supposedly very roughly an average of 100 IU VitA / 100g, i.e. 800g sprouts to reach the RDA.

Starchy root vegetables are extremely variable with potatoes and cassava being grossly deficient in VitA, and sweet potatoes and yams having massive 6,000% sufficiencies of Vit A, on a per 2500 kcal basis.

Fruits on average have a sufficiency of VitA on a per 2500 kcal basis, but there is a wide variation, from grossly deficient to massively sufficient, with redness of colour (which includes orange and purple coloration) being fairly reliable predictors of high VitA content, but yellowness of colour much less so (suggesting differences within the sweet potatoes also, by colour?...).

Green leaf, flower and shoot vegetables are extremely variable also, with 500 g of cauliflower and bamboo shoots providing nearly no VitA, and 500 g of the standard Bullethead cabbage and broccoflower providing roughly only 1/14 of the VitA RDA. 500 g of Kale provides 400% sufficiency of VitA.

Squash and zucchini are similarly variable with 500 g of summer squash exceeding the RDA in VitA, whereas 500 g of winter squash provided less than 1/10 of the RDA of VitA. 500 g of summer zucchini provided roughly 1/3 of the VitA RDA.

The meaning of the food vitamin A data I interpret to be that red and orange/purple fruits and (i.e., fruit-) vegetables should be a regular substantial part of the diet, and that 500 g of a good variety of green vegetables must be had to make up shortfalls in red/orange/purple fruit unless it is known that the particular green vegetable(s) eaten are those high in VitA.

However, probably almost any deeply dark green leafy vegetable is likely to be able to do so on its own, our common western cabbage (not the Savoy or purple cabbages, note) being the culprit dragging the averages down, notwithstanding its great usefulness overall as an affordable and readily available green(ish) leaf vegetable staple food.
Vitamin A being a fat soluble vitamin, dietary fat eaten in the same meal will aid absorption from the digestive tract, and there is some substantial capacity for good body stores to suffice for probably a small number of months of dietary shortfall of vitamin A with little or no compromise of health.

4.2.12 Vitamin E (VitE) (RDA 15 micrograms (ug) ∝ (alpha)-tocopherol / day):

Human breast milk has a sufficiency of 130% of vitamin E on a per 2500 kcal basis.

Cow and goat milk are both very deficient in VitE at 1.7 and 3 mg / 2500 kcal respectively.

Land animal flesh is substantially deficient in VitE at an average less than 1/3 of RDA per 2,500 kcal.
Marine fish flesh on average is sufficient in VitE on a per 2500 kcal basis.
Marine mollusk and crustacean flesh on average has a 200% sufficiency of VitE on a per 2500 kcal basis.

Eggs have a sufficiency of VitE on a per 2500 kcal basis.

Legumes, all in the unsprouted condition, are substantially deficient in VitE at an average 1/3 of RDA per 2,500 kcal.
Peanuts are an exception at 200% of RDA on a per 2500 kcal basis.
Sprouting increases the VitE levels of legumes, but perhaps not by much.

Seeds, all in the unsprouted condition, are very variable in VitE – pumpkin seeds are very deficient at only 1/6 of RDA on a per 2500 kcal basis, sesame seeds/ tahini are slightly deficient at 2/3 of RDA on a per 2500 kcal basis, and sunflower seeds are 1,000% sufficient on a per 2500 kcal basis.
Sprouting increases the VitE levels of seeds, but perhaps not by much.

Nuts on average have a 180% sufficiency of VitE on a 2500 kcal basis, all in the unsprouted condition, but there is a great variability.
Macadamia nuts are very deficient in VitE at only 1/7 of RDA on a per kcal basis, walnuts having only 1/5, pecans only 1/3, cashews only 1/3, and pistachios only 2/3 of RDA on a per 2500 kcal basis; whereas Brazil nuts have a sufficiency of 140%, hazelnuts have a sufficiency of 400%, and almonds have a sufficiency of 700% of VitE RDA on a per 2500 kcal basis.
Sprouting increases the VitE levels of nuts, but perhaps not by much.

Grains, all in the unsprouted condition, are substantially deficient in VitE at only 1/3 of RDA on a per 2500 kcal basis. Noteworthy is that rice is very deficient at only 1/15 of VitE RDA on a per 2500 kcal basis.
Sprouting increases the VitE levels of grains, but perhaps not by much.

Starchy root vegetables are very variable in VitE, with potatoes and cassava being very or substantially deficient in VitE at only 1/10 and 1/4 of VitE RDA on a per 2500 kcal basis, and sweet potatoes and yams having sufficiencies of 130% and 200% respectively of VitE RDA on a per 2500 kcal basis.

Fruits on average have only a marginal deficiency of VitE at 4/5 of RDA on a per 2500 kcal basis, but there is a wide variation, from very deficient pineapples, banana and honeydew melon at around 1/7 of RDA on a per kcal basis; most of the fruits being substantially deficient but higher than that; and peaches at 300%, pawpaw at 300%, mangoes at 300%, and plums at 110% of RDA on a per 2500 kcal basis.
Note that the high VitE fruits noted here represent 3 different botanical groups and 3 different geo-climatic area types.

500 g of green leaf, flower and shoot vegetables on average provides only 1/10 of the VitE RDA.

500 g of squash and zucchini provides only 1/30 of the VitE RDA.

The meaning of the food vitamin E data I interpret to be that to provide a good probability of the diet providing a good supply of vitamin E, on average, a good mixture (on a per kcal basis) of different types of seeds, nuts, tubers and fruit should comprise the large majority of the diet.
These could be substituted (on a per kcal basis) by marine animal foods, or eggs, regards VitE, but this does not consider other factors, which generally suggest that such substitution would be best kept to a low proportion of kcal. (see below)

4.2.13 Selenium (Se) (RDA 55 micrograms (ug) /day) (Safe Upper limit 400 ug / day):

Human breast milk is reported to have 25-120 micrograms Se / 2,500 kcal. As this range is probably drawn from a population, the USA, that has documented areas where selenium toxicity from excessive intake occurs in both cattle and humans, with the humans also eating the cattle at an intake higher than almost anywhere else in the world, and as meats tend to accumulate selenium, and as the safe upper limit has been set at 400 micrograms Se / day, and as the vested interests of the meat companies have also possibly had an effect in inflating the selenium RDA, and as the vested interests of the meat companies have also possibly had an effect in increasing the safe upper limit (which is a little less than half the intake level at which overt pathological changes in hair and fingernail tissues has been documented) it is quite possible that the RDA of 55 micrograms Se is inflated above what should really be suggested.
However, in this regard, note that the Se intake in Keshan, China, where the Keshan disease cardiomyopathy was documented, was 3-22 micrograms/day, and that over several months of repletion with Se (the implication is that repletion was titrated, with an increasing dose until the plateau criteria was met, or, that different doses were used) glutathione peroxidase activity plateaued (though note that this was the maximum activity level, which is not necessarily physiological normality) at similar levels in men at a dose of 40 micrograms Se / day – this was raised a fraction to give the Se RDA (Reilly 1991).

Cow milk has a 300% sufficiency of selenium on a per 2500 kcal basis and goat milk is very nearly sufficient and probably sufficient in reality.

Land animal flesh on average approaches the safe upper limit for Se on a per 2500 kcal basis with a sufficiency of 550% at 280 ug.
Marine fish flesh on average exceeds the safe upper limit for Se on a per 2500 kcal basis with a sufficiency of 1,400% at 730 ug.
Marine mollusks and crustaceans on average exceed the safe upper limit of Se on a per 2500 kcal basis with a sufficiency of 2,200% at 1,200 ug.

Eggs on average exceed the safe upper limit of Se on a per 2500 kcal basis with a sufficiency of 1,000% at 550 ug.

Legumes on average are marginally deficient in Se on a per 2500 kcal basis with 45 ug Se / 2,500 kcal, but in reality this is probably quite sufficient.

Seeds on average have a 200% sufficiency of Se on a per 2500 kcal basis, but this is quite variable, with sunflower seeds having a 400% sufficiency, and both pumpkin seeds and sesame/tahini being deficient (probably only marginally in reality) in Se with only 1/2 of the RDA per 2,500 kcal.
Nuts must be considered with Brazil nuts separate from the others here.
Brazil nuts (on average) exceed the safe upper limit of Se on a per 2500 kcal basis with a sufficiency of 9,000% at 5,100 ug per 2,500 kcal. Put another way, 1/10 of calories as Brazil nuts would exceed the safe upper limit for Se, and 1/5 of calories as Brazil nuts would cause overt pathological changes visible in hair and fingernails in some people.
Nuts other than Brazil nuts are deficient (probably only marginally in reality) in Se with only 1/2 of the RDA per 2,500 kcal.
A calorically equal mixture of all eight nuts including Brazil nuts would, at 730 ug Se / 2,500 kcal, still well exceed the safe upper limit for Se, and approach the level at which excess Se causes overt pathological changes visible in hair and fingernails in some people.

Grains on average have a 500% sufficiency of Se on a per 2500 kcal basis, but there is a substantial variability, and those exceeding the upper safe limit on a per kcal basis include oats at 460 ug / 2,500 kcal, wheat at 520 ug / 2,500 kcal, and the middle and high end of the brown rice range at 540 and 870 ug Se / 2,500 kcal exceed the safe upper limit of Se and reach the level at which excess Se causes overt pathological changes visible in hair and fingernails in some people. Corn is the exception, being very deficient in Se at only 1/10 of the RDA per 2,500 kcal.

Starchy root vegetables are substantially deficient in Se, at only 1/5 of the RDA per 2,500 kcal.

Fruits on average are deficient in Se, at only 1/3 of the RDA per 2,500 kcal, but in reality this is probably near to sufficiency, though the range of fruit Se contents includes values probably substantially deficient even in reality.

500g of green leaf, flower or shoot vegetables on average provides only 1/10 of the RDA for Se.

500g of squash or zucchini provides approximately nil Se.

The meaning of the food selenium data I interpret to be that the addition of roughly equal amounts of legumes, nuts, seeds and cereal grains to an amount of fruit and/or starchy root vegetables that meets most of the dietary caloric requirement, and 500 g of green leaf, flower and shoot vegetables would provide a nicely sufficient selenium intake well short of the upper safe limit and toxic levels.

Brazil nuts should not constitute more than 1/20 of calories.

Land animal flesh on average should not constitute more than roughly 1/4 of calories.
Marine fish flesh on average should not constitute more than roughly 1/6 of calories.
Marine mollusks and crustaceans on average should not constitute more than roughly 1/10 of calories.

Squash and zucchini seem to be generally nutritionally inferior to green leaf, flower and shoot vegetables.

4.2.14 Thiamin (VitB1) (RDA 1.1 mg / day) ; Riboflavin (VitB2) (RDA 1.2 mg / day) ; Niacin (VitB3) (RDA 15 mg / day) & Pyridoxal/ine/amine (VitB6) (RDA 1.3 mg / day):

By some coincidence, perhaps for reasons including that they are all water soluble and were the first four water soluble vitamins isolated, perhaps because they are more closely interrelated/ connected in biochemistry than various other nutrients are, these four nutrients are similar enough to each other in both their contents in foods and food groups, and in their sufficiency or deficiency according to the RDA/AI on a per 2,500 kcal basis, that here I will deal with them all together, by food group (and commenting on isolated foods as occasionally appropriate), rather than singly as the nutrients are dealt with in sections 4.3, 4.4 and 4.5 above.

Human breast milk is a little low in VitB1 (0.75 mg / 2,500 kcal); substantially low in VitB6 (0.35-1.1 mg / 2,500 kcal); and ranges into a low content of VitB3 (6.5-20 mg / 2,500 kcal), according to the RDAs. Absorption of these vitamins from human breast milk is probably greater than it is from foods, so it seems that the RDAs are probably roughly correct.
Human breast milk has a 125-200% sufficiency of VitB2 on a per 2500 kcal basis.

An average of cow and goat milk, which are quite similar to each other in their contents of these four nutrients per 2,500 kcal, has:
A 150% sufficiency of VitB1, a 500% sufficiency of VitB2, only 1/2 of VitB3, and a sufficiency of VitB6, on a per 2,500 kcal basis.

Land animal flesh on average has a marginal deficiency (beef more so at only 1/2) of VitB1, a 200% sufficiency of VitB2, a 500% sufficiency of VitB3, and a 200% sufficiency of VitB6, on a per 2,500 kcal basis.

Marine fish flesh on average has a 170% sufficiency of VitB1, a sufficiency of VitB2, a 600% sufficiency of VitB3, and only 3/5 of VitB6, on a per 2,500 kcal basis.

Marine mollusk and crustacean flesh on average has a 180% sufficiency of VitB1, a 350% sufficiency of VitB2, a 350% of VitB3, and a 250% sufficiency of VitB6, on a per 2,500 kcal basis.

Eggs have a very marginal deficiency (probably a sufficiency) of VitB1, a 700% sufficiency of VitB2, only 1/10 of VitB3 RDA, and a 1,500% sufficiency of VitB6, on a per 2,500 kcal basis.

Legumes on average have a 350% sufficiency of VitB1, a sufficiency of VitB2, a 150% sufficiency of VitB3, and a 180% sufficiency of VitB6, on a per 2,500 kcal basis.

Seeds on average have a 350% sufficiency of VitB1, a sufficiency of VitB2 (excepting sesame/tahini at 1/2), a 150% sufficiency of VitB3 (excepting pumpkin seeds at 1/2), and a 180% sufficiency of VitB6 (excepting sesame/tahini and pumpkin seeds at 1/2), on a per 2,500 kcal basis.

Nuts on average have a 180% sufficiency of VitB1, a deficiency of 1/2 of VitB2 (excepting almonds at 250%), a deficiency of VitB3 (have only 1/3, excepting almonds, which are sufficient), and a sufficiency of VitB6 (notably pistachios at 400% sufficiency, and excepting Brazil nuts at 1/4), on a per 2,500 kcal basis.

(continued below)

Cereal grains on average have a 350% sufficiency of VitB1 (it seems the white rice of the data was enriched in B1, given the famed beriberi disease historically associated with it…), a 200% sufficiency of VitB2 (excepting some (i.e. white) rice at having only 1/6), a 200% sufficiency of VitB3 (excepting oats at having only 3/5), and a 150% sufficiency of VitB6, on a per 2,500 kcal basis.

Starchy root vegetables on average have a 150% sufficiency of VitB1; a deficiency of VitB2 for yams and cassava having 1/2 or less, and a sufficiency for potatoes and sweet potatoes; a 140% sufficiency of VitB3; and a 400% sufficiency of VitB6, on a per 2,500 kcal basis.

Fruits on average have a 180% sufficiency of VitB1, a 130% sufficiency of VitB2, a 130% sufficiency of VitB3 (excepting apples having only 1/5 and persimmons having only 1/10), and a 250% sufficiency of VitB6, on a per 2,500 kcal basis.

500 grams of green leaf, flower and shoot vegetables on average provides 1/4 of the VitB1 RDA, 1/4 of the VitB2 RDA, 1/7 of the VitB3 RDA, and 1/2 of the VitB6 RDA.

500 grams of squash and zucchini on average provides 1/5 of the VitB1 RDA, 1/6 of the VitB2 RDA, 1/6 of the VitB3 RDA, and 2/5 of the VitB6 RDA.

The meaning of the food thiamin (VitB1), riboflavin (VitB2), niacin (VitB3) and pyridoxal/ine/amine (VitB6) data I interpret to be that the addition of roughly equal amounts of legumes, nuts, seeds and cereal grains, to an amount of fruit and/or starchy root vegetables that meets most of the dietary caloric requirement, and 500 g of green leaf, flower and shoot vegetables would provide a very sufficient intake of all four of these vitamins. Even including the nuts, which are moderately deficient (compared to the RDA) on average in riboflavin (VitB2) and niacin (VitB3), and also have moderate deficiencies of thiamin (VitB1) and pyridox(al/ine/amine) (VitB6) in many individual nut species within the average values, the ratio of these food groups could be varied considerably from this, and the diet on average would provide a very sufficient intake (exceeding the RDAs) of thiamin (VitB1), riboflavin (VitB2), niacin (VitB3) and pyridoxal/ine/amine (VitB6).
It is quite noteworthy that the fruits on their own on average on a per 2500 kcal basis provide a very sufficient amount of thiamin (VitB1), riboflavin (VitB2), niacin (VitB3) and pyridoxal/ine/amine (VitB6). This came as a surprise even to myself, even fairly aware that I am of the extent to which much nonsense has been very ignorantly, foolishly and wrongly said about various plant foods by various meat-habituated vocalisers/writers, such as the many misconceptions and wrong theories that I have savagely (and very rightly so…) debunked in this work here without fear or favor of network consequences, reveling in the truth, like a scientist should be very much in the habit of.
Also, that 500 g of green leaf, flower and shoot vegetables on average supply a very significant and useful fraction of the RDAs for thiamin (VitB1), riboflavin (VitB2), niacin (VitB3) and pyridoxal/ine/amine (VitB6).

4.3 Calcium (Ca), Phosphorus (P) & Magnesium (Mg), and Fiber Content, and Weight, per 2500kcal of Foods, and their sufficiencies or deficiencies related to the Recommended Daily Allowances (RDAs) / Adequate Intakes (AIs)

Table 7: Calcium (Ca), Phosphorus (P) & Magnesium (Mg), and Fiber Content, and Weight, per 2500kcal of Foods, and RDAs /AIs

Animal Flesh	**Ca**	**Phos**	**Mg**	**Fiber**	
Units	**(mg)**	**(mg)**	**(mg)**	**(g)**	
Meat (5% fat ground beef, broiled):	110	3,000	320	0	/ 2500kcal (1.5kg)
Meat (15% fat ground beef, broiled):	180	2,000	210	0	/ 2500kcal (1.0kg)
Ham (lean, roasted):	115	3,600	350	0	/ 2500kcal (1.6kg)
Lamb (lean, roasted):	110	2,700	340	0	/ 2500kcal (1.3kg)
Chicken Meat (all, fried):	195	2,300	310	0	/ 2500kcal (1.1kg)
Average Land Animal Flesh:	**140**	**2,700**	**310**	**0**	**/ 2500kcal (1.3kg)**
Tuna (baked/broiled):	140	4,400	870	0	/ 2500kcal (1.4kg)
Salmon (fillet, baked/broiled):	105	3,500	510	0	/ 2500kcal (1.4kg)
Sea Bass (fillet, baked/broiled):	260	5,000	1,100	0	/ 2500kcal (2.0kg)
Snapper (fillet, baked/broiled):	780	3,900	720	0	/ 2500kcal (1.9kg)
Sole/Flounder (fillet, broiled):	390	6,100	1,200	0	/ 2500kcal (2.1kg)
Striped Mullet (fillet, baked):	520	4,100	550	0	/ 2500kcal (1.7kg)
Average Sea Fish Flesh:	**365**	**4,500**	**820**	**0**	**/ 2500kcal (1.8kg)**
Squid (calamari, baked):	720	4,800	720	0	/ 2500kcal (1.8kg)
Octopus (raw):	1,600	5,900	960	0	/ 2500kcal (3.0kg)
Crab (leg, boiled):	1,500	7,200	1,600	0	/ 2500kcal (2.6kg)
Prawns (steamed):	1,000	3,400	800	0	/ 2500kcal (2.5kg)
Clams (canned):	1,600	5,700	310	0	/ 2500kcal (1.7kg)
Oysters (raw):	1,700	5,000	1,700	0	/ 2500kcal (3.7kg)
Mussels (steamed/boiled):	480	4,100	530	0	/ 2500kcal (1.5kg)
Av. Sea Mollusk/Crustac'n Flesh:	**1,200**	**5,200**	**950**	**0**	**/ 2500kcal (2.4kg)**
Average Animal Flesh:	**630**	**4,300**	**730**	**0**	**/ 2500kcal (1.9kg)**
Recommended Daily Allowance:	**1,000**	**700**	**360**	**32**	

Note: The values given are for grown, non-pregnant, non-lactating adults under the age of 50, as given in Whitney and Rolfes (2008).

Egg (whole (no shell), raw):	**Ca**	**Phos**	**Mg**	**Fiber**	
Units	**(mg)**	**(mg)**	**(mg)**	**(g)**	
	920	**3,300**	**201**	**0**	**/2500kcal(31of, 1.6kg)**
Recommended Daily Allowance:	**1,000**	**700**	**360**	**32**	

Milk:	**Ca**	**Phos**	**Mg**	**Fiber**	
Units	**(mg)**	**(mg)**	**(mg)**	**(g)**	
Milk (cow, whole):	4,700	3,800	410	0	/ 2500kcal (4.2kg/L)
Milk (goat):	4,900	4,000	510	0	/ 2500kcal (3.6kg/L)
Average Cow/Goat Milk:	**4,800**	**3,900**	**460**	**0**	**/ 2500kcal (3.9kg/L)**
Recommended Daily Allowance:	**1,000**	**700**	**360**	**32**	

Milk (human, mature (after day10)):	**Ca**	**Phos**	**Mg**	**Fiber**	
Units	**(mg)**	**(mg)**	**(mg)**	**(g)**	
	830	**480**	**120**	**0**	**/2500kcal (3.7kg/L)**
Recommended Daily Allowance:	**1,000**	**700**	**360**	**32**	

Legumes:	**Ca**	**Phos**	**Mg**	**Fiber**	
Units	**(mg)**	**(mg)**	**(mg)**	**(g)**	
Soy beans (cooked):	1,500	3,500	1,200	85	/ 2500kcal (1.4kg)
Lentils (cooked):	410	3,900	770	170	/ 2500kcal (2.2kg)
Black beans (boiled):	510	2,700	1,300	170	/ 2500kcal (1.9kg)
Red kidney beans (canned):	560	2,800	890	140	/ 2500kcal (3.0kg)
White beans (boiled):	1,300	----	1,200	180	/ 2500kcal (1.8kg)
Navy beans (boiled):	1,200	2,600	940	190	/ 2500kcal (1.8kg)
Green Peas (boiled):	820	3,500	1,200	150	/ 2500kcal (3.0kg)
Adzuki beans (boiled):	540	----	1,000	140	/ 2500kcal (2.0kg)
Chickpeas (cooked):	740	2,600	730	110	/ 2500kcal (1.5kg)
Average Legume (cooked/soaked):	**840**	**3,100**	**1,000**	**150**	**/ 2500kcal (2.1kg)**
Peanuts (P. butter = raw):	410	2,200	720	35	/2500kcal (0.43kg)
Recommended Daily Allowance:	**1,000**	**700**	**360**	**32**	

Seeds:	**Ca**	**Phos**	**Mg**	**Fiber**	
Units	**(mg)**	**(mg)**	**(mg)**	**(g)**	
Sunflower seeds (dry = raw):	510	3,100	1,600	45	/ 2500kcal (0.42kg)
Pumpkin seeds (roasted):	310	520	1,500	--	/ 2500kcal (0.56kg)
Sesame seeds/ Tahini:	590	3,300	1,400	29	/ 2500kcal (0.44kg)
Average Seeds (dry/raw):	**470**	**(NA)2,300**	**1,500**	**37**	**/ 2500kcal (0.47kg)**
Recommended Daily Allow'ce:	**1,000**	**700**	**360**	**32**	

Nuts:	Ca	Phos	Mg	Fiber	
Units	**(mg)**	**(mg)**	**(mg)**	**(g)**	
Pistachios:	470	2,200	540	45	/ 2500kcal (0.45kg)
Almonds:	1,100	2,100	1,200	50	/ 2500kcal (0.43kg)
Cashews (dry roasted):	200	2,100	1,100	15	/ 2500kcal (0.43kg)
Hazelnuts:	450	1,200	650	40	/ 2500kcal (0.40kg)
Walnuts:	370	1,300	600	25	/ 2500kcal (0.40kg)
Brazil nuts (dried = raw?):	610	----	1,400	30	/ 2500kcal (0.38kg)
Pecan nuts:	250	1,000	440	35	/ 2500kcal (0.36kg)
Macadamia nuts (raw):	290	----	450	30	/2500kcal (0.35kg)
Average Nuts (raw/dry):	**470**	**1,600**	**800**	**34**	**/ 2500kcal (0.40kg)**
Recommended Daily Allowance:	**1,000**	**700**	**360**	**32**	

Grains:	Ca	Phos	Mg	Fiber	
Units	**(mg)**	**(mg)**	**(mg)**	**(g)**	
Oats (rolled):	340	3,100	960	65	/ 2500kcal (0.65kg)
Wheat (flour, whole):	250	2,500	1,000	90	/ 2500kcal (0.74kg)
Millet (cooked, 70% water): 2.1kgwet	60(?)	2,100	930	25	/ 2500kcal (0.63dry)
Barley (whole, cooked, wet):1.8kgwet	240	2,100	410	130	/ 2500kcal (0.63dry)
Rice (brown, cooked, wet): 2.2kgwet	230	1,800	970	45	/ 2500kcal (0.68dry)
Rice (white, cooked, wet):1.9kgwet	62-200	770	100-250	10	/ 2500kcal (0.55dry)
Av. Grains (4 wet, 6 dry):2.0kg wet	**210**	**2,100**	**740**	**60**	**/ 2500kcal (0.65 dry)**
Corn (yellow, boiled):	55	2,400	730	55	/ 2500kcal (2.3kg)
Av. Grains (all 7 here):	**190**	**2,100**	**740**	**60**	**/ 2500kcal**
Recommended Daily Allowance:	**1,000**	**700**	**360**	**32**	

Starchy Root Vegetables:	Ca	Phos	Mg	Fiber	
Units	**(mg)**	**(mg)**	**(mg)**	**(g)**	
Potatoes (baked, with skin):	400	1,900	760	55	/ 2500kcal (2.9kg)
Sweet potatoes (baked):	1,000	1,500	750	95	/ 2500kcal (2.8kg)
Yams (orange, peeled, boiled):	890	1,100	590	80	/ 2500kcal (3.3kg)
Cassava:	240	380	320	23	/ 2500kcal (1.6kg)
Av. Starchy Root Veg's (undried):	**630**	**1,200**	**605**	**65**	**/ 2500kcal (2.6kg)**

Note: Most of the potato's fiber was in the skin.

Recommended Daily Allowance:	**1,000**	**700**	**360**	**32**

Carrots (raw):	Ca	Phos	Mg	Fiber	
Units	**(mg)**	**(mg)**	**(mg)**	**(g)**	
	2,000	**2,100**	**750**	**170**	**/ 2500kcal (6.0kg)**
	170	**180**	**60**	**15**	**210kcal (500g)**
Recommended Daily Allowance:	**1,000**	**700**	**360**	**32**	

Fruit:	Ca	Phos	Mg	Fiber	
Units	**(mg)**	**(mg)**	**(mg)**	**(g)**	
Apricot:	730	1,200	590	150	/ 2500kcal (5.2kg)
Cantaloupe/Rockmelon:	650	1,100	880	45	/ 2500kcal (7.3kg)
Peach:	390	1,300	590	65	/ 2500kcal (6.4kg)
Watermelon:	600	920	820	55	/ 2500kcal (8.3kg)
Oranges:	2,100	740	530	120	/ 2500kcal (5.3kg)
Pawpaw:	1,500	320	630	110	/ 2500kcal (6.4kg)
Honeydew melon (cubes)	410	780	670	40	/ 2500kcal (7.0kg)
Plums:	330	920	420	85	/ 2500kcal (5.5kg)
Jackfruit:	900	----	980	40	/2500kcal (2.7kg)
Pear (fresh):	390	470	310	130	/ 2500kcal (4.3kg)
Pineapple (raw):	690	440	620	60	/ 2500kcal (5.0kg)
Bananas:	140	620	760	70	/ 2500kcal (2.8kg)
Dates:	340	550	380	70	/ 2500kcal (0.9kg)
Grapes (seedless):	370	740	290	35	/ 2500kcal (3.7kg)
Mango:	390	430	350	75	/ 2500kcal (3.8kg)
Persimmon:	150(?)-550	323	170	70	/ 2500kcal (1.9kg)
Apples (average (un)dried):	250	520	240	100	/ 2500kcal (4.7kg)
Average Fruit (undried):	**610**	**720**	**550**	**80**	**/ 2500kcal (4.8kg)**

Note: Apple skin and flesh similar.

Recommended Daily Allowance:	1,000	700	360	32

Avocado (fatty fruit):	Ca	Phos	Mg	Fiber	
Units	**(mg)**	**(mg)**	**(mg)**	**(g)**	
	190	**820**	**450**	**100(?)**	**/2500kcal (1.6kg)**

Green Leaf, Flower & Shoot Veg:	Ca	Phos	Mg	Fiber	
Units	**(mg)**	**(mg)**	**(mg)**	**(g)**	
Kale (raw):	670	----	170	10	/(14gProt, 160kcal)/500g
Broccoli (raw):	230	----	100	13	/(14gProt, 160kcal)/500g
Broccoflower (raw):	160	320	25	15	/(15gProt, 160kcal)/500g
Cauliflower (raw):	110	220	80	10	/(10gProt, 120kcal)/500g
Lettuce:	90	100	35	9	/(10gProt, 70kcal)/500g
Cabbage (Savoy):	170	----	140	16	/(10gProt, 145kcal)/500g
Chinese Cabbage:	460	----	55	5	/(8gProt, 60kcal)/500g
Bamboo shoots (boiled):	50	----	15	6	/(7gProt, 60kcal)/500g
Cabbage (Red):	230	150	80	7	/(7gProt, 160kcal)/500g
Cabbage (standard):	200	130	60	15	/(7gProt, 120kcal)/500g
Av. Leaf, Flower, Shoot Veg's:	**236**	**180**	**75**	**10**	**/(10gProt,100kcal)/500g**

Recommended Daily Allow.:	1,000	700	360	32

Squash and Zucchini:	Ca	Phos	Mg	Fiber	
Units	**(mg)**	**(mg)**	**(mg)**	**(g)**	
Squash, summer, all 5 varieties (boiled):	130	----	120	7	/ 100kcal (500g)
Squash, winter, all 5 varieties (baked):	110	----	65	14	/ 155kcal (500g)
Zucchini, summer (boiled):	65	190	110	8	/ 80kcal (500g)
Average Squash and Zucchini:	**100**	**(NA)190**	**100**	**10**	**/ 110kcal (500g)**

Note: the zucchini Phos value was for raw zucchini.

Recommended Daily Allowance:	1,000	700	360	32

For convenience, the average values of the different food groups collated above (Table 7) are listed here below in Table 8, with a few common oddments appropriately placed:

Table 8: Average Calcium (Ca), Phosphorus (P) & Magnesium (Mg), and Fiber Content, and Weight, per 2500kcal of Food Groups, and RDAs /AIs

Mineral/Vitamin:	**Ca**	**Phos**	**Mg**	**Fiber**
Units:	**(mg)**	**(mg)**	**(mg)**	**(g)**
Recommended Daily Allowance:	**1,000**	**700**	**360**	**32**

Note: The values given are for grown, non-pregnant, non-lactating adults under the age of 50, as given in Whitney and Rolfes (2008).

Average Land Animal Flesh:	**140**	**2,700**	**310**	**0**	**/ 2500kcal (1.3kg)**
Average Sea Fish Flesh:	**365**	**4,500**	**820**	**0**	**/ 2500kcal (1.8kg)**
Av. Sea Mollusk/Crustac'n Flesh:	**1,200**	**5,200**	**950**	**0**	**/ 2500kcal (2.4kg)**
Average Animal Flesh:	**630**	**4,300**	**730**	**0**	**/ 2500kcal (1.9kg)**
Egg (whole (no shell), raw):	**920**	**3,300**	**201**	**0**	**/2500kcal(31of, 1.6kg)**
Average Cow/Goat Milk:	**4,800**	**3,900**	**460**	**0**	**/ 2500kcal (3.9kg/L)**
Human Milk (mature (day10+)):	**830**	**480**	**120**	**0**	**/2500kcal (3.7kg/L)**
Average Legume (cooked/soaked):	**840**	**3,100**	**1,000**	**150**	**/ 2500kcal (2.1kg)**
Peanuts (P. butter = raw):	410	2,200	720	35	/2500kcal (0.43kg)
Average Seeds (dry/raw):	**470**	**(NA)2,300**	**1,500**	**37**	**/ 2500kcal (0.47kg)**
Average Nuts (raw/dry):	**470**	**1,600**	**800**	**34**	**/ 2500kcal (0.40kg)**
Av. Grains (4 wet, 6 dry):2.0kg wet	**210**	**2,100**	**740**	**60**	**/ 2500kcal (0.65 dry)**
Corn (yellow, boiled):	55	2,400	730	55	/ 2500kcal (2.3kg)
Av. Grains (all 7 here):	**190**	**2,100**	**740**	**60**	**/ 2500kcal**
Av. Starchy Root Veg's (undried):	**630**	**1,200**	**605**	**65**	**/ 2500kcal (2.6kg)**

Note: Most of the potato's fiber was in the skin. Cassava is unrepresentatively deficient in all four nutrients.

Carrots (raw):	**2,000**	**2,100**	**750**	**170**	**/ 2500kcal (6.0kg)**
	170	**180**	**60**	**15**	**/ 210kcal (500g)**
Average Fruit (undried):	**610**	**720**	**550**	**80**	**/ 2500kcal (4.8kg)**

Note: Apple skin and flesh similar. Oranges & pawpaw 2X anything else Ca. Pawpaw lowest Phos.

Avocado (fatty fruit):	**190**	**820**	**450**	**100(?)**	**/2500kcal (1.6kg)**
Av. Leaf, Flower, Shoot Veg's:	**236**	**180**	**75**	**10**	**/(10gProt,100kcal)/500g**

Note: Bamboo shoots unrepresentatively low in all three nutrients.

Average Squash and Zucchini:	**100**	**(NA)190**	**100**	**10**	**/ 110kcal (500g)**

Note: the zucchini Phos value only available for (raw) zucchini.

Mineral/Vitamin:	**Ca**	**Phos**	**Mg**	**Fiber**
Units:	**(mg)**	**(mg)**	**(mg)**	**(g)**
Recommended Daily Allowance:	**1,000**	**700**	**360**	**32**

Calcium (Ca) (RDA 1000 mg / day):

Lean land animal flesh is markedly deficient in calcium, and considering that only 50% of calories may be supplied as protein without metabolic toxicity being experienced as illness, so is marine fish flesh, supplying, respectively, at 50% of calories, 70 and 180 mg of the 1000 mg RDA.

A noteworthy curiosity that I've never seen or heard remarked on is that land animal fat has much more calcium than lean meat (which is not chemically surprising given the known ability of cationic calcium (Ca^{2+}) to join to the anionic end of fatty acids (FA^-)), such that 1kg of fat contains roughly 1000 mg of calcium along with 9000 kcal of energy. So Eskimos, if marine mammal fat calcium content is the same, which is not at all certain (it's probably greater, in accord with marine fish flesh), or arctic land animal eating hunter-gatherer peoples, from 1,500 kcal (roughly half their daily requirement) of fat might get only very roughly 175 mg of Ca and maybe another 25mg of Ca if they ate lean meat for the other half of their caloric requirement – the maximum Ca available even if only enough meat was eaten to supply protein needs (which does not address what *other* nutrients might require more than that small amount (roughly 200 grams) of lean meat to be eaten in a context of there being little or no plant foods in the diet) and all the energy was to come from fat might be roughly 300 mg – that these people are not reported (the reporting is probably very lacking/deficient in any case) to suffer from osteoporosis is suggestive that the RDA of 1000 mg Ca per day is too high.

As marine fish flesh is higher in calcium than land animal flesh, it may well be that marine fish fat is similarly higher, which would put fatty (more likely to be from cold waters) marine fish, consumed such that fat was a large fraction of the intake of them, close to being sufficient in calcium – and the same might apply to marine mammals, as the seawater environment (Seawater's high calcium content) might have something to do with the other land-marine difference.

On average marine mollusk and crustacean flesh is sufficient in calcium on a per 2500 kcal basis.

Whole egg is marginally deficient in calcium on a per 2500 kcal basis according to the RDA, but is very probably sufficient in reality, the RDA very probably being excessive as discussed below.

That human breast milk would be deemed a calcium deficient food, supplying 830 mg of the 1000 mg RDA on a per 2500 kcal basis, given that an infant must be forming some times over the amount of bone per body weight than an adult, even though absorption from breast milk by an infant is probably greater than absorption from foods by adults, seems to me a fairly strong indication that the adult RDA is in fact excessive – probably due to the vested interests of the dairy industry and the research it funds, and due to faulty science that does not relate calcium intake to other important factors having a bearing on i.e. osteoporosis, such as weight-bearing exercise, Ca/P ratio, protein intake, sodium intake, and coffee.
The older World Health Organisation (WHO) levels of around 400 mg per day seem to me to be more correct in the context of those other factors being as they ought – and bear in mind that tissue calcification has roles in disease pathologies that are not yet fully understood (calcification of atherosclerotic arterial lesions is becoming used as a marker of risk in atherosclerotic disease imaging), and it is probable that excessive calcium has bad effects, as calcium is by no means a very micro nutrient, and the more macro a nutrient is, the greater the likelihood that a proportionate excess of it will have a bad effect (think about carbohydrate, fat, sugars and sodium, for example).

That cow and goat milk have 5 times the amount of calcium per unit of energy than human breast milk seems to me that for the purposes of the human diet they are unnecessarily high and perhaps undesirably excessive in calcium, depending on the sufficiency of the rest of the diet. This seems likely to me to be related to the vested interests of the dairy companies being involved in falsely inflating the RDA for calcium, in order to maintain a market for the product, which might otherwise be rightly seen as unnecessary or even harmful in the diet.

Legumes at 840 mg Ca / 2500 kcal are deemed deficient in Ca by the RDA, but I strongly suspect that in reality they are very adequate, due to the RDA being artificially inflated as discussed above.

Nuts and seeds at 470 mg Ca / 2500 kcal are deemed deficient in Ca by the RDA, but I suspect that in reality they are adequate, due to the RDA being artificially inflated as discussed above.

(Cereal) Grains at 200 mg Ca / 2500 kcal are deficient in Ca, but not as lacking as lean meat, and their addition to the diet as a caloric replacement for some fraction of fruit would still leave the overall meal/diet very probably quite sufficient in calcium.
Corn is the exception at 55 mg Ca / 2500 kcal, raising the possibility that this low calcium content is part of the association of dental caries with cooked corn as a predominant staple of diet, along with the deposition of carbohydrate along the gum line and on or between teeth.

Starchy root vegetables at 630 mg Ca / 2500 kcal / 2.6 kg are deemed deficient in Ca by the RDA, but I strongly suspect that in reality they are adequate, due to the RDA being artificially inflated as discussed above.

Carrots are root vegetables but not starchy enough to be considered as starchy root vegetables – 1.5 kg would supply 500 mg Ca (which would probably in reality be adequate, due to the RDA being artificially inflated as discussed above), but only 625 kcal, and would be too difficult for even a carrot-liking committed raw food eater to enjoy chewing without grating or some other processing – however, their inclusion in the diet would not negatively impact the diet's calcium sufficiency.

Fruits at 610 mg Ca / 2500 kcal are deemed deficient in Ca by the RDA, but I strongly suspect that in reality they are very adequate, due to the RDA being artificially inflated as discussed above. The weight to be eaten for this is 4.8 kg but with other more energy-dense foods, such as the protein-plus-fat/carbohydrate sources this would reduce. Note that for humans using 2500 kcal of energy an average water intake of roughly 3 litres / day is considered desirable, such as might come from roughly 3.5 kg of fruit.

Green leaf, flower and shoot vegetables at 240 mg Ca / 500 gram / 10 gram protein / negligible kcal are quite a good source of Ca in a balanced meal/diet.

Squash and zucchini are deficient in calcium.

The meaning of the food calcium (Ca) data I interpret to be that it is significant that 500 grams of green leaf, flower and shoot vegetables eaten along with enough fruit to meet caloric requirements would give 850 mg Ca per day, which I strongly suspect in reality would be very adequate for the very large majority of the human population(s), due to the RDA being artificially inflated as discussed above.

This could provide a little too much water for comfort, but replacement of some fraction of the fruit with tubers, legumes, nuts and seeds, and even grains would still very probably provide a sufficient intake of calcium.

Phosphorus (P) (RDA 700 mg / day):

All of the animal foods, with the very dubious exception of human breast milk, are sufficient in phosphorus.
Legumes, seeds, nuts, grains, starchy root vegetables and fruit are, on group average, sufficient in phosphorus.
Green leaf, flower and shoot vegetables are a good source of P in a balanced meal/diet.

Squash and zucchini are deficient in phosphorus.

Calcium/Phosphorus ratio (Ca/P):

Mature (at 10 days of suckling) human breast milk is roughly 2:1 in Ca/P.
Land animals and marine fish meat on average are roughly 1:9
Marine mollusks and crustaceans on average are roughly 1:5
Eggs 1:3.5
Cow&Goat milk 1.25:1
Legumes 1:1.3
Seeds 1:5
Nuts 1:3.5
Grains 1:10
Fruits 1:1
Green leaf, flower and shoot vegetables 1.3:1

The meaning of the food Ca/P ratio data I interpret to be that perhaps the most meaningful comment that could be made on this data is that with the exception of dairy milk (which has 4 times the protein/energy of human breast milk and therefore should not make up a large proportion of the diet) the dietary proportionation of food groups that would make the dietary Ca/P ratio most consistent with both the RDA Ca/P and the human breast milk Ca/P would be to obtain most of the calories through fruit, to have also at least 500 g of green leaf, flower or shoot vegetables, and to have a relatively small proportion (all together) of the other foods.

Magnesium (Mg) (RDA 360 mg / day):

Human breast milk is very dubiously deficient in Mg at 120 mg/ 2500 kcal on the basis of the adult RDA / kcal.

Meat is marginally deficient at 310 mg / 2500 kcal and eggs deficient at 200 mg / 2500 kcal, compared to the RDA of 360 mg, but on consideration of the lower level in human breast milk, although there is not at all any guarantee that the ratios of nutrients required for an infant are the same as those required for adults (consider bone hardness differences, in which Mg is involved, for example), it might be possible to consider both these foods tentatively actually sufficient in Mg. All other food groups are on average sufficient in Mg on a per 2500 kcal basis, some to the extent of being possibly excessively so.

The meaning of the food magnesium data I interpret to be that perhaps the most meaningful comment that could be made on this data is that if we take the actual

optimal requirements as being probably on average no higher than the RDA (as the breast milk level at 120 mg / 2500 kcal is much lower), then seafoods, cow/goat milk, legumes and seeds (very roughly around 1000 mg / 2500 kcal) should be (all together) no more than a lesser proportion of the diet.

<u>Fiber (RDA 32 g / day):</u>

All of the animal foods; flesh, eggs and milks; are absolutely deficient in fiber, having none.

It is the opinion of many orthodox dieticians and nutritionists that 80 g of fiber / day would be better target than the RDA here of 32 g, and fruit and green leaf, flower and shoot vegetables seems the best way of meeting or (perhaps to even greater benefit) exceeding that amount.

4.4 Vitamin B12 (VitB12), Folate, Iron (Fe), and Zinc (Zn) Content, and Weight, per 2500 kcal of Foods, and their sufficiencies or deficiencies related to the Recommended Daily Allowances (RDAs) / Adequate Intakes (AIs)

Table 9: Iron (Fe), Zinc (Zn), Folate, and Vitamin B12 content, and weight, per 2500 kcal of foods, & RDAs

Animal Flesh	Fe	Zn	Fol	VitB12	
Units	**(mg)**	**(mg)**	**(ug)**	**(ug)**	
Meat (5% fat ground beef, broiled):	41	95	105	35	/ 2500kcal (1.5kg)
Meat (15% fat ground beef, broiled):	26	64	90	25	/ 2500kcal (1.0kg)
Ham (lean, roasted):	15	41	70	10	/ 2500kcal (1.6kg)
Lamb (lean, roasted):	28	65	310	35	/ 2500kcal (1.3kg)
Chicken Meat (all, fried):	15	25	80	5	/ 2500kcal (1.1kg)
Average Land Animal Flesh:	**25**	**58**	**130**	**20**	**/ 2500kcal (1.3kg)**
Tuna (baked/broiled):	18	11	32	150	/ 2500kcal (1.4kg)
Salmon (fillet, baked/broiled):	14	12	400	40	/ 2500kcal (1.4kg)
Sea Bass (fillet, baked/broiled):	8	10	120	6	/ 2500kcal (2.0kg)
Snapper (fillet, baked/broiled):	5	8	115	70	/ 2500kcal (1.9kg)
Sole/Flounder (fillet, broiled):	7	13	185	55	/ 2500kcal (2.1kg)
Striped Mullet (fillet, baked):	23	14	160	5	/ 2500kcal (1.7kg)
Average Sea Fish Flesh:	**12**	**11**	**170**	**55**	**/ 2500kcal (1.8kg)**
Squid (calamari, baked):	15	34	105	25	/ 2500kcal (1.8kg)
Octopus (raw):	161	51	483	610	/ 2500kcal (3.0kg)
Crab (leg, boiled):	19	195	1,300	300	/ 2500kcal (2.6kg)
Prawns (steamed):	80	34	115	35	/ 2500kcal (2.5kg)
Clams (canned):	470	46	490	1,700	/ 2500kcal (1.7kg)
Oysters (raw):	244	3,300	370	710	/ 2500kcal (3.7kg)
Mussels (steamed/boiled):	100	39	1,100	350	/ 2500kcal (1.5kg)
Av. Sea Mollusk/Crustacean Flesh:	**155**	**530**	**565**	**530**	**/ 2500kcal (2.4kg)**
Average Animal Flesh:	**70**	**225**	**310**	**230**	**/ 2500kcal (1.9kg)**
Recommended Daily Allowance:	**8m18f**	**9.5**	**400**	**2.4**	

Note: The RDA given here for Zn is the average of male and female values as it is obvious that body weight is the factor for the small difference between them. The values given are for grown, non-pregnant, non-lactating adults under the age of 50 y, as given in Whitney and Rolfes (2008).

Egg (whole (no shell), raw):	Fe	Zn	Fol	VitB12	
Units	**(mg)**	**(mg)**	**(ug)**	**(ug)**	
	32	**20**	**830**	**25**	**/2500kcal (31of, 1.6kg)**
Recommended Daily Allowance:	**8m18f**	**9.5**	**400**	**2.4**	

Milk:	**Fe**	**Zn**	**Fol**	**VitB12**	
Units	**(mg)**	**(mg)**	**(ug)**	**(ug)**	
Milk (cow, whole):	1.7	17	205	20	/ 2500kcal (4.2kg/L)
Milk (goat):	1.5	10	30	3	/ 2500kcal (3.6kg/L)
Average Cow/Goat Milk:	**1.6**	**14**	**120**	**12**	**/ 2500kcal (3.9kg/L)**
Recommended Daily Allowance:	**8m18f**	**9.5**	**400**	**2.4**	

Milk (human, mature (after day10)):	**Fe**	**Zn**	**Fol**	**VitB12**	
Units	**(mg)**	**(mg)**	**(ug)**	**(ug)**	
	2.2	**7.4**	**400**	**1.8-3.7**	**/2500kcal (3.7kg/L)**
Recommended Daily Allowance:	**8m18f**	**9.5**	**400**	**2.4**	

Legumes:	**Fe**	**Zn**	**Fol**	**VitB12**	
Units	**(mg)**	**(mg)**	**(ug)**	**(ug)**	
Soy beans (cooked):	75	17	780	0	/ 2500kcal (1.4kg)
Lentils (cooked):	70	27	3,900	0	/ 2500kcal (2.2kg)
Black beans (boiled):	40	21	2,800	0	/ 2500kcal (1.9kg)
Red kidney beans (canned):	60	21	2,600	0	/ 2500kcal (3.0kg)
White beans (boiled):	66	25	2,400	0	/ 2500kcal (1.8kg)
Navy beans (boiled):	42	19	2,500	0	/ 2500kcal (1.8kg)
Green Peas (boiled):	45	37	1,700	0	/ 2500kcal (3.0kg)
Adzuki beans (boiled):	39	35	2,400	0	/ 2500kcal (2.0kg)
Chickpeas (cooked):	44	23	2,600	0	/ 2500kcal (1.5kg)
Average Legume (cooked/soaked):	**53**	**25**	**2,400**	**0**	**/ 2500kcal (2.1kg)**
Peanuts (P. butter = raw):	8.0	12	390	0	/2500kcal (0.43kg)
Recommended Daily Allowance:	**8m18f**	**9.5**	**400**	**2.4**	

Seeds:	**Fe**	**Zn**	**Fol**	**Vitb12**	
Units	**(mg)**	**(mg)**	**(ug)**	**(ug)**	
Sunflower seeds (dry = raw):	30	22	1,000	0	/ 2500kcal (0.42kg)
Pumpkin seeds (roasted):	18	58	55	0	/ 2500kcal (0.56kg)
Sesame seeds/ Tahini:	26	44	410	0	/ 2500kcal (0.44kg)
Average Seeds (dry/raw):	**25**	**41**	**490**	**0**	**/ 2500kcal (0.47kg)**
Recommended Daily Allowance:	**8m18f**	**9.5**	**400**	**2.4**	

Nuts:	**Fe**	**Zn**	**Fol**	**VitB12**	
Units	**(mg)**	**(mg)**	**(ug)**	**(ug)**	
Pistachios:	19	10	220	0	/ 2500kcal (0.45kg)
Almonds:	18	14	120	0	/ 2500kcal (0.43kg)
Cashews (dry roasted):	26	23	300	0	/ 2500kcal (0.43kg)
Hazelnuts:	19	10	450	0	/ 2500kcal (0.40kg)
Walnuts:	11	12	370	0	/ 2500kcal (0.40kg)
Brazil nuts (dried = raw?):	9	15	85	0	/ 2500kcal (0.38kg)
Pecan nuts:	9	16	80	0	/ 2500kcal (0.36kg)
Macadamia nuts (raw):	13	5	40	0	/2500kcal (0.35kg)
Average Nuts (raw/dry):	**16**	**13**	**210**	**0**	**/ 2500kcal (0.40kg)**

Recommended Daily Allowance: 8m18f 9.5 400 2.4

Grains:	**Fe**	**Zn**	**Fol**	**VitB12**	
Units	**(mg)**	**(mg)**	**(ug)**	**(ug)**	
Oats (rolled):	28	20	105	0	/ 2500kcal (0.65kg)
Wheat (flour, whole):	29	21	330	0	/ 2500kcal (0.74kg)
Millet (cooked, 70% water): 2.1kgwet	13	19	400	0	/ 2500kcal (0.63dry)
Barley (whole, cooked, wet):1.8kgwet	19	15	150	0	/ 2500kcal (0.63dry)
Rice (brown, cooked, wet): 2.2kgwet	9	14	95(?)	0	/ 2500kcal (0.68dry)
Rice (white, cooked, wet):1.9kgwet	34	10	1,100(?)	0	/ 2500kcal (0.55dry)
Av. Grains (4 wet, 6 dry):2.0kg wet	**22**	**16**	**360**	**0**	**/ 2500kcal (0.65 dry)**
Corn (yellow, boiled):	14	11	1,100	0	/ 2500kcal (2.3kg)
Av. Grains (all 7 here):	**21**	**15**	**470**	**0**	**/ 2500kcal**

Recommended Daily Allowance: 8m18f 9.5 400 2.4

Starchy Root Vegetables:	**Fe**	**Zn**	**Fol**	**VitB12**	
Units	**(mg)**	**(mg)**	**(ug)**	**(ug)**	
Potatoes (baked, with skin):	29	9?	760	0	/ 2500kcal (2.9kg)
Sweet potatoes (baked):	19	9?	170	0	/ 2500kcal (2.8kg)
Yams (orange, boiled):	24	7	200	0	/ 2500kcal (3.3kg)
Cassava:	4	5	270	0	/ 2500kcal (1.6kg)
Av. Starchy Root Veg's (undried):	**19**	**8**	**350**	**0**	**/ 2500kcal (2.6kg)**

Note: Almost all of the potato's Fe was in the skin

Recommended Daily Allowance: 8m18f 9.5 400 2.4

Carrots (raw):	**Fe**	**Zn**	**Fol**	**VitB12**	
Units	**(mg)**	**(mg)**	**(ug)**	**(ug)**	
	17	**17**	**1,200**	**0**	**/ 2500kcal (6.0kg)**
	1.4	**1.4**	**100**	**0**	**/ 210kcal (500g)**

Recommended Daily Allowance: 8m18f 9.5 400 2.4

Fruit:	**Fe**	**Zn**	**Fol**	**VitB12**	
Units	**(mg)**	**(mg)**	**(ug)**	**(ug)**	
Apricot:	15	15	440	0	/ 2500kcal (5.2kg)
Cantaloupe/Rockmelon:	14	14	1,600	0	/ 2500kcal (7.3kg)
Peach:	13	13	260	0	/ 2500kcal (6.4kg)
Watermelon:	22	11	270	0	/ 2500kcal (8.3kg)
Oranges:	6	3	1,600	0	/ 2500kcal (5.3kg)
Pawpaw:	6	4	2,400	0	/ 2500kcal (6.4kg)
Honeydew melon (cubes)	12	8	1,300	0	/ 2500kcal (7.0kg)
Plums:	8	8	250	0	/ 2500kcal (5.5kg)
Jackfruit:	16	11	370	0	/2500kcal (2.7kg)
Pear (fresh):	8	5	310	0	/ 2500kcal (4.3kg)
Pineapple (raw):	14	6	900	0	/ 2500kcal (5.0kg)
Bananas:	7	5	570	0	/ 2500kcal (2.8kg)
Dates:	8	2	170	0	/ 2500kcal (0.89kg)
Grapes (seedless):	15	'3'	75	0	/ 2500kcal (3.7kg)
Mango:	5	1	540	0	/ 2500kcal (3.8kg)
Persimmon:	3-48(?)	2	150	0	/ 2500kcal (1.9kg)
Apples (average (un)dried):	7	3	70	0	/ 2500kcal (4.7kg, undr)
Average Fruit (undried):	**13**	**7**	**670**	**0**	**/ 2500kcal (4.8kg)**

Recommended Daily Allowance:	**8m18f**	**9.5**	**400**	**2.4**

Avocado (fatty fruit):	**Fe**	**Zn**	**Fol**	**VitB12**	
Units	**(mg)**	**(mg)**	**(ug)**	**(ug)**	
	8.5	**10**	**1,300**	**0**	**/2500kcal (1.6kg)**

Green Leaf, Flower & Shoot Veg:	**Fe**	**Zn**	**Fol**	**VitB12**	
Units	**(mg)**	**(mg)**	**(ug)**	**(ug)**	
Kale (raw):	8.5	2	140	0	/(14gProt, 160kcal)/500g
Broccoli (raw):	3.5	2	310	0	/(14gProt, 160kcal)/500g
Broccoflower (raw):	3.5	2.5	280	0	/(15gProt, 160kcal)/500g
Cauliflower (raw):	2	1	280	0	/(10gProt, 120kcal)/500g
Lettuce:	2	1	150	0	/(10gProt, 70kcal)/500g
Cabbage (Savoy):	2	1.5	400	0	/(10gProt,145kcal)/500g
Chinese Cabbage:	5	1	205	0	/(8gProt, 60kcal)/500g
Bamboo shoots (boiled):	1	2.5	13	0	/(7gProt, 60kcal)/500g
Cabbage (Red):	4	1.5	90	0	/(7gProt,160kcal)/500g
Cabbage (standard):	2	1	210	0	/(7gProt,120kcal)/500g
Av. Leaf, Flower, Shoot Veg's:	**3.5**	**1.5**	**210**	**0**	**/(10gProt,100kcal)/500g**

Note: The broccoflower Fe value was taken from Whitney and Rolfes (2008) as the Insel, Turner and Ross (2007) value was too low to be consistent with the values amongst both books for broccoli and cauliflower, from which broccoflower is derived.

Recommended Daily Allowance:	**8m18f**	**9.5**	**400**	**2.4**

Squash and Zucchini:	**Fe**	**Zn**	**Fol**	**VitB12**	
Units	**(mg)**	**(mg)**	**(ug)**	**(ug)**	
Squash, summer, all 5 varieties (boiled):	2	2	100	0	/ 100kcal (500g)
Squash, winter, all 5 varieties (baked):	2	1	100	0	/ 155kcal (500g)
Zucchini, summer (boiled):	2	1	60	0	/ 80kcal (500g)
Average Squash and Zucchini:	**2**	**1.5**	**85**	**0**	**/ 110kcal (500g)**

Recommended Daily Allowance:	**8m18f**	**9.5**	**400**	**2.4**

For convenience, the average values of the different food groups collated above (Table 9) are listed here below in Table 10, with a few common oddments relocated underneath:

Table 10: Average Iron (Fe), Zinc (Zn), Folate and Vitamin B12 Content, and Weight, per 2500 kcal of Food Groups, & RDAs

Mineral/Vitamin:	Fe	Zn	Fol	VitB12	
Units:	(mg)	(mg)	(ug)	(ug)	
Recommended Daily Allowance:	8m18f	9.5	400	2.4	
Note:					
Average Land Animal Flesh:	25	58	130	20	/ 2500kcal (1.3kg)
Average Sea Fish Flesh:	12	11	170	55	/ 2500kcal (1.8kg)
Av. Sea Mollusk/Crustacean Flesh:	155	530	565	530	/ 2500kcal (2.4kg)
Average Animal Flesh:	70	225	310	230	/ 2500kcal (1.9kg)
Egg (whole (no shell), raw): 31of	32	20	830	25	/2500kcal(1.6kg)
Average Cow/Goat Milk:	1.6	14	120	12	/2500kcal (3.9kg/L)
Human Milk (mature (day10+)):	2.2	7.4	400	1.8-3.7	/2500kcal (3.7kg/L)
Average Legume (cooked/soaked):	53	25	2,400	0	/ 2500kcal (2.1kg)
Peanuts (P. butter = raw):	8.0	12	390	0	/2500kcal (0.43kg)
Average Seeds (dry/raw):	25	41	490	0	/2500kcal (0.47kg)
Average Nuts (raw/dry):	16	13	210	0	/2500kcal (0.40kg)
Av. Grains (4 wet, 6 dry):2.0kg wet:	22	16	360	0	/2500kcal (0.65dry)
Corn (yellow, boiled):	14	11	1,100	0	/ 2500kcal (2.3kg)
Av. Grains (all 7 here):	21	15	470	0	/ 2500kcal
Av. Starchy Root Veg's (undried):	19	8	350	0	/ 2500kcal (2.6kg)
Notes: 1. Zn and folate are very close to the RDA. 2. Almost all of the potato's Fe was in the skin					
Carrots (raw):	17	17	1,200	0	/ 2500kcal (6.0kg)
	1.4	1.4	100	0	/ 210kcal (500g)
Average Fruit (undried):	13	7	670	0	/ 2500kcal (4.8kg)
Avocado (fatty fruit):	8.5	10	1,300	0	/2500kcal (1.6kg)
Av. Gr. Leaf, Flwr, Sht Veg: 100kcal	3.5	1.5	210	0	/(10gProt,)/500g
Average Squash and Zucchini:	2	1.5	85	0	/ 110kcal (500g)
Mineral/Vitamin:	Fe	Zn	Fol	VitB12	
Units:	(mg)	(mg)	(ug)	(ug)	
Recommended Daily Allowance:	8m18f	9.5	400	2.4	

Vitamin B12 (RDA 2.4 micrograms (ug) / day):

Human breast milk has roughly 3 micrograms VitB12 / 2500 kcal. The human RDA is 2.4 micrograms / day.

All other animal foods far surpass this level:
Cow/goat milk has 12 micrograms VitB12 / 2500 kcal,
Land animal flesh on average has 20 micrograms VitB12 / 2500 kcal,
Marine fish flesh on average has 55 micrograms VitB12 / 2500 kcal,
Marine mollusks and crustaceans on average have 530 micrograms VitB12 / 2500 kcal,
Eggs have 25 micrograms VitB12 / 2500 kcal (though reportedly only 10% available vs. 50% for flesh foods, which would bring the amount available per kcal back to something probably very close to that from human breast milk),
Insects vary widely – some have no B12, most (?) sufficient B12, termites a great excess.

All plant foods reportedly have nil intrinsic B12, though some variable and small amount may accrue from adherence from the environment (incl. insects and int./ext. bacteria).

Some seaweeds, green and purple lavers, have B12 useful to humans, others, like blue-green algae ("Spirulina" for example, I think) have inactive B12 analogues that may be worse than useless as they occupy B12 sites in the body or digestive tract (although the digestive tract reportedly has a mechanism that prevents some of these being absorbed).

The meaning of the food vitamin B12 data I interpret to be that as the human infant is making more neural tissue (absolutely, not only relative to its bodyweight) and other things that adult humans are, but needs absolutely less energy (1000 kcal / day for a 10 kg infant, for example), and therefore receives only roughly 1 microgram per day from the breast milk it drinks in meeting its lower absolute energy needs, it seems logical that adult human requirements, at similar absorption + intestinal production (?), should be (much?) less than 1 microgram of vit. B12 / day.
This would be met by only a very small amount of animal matter, and could probably all be met by the amount of insect matter, larvae and adults, that apes eat incidentally in plant matter eaten, w/wo that eaten deliberately in focused insect eating. Note that RDAs are set to accommodate the very large majority of the population including those with reduced absorption, transport, storage and use of a particular nutrient.

Folate (RDA 400 micrograms / day):

Human breast milk has about 400 micrograms folate / 2500 kcal.
Cow and goat milk have much less at about 200 and 30 micrograms folate / 2500 kcal respectively, despite the fact that they naturally eat food that would supply much more folate than would the diet of carnivores.
Folate supplementation for neural tube defect (NTD) prevention is one of the few instances of supplemented isolated nutrients being substantially effective in improving a serious human health condition at the population (-representative) level.
These facts, together with those immediately below, suggest that humans are a very high-folate requiring animal compared to both carnivores and other mammals.

Land animal flesh and marine fish flesh are both quite deficient in folate at 130 and 170 micrograms folate / 2500 kcal respectively, particularly when considering that for humans only 50% or less of the calories may be eaten as protein without metabolic toxicity being experienced as illness – for 1250 kcal (half the daily energy requirement) the intakes would be 65 and 85 micrograms folate / 2500 kcal respectively, compared to the RDA of 400 micrograms, and that folate being a water-soluble, and therefore not a fat-soluble, vitamin, the fat of any carcass will contain less folate than the lean meat (consistent with the nutrient data), so fat eating would not make up for the deficiency. Note that there is no mechanism for storing folate in fat as there is for ionic calcium (which is very water soluble, though perhaps it is taken up into fatty tissue already joined to fatty acid (the 'calcium salt' of the fatty acid), nor for the storage of folate in a similar fashion as vitamin B12 (also water soluble) is stored (in liver).

On average marine mollusk and crustacean flesh is sufficient in folate per 2500 kcal.

Eggs are 200% sufficient on a per 2500 kcal basis in folate.

Legumes are 600% sufficient on average on a per 2500 kcal basis in folate. Therefore consuming roughly enough legumes to meet protein requirements would meet the RDA for folate, and provide roughly 1/4 of the daily energy requirement.

Seeds are sufficient on a per kcal basis in folate on average for the only three types considered, but pumpkin seeds were very deficient. Consuming roughly enough seeds to meet protein requirements would provide roughly 1/2 (but only 1/30 for pumpkin seeds) of the RDA for folate, and roughly 1/4 of the daily energy requirement.

Nuts are deficient on a per 2500 kcal basis in folate on average, at 210 micrograms folate / 2500 kcal. Consuming roughly enough nuts to meet protein requirements would provide roughly 1/8 of the RDA for folate, and roughly 1/4 of the daily energy requirement.

Grains (without corn) are marginally deficient on a per 2500 kcal basis in folate on average, at 360 micrograms folate / 2500 calories. Consuming roughly enough grain to meet protein requirements would provide marginally less than the RDA for folate (which would very probably be sufficient for the very large majority of the population(s), and roughly the entire daily energy requirement. Corn has a roughly 300% sufficiency in folate on a per 2500 kcal basis.

Starchy root vegetables are marginally deficient on a per 2500 kcal basis in folate on average, at 350 micrograms folate / 2500 calories. Consuming roughly enough starchy root vegetables to meet protein requirements would provide marginally less than the RDA for folate (which would very probably be sufficient for the very large majority of the population(s), and roughly the entire daily energy requirement.
Cassava is markedly deficient in many nutrients, including folate, unless the data or assays are faulty.

Fruit is 170% sufficient on a per 2500 kcal basis in folate on average at 670 micrograms folate / 2500 kcal. Consuming roughly enough fruit to meet the average daily water requirement of 3 litres would provide roughly all of the folate RDA, roughly 1/2 of the protein requirements, and roughly 2/3 of the daily energy requirement.

Eating 500 grams of green leaf, flower and shoot vegetables per day will provide 1/2 of the folate RDA on average, roughly 1/7 of the average daily water requirements, and roughly 1/6 of the daily protein requirements.

Eating 500 grams of squash or zucchini per day will provide roughly 1/4 of the folate RDA on average.

In considering calcium, calcium/phosphorus ratio, fiber, folate, protein, water and energy together, it seems that any more (as a replacement for any other of the food groups noted here) than the very small amount of animal food required for meeting

vitamin B12 requirements would have a detrimental effect (with reference to the RDAs) on a diet composed of mostly (roughly 3 kg, such as to roughly meet the average daily water requirements) fruit and starchy root vegetables, having at least 500 grams of green leaf, flower or shoot vegetables, and roughly 200 grams dry weight of combined legumes, seeds, nuts and cereal grains. The same seems to apply even if these proportions are varied quite a bit.

Zinc (Zn) (RDA 9.5 mg / day):

Human breast milk has roughly 7.4 millegrams Zn / 2500 kcal, which is (very dubiously, though absorption will be higher) deficient on a kcal basis according to the RDA. The human RDA is 9.5 millegrams / day.

As the human infant is making many times more proteinaceous tissue (relative to its bodyweight) than adult humans are, it seems logical that adult human requirements, in the absence of some malfunction, should need only a fraction of 7.4 millegrams of Zn per day. It seems probable to me that the vested interests of the meat companies, via the research that they fund, in conjunction with the cherished dietary habits of sundry authorities, has been responsible for the artificial inflation of the actual human requirement for Zn.

Land animal flesh has an average 600% sufficiency of Zn, marine fish has a sufficiency, and marine mollusks and crustaceans have a 5000% sufficiency (raising questions of possible long-term toxicity) of Zn on per 2500 kcal basis.

Cow and goat milk are sufficient in Zn on a per 2500 kcal basis.

Legumes have a 250% sufficiency of Zn, seeds have a 400% sufficiency of Zn, nuts a sufficiency of Zn, and grains a 150% sufficiency, on average on a per 2500 kcal basis. (Phytic acid, sprouting, vitamin C, cooking, and probably other factors potentially significantly modify the availability of Zn from food.)

Starchy root vegetables are marginally deficient in Zn, at 8 mg / 2500 kcal, according to the RDA of 9.5 mg / 2500 kcal, but it is very probable that in reality this amount is sufficient for the very large majority of the human population(s).
Cassava is markedly deficient in many nutrients, including Zn, unless the data or assays are faulty.

Fruits are marginally deficient in Zn, at 7 mg / 2500 kcal, according to the RDA of 9.5 mg / 2500 kcal, but it is very probable that in reality this amount is sufficient for the large majority of the human population(s).

500 g of green leaf, flower and shoot vegetables on average provide only a surprisingly (given the 10 g of protein provided) low 1.5 mg Zn. Nevertheless, this is still a useful lesser fraction of the Zn requirement from a food group that ought be only a fraction of diet and is not considered as contributing to the meeting of caloric requirements.
500 g of squash and zucchinis likewise, but less surprisingly, provide 1.5 mg Zn.

Iron (Fe) (RDA 8 mg / day for males; 18 mg / day for menstruating females):

Human breast milk has roughly 2.2 millegrams Fe / 2500 kcal, which is (very dubiously, though absorption will be higher) deficient on a per kcal basis according to the RDA. The human RDA is 8(male)/18(female) millegrams / day.

As the human infant is making many times more blood and other tissue (relative to its bodyweight) than adult humans are, it seems logical that adult human requirements, in the absence of some malfunction, leaving aside menstrual losses of Fe in blood, should need only about 2.2 millegrams of Fe per day. It seems probable to me that the vested interests of the meat companies, via the research that they fund, in conjunction with the cherished dietary habits of sundry authorities, has been responsible for the artificial inflation of the actual human requirements for Fe.

Land animal flesh on average is sufficient in Fe on a per kcal basis.
Marine fish flesh on average is sufficient in Fe for males but slightly deficient in Fe for females, on a per 2500 kcal basis.
Marine mollusk and crustacean flesh on average is 900% sufficient in Fe (raising questions of possible long-term toxicity), on a per 2500 kcal basis.

Eggs have a 160% sufficiency of Fe on a per 2500 kcal basis.

Cow and goat milk are greatly deficient in Fe on a per kcal basis, providing less than 1/10 of pre-menopausal female RDA. (Roughly similar to human breast milk)

Legumes on average have a 250% sufficiency of Fe on a per 2500 kcal basis.

Seeds on average are sufficient in Fe on a per 2500 kcal basis.

Nuts on average are sufficient in Fe for males but marginally deficient for females, according to the RDA, but in reality are probably sufficient for the large majority of the population(s), on a per 2500 kcal basis.

Grains on average are sufficient in Fe on a per 2500 kcal basis. (Phytic acid, sprouting, vitamin C, cooking, and probably other factors potentially significantly modify the availability of Fe from food.)

Starchy root vegetables on average are sufficient in Fe on a per 2500 kcal basis.
Cassava is markedly deficient in many nutrients, including Fe, unless the data or assays are faulty.

Fruits on average are sufficient for males in Fe, but slightly deficient for pre-menopausal females in Fe, on a per 2500 kcal basis, but the reality is probably that fruits on average are actually sufficient for the large majority of the human population(s), particularly given the lack of phytic acid and the presence of vitamin C.

500 g of green leaf, flower and shoot vegetables on average provide only 3.5 mg Fe. Nevertheless, this is still a useful lesser fraction of the Fe requirement from a food group that ought be only a fraction of diet and is not considered as contributing to the meeting of caloric requirements.
500 g of squash and zucchinis provide 2 mg Fe.

The meaning of the food data (for calcium (Ca), phosphorus (P), their ratio Ca/P, magnesium (Mg), fiber, iron (Fe), zinc (Zn), folate and vitamin B12) considered so far in this chapter I interpret to be that in considering calcium, calcium/phosphorus ratio, fiber, folate, protein, water, and energy, to which now are added zinc and iron, all together, it still seems that any more (as a replacement for any other of the food groups noted here) than the very small amount of animal food (and certainly of milk with its great deficiency of iron) required for meeting vitamin B12 requirements would have a detrimental effect (with reference to the RDAs) on a diet composed of

mostly (roughly 3 kg, such as to roughly meet the average daily water requirements) fruit and starchy root vegetables, having at least 500 grams of green leaf, flower or shoot vegetables, and roughly 200 grams dry weight of combined legumes, seeds, nuts and cereal grains. The same seems to apply even if these proportions are varied quite a bit.

4.5 Vitamin C, Vitamin A, Vitamin E, and Selenium (Se) Content, and Weight, per 2500 kcal of Foods, and their sufficiencies or deficiencies related to the Recommended Daily Allowances (RDAs) / Adequate Intakes (AIs)

Table 11: VitC, VitA, VitE & Se content, and weight, per 2500 kcal of foods, & RDAs

Animal Flesh	**VitC**	**VitA**	**VitE**	**Se**	
Units	**(mg)**	**(ug RetEq)**	**(mg ∝)**	**(ug)**	
Meat (5% fat ground beef, broiled):	0	0	5.5	200	/ 2500kcal (1.5kg)
Meat (15% fat ground beef, broiled):	0	0	4	160	/ 2500kcal (1.0kg)
Ham (lean, roasted):	0	0	4.5	340	/ 2500kcal (1.6kg)
Lamb (lean, roasted):	0	0	3	350	/ 2500kcal (1.3kg)
Chicken Meat (all, fried):	0	200	5	370	/ 2500kcal (1.1kg)
Average Land Animal Flesh:	**0**	**(NA)40**	**4.5**	**280**	**/ 2500kcal (1.3kg)**
Tuna (baked/broiled):	0	10,000	17	840	/ 2500kcal (1.4kg)
Salmon (fillet, baked/broiled):	0	180	17	>190	/ 2500kcal (1.4kg)
Sea Bass (fillet, baked/broiled):	0	1,300	12	950	/ 2500kcal (2.0kg)
Snapper (fillet, baked/broiled):	35	690	13	940	/ 2500kcal (1.9kg)
Sole/Flounder (fillet, broiled):	0	290	13	740	/ 2500kcal (2.1kg)
Striped Mullet (fillet, baked):	20	700	20	---	/ 2500kcal (1.7kg)
Average Sea Fish Flesh:	**(NA)9**	**(NA)2,200**	**15**	**730**	**/ 2500kcal (1.8kg)**
Note: Tuna VitA >7x anything else					
Squid (calamari, baked):	100	730	35	1,200	/ 2500kcal (1.8kg)
Octopus (raw):	160	1,400	40	1,400	/ 2500kcal (3.0kg)
Crab (leg, boiled):	190	230	23	980	/ 2500kcal (2.6kg)
Prawns (steamed):	55	1,700	34	1,000	/ 2500kcal (2.5kg)
Clams (canned):	370	3,000	11	1,100	/ 2500kcal (1.7kg)
Oysters (raw):	85	250	31	1,300	/ 2500kcal (3.7kg)
Mussels (steamed/boiled):	210	1,300	21	1,300	/ 2500kcal (1.5kg)
Av. Sea Mollusk/Crustac'n Flesh:	**170**	**1,200**	**28**	**1,200**	**/ 2500kcal (2.4kg)**
Average Animal Flesh:	**(NA)70**	**(NA)1,200**	**17**	**790**	**/ 2500kcal (1.9kg)**
Recommended Daily Allowance:	**82**	**800**	**15**	**55**	

The values given are for grown, non-pregnant, non-lactating adults under the age of 50, as given in Whitney and Rolfes (2008).

Egg (whole (no shell), raw):	**VitC**	**VitA**	**VitE**	**Se**	
Units	**(mg)**	**(ugRetEq)**	**(mg ∝)**	**(ug)**	
	0	**2,500**	**17**	**550**	**/2500kcal(31of, 1.6kg)**
Recommended Daily Allowance:	**82**	**800**	**15**	**55**	

Milk:	VitC	VitA	VitE	Se	
Units	**(mg)**	**(ugRetEq)**	**(mg ∝)**	**(ug)**	
Milk (cow, whole):	0	1,200	1.7	150	/ 2500kcal (4.2kg/L)
Milk (goat):	45	2,100	3.0	50	/ 2500kcal (3.6kg/L)
Average Cow/Goat Milk:	**(NA)20**	**1,600**	**2.4**	**100**	**/ 2500kcal (3.9kg/L)**
Recommended Daily Allowance:	**82**	**800**	**15**	**55**	

Milk (human, mature (after day10)):	VitC	VitA	VitE	Se	
Units	**(mg)**	**(ugRetEq)**	**(mg ∝)**	**(ug)**	
	370	**1,700**	**20**	**25-120**	**/2500kcal (3.7kg/L)**
Recommended Daily Allowance:	**82**	**800**	**15**	**55**	

Legumes:	VitC	VitA	VitE	Se	
Units	**(mg)**	**(ugRetEq)**	**(mg ∝)**	**(ug)**	
Soy beans (cooked):	25	65	5	110	/ 2500kcal (1.4kg)
Lentils (cooked):	35	20	2	60	/ 2500kcal (2.2kg)
Black beans (boiled):	0	10	1	20	/ 2500kcal (1.9kg)
Red kidney beans (canned):	20	0	16	15	/ 2500kcal (3.0kg)
White beans (boiled):	0	0	- -	25	/ 2500kcal (1.8kg)
Navy beans (boiled):	20	0	1	50	/ 2500kcal (1.8kg)
Green Peas (boiled):	10	40	3.5	30	/ 2500kcal (3.0kg)
Adzuki beans (boiled):	0	0	--	25	/ 2500kcal (2.0kg)
Chickpeas (cooked):	20	30	5.5	55	/ 2500kcal (1.5kg)
Average Legume (cooked/soaked):	**15**	**20**	**5**	**45**	**/ 2500kcal (2.1kg)**
Peanuts (P. butter = raw):	0	0	27	30	/2500kcal (0.43kg)
Recommended Daily Allowance:	**82**	**800**	**15**	**55**	

Seeds:	VitC	VitA	VitE	Se	
Units	**(mg)**	**(ugRetEq)**	**(mg ∝)**	**(ug)**	
Sunflower seeds (dry = raw):	5	30	150	230	/ 2500kcal (0.42kg)
Pumpkin seeds (roasted):	5	35	2.5	25	/ 2500kcal (0.56kg)
Sesame seeds/ Tahini:	0	30	9	25	/ 2500kcal (0.44kg)
Average Seeds (dry/raw):	**5**	**30**	**(NA)55**	**(NA)90**	**/ 2500kcal (0.47kg)**
Recommended Daily Allowance:	**82**	**800**	**15**	**55**	

Nuts:	**VitC**	**VitA**	**VitE**	**Se**	
Units	**(mg)**	**(ugRetEq)**	**(mg ∝)**	**(ug)**	
Pistachios:	15	250	11	40	/ 2500kcal (0.45kg)
Almonds:	0	5	110	10	/ 2500kcal (0.43kg)
Cashews (dry roasted):	0	0	4	75	/ 2500kcal (0.43kg)
Hazelnuts:	25	5	60	15	/ 2500kcal (0.40kg)
Walnuts:	5	10	3	20-70	/ 2500kcal (0.40kg)
Brazil Nuts (dried = raw?):	0(2)	0	22	5,100	/ 2500kcal (0.38kg)
Pecan Nuts:	5	20	5	15	/ 2500kcal (0.36kg)
Macadamia Nuts (raw):	5	0	2	10	/2500kcal (0.35kg)
Average Nuts (raw/dry):	**5**	**35**	**(NA)27**	**(NA30/760)**	**/2500kcal (0.40kg)**

Notes: The Se average is w/o Brazil Nuts / w B.N (B.N. are 5,000). Vit E highly variable.

Recommended Daily Allowance:	**82**	**800**	**15**	**55**

Grains:	**VitC**	**VitA**	**VitE**	**Se**	
Units	**(mg)**	**(ugRetEq)**	**(mg ∝)**	**(ug)**	
Oats (rolled, baked):	0	65	5	460	/ 2500kcal (0.65kg)
Wheat (flour, whole):	0	5	6	520	/ 2500kcal (0.74kg)
Millet (cooked, 70% water): 2.1kgwet	0	10	6	20	/ 2500kcal (0.63dry)
Barley (whole, cooked, wet):1.8kgwet	0	0	11	25-180	/ 2500kcal (0.63dry)
Rice (brown, cooked, wet): 2.2kgwet	0	0	1	220-870	/ 2500kcal (0.68dry)
Rice (white, cooked, wet):1.9kgwet	0	0	1	150	/ 2500kcal (0.55dry)
Av. Grains (4 wet, 6 dry):2.0kg wet	**0**	**10**	**5**	**300**	**/2500kcal (0.65dry)**
Corn (yellow, boiled):	150	600	2.8	5	/ 2500kcal (2.3kg)
Av. Grains (all 7 here):	**(NA)20**	**(NA)70**	**(NA)5**	**(NA)260**	**/ 2500kcal**

Notes: Only corn has VitC. Corn has 10x anything else VitA. Corn Se is only 5.

Recommended Daily Allowance:	**82**	**800**	**15**	**55**

Starchy Root Vegetables:	**VitC**	**VitA**	**VitE**	**Se**	
Units	**(mg)**	**(ugRetEq)**	**(mg ∝)**	**(ug)**	
Potatoes (baked, with skin):	250	55	1.3	10	/ 2500kcal (2.9kg)
Sweet potatoes (baked):	530	55,000	19	5	/ 2500kcal (2.8kg)
Yams (orange, boiled):	420	50,000	31	15	/ 2500kcal (3.3kg)
Cassava:	210	20	3.4	10	/ 2500kcal (1.6kg)
Av. Starchy Root Veg's (undried):	**350**	**NotApplic**	**14(NA)**	**10**	**/ 2500kcal (2.6kg)**

Notes: Cassava & Potato v. deficient VitA; Sweet P & orange Yams v. high (40xRDA). Cassava and Potato deficient (av. 1/6th RDA) VitE.

Recommended Daily Allowance:	**82**	**800**	**15**	**55**

Carrots (raw):	**VitC**	**VitA**	**VitE**	**Se**	
Units	**(mg)**	**(ugRetEq)**	**(mg ∝)**	**(ug)**	
	330	**100,000**	**42**	**10**	**/ 2500kcal (6.0kg)**
	30	**8,300**	**5**	**0(1)**	**/ 208kcal (500g)**
Recommended Daily Allowance:	**82**	**800**	**15**	**55**	

Fruit:	VitC	VitA	VitE	Se	
Units	(mg)	(ugRetEq)	(mg ∝)	(ug)	
Apricot:	590	10,000	4.5	5	/ 2500kcal (5.2kg)
Cantaloupe/Rockmelon:	2,500	25,000	4.5	30	/ 2500kcal (7.3kg)
Peach:	390	2,000	46	10	/ 2500kcal (6.4kg)
Watermelon:	650	4,600	5.5	35	/ 2500kcal (8.3kg)
Oranges:	2,800	1,200	9	30	/ 2500kcal (5.3kg)
Pawpaw:	3,900	7,000	46	35	/ 2500kcal (6.4kg)
Honeydew melon (cubes)	1,300	410	2	45	/ 2500kcal (7.0kg)
Plums:	500	1,800	17	0	/ 2500kcal (5.5kg)
Jackfruit:	180	400	--	15	/2500kcal (2.7kg)
Pear (fresh):	180	80	5	5	/ 2500kcal (4.3kg)
Pineapple (raw):	2,400	150	1.5	5	/ 2500kcal (5.0kg)
Bananas:	240	170	2.5	30	/ 2500kcal (2.8kg)
Dates:	5	5	0.5	25	/ 2500kcal (0.89kg)
Grapes (seedless):	370	220	7.5	5	/ 2500kcal (3.7kg)
Mango:	1,100	1,500	43	25	/ 2500kcal (3.8kg)
Persimmon:	140-1,300	30,000	--	11	/ 2500kcal (1.9kg)
Apples (average (un)dried):	210	280	7	5	/ 2500kcal (4.7kg)
Average Fruit (undried):	**1,100**	**(NA)3,200**	**(NA)13**	**18**	**/ 2500kcal (4.8kg)**

Notes: Vit A highly variable. Vit E is highly variable. Se is quite variable.
Dates are very deficient in Vit C, VitA and VitE.

Recommended Daily Allowance:	**82**	**800**	**15**	**55**

Avocado (fatty fruit):	VitC	VitA	VitE	Se	
Units	(mg)	(ugRetEq)	(mg ∝)	(ug)	
	160	**220**	**33**	**5**	**/2500kcal (1.6kg)**

Green Leaf, Flower & Shoot Veg:	VitC	VitA	VitE	Se	
Units	(mg)	(ugRetEq)	(ug ∝)	(ug)	
Kale (raw):	600	3,800	--	5	/(14gProt, 160kcal)/500g
Broccoli (raw):	440	150	4	10	/(14gProt, 160kcal)/500g
Broccoflower (raw):	440	40	0.5	5	/(15gProt, 160kcal)/500g
Cauliflower (raw):	230	5	0.5	5	/(10gProt, 120kcal)/500g
Lettuce:	20	250	1	0	/(10gProt, 70kcal)/500g
Cabbage (Savoy):	160	250	1	5	/(10gProt,145kcal)/500g
Chinese Cabbage:	130	1,100	0.5	0	/(8gProt, 60kcal)/500g
Bamboo shoots (boiled):	0(?)	0(?)	3	0	/(7gProt, 60kcal)/500g
Cabbage (Red):	290	560	0.5	5	/(7gProt,160kcal)/500g
Cabbage (standard):	190	50	0.5	0	/(7gProt,120kcal)/500g
Av. Leaf, Flower, Shoot Veg's:	**240**	**(NA)620**	**1.5**	**5**	**/(10gProt,100kcal)/500g**

Note: Lettuce deficient in Vit C. Kale 3+x anything else VitA, only Kale and Chinese Cabbage sufficient in VitA.

Recommended Daily Allow.:	**82**	**800**	**15**	**55**

Squash and Zucchini:	VitC	VitA	VitE	Se	
Units	(mg)	(ugRetEq)	(ug ∝)	(ug)	
Squash, summer, all 5 varieties (boiled):	30	55(?)	0.5	0	/ 100kcal (500g)
Squash, winter, all 5 varieties (baked):	50	1,300(?)	0.5	0	/ 155kcal (500g)
Zucchini, summer (boiled):	25	280	0.5	0	/ 80kcal (500g)
Average Squash and Zucchini:	**35**	**NotApplic**	**0.5**	**0**	**/ 110kcal (500g)**

Note: Only winter Squash sufficient in Vit A, 5+x anything else, summer Sq. <1/10 RDA.

Recommended Daily Allowance:	**82**	**800**	**15**	**55**

For convenience, the average values of the different food groups collated above are listed here below in Table 12, with a few common oddments relocated underneath:

Table 12: Average VitC, VitA, VitE & Se, and weight, per 2500 kcal, of food groups, and RDAs

Mineral/Vitamin:	**VitC**	**VitA**	**VitE**	**Se**	
Units:	**(mg)**	**(ugRetEq)**	**(ug ∝)**	**(ug)**	
Recommended Daily Allowance:	**82**	**800**	**15**	**55**	
For grown, non-pregnant, non-lactating adults <50y (Whitney and Rolfes, 2008).					
Average Land Animal Flesh:	**0**	**(NA)40**	**4.5**	**280**	**/ 2500kcal (1.3kg)**
Note: Only the chicken had any VitA					
Average Sea Fish Flesh:	**(NA)9**	**(NA)2,200**	**15**	**730**	**/ 2500kcal (1.8kg)**
Note: Tuna >7x anything else VitA					
Av. Sea Mollusk/Crustac'n Flesh:	**170**	**1,200**	**28**	**1,200**	**/ 2500kcal (2.4kg)**
Note: VitA range oysters 230 – clams 3,000.					
Average Animal Flesh:	**(NA)70**	**(NA)1,200**	**17**	**790**	**/ 2500kcal (1.9kg)**
Egg (whole (no shell), raw): (31of)	**0**	**2,500**	**17**	**550**	**/2500kcal (1.6kg)**
Average Cow/Goat Milk:	**(NA)20**	**1,600**	**2.4**	**100**	**/2500kcal (3.9kg/L)**
Human Milk (mature (day10+)):	**370**	**1,700**	**20**	**25-120**	**/2500kcal (3.7kg/L)**
Average Legume (cooked/soaked):	**15**	**20**	**5**	**45**	**/ 2500kcal (2.1kg)**
Peanuts (P. butter = raw):	0	0	27	30	/2500kcal (0.43kg)
Average Seeds (dry/raw):	**5**	**30**	**(NA)55**	**(NA)90**	**/ 2500kcal (0.47kg)**
Notes: VitE range Pumpkin s 2.5 - sunflower s 150. Se range pumpk/ses 25 – sunfl 230					
Average Nuts (raw/dry):	**5**	**35**	**(NA)27**	**(NA30/760)**	**/2500kcal (0.40kg)**
Notes: The Se average is w/o Brazil Nuts / w B.N (B.N. are 5,000). Vit E range 2-110.					
Av. Grains (4 wet, 6 dry):2.0kg wet	**0**	**10**	**5**	**300**	**/2500kcal (0.65dry)**
Corn (yellow, boiled):	150	600	2.8	5	/ 2500kcal (2.3kg)
Av. Grains (all 7 here):	**(NA)20**	**(NA)70**	**(NA)5**	**(NA)260**	**/ 2500kcal**
Notes: Only corn has VitC. Corn 10x any other VitA. Corn Se is only 5. VitE range 1-11.					
Av. Starchy Root Veg's (undried):	**350**	**NotApplic**	**14(NA)**	**10**	**/ 2500kcal (2.6kg)**
Notes: Cassava & Potato v. deficient VitA; Sweet P & orange Yams v. high (40xRDA). Cassava and Potato deficient (av. $1/6^{th}$ RDA) VitE.					
Carrots (raw):	**330**	**100,000**	**42**	**10**	**/ 2500kcal (6.0kg)**
	30	**8,300**	**5**	**0(1)**	**/ 208kcal (500g)**
Average Fruit (undried):	**1,100**	**(NA)3,200**	**(NA)13**	**18**	**/ 2500kcal (4.8kg)**
Notes: Vit A range 5/80-10,000. Vit E range 2-46. Se range 0(?)-45. Dates are very deficient in Vit C(5), VitA(5) and VitE(0.5).					
Avocado (fatty fruit):	**160**	**220**	**33**	**5**	**/2500kcal (1.6kg)**
Av. Leaf, Flower, Shoot Veg's:	**240**	**(NA)620**	**1.5**	**5**	**/(10gProt,100kcal)/500g**
Note: Lettuce Vit C 20. Kale >3x anything else VitA, only Kale and Chinese Cabbage sufficient in VitA.					
Average Squash and Zucchini:	**35**	**NotApplic**	**0.5**	**0**	**/ 110kcal (500g)**
Note: Only winter Squash sufficient in Vit A, 5+x anything else, summer Sq. <1/10 RDA.					
Recommended Daily Allowance:	**82**	**800**	**15**	**55**	

Vitamin C (VitC) (RDA 82 mg / day):

Human breast milk has a sufficiency of 450% of vitamin C per 2500 kcal. This relatively high level may relate to the need for more vitamin C for connective tissue synthesis. Cow and goat milk are both deficient in VitC at 0 and 45 mg / 2500 kcal respectively, though the 0 value seems dubious. It seems odd that such relatively quickly growing animals with their milk protein/energy ratio some times over that of humans would not need more VitC for the infant – maybe cows make their own VitC, and maybe calves eat grass very early and during suckling…
In any case, it seems to me that the figures suggest that the RDA of 82 mg VitC / day may be set lower than is really required for optimal health, including regards antioxidant function, and other fairly credible authorities share this opinion.

Land animal flesh is grossly deficient in VitC.
Marine fish flesh on average is very deficient in VitC.
Marine mollusk and crustacean flesh on average is sufficient in VitC per 2500 kcal.

Eggs are absolutely deficient in VitC, having none.

Legumes, seeds, nuts and grains (other than corn), all in the unsprouted condition, are very deficient in VitC. Sprouting increases the Vit C levels of such foods, for example supposedly to an average of 17mg/100g (wet weight) for legumes or i.e., 500 g sprouts to meet the RDA.

Starchy root vegetables on average have a 400% sufficiency of VitC per 2500 kcal.

Fruits on average have a 1,300% sufficiency of VitC on a per 2500 kcal basis.

Eating 500 grams of green leaf, flower and shoot vegetables per day will provide a 300% sufficiency of VitC.

Squash and zucchini are deficient in VitC.

The meaning of the food vitamin C data I interpret to be that:
Taken together with the data named above, it seems safe to say that much more green leaf, flower and shoot vegetables should be eaten than squash and zucchini. Also it seems safe to say that on the basis of vitamin C content alone, the human diet generally should derive most of its calories from fruits and/or starchy root vegetables, and that this does not contradict the optimum proportion of diet of these foods suggested by consideration of the other nutrients done above.

Vitamin A (VitA) (RDA 800 micrograms / day):

Human breast milk has a sufficiency of 200% of vitamin A on a per 2500 kcal basis. This relatively high level may relate to the increased need in infancy for more vitamin A for tissue formation and protection, and growth.

Cow and goat milk also have a sufficiency of 200% of VitA on a per 2500 kcal basis.

Land animal flesh (other than liver) is grossly deficient in VitA.
Marine fish flesh on average is sufficient in VitA, but there is great variability, from 1/4 to 10 times the RDA per 2,500 kcal.
Marine mollusk and crustacean flesh on average is sufficient in VitA, but there is great variability, from 1/4 to 4 times the RDA per 2,500 kcal.

Eggs have a 300% sufficiency of VitA on a per 2500 kcal basis.

Legumes, seeds, nuts and grains (other than corn), all in the unsprouted condition, are very deficient in VitA on a per 2500 kcal basis. Sprouting increases the VitA levels of legumes, but the result is still supposedly very roughly an average of only 100IU VitA / 100g, or 800g sprouts needed to reach the RDA.

Starchy root vegetables are extremely variable with potatoes and cassava being grossly deficient in VitA, and sweet potatoes and yams having massive 6,000% sufficiencies of Vit A, on a per 2500 kcal basis.

Fruits on average have a sufficiency of VitA on a per 2500 kcal basis, but there is a wide variation, from grossly deficient to massively sufficient, with redness of colour (which includes orange and purple coloration) being a fairly reliable predictor of high VitA content, but yellowness of colour much less so (suggesting differences within the sweet potatoes also, by colour?...).

Green leaf, flower and shoot vegetables are extremely variable also, with 500 g of cauliflower or bamboo shoots providing nearly no VitA, and 500 g of the standard Bullethead cabbage and broccoflower providing roughly only 1/14 of the VitA RDA. 500 g of Kale provides 400% sufficiency of VitA.

Squash and zucchini are similarly variable with 500 g of summer squash exceeding the RDA in VitA, whereas 500 g of winter squash provided less than 1/10 of the RDA of VitA. 500 g of summer zucchini provided roughly 1/3 of the VitA RDA.

The meaning of the food vitamin A data I interpret to be that red and orange/purple fruits and (i.e., fruit-) vegetables should be a regular substantial part of the diet, and that 500 g of a good variety of green vegetables must be had to make up shortfalls in red/orange/purple fruit unless it is known that the particular green vegetable(s) eaten are those high in VitA. However, probably almost any deeply dark green leafy vegetable is likely to be able to do so on its own, our common western cabbage (not the Savoy or purple cabbages, note) being the culprit dragging the averages down, notwithstanding its great usefulness overall as an affordable and readily available green(ish) leaf vegetable staple food.
Vitamin A being a fat soluble vitamin, dietary fat eaten in the same meal will aid absorption from the digestive tract, and there is some substantial capacity for good body stores to suffice for probably a small number of months of dietary shortfall of vitamin A with little or no compromise of health.

<u>Vitamin E (VitE) (RDA 15 micrograms (ug) $\propto$ (alpha)-tocopherol / day):</u>

Human breast milk has a sufficiency of 130% of vitamin E on a per 2500 kcal basis.

Cow and goat milk are both very deficient in VitE at 1.7 and 3 mg / 2500 kcal respectively.

Land animal flesh is substantially deficient in VitE at an average less than 1/3 of RDA per 2,500 kcal.
Marine fish flesh on average is sufficient in VitE on a per 2500 kcal basis.
Marine mollusk and crustacean flesh on average have a 200% sufficiency of VitE on a per 2500 kcal basis.

Eggs have a sufficiency of VitE on a per kcal basis.

Legumes, all in the unsprouted condition, are substantially deficient in VitE at an average 1/3 of RDA per 2,500 kcal.
Peanuts are an exception at 200% of RDA on a per 2500 kcal basis.
Sprouting increases the VitE levels of legumes, but perhaps not by much.

Seeds, all in the unsprouted condition, are very variable in VitE – pumpkin seeds are very deficient at only 1/6 of RDA on a per 2500 kcal basis, sesame seeds/ tahini are slightly deficient at 2/3 of RDA on a per kcal basis, and sunflower seeds are 1,000% sufficient on a per 2500 kcal basis. The massive differences make me question the assay accuracy.
Sprouting increases the VitE levels of seeds, but perhaps not by much.

Nuts on average have a 180% sufficiency of VitE on a per 2500 kcal basis, all in the unsprouted condition, but there is a great variability.
Macadamia nuts are very deficient in VitE at only 1/7 of RDA on a per kcal basis, walnuts having only 1/5, pecans only 1/3, cashews only 1/3, and pistachios only 2/3 of RDA on a per 2500 kcal basis; whereas Brazil nuts have a sufficiency of 140%, hazelnuts have a sufficiency of 400%, and almonds have a sufficiency of 700% of VitE RDA on a per 2500 kcal basis.
Sprouting increases the VitE levels of nuts, but perhaps not by much.

Grains, all in the unsprouted condition, are substantially deficient in VitE at only 1/3 of RDA on a per 2500 kcal basis. Noteworthy is that rice is very deficient at only 1/15 of VitE RDA on a per 2500 kcal basis.
Sprouting increases the VitE levels of such grains, but perhaps not by much.

Starchy root vegetables are very variable in VitE, with potatoes and cassava being very or substantially deficient in VitE at only 1/10 and ¼ respectively of VitE RDA on a per 2500 kcal basis, and sweet potatoes and yams having sufficiencies of 130% and 200% respectively of VitE RDA on a per 2500 kcal basis.

Fruits on average have only a marginal deficiency of VitE at 4/5 of RDA on a per 2500 kcal basis, but there is a wide variation, from very deficient pineapples, banana and honeydew melon at around 1/7 of RDA on a per 2500 kcal basis; most of the fruits being substantially deficient but higher than that; and peaches at 300%, pawpaw at 300%, mangoes at 300%, and plums at 110% of RDA on a per 2500 kcal basis.
Note that the high VitE fruits noted here represent 3 different botanical groups and 3 different geo-climatic area types.

500 g of green leaf, flower and shoot vegetables on average provides only 1/10 of the VitE RDA.

500 g of squash and zucchini provides only 1/30 of the VitE RDA.

The meaning of the food vitamin E data I interpret to be that to provide a good probability of the diet providing a good supply of vitamin E, on average, a good mixture (on a per kcal basis) of different types of seeds, nuts, tubers and fruit should comprise the large majority of the diet. These could be substituted (on a per kcal basis) by marine animal foods, or eggs, regards VitE, but this does not consider other factors, which generally suggest that such substitution would be best kept to a low proportion of kcal. (see below)

(continues below)

Selenium (Se) (RDA 55 micrograms (ug) /day) (Safe Upper limit 400 ug / day):

Human breast milk is reported to have 25-120 micrograms Se / 2,500 kcal. As this range is probably drawn from a population, the USA, that has documented areas where selenium toxicity from excessive intake occurs in both cattle and humans, with the humans also eating the cattle at an intake higher than almost anywhere else in the world, and as meats tend to accumulate selenium, and as the safe upper limit has been set at 400 micrograms Se / day, and as the vested interests of the meat companies have also possibly had an effect in inflating the selenium RDA, and as the vested interests of the meat companies have also possibly had an effect in increasing the safe upper limit (Which is a little less than half the intake level at which overt pathological changes in hair and fingernail tissues has been documented) it is quite possible that the RDA of 55 micrograms Se is inflated above what should really be suggested.
However, in this regard, note that the Se intake in Keshan, China, where the Keshan disease cardiomyopathy was documented, was 3-22 micrograms/day, and that over several months of repletion with Se (the implication is that repletion was titrated, with an increasing dose until the plateau criteria was met, or, that different doses were used) glutathione peroxidase activity plateaued (though note that this was the maximum activity level, which is not necessarily physiological normality) at similar levels in men at a dose of 40 micrograms Se / day – this was raised a fraction to give the Se RDA (Reilly 1991).

Cow milk has a 300% sufficiency of selenium on a per 2500 kcal basis and goat milk is very nearly sufficient and probably sufficient in reality.

Land animal flesh on average approaches the safe upper limit for Se on a per 2500 kcal basis with a sufficiency of 550% at 280 ug.
Marine fish flesh on average exceeds the safe upper limit for Se on a per 2500 kcal basis with a sufficiency of 1,400% at 730 ug.
Marine mollusks and crustaceans on average exceed the safe upper limit of Se on a per 2500 kcal basis with a sufficiency of 2,200% at 1,200 ug.

Eggs on average exceed the safe upper limit of Se on a per 2500 kcal basis with a sufficiency of 1,000% at 550 ug.

Legumes on average are marginally deficient in Se on a per 2500 kcal basis with 45 ug Se / 2,500 kcal, but in reality this is probably quite sufficient.

Seeds on average have a 200% sufficiency of Se on a per 2500 kcal basis, but this is quite variable, with sunflower seeds having a 400% sufficiency, and both pumpkin seeds and sesame/tahini being deficient (probably only marginally in reality) in Se with only 1/2 of the RDA per 2,500 kcal.

Nuts must be considered with Brazil nuts separate from the others, here.
Brazil nuts (on average) exceed the safe upper limit of Se on a per 2500 kcal basis with a sufficiency of 9,000% at 5,100 ug per 2,500 kcal. Put another way, 1/10 of calories as Brazil nuts would exceed the safe upper limit for Se, and 1/5 of calories as Brazil nuts would cause overt pathological changes visible in hair and fingernails in some people.
Nuts other than Brazil nuts are deficient (probably only marginally in reality) in Se with only 1/2 of the RDA per 2,500 kcal.
A calorically equal mixture of all eight nuts including Brazil nuts would, at 730 ug Se / 2,500 kcal, still well exceed the safe upper limit for Se, and approach the level at which excess Se causes overt pathological changes visible in hair and fingernails in some people. Of course it is not common or recommended that all energy would come from mixed nuts.

Grains on average have a 500% sufficiency of Se on a per 2500 kcal basis, but there is a substantial variability, and those exceeding the upper safe limit on a per 2500 kcal basis include oats at 460 ug / 2,500 kcal, wheat at 520 ug / 2,500 kcal, and the middle and high end of the brown rice range at 540 and 870 ug Se / 2,500 kcal exceed the safe upper limit of Se and reach the level at which excess Se causes overt pathological changes visible in hair and fingernails in some people. Corn is the exception, being very deficient in Se at only 1/10 of the RDA per 2,500 kcal.

Starchy root vegetables are substantially deficient in Se, at only 1/5 of the RDA per 2,500 kcal.

Fruits on average are deficient in Se, at only 1/3 of the RDA per 2,500 kcal, but in reality this is probably near to sufficiency, though the range of fruit Se contents includes values probably substantially deficient even in reality.

500g of green leaf, flower or shoot vegetables on average provides only 1/10 of the RDA for Se.

500g of squash or zucchini provides approximately nil Se.

The meaning of the food selenium data I interpret to be that the addition of roughly equal amounts of legumes, nuts, seeds and cereal grains to an amount of fruit and/or starchy root vegetables that meets most of the dietary caloric requirement, and 500 g of green leaf, flower and shoot vegetables would provide a nicely sufficient selenium intake well short of the upper safe limit and toxic levels.
Brazil nuts should not constitute more than 1/20 of calories.
Land animal flesh on average should not constitute more than roughly 1/4 of calories.
Marine fish flesh on average should not constitute more than roughly 1/6 of calories.
Marine mollusks and crustaceans on average should not constitute more than roughly 1/10 of calories.
Note that consideration of other nutrients indicates that certainly the 1/4 and often the 1/6 fractions given here would be too high.

Squash and zucchini seem to be generally nutritionally inferior to green leaf, flower and shoot vegetables.

4.6 Thiamin (VitB1), Riboflavin (VitB2), Niacin (VitB3) & Pyridox(al/ine/amine) (VitB6) Content, and Weight, per 2500 kcal of Foods, and their sufficiencies or deficiencies related to the Recommended Daily Allowances (RDAs) / Adequate Intakes (AIs)

Table 13: Thiamin (VitB1), Riboflavin (VitB2), Niacin (VitB3) & Pyridoxal/ine/amine (VitB6) content per 2500 kcal weight of foods, & RDAs/AIs

Animal Flesh	B1	B2	B3	B6	
Units	(mg)	(mg)	(mg)	(mg)	
Meat (5% fat ground beef, broiled):	0.5	2.3	50	2.0	/ 2500kcal (1.5kg)
Meat (15% fat ground beef, broiled):	0.5	2.3	50	2.0	/ 2500kcal (1.0kg)
Ham (lean, roasted):	1.0	3.0	70	1.7	/ 2500kcal (1.6kg)
Lamb (lean, roasted):	1.0	3.2	70	1.7	/ 2500kcal (1.3kg)
Chicken Meat (all, fried):	0.9	2.1	120	6.0	/ 2500kcal (1.1kg)
Average Land Animal Flesh:	**0.8**	**2.6**	**72**	**2.7**	**/ 2500kcal (1.3kg)**
Tuna (baked/broiled):	0.5-7.5	1.1-1.6	200	1-20	/ 2500kcal (1.4kg)
Salmon (fillet, baked/broiled):	0.4-1.5	1.6	110	5-10	/ 2500kcal (1.4kg)
Sea Bass (fillet, baked/broiled):	2.1	0.7	50	7	/ 2500kcal (2.0kg)
Snapper (fillet, baked/broiled):	0.9	0(?)	7	9	/ 2500kcal (1.9kg)
Sole/Flounder (fillet, broiled):	1.3	1.5	45	4	/ 2500kcal (2.1kg)
Striped Mullet (fillet, baked):	1.8	1.8	110	9	/ 2500kcal (1.7kg)
Average Sea Fish Flesh:	**1.8**	**1.2**	**87**	**8**	**/ 2500kcal (1.8kg)**

Notes: Tuna canned-raw; Salmon smoked-raw. Can/smoking reduce B1 & B6 2-20fold.

Squid (calamari, baked):	0.5	6	35	0.7	/ 2500kcal (1.8kg)
Octopus (raw):	0.7	1	60	10	/ 2500kcal (3.0kg)
Crab (leg, boiled):	1.1-2.3	1.8-4.5	34-82	4	/ 2500kcal (2.6kg)
Prawns (steamed):	0.6	0.6	65	3	/ 2500kcal (2.5kg)
Clams (canned):	2.4	7	55	2	/ 2500kcal (1.7kg)
Oysters (raw):	1.8-4	2.5-7	60	2	/ 2500kcal (3.7kg)
Mussels (steamed/boiled):	4.5	6	45	1.4	/ 2500kcal (1.5kg)
Av. Sea Mollusk/Crustacean Flesh:	**1.9**	**4.1**	**55**	**3.3**	**/ 2500kcal (2.4kg)**
Average Animal Flesh:	**1.6**	**7**	**70**	**4.7**	**/ 2500kcal (1.9kg)**
Recommended Daily Allowance:	**1.1**	**1.2**	**15**	**1.3**	

Note: The values given are for grown, non-pregnant, non-lactating adults under the age of 50, as given in Whitney and Rolfes (2008).

Egg (whole (no shell), raw):	B1	B2	B3	B6	
Units	(mg)	(mg)	(mg)	(mg)	
	1.0	**8**	**1.0**	**20**	**/2500kcal(31of, 1.6kg)**

Note: The calculated 1.0 is only entitled to significance of the .0 at the level of 0.3 and that only for the figure given in the table, derived from an unknown number of samples, maybe only one.

Recommended Daily Allowance:	**1.1**	**1.2**	**15**	**1.3**

Milk:	B1	B2	B3	B6	
Units	**(mg)**	**(mg)**	**(mg)**	**(mg)**	
Milk (cow, whole):	1.7	8	4	1.4	/ 2500kcal (4.2kg/L)
Milk (goat):	1.6	4.9	10	1.6	/ 2500kcal (3.6kg/L)
Average Cow/Goat Milk:	**1.6**	**6.5**	**7**	**1.5**	**/ 2500kcal (3.9kg/L)**
Recommended Daily Allowance:	**1.1**	**1.2**	**15**	**1.3**	

Milk (human, mature (after day10)):	B1	B2	B3	B6	
Units	**(mg)**	**(mg)**	**(mg)**	**(mg)**	
	0.75	**1.5-2.3**	**6.5-20**	**0.35-1.1**	**/2500kcal (3.7kg/L)**
Recommended Daily Allowance:	**1.1**	**1.2**	**15**	**1.3**	

Legumes:	B1	B2	B3	B6	
Units	**(mg)**	**(mg)**	**(mg)**	**(mg)**	
Soy beans (cooked):	3	3.5	6-20	1.0	/ 2500kcal (1.4kg)
Lentils (cooked):	3.5	1.8	25	4	/ 2500kcal (2.2kg)
Black beans (boiled):	4.5	1.1	9	1.1	/ 2500kcal (1.9kg)
Red kidney beans (canned):	3	2.5	13	2.3	/ 2500kcal (3.0kg)
White beans (boiled):	4	1.0	5	2.2	/ 2500kcal (1.8kg)
Navy beans (boiled):	4	1.2	12	2.4	/ 2500kcal (1.8kg)
Green Peas (boiled):	6.5	2.9	40	2.9	/ 2500kcal (3.0kg)
Adzuki beans (boiled):	2.2	1.2	14	1.9	/ 2500kcal (2.0kg)
Chickpeas (cooked):	1.7	0.9	8	2.0	/ 2500kcal (1.5kg)
Average Legume (cooked/soaked):	**3.6**	**1.8**	**15.5**	**2.2**	**/ 2500kcal (2.1kg)**
Peanuts (P. butter = raw):	2.8	0.5	53	1.4	/2500kcal (0.43kg)
Recommended Daily Allowance:	**1.1**	**1.2**	**15**	**1.3**	

Seeds:	B1	B2	B3	B6	
Units	**(mg)**	**(mg)**	**(mg)**	**(mg)**	
Sunflower seeds (dry = raw):	6	1.4	35	5.5	/ 2500kcal (0.42kg)
Pumpkin seeds (roasted):	1.0	1.5	8	0.4	/ 2500kcal (0.56kg)
Sesame seeds/ Tahini:	1.1-6.5	0.6	24	0.6	/ 2500kcal (0.44kg)
Average Seeds (dry/raw):	**3.6**	**1.2**	**22**	**2.2**	**/ 2500kcal (0.47kg)**
Note: B6 range 0.4/0.5(pumpkin/sesame)-5.5(sunflower)					
Recommended Daily Allowance:	**1.1**	**1.2**	**15**	**1.3**	

Nuts:	**B1**	**B2**	**B3**	**B6**	
Units	**(mg)**	**(mg)**	**(mg)**	**(mg)**	
Pistachios:	3.7	0.7	6	5.5	/ 2500kcal (0.45kg)
Almonds:	0.8	3	16	0.5	/ 2500kcal (0.43kg)
Cashews (dry roasted):	0.9	0.9	6	1.1	/ 2500kcal (0.43kg)
Hazelnuts:	1.8	0.4	6	2.2	/ 2500kcal (0.40kg)
Walnuts:	0.1-1.0	0.5	4	2.0	/ 2500kcal (0.40kg)
Brazil nuts (dried = raw?):	2.3	0.1	1	0.3	/ 2500kcal (0.38kg)
Pecan nuts:	2.4	0.5	4	0.8	/ 2500kcal (0.36kg)
Macadamia nuts (raw):	4	0.5	9	0.9	/2500kcal (0.35kg)
Average Nuts (raw/dry):	**2.0**	**0.8**	**6**	**(NA)1.7**	**/ 2500kcal (0.40kg)**
Note: VitB6 range 0.3-5.5					
Recommended Daily Allowance:	**1.1**	**1.2**	**15**	**1.3**	

Grains:	**B1**	**B2**	**B3**	**B6**	
Units	**(mg)**	**(mg)**	**(mg)**	**(mg)**	
Oats (rolled):	12	2.2	9	1.5	/ 2500kcal (0.65kg)
Wheat (flour, whole):	3	1.5	45	2.5	/ 2500kcal (0.74kg)
Millet (cooked, 70% water): 2.1kgwet	2	1.6	30	2.3	/ 2500kcal (0.63dry)
Barley (whole, cooked, wet):1.8kgwet	1.5	1.0	40	2.3	/ 2500kcal (0.63dry)
Rice (brown, cooked, wet): 2.2kgwet	1.2-2	2.1-0.3	30	3	/ 2500kcal (0.68dry)
Rice (white, cooked, wet):1.9kgwet	3(?)	0.2	30	1.7	/ 2500kcal (0.55dry)
Av. Grains (4 wet, 6 dry):2.0kg wet	**4**	**(NA)1.4**	**30**	**2.2**	**/ 2500kcal (0.65 dry)**
Note: Brown rice B1 & B2 are Wild-(Long/Med Grain).					
Note: White rice appears VitB1-enriched here, considering historical beriberi association					
Corn (yellow, boiled):	5	1.4	37	1.1	/ 2500kcal (2.3kg)
Av. Grains (all 7 here):	**4**	**1.4**	**31**	**2.0**	**/ 2500kcal**
Recommended Daily Allowance:	**1.1**	**1.2**	**15**	**1.3**	

Starchy Root Vegetables:	**B1**	**B2**	**B3**	**B6**	
Units	**(mg)**	**(mg)**	**(mg)**	**(mg)**	
Potatoes (baked, with skin):	1.6	1.2	40	8	/ 2500kcal (2.9kg)
Sweet potatoes (baked):	1.8	1.6	18	5	/ 2500kcal (2.8kg)
Yams (orange, boiled):	1.9	0.3	12	4.5	/ 2500kcal (3.3kg)
Cassava:	1.2	0.6	13	1.4	/ 2500kcal (1.6kg)
Av. Starchy Root Veg's (undried):	**1.6**	**(NA)0.9**	**(NA)21**	**4.5**	**/ 2500kcal (2.6kg)**
Note: Potato skin B2 = 3 x flesh B2. VitB2 range 0.3 Cassava – 1.6 Sweet potatoes					
Recommended Daily Allowance:	**1.1**	**1.2**	**15**	**1.3**	

Carrots (raw):	**B1**	**B2**	**B3**	**B6**	
Units	**(mg)**	**(mg)**	**(mg)**	**(mg)**	
	4	**4**	**60**	**8**	**/ 2500kcal (6.0kg)**
	0.3	**0.3**	**5**	**0.7**	**/ 210kcal (500g)**
Recommended Daily Allowance:	**1.1**	**1.2**	**15**	**1.3**	

Fruit:	**B1**	**B2**	**B3**	**B6**	
Units	**(mg)**	**(mg)**	**(mg)**	**(mg)**	
Apricot:	1.5	1.9	30	2.6	/ 2500kcal (5.2kg)
Cantaloupe/Rockmelon:	3	0.9	55	4.6	/ 2500kcal (7.3kg)
Peach:	1.3	1.7	50	1.3	/ 2500kcal (6.4kg)
Watermelon:	2.2	1.1	14	3.3	/ 2500kcal (8.3kg)
Oranges:	4.4	2.0	14.5	2.8	/ 2500kcal (5.3kg)
Pawpaw:	1.8	1.8	20	0.9	/ 2500kcal (6.4kg)
Honeydew melon (cubes)	2.3	0.8	30	5.5	/ 2500kcal (7.0kg)
Plums:	1.7	1.7	20	1.7	/ 2500kcal (5.5kg)
Jackfruit:	0.6	2.9	11	2.9	/2500kcal (2.7kg)
Pear (fresh):	0.5	1.0	7	1.0	/ 2500kcal (4.3kg)
Pineapple (raw):	4	1.9	25	6	/ 2500kcal (5.0kg)
Bananas:	0.7	1.9	18	10	/ 2500kcal (2.8kg)
Dates:	0.4	0.4	11	1.4	/ 2500kcal (0.89kg)
Grapes (seedless):	2.8	2.2	9	4	/ 2500kcal (3.7kg)
Mango:	2.3	1.9	22	5	/ 2500kcal (3.8kg)
Persimmon:	1.1	0(?)	2	--	/ 2500kcal (1.9kg)
Apples (average (un)dried):	0.7	1.0	4	1.7	/ 2500kcal (4.7kg, undr)
Av. Fruit (undried):	**1.9**	**1.6**	**20**	**3.4**	**/ 2500kcal (4.8kg):**

Recommended Daily Allowance:	**1.1**	**1.2**	**15**	**1.3**

Avocado (fatty fruit):					
Units	**(mg)**	**(mg)**	**(mg)**	**(mg)**	
	0.4-1.0	**0.7-2.1**	**14-29**	**1.4-4**	**/2500kcal (1.6kg)**

Note: Avocado data is Florida-California.

Green Leaf, Flower & Shoot Veg:	**B1**	**B2**	**B3**	**B6**	
Units	**(mg)**	**(mg)**	**(mg)**	**(mg)**	
Kale (raw):	0.5	0.6	5	1.3	/(14gProt, 160kcal)/500g
Broccoli (raw):	0.3	0.5	3	0.9	/(14gProt, 160kcal)/500g
Broccoflower (raw):	0.3	0.5	4	1.1	/(15gProt, 160kcal)/500g
Cauliflower (raw):	0.3	0.3	3	1.1	/(10gProt, 120kcal)/500g
Lettuce:	0.2	0.1	0.5	0.2	/(10gProt, 70kcal)/500g
Cabbage (Savoy):	0.4	0.1	1.5	0.9	/(10gProt,145kcal)/500g
Chinese Cabbage:	0.1	0.3	2	0.8	/(8gProt, 60kcal)/500g
Bamboo shoots (boiled):	0.1	0.2	1	0.6	/(7gProt, 60kcal)/500g
Cabbage (Red):	0.3	0.4	2	1.0	/(7gProt,160kcal)/500g
Cabbage (standard):	0.3	0.1	1	0.6	/(7gProt,120kcal)/500g
Av. Leaf, Flower, Shoot Veg's:	**0.3**	**0.3**	**2**	**0.8**	**/(10gProt,100kcal)/500g**

Recommended Daily Allow.:	**1.1**	**1.2**	**15**	**1.3**

Squash and Zucchini:	**B1**	**B2**	**B3**	**B6**	
Units	**(mg)**	**(mg)**	**(mg)**	**(mg)**	
Squash, summer, all 5 varieties (boiled):	0.2	0.2	2.5	0.3	/ 100kcal (500g)
Squash, winter, all 5 varieties (baked):	0.1	0.3	2.5	0.8	/ 155kcal (500g)
Zucchini, summer (boiled):	0.2	0.2	2	0.4	/ 80kcal (500g)
Average Squash and Zucchini:	**0.2**	**0.2**	**2.5**	**0.5**	**/ 110kcal (500g)**

Recommended Daily Allowance:	**1.1**	**1.2**	**15**	**1.3**

For convenience, the average values of the different food groups collated above are listed here below in Table 14, with a few common oddments relocated underneath:

Table 14: Average Thiamin (VitB1), Riboflavin (VitB2), Niacin (VitB3) & Pyridoxal/ine/amine (VitB6) content per 2500 kcal weight of foods, & RDAs/AIs

Mineral/Vitamin:	B1	B2	B3	B6	
Units	**(mg)**	**(mg)**	**(mg)**	**(mg)**	
Recommended Daily Allowance:	**1.1**	**1.2**	**15**	**1.3**	
Note: The values given are for grown, non-pregnant, non-lactating adults under the age of 50, as given in Whitney and Rolfes (2008).					
Average Land Animal Flesh:	**0.8**	**2.6**	**72**	**2.7**	**/ 2500kcal (1.3kg)**
Average Sea Fish Flesh:	**1.8**	**1.2**	**87**	**8**	**/ 2500kcal (1.8kg)**
Notes: Tuna canned-raw; Salmon smoked-raw. Can/smoking reduce B1 & B6 2-20fold.					
Av. Sea Mollusk/Crustacean Flesh:	**1.9**	**4.1**	**55**	**3.3**	**/ 2500kcal (2.4kg)**
Average Animal Flesh:	**1.6**	**7**	**70**	**4.7**	**/ 2500kcal (1.9kg)**
Egg (whole (no shell), raw): (31of)	**1.0**	**8**	**1.5**	**20**	**/2500kcal, 1.6kg)**
Note: The calculated 1.0 is only entitled to significance of the .0 at the level of 0.3.					
Average Cow/Goat Milk:	**1.6**	**6.5**	**7**	**1.5**	**/ 2500kcal (3.9kg/L)**
Human Milk (mature (day10+)):	**0.75**	**1.5-2.3**	**6.5-20**	**0.35-1.1**	**/2500kcal(3.7kg/L)**
Average Legume (cooked/soaked):	**3.6**	**1.8**	**15.5**	**2.2**	**/ 2500kcal (2.1kg)**
Peanuts (P. butter = raw):	2.8	0.5	53	1.4	/2500kcal (0.43kg)
Average Seeds (dry/raw):	**3.6**	**1.2**	**22**	**2.2**	**/ 2500kcal (0.47kg)**
Note: B6 range 0.4/0.5(pumpkin/sesame)-5.5(sunflower)					
Average Nuts (raw/dry):	**2.0**	**0.8**	**6**	**(NA)1.7**	**/2500kcal (0.40kg)**
Notes: VitB6 range Brazil N 0.3 – Pistachio 5.5. Brazil N B2 = 0.1, Almond B2 = 3. Walnut range 0.1-1.0					
Av. Grains (4 wet, 6 dry):2.0kg wet	**4**	**(NA)1.4**	**30**	**2.2**	**/ 2500kcal (0.65 dry)**
Note: White Rice B2 = 0.2, non-Wild (Long/Med Grain) Brown rice B2 = 0.3.					
Note: White rice appears VitB1-enriched here considering historical beriberi association					
Corn (yellow, boiled):	5	1.4	37	1.1	/ 2500kcal (2.3kg)
Av. Grains (all 7 here):	**4**	**1.4**	**31**	**2.0**	**/ 2500kcal**
Av. Starchy Root Veg's (undried):	**1.6**	**(NA)0.9**	**(NA)21**	**4.5**	**/ 2500kcal (2.6kg)**
Note: Potato skin B2 = 3 x flesh B2. VitB2 range 0.3 Cassava – 1.6 Sweet potatoes					
Carrots (raw):	**4**	**4**	**60**	**8**	**/ 2500kcal (6.0kg)**
	0.3	**0.3**	**5**	**0.7**	**/ 210kcal (500g)**
Av. Fruit (undried):	**1.9**	**1.6**	**20**	**3.4**	**/ 2500kcal (4.8kg)**
Note: B3 range Persimmon 2, apples 4 – Peach 50, Cantaloupe/Rockmelon 55					
Avocado (fatty fruit):	0.4-1.0	0.7-2.1	14-29	1.4-4	/2500kcal (1.6kg)
Note: Avocado data is Florida-California.					
Av. Leaf, Flower, Shoot Veg's:	**0.3**	**0.3**	**2**	**0.8**	**/(10gProt,100kcal)/500g**
Note: Lettuce at or near the low end of the ranges for all four nutrients, Kale at the high.					
Average Squash and Zucchini:	**0.2**	**0.2**	**2.5**	**0.5**	**/ 110kcal (500g)**
Recommended Daily Allowance:	**1.1**	**1.2**	**15**	**1.3**	

Thiamin (VitB1) (RDA 1.1 mg / day) ; Riboflavin (VitB2) (RDA 1.2 mg / day) ; Niacin (VitB3) (RDA 15 mg / day) & Pyridoxal/ine/amine (VitB6) (RDA 1.3 mg / day):

By some coincidence, perhaps for reasons including that they are all water soluble and were the first four water soluble vitamins isolated, perhaps because they are more closely interrelated/ connected in biochemistry than various other nutrients are, these four nutrients are similar enough to each other in both foods and food groups, in their sufficiency or deficiency according to the RDA/AI on a per 2,500 kcal basis, that here I will deal with them all together, by food group (and commenting on isolated foods as occasionally appropriate), rather than singly as the nutrients are dealt with in sections 4.3, 4.4 and 4.5 above.

Human breast milk is a little low in VitB1 (0.75 mg / 2,500 kcal); sufficient in VitB2; substantially low in VitB6 (0.35-1.1 mg / 2,500 kcal); and ranges into a low content of VitB3 (6.5-20 mg / 2500 kcal), on a per 2500 kcal basis, according to the RDAs. Absorption of these vitamins from human breast milk is probably greater than it is from foods, so it seems that the RDAs are probably roughly correct.

An average of cow and goat milk, which are quite similar to each other in their contents of these four nutrients per 2,500 kcal, has:
A 150% sufficiency of VitB1, a 500% sufficiency of VitB2, only 1/2 of VitB3, and a sufficiency of VitB6, on a per 2,500 kcal basis.

Land animal flesh on average has a marginal deficiency (beef more so at 1/2) of VitB1, a 200% sufficiency of VitB2, a 500% sufficiency of VitB3, and a 200% sufficiency of VitB6, on a per 2,500 kcal basis.

Marine fish flesh on average has a 170% sufficiency of VitB1, a sufficiency of VitB2, a 600% sufficiency of VitB3, and only 3/5 of VitB6, on a per 2,500 kcal basis.

Marine mollusk and crustacean flesh on average has a 180% sufficiency of VitB1, a 350% sufficiency of VitB2, a 350% of VitB3, and a 250% sufficiency of VitB6, on a per 2,500 kcal basis.

Eggs have a very marginal deficiency (probably a sufficiency) of VitB1, a 700% sufficiency of VitB2, only 1/10 of VitB3 RDA, and a 1,500% sufficiency of VitB6, on a per 2,500 kcal basis.

Legumes on average have a 350% sufficiency of VitB1, a sufficiency of VitB2, a 150% sufficiency of VitB3, and a 180% sufficiency of VitB6, on a per 2,500 kcal basis.

Seeds on average have a 350% sufficiency of VitB1, a sufficiency of VitB2 (excepting sesame/tahini at 1/2), a 150% sufficiency of VitB3 (excepting pumpkin seeds at 1/2), and a 180% sufficiency of VitB6 (excepting sesame/tahini and pumpkin seeds at 1/2), on a per 2,500 kcal basis.

Nuts on average have a 180% sufficiency of VitB1, a deficiency of 1/2 of VitB2 (excepting almonds at 250%), a deficiency of VitB3 (have only 1/3, excepting almonds, which are sufficient), and a sufficiency of VitB6 (notably pistachios at 400% sufficiency, and excepting Brazil nuts at 1/4), on a per 2,500 kcal basis.

Cereal grains on average have a 350% sufficiency of VitB1, a 200% sufficiency of VitB2 (excepting some (i.e. white) rice at having only 1/4), a 200% sufficiency of VitB3 (excepting oats at having only 3/5), and a 150% sufficiency of VitB6, on a per 2,500 kcal basis. (White rice appears to be VitB1-enriched here, as historically it caused beriberi.)

Starchy root vegetables on average have a 150% sufficiency of VitB1; a deficiency of VitB2 for yams and cassava having 1/2 or less, and a sufficiency for potatoes and sweet potatoes; a 140% sufficiency of VitB3; and a 400% sufficiency of VitB6, on a per 2,500 kcal basis.

Fruits on average have a 180% sufficiency of VitB1, a 130% sufficiency of VitB2, a 130% sufficiency of VitB3 (excepting apples having only 1/5 and persimmons having only 1/10), and a 250% sufficiency of VitB6, on a per 2,500 kcal basis.

500 g of green leaf, flower and shoot vegetables on average provides 1/4 of the VitB1 RDA, 1/4 of the VitB2 RDA, 1/7 of the VitB3 RDA, and 1/2 of the VitB6 RDA.

500 grams of squash and zucchini on average provides 1/5 of the VitB1 RDA, 1/6 of the VitB2 RDA, 1/6 of the VitB3 RDA, and 2/5 of the VitB6 RDA.

The meaning of the food thiamin (VitB1), riboflavin (VitB2), niacin (VitB3) and pyridoxal/ine/amine (VitB6) data I interpret to be that the addition of roughly equal amounts of legumes, nuts, seeds and cereal grains, to an amount of fruit and/or starchy root vegetables that meets most of the dietary caloric requirement, and 500 g of green leaf, flower and shoot vegetables would provide a very sufficient intake of all four of these vitamins. Even including the nuts, which are moderately deficient (compared to the RDAs, on a 2500 kcal basis) on average in riboflavin (VitB2) and niacin (VitB3) and also have moderate deficiencies of thiamin (VitB1) and pyridox(al/ine/amine) (VitB6) in many individual nut species within the average values, the ratio of these food groups could be varied considerably from this, and the diet on average would provide a very sufficient intake (exceeding the RDAs) of thiamin (VitB1), riboflavin (VitB2), niacin (VitB3) and pyridoxal/ine/amine (VitB6).
It is quite noteworthy that the fruits on their own on average on a per 2500 kcal basis provide a very sufficient amount of thiamin (VitB1), riboflavin (VitB2), niacin (VitB3) and pyridoxal/ine/amine (VitB6). This came as a surprise even to myself, even fairly aware as I am of the extent to which much nonsense has been very ignorantly, foolishly and wrongly said about various plant foods by various meat-habituated vocalisers/writers, such as the many misconceptions and wrong theories that I have savagely (and very rightly so…) debunked in this work here without fear or favor of network consequences, reveling in the truth, like a scientist ought.
Also, that 500 g of green leaf, flower and shoot vegetables on average supply a very significant and useful fraction of the RDAs for thiamin (VitB1), riboflavin (VitB2), niacin (VitB3) and pyridoxal/ine/amine (VitB6).

References for chapter 4

Insel P, Turner RE, Ross D (2007) Nutrition 3rd Ed; American Dietetic Association; Jones and Bartlett, Boston, London.

Reilly C (1991) Selenium. In, Sadler MJ, Strain JJ, Caballero B (1991) Encyclopedia of Human Nutrition. Academic Press, Sydney.

Shils ME, Shike M, Ross AC, Caballero B, Cousins RJ (2006) (Eds) Modern Nutrition in Health and Disease (10th Ed), Lippincott Williams and Wilkins, New York, London.

Whitney E, Rolfes SR (2008) Understanding Nutrition (11th Ed) Thomson Wadsworth, Australia, UK, USA.

Chapter 5. Essential amino acid complementarities among food groups

5.1 Introduction p343

5.2 Summary of chapter 5 p344

5.3 Table 15: The lowest and highest essential amino acids of foods relative to the proportion required by the human body p345

5.4 Interpretation of the data p348

References for chapter 5 p349

5.1 Introduction: In the following table the data is taken from the Protein Quality scale provided by the SelfNutritionData website at: http://nutritiondata.self.com/
And I acknowledge and thank the site's producers and maintainers.

The Protein Quality scale provided by the SelfNutritionData website gives the essential amino acids as being 1,2,3,4 or 5 out of 5, 5 out of 5 being the category where the particular essential amino acid is present in its highest proportion relative to the proportion required by the human body, and 1 out of 5 being the lowest category.
It seems, however, due to more foods than expected having 5/5 for all essential amino acids, that the algorithm for this does not allow excessive richness (only deficiency) in any essential amino acid to be visible - possibly if *non*-essential amino acids are present in high enough proportion relative to the essential amino acids, then it becomes more likely that all essential amino acid(s) may receive a rating of 5/5, and no higher rating be available to signify excess of any particular essential amino acid available for complementation of any other food's essential amino acid proportions – such a compromise in the graphic depiction (internal sufficiency of the food's essential amino acid contents on their own, versus the ability of the food to complement any particular other food's essential amino acid contents) would be sensible, but better detailed on the site, difficult though it might be for the general public to understand.

In the table below only the essential amino acids in the lowest category (1/5, 2/5, 3/5 or 4/5), those the particular food is most deficient in, and those in the highest category (5/5), those the particular food is most sufficient in, relative to the proportions required by the human body, are given, as those essential amino acids in intermediate categories are generally fairly irrelevant to the investigation here, which is to address the question of how the essential amino acid proportions of particular foods or food groups can complement each other when eaten together to improve the overall proportion at a particular meal (amino acids are generally not stored, and other than when muscles atrophy from disuse and amino acids may come back into the bloodstream, which is probably not an insignificant process in various illnesses and injuries, so the question must be examined at the level of the meal, rather than at the level of the diet).

The 'amino acid scores' (aas__) provided by the SelfNutritionData website give some summary measure of the completeness of the essential amino acids in a particular food relative to the proportions required by the human body. I do not yet know the exact algorithm for determining them, but they seem generally in good agreement with the data provided by the website in the Protein Quality scale – in general the aas will be heavily influenced (reduced) by the lowest proportion relative to the proportions required by the human body that any essential amino acid is found in – for example one 5/5 and the all the rest 4/5 would score much higher than one 1/5 and all the rest 5/5.

5.2 Summary of chapter 5

Animal foods require no dietary complementation of their essential amino acid profile by other food groups.

For almost any conceivable practical purpose, legumes require no dietary complementation of their essential amino acid profile by other food groups.

Nuts, seeds and grains are generally mildly or moderately low in lysine. Some nuts are also low in (methionine + cysteine).

Generally, fruits as a group may not be relied on to complement any other food group's essential amino acid profile, as different fruits vary so widely in their amino acid profile.

In a diet made up calorically of mostly fruit or starchy root vegetables in almost any proportion, with the caloric balance coming from almost any proportion of legumes, seeds or grains, with or without nuts, along with 500 grams of green leaf, flower or shoot vegetables, there is almost no chance of there being any significant dietary shortage of any essential amino acid.
The only combination of food groups likely to result in a significant dietary shortage of any essential amino acid would be an unfortunate very unusual exclusive combination of one type of only certain nuts (i.e., macadamia nuts in particular, extremely low ('0/5') in lysine and very low ('1/5') in (methionine + cysteine)) with certain fruits (i.e., pawpaw/ papaya, mandarins, apricots, very low ('1/5') in (methionine + cysteine), with or without green leaf, flower or shoot vegetables.

5.3 Table 15: The lowest and highest essential amino acids of foods relative to the proportion required by the human body

Animal Flesh

Beef:		All essential amino acids 5/5 aas94
Ham:		All essential amino acids 5/5 aas135
Lamb:		All essential amino acids 5/5 aas141
Chicken Meat (all):		All essential amino acids 5/5 aas136
Tuna:		All essential amino acids 5/5 aas148
Salmon:		All essential amino acids 5/5 aas148
Sea Bass:		
Snapper:		All essential amino acids 5/5 aas148
Sole/Flounder:		
Striped Mullet:		
Squid:		All essential amino acids 5/5 aas107
Octopus:		All essential amino acids 5/5 aas107
Crab:		All essential amino acids 5/5 aas113
Prawns:		All essential amino acids 5/5 aas113
Clams:		All essential amino acids 5/5 aas107
Oysters:		All essential amino acids 5/5 aas106
Mussels:		All essential amino acids 5/5 aas106

Generalisations for Animal Flesh: All animal flesh has all essential amino acids present in approximately the proportions required by the human body.

Egg (whole (no shell), raw): All essential amino acids 5/5 - all essential amino acids present in approximately the proportions required by the human body; aas 136.

Milk:

Milk (cow, whole):		All essential amino acids 5/5 aas137
Milk (goat):		All essential amino acids 5/5 aas139

Generalisations for Cow/Goat Milk: Both cow and goat milk have all essential amino acids present in approximately the proportions required by the human body. This is assumed to apply also for human breast milk and infants.

Legumes:

Soy beans:		All essential amino acids 5/5 aas132
Lentils:	Lowest (Met + Cys) 4/5;	all others are 5/5 aas96
Black beans:		All essential amino acids 5/5 aas104
Red kidney beans:		All essential amino acids 5/5 aas104
White beans:		All essential amino acids 5/5 aas104
Navy beans:	Lowest (Met + Cys) 4/5;	all others are 5/5 aas82
Green Peas:		All essential amino acids 5/5 aas102
Adzuki beans:	Lowest (Met + Cys) 4/5;	all others are 5/5 aas79
Chickpeas:		All essential amino acids 5/5 aas106

Generalisations for Legumes:
The legumes on average are only very mildly low in their proportions of the essential amino acids (methionine + cysteine), relative to the proportions required by the human body. Soybeans not at all.

Peanuts:	Lowest Lysine 4/5;	all others 5/5 aas70

Seeds:

Sunflower seeds:	Lowest Lysine 4/5;	all others 5/5 aas88
Pumpkin seeds:		All essential amino acids 5/5 aas136
Sesame seeds/ Tahini:	Lowest Lysine 3/5;	all others 5/5 aas63

Generalisations for Seeds:

The seeds on average (not a valid procedure here, really) are only mildly low in their proportions of the essential amino acid lysine, relative to the proportions required by the human body. Pumpkin seeds have a very optimal essential amino acid content, whereas sesame seeds/ tahini are substantially low in lysine.

Nuts:

Pistachios:		All essential amino acids 5/5 aas109
Almonds:	Lowest Lys, (Met + Cys) 3/5;	all others 5/5 aas54
Cashews:		All essential amino acids 5/5 aas100
Hazelnuts:	Lowest Lys 3/5;	all others 5/5 aas55
Walnuts:	Lowest Lys 3/5;	all others 5/5 aas55
Brazil nuts:	No data available	
Pecan nuts:	Lowest Lys 3/5;	all others 5/5 aas61
Macadamia nuts:	Lowest Lys 0/5, (Met + Cys) 1/5;	all others 5/5 aas4

Generalisations for Nuts:

The nuts generally (with the exception of pistachios and cashews) are substantially low in their proportions of the essential amino acid lysine, some also in (methionine + cysteine), relative to the proportions required by the human body. Macadamia nuts are extremely deficient in lysine and (methionine + cysteine).

Cereal Grains:

Oats:	Lowest Lys 4/5;	all others are 5/5 aas81
Wheat (whole):	Lowest Lys 3/5;	all others are 5/5 aas54
Millet:	Lowest Lys 2/5;	all others are 5/5 aas38
Barley:	Lowest Lys 4/5;	all others are 5/5 aas81
Rice (brown):	Lowest Lys 4/5;	all others are 5/5 aas75
Rice (white):	Lowest Lys 4/5;	all others are 5/5 aas71
Corn (yellow):	Lowest Lys 3/5;	all others are 5/5 aas55

Generalisation for Grains:

The cereal grains are all low in their proportions of the essential amino acid lysine, relative to the proportions required by the human body. Rice, barley and oats only mildly so; corn and wheat moderately so, and millet substantially so.

Starchy Root Vegetables:

Potatoes (with skin):		All essential amino acids 5/5 aas112
Sweet potatoes:	Lowest Lys 4/5;	all others are 5/5 aas82
Yams:	Lowest Lys 4/5;	all others are 5/5 aas76
Cassava:	Lowest Leu, Lys (Phe + Tyr) 3/5;	Trp, (Met + Cys) are 5/5 aas52

Generalisations for Starchy Root Veg's:

With the exception of cassava, which is from the data nutritionally inferior in many other ways also, the starchy root vegetables (other than potatoes, which are not low in any essential amino acids) are only mildly low in their proportions of the essential amino acid lysine relative to the proportions required by the human body.

Carrots (raw): Lowest (Met + Cys) 4/5; all others 5/5 aas82

Fruit:

Apricot: Lowest (Met + Cys) 1/5; all others 5/5
Peach: Lowest Lys, Thr, Leu 3/5; (Met + Cys), Trp 5/5 aas54
Plums: Lowest Leu Lys 2/5; Trp 5/5 aas39
Cantaloupe/Rockmelon: Lowest Trp 2/5; Val, Hys, Iso, (Phe + Tyr) 5/5 aas34
Watermelon: Low. (Met + Cys), Leu, Hys 3/5; Trp, Thr, Lys, Iso, (Phe + Tyr) 5/5 aas52
Honeydew melon: Low. Leu, Lys, Hys 3/5; Trp, Val, Iso (Phe + Tyr) 5/5 aas51
Oranges: Lowest Leu 3/5; (Phe + Tyr) 5/5 aas58
Mandarines: Low. (Met + Cys) 1/5; Thr, Iso, Lys, (Phe + Tyr), Val, Hys only 4/5? aas20
Apples: Lowest (Met + Cys) 2/5; Leu, Lys, Val, Hys 5/5 aas30
Pear: Lowest Hys 1/5 (Met + Cys) 2/5; Leu, Thr, Val, Iso 5/5 aas29
Pawpaw: Lowest (Met + Cys) 1/5; Trp 5/5 aas7
Jackfruit: data not available
Pineapple: Lowest Leu 4/5; all others 5/5 aas81
Bananas: Lowest (Met + Cys) 3/5; all others 5/5 aas62
Dates: Lowest Thr, Leu, Lys, (Phe + Tyr) 3/5; (Met + Cys), Val 5/5 aas53
Grapes: Lowest Iso, Leu, Lys 2/5; (Met + Cys), Thr 5/5 aas32
Mango: Lowest (Met + Cys) 2/5; all others 5/5 aas31
Persimmon: All essential amino acids 5/5 aas112

Generalisations for Fruit:

Different fruits seem to differ greatly in their proportions of the essential amino acids relative to the proportions required by the human body.
As 7/16 fruits are low in methionine and cysteine, and 5/16 are high in these, far from there being an average statement that might be made, there is more of a bimodal distribution – high or low but not in between.
There is some tendency to being high in valine (10/16) and phenylalanine + tyrosine (9/16). There is some tendency to being low in leucine (8/16).
Avocado (fatty fruit): All essential amino acids 5/5 aas129

Green Leaf, Flower & Shoot Veg:

Kale (raw): All essential amino acids 5/5 aas92
Broccoli (raw): Lowest Leu, (Met + Cys) 4/5; all others 5/5 aas72
Broccoflower (raw): No data available
Cauliflower (raw): All essential amino acids 5/5 aas103
Lettuce: Lowest (Met+Cys) (2-5)/5; others but Lys, Leu, Hys ((3-4/5)/5) (4-5)/5 aas44-90
Cabbage (Savoy): Lowest (Met + Cys) 4/5; all others 5/5 aas74
Chinese Cabbage: Lowest (Met + Cys) 3/5; all others 5/5 aas68
Bamboo shoots: Lowest (Met + Cys), (Phe + Tyr) 4/5; all others 5/5 aas73
Cabbage (Red): Lowest Leu, Lys 3/5; all others but (Met + Cys), (Phe + Tyr)5/5 aas58
Cabbage: Lowest Leu, Lys 3/5; all others but (Met + Cys), (Phe + Tyr)5/5 aas58

Generalisations for Leaf, Flower, Shoot Veg's:

There is a strong tendency for green leaf, flower and shoot vegetables to be substantially low in methionine and cysteine, relative to the proportions required by the human body. Lysine and Leucine are also often moderately low.

Squash and Zucchini:

Squash, summer, all 5 varieties: Lowest Thr 4/5; all others 5/5 aas86
Squash, winter, all 5 varieties: All essential amino acids 5/5 aas109
Zucchini, summer: Lowest Thr 4/5; all others 5/5 aas89

Generalisations for Squash and Zucchini:

Consistent with the very high rating for pumpkin and squash seeds (see above, and squash and zucchini by definition include immature seeds), relative to the proportions required by the human body, squash and zucchini are generally only mildly low in threonine.

5.4 Interpretation of the data:

Generalisations for the food groups:

Generalisations for Animal Flesh: All animal flesh has all essential amino acids present in approximately the proportions required by the human body.

Egg (whole (no shell), raw): (Chicken) eggs have all essential amino acids present in approximately the proportions required by the human body.

Generalisations for Cow/Goat Milk: Both cow and goat milk have all essential amino acids present in approximately the proportions required by the human body. This is assumed to apply also for human breast milk and infants.

Generalisations for Legumes: The legumes on average are only very mildly low in their proportions of the essential amino acids methionine + cysteine, relative to the proportions required by the human body. Soybeans not at all.
Peanuts: Lowest Lysine 4/5; all others 5/5 aas70

Generalisations for Seeds: The seeds on average (not a valid procedure here, really) are only mildly low in their proportions of the essential amino acid lysine, relative to the proportions required by the human body. Pumpkin seeds have a very optimal essential amino acid content, whereas sesame seeds/ tahini are substantially low in lysine.

Generalisations for Nuts: The nuts generally (with the exception of pistachios and cashews) are substantially low in their proportions of the essential amino acid lysine, some also in (methionine + cysteine), relative to the proportions required by the human body. Macadamia nuts are extremely deficient in lysine and (methionine + cysteine).

Generalisation for Grains:
The cereal grains are all low in their proportions of the essential amino acid lysine, relative to the proportions required by the human body. Rice, barley and oats only mildly so; corn and wheat moderately so, and millet substantially so.

Generalisations for Starchy Root Veg's: With the exception of cassava, which is nutritionally inferior in other ways also, the starchy root vegetables (other than potatoes, which are not low in any essential amino acids) are only mildly low in their proportions of the essential amino acid lysine relative to the proportions required by the human body.
Carrots: Lowest (Met + Cys) 4/5; all others 5/5 aas82

Generalisations for Fruit: Different fruits seem to differ greatly in their proportions of the essential amino acids relative to the proportions required by the human body.
As 7/16 fruits are low in methionine and cysteine, and 5/16 are high in these, far from there being an average quantification that might be made, there is more of a bimodal distribution – high or low but not in between.
There is some tendency to being high in valine (10/16) and phenylalanine + tyrosine (9/16).
There is some tendency to being low in leucine (8/16).
Avocado (fatty fruit): All essential amino acids 5/5 aas129

Generalisations for Leaf, Flower, Shoot Veg's: There is a strong tendency for green leaf, flower and shoot vegetables to be substantially low in (methionine + cysteine), relative to the proportions required by the human body.
Lysine and Leucine are also often moderately low.

Generalisations for Squash and Zucchini:
Consistent with the very high rating for pumpkin and squash seeds (see above, and squash and zucchini by definition include immature seeds), relative to the proportions required by the human body, squash and zucchini are generally only mildly low in threonine.

Interpretation:

Animal foods require no dietary complementation of their essential amino acid profile by other food groups.

For almost any conceivable practical purpose, legumes require no dietary complementation of their essential amino acid profile by other food groups.

Nuts, seeds and grains are generally mildly or moderately low in lysine. Some nuts are also low in (methionine + cysteine).

Generally, fruit may not be relied on to complement any other food group's essential amino acid profile, as different fruits vary so widely in their amino acid profile.

In a diet made up calorically of mostly fruit or starchy root vegetables in almost any proportion, with the caloric balance coming from almost any proportion of legumes, seeds or grains, with or without nuts, along with 500 grams of green leaf, flower or shoot vegetables, there is almost no chance of there being any significant dietary shortage of any essential amino acid.
The only combination of food groups likely to result in a significant dietary shortage of any essential amino acid would be an unfortunate very unusual exclusive combination of one type of only certain nuts (i.e., macadamia nuts in particular, extremely low ('0/5') in lysine and very low ('1/5') in (methionine + cysteine)) with certain fruits (i.e., pawpaw/ papaya, mandarins, apricots, very low ('1/5') in (methionine + cysteine), with or without green leaf, flower or shoot vegetables.

Cooking may create Maillard Products from lysine + sugar, which then renders that lysine unabsorbable, making lysine even more so the limiting amino acid that it is in for example grains, nuts and seeds, and increasing the slight extent to which lysine is the limiting amino acid for animal foods.

Reference for chapter 5

SelfNutritionData website at: http://nutritiondata.self.com/

Chapter 6. The difference between 'natural' and 'optimal' diets: natural statistical distributions of optimality – perfection is unusual; the average is far from perfect

The term 'natural' as applied to foods is one that has varied connotations to various people – some think of natural food as being food that is not subjected to any or much processing, some think of it as being 'whole' foods, some think of raw foods, and others think about what is evolutionarily the natural food of some species. Here in this section, the meaning is taken to be what is evolutionarily the natural food of humans, and the other meanings of 'natural' will be addressed in other sections.

Very or most often some characteristic of nature will have a variation that when represented on a graph fits a 'normal' or Gaussian distribution (the Bell (shaped) Curve), for example height and IQ, or some other distribution curve such as the Chi-squared (a lopsided or asymmetrical bellish shape or even a bimodal (two bell or bellish shapes offset overlying each other) distribution curve.

Now, nature is not generally some sort of paradise or utopia for the organisms living in it. It is a somewhat chaotic system (with a Gaussian or some other such distribution(s) of orderliness in it) in which at any one time a large fraction of the organisms in it is faced with some sort of imperfection in some aspect of its life – predation, starvation, other violence from members of their own or other species, weather, physical accidents, infection, parasitization, cancer, population explosion, extinction or other unfavorable increases or decreases. And, it is logical to assume that the quality or suitability of food that is being eaten has some similar level of variation to it.

For example, Galapagos Island Tortoises, normally an almost exclusively grass-eating animal, have been recorded as turning to eating the corpses of their dead fellows in one period of starvation due to their population having increased for some reason such that nearly all the grass was eaten up and not enough was left growing up to meet nutritional requirements.

So, what happens in nature is that quite often, in various circumstances, the sort of food that an individual or population is eating is not actually what would be the best for their health, it is what they are eating in order to survive. In other words, 'fallback foods', as discussed above.
And population explosions are probably in general the greatest cause of hunger and the eating by a population of foods that are not best for the health of members of their particular species.
Environmental and habitat change would be other such causes.

Note that the drive in nature, including regarding the criteria for any eating behavior to survive evolutionary selection, is generally that it enable the organism to live long enough to reproduce such that its young can grow to reproduce also, and so on – nature does not in general 'care' whether an animal lives to an advanced age and dies there of pure old age free of any disease, and death at an advanced age totally free of any disease other than pure aging is far and away the exception, not the rule – it is very, very rare.

Foods eaten in natural circumstances are not necessarily the best for the health of the animal eating them, and may be far from it.

Evolution generally selects for organisms to have some flexibility in the foods they can eat for survival and reproduction – this does not indicate that all foods eaten in surviving and reproducing, or the proportions of them, are the best for the long-term health of individuals of that species, only that they will permit survival of that species.

Chapter 7. Hunter-gatherers and/or gatherer-hunters

7.1 Summary of chapter 7 p352

7.2 Presentism: over-extrapolation of evidence towards conformity with our own lives and lives of our times (Walker 2007) p354

7.3 Survey of 104 gatherer-hunter groups; dubious dietary data; meat nutritionally secondary; plant food nutritionally primary; extreme habitat exceptions; seasonal exceptions; insects; infanticide; cannibalism; group size; division of labour (Hayden 1981); hunting as an escape from drudgery by males; hunting as a proxy display of martial power, and meat as its symbol; maintenance of territory; meat sharing in bonding; cuckoldry, prostitution, and pimping; far too little mention of defense needs p355

7.4 Californian and Australian gatherer-hunters; diet mostly vegetarian; egalitarian meat sharing (Gould 1981) p361

7.5 Kalahari G/Wi gatherer-hunters; diet mostly plant foods; seasons; water from melons; seasonal plant food plenty allows male freedom from gathering; nitrogen metabolism endproducts diuretic; power scavenging; hide acquisition and use; hunting ages and activities; male prestige of hunting; high symbolic value of meat; substantially/ sufficiently egalitarian sharing of meat (Silberbauer 1981) p362

7.6 Mbuti pygmies, Zaire; interaction with agricultural civilization; diet seasonality; hunting; meat-sharing protocol includes all group members; prestige of elephant killing (Harako 1981); big-game hunting as a proxy display of martial power; no mention of infanticide, warfare, cannibalism, or disease p365

7.7 Hadza of Africa; seasonal food types; tubers, wild vs. cultivated (Schoeninger 2007) p368

7.8 Australian Aborigines; diet and season; tubers; native millet, a summer staple (O'Dea 1991) fixation on animal hunting and failure to address well-documented plant food use as proportion of diet; failure to address fallback foods; hunting as a proxy display of martial power; territory acquisition and defense p369

7.9 Hadza of Africa; large prey as common goods with extensive sharing and continuity of supply (Hawkes, O'Connell, Blurton Jones 1991); failure to consider big-game hunting as a proxy display of martial power with value within the group as a mating and power display and as an outlet for competitive and aggressive urges, and between groups in the acquisition and defense of territory p371

7.10 Amazonian Araya, Parakana, Arawete (neighboring), and Mayoruna natives; different carbohydrate-rich staples and tooth decay; arbitrary prey type selection and avoidance and cultural distancing; fat- and protein-rich palm larvae (Milton 1991); high dietary proportion of animal food (including human-like monkeys) associated with high degree of inter-tribal warfare and killing; tenability of arbitrary prey type selection indicates energy- not protein-limitation of available foods p372

References for chapter 7 p374

7.1 Summary of chapter 7

[In the past, and to some extent carrying over to the present in the (mis)conceptions of the public and some academics, there has been a focus on the big-game hunting of those traditional 'hunter-gatherer' (more correctly gatherer-hunter) groups studied, due in no small part to an ethnocentric male-centric investigator base, with notions of manhood tied to savagely dispatching the hapless wildlife rife.

It seems, rather, that other than in the most extreme of habitats such as in the arctic, or in some other particular ones such as some islands and coasts, substantially more plant food (more than 70%) than animal food is generally eaten. Hence gatherer-hunters is usually a more correct term than hunter-gatherers.

And, this needs to be considered in the context that the areas occupied by present-day and historically recent gatherer-hunters and hunter-gatherers are areas that are more marginal (less rich or suited) to supplying the optimal foods for humans, because the richest or most suitable areas have been taken over by agriculture.
A particular aspect of this that seems rather obvious once considered, but that receives little mention, is that agriculture is largely based on growing cereal grains, and that therefore, land remaining after agriculture has taken over lands more suitable for the growing of those cereal grains is likely to be less suitable for the growing of cereal grains, and cereal grains are therefore less likely to be substantially represented in the diet of the peoples living in those remaining lands, and therefore their diets will underrepresent (and maybe greatly) the extent to which cereal grains were a part of our most recent prehistorical gatherer-hunter ancestors' diets.
And a similar consideration may apply to the remaining dietary 30% as animal foods – in richer areas, in preagricultural times, the proportion may have been considerably less than this.

Aside from a marginal-rich continuum, some places are more merely *different* from others in the foodstuffs (and therefore the nutrients) they supply, and there is a fair bit of variety in the diets of studied gatherer-hunters and hunter-gatherers.
This is not at all to be taken to mean that all such diets are equally good for all humans. There are very likely to be significantly beneficial changes that could be made to most of these diets.
And conversely, for peoples whose ancestry has a long line in a particular place, eating particular foods such as to give a particular dietary nutrient profile, the optimal diets are likely to be a little dissimilar, though we should be careful not to overestimate the probability of genetic differences arising from such population dietary histories. (see also chapter 3 above)

It seems however, that the substantial use of tubers as a foodstuff has been very present in our most recent prehistorical gatherer-hunter ancestors' diets, as it remains so in studied present-day or recent historical gatherer-hunter groups. The same can be said of fruits, legumes, nuts, oily and other non-cereal seeds, insects and their larvae, and some leaves and other vegetative plant parts (i.e. shoots).

On energetic considerations, it seems that the calories returned from the meat acquired by hunting are less than the calories returned from plant foods acquired by gathering. It seems that there are two possible situations:
1. The plentifulness of plant foods actually allow the nutritionally energetically wasteful hunting to be indulged in (see just below), and when the plant foods are in short supply the men reduce or stop hunting and put more effort into gathering.

2. If plant foods are in such short supply that the whole group together cannot gather enough to meet their needs, then hunting is increased. Here it needs to be kept in mind that more than 50% of calories as protein is overtly toxic to humans and illness is felt – and accordingly, in some conditions of great shortage of plant foods, prey that does not have enough fat is discarded even if the hunters are in a state of hunger.

Claims of meat being more energy-dense than plant foods are palpably absurd: raw meat is about 70% water; raw nuts and oily seeds have very little water, the large majority of the weight being taken up by about 20% protein on average, and the another 70-odd% being fully fat or carbohydrate – so obviously raw nuts (600 kcal / 100 g) (or oily seeds (600 kcal / 100 g), and also cereal grains (380 kcal / 100 g) for that matter…) are more energy dense than raw meat (190 kcal / 100 g, lean; 230 kcal / 100 g, regular) - the maths here is at about a primary/ elementary school level – just open up a standard food nutrient content tables and do the primary/ elementary school maths...

Furthermore, note that high-protein diets are reported to increase the cost of digestion by about 30% compared to high-fat diets – it may be more accurate to state that digestion, absorption, transport and metabolism of protein that is in excess of minimum requirements for body proteins, so as to derive energy from it, has an energetic cost 30% more than the equivalent calories derived from fat and carbohydrate, but I will not clarify this further, and the general point of there being some substantial excess metabolic cost for acquiring energy from protein rather than from fat or carbohydrate is probably enough for the large majority of the readers of this' purposes.

Also, the excess nitrogenous waste resulting from the breakdown of dietary protein for energy is diuretic – that is, it causes an increased excretion of water as urine. This is undesirable in habitats where water is in short supply.

Often, it is not protein or energy substrates (carbohydrate or fat) that is the limiting macronutrient for gatherer-hunters, but water, though we are not in the habit of labeling it as a nutrient. There are some circumstances where a major focus of gatherer-hunters is on gathering enough water-rich foods, a notable example being melons.
The diets of present-day apes generally supply all of their water requirements.

Humans need very little protein in comparison with their energy needs.
Quantifications of this include that with purified amino acids supplied in the right proportions in experimental diets as little as 20 grams of amino acids per day per average sized male have been sufficient to maintain positive nitrogen balance (to prevent reduction of the body's muscle mass and other protein components), and that generally accepted requirements for providing a good margin of safety to the whole population have been around 0.6 grams of food protein per kg of body weight (about 40 grams of food protein per day for a 60kg human.
Also, that human breast milk protein/energy ratios are a mere fraction of those of other mammals for which it is known (see also section 3.5 of chapter 3 above)

It seems fairly likely to me that present-day/modern gatherer-hunter men (mostly men, but also other group members for some things) don't mind that hunting is often nutritionally energetically wasteful, because:

1. They are released from the more mundane and unexciting drudgeries of plant food collecting (of surplus over and above what plant food they eat while looking around for animals to hunt…);

2. They have an outlet for competitive, dominance and aggressive urges;

3. They have a mechanism for proxy displays of martial power to all three of: neighboring group members; other males within their own group; and females; which is safer than actual combat in the context of potentially lethal weaponization.

4. Males with less martial power have a mechanism, via cuckoldry against them deliberately left unknown to them due to them being out hunting, to attempt (or not, to the extent that it was more a psychological displacement activity to remove one from the potential danger of conflict due to food contribution imbalance or cuckoldry prevention or discovery) to balance the imbalance in meat contributions with the more successful (on average more martially powerful by proxy of greater hunting success) males, to exchange sexual access to their wives for reduced risk in potentially fatal weaponized combat with the cuckolders, a share in the food (i.e. meat) greater than that proportional to their own contribution, and a share in the martial power of those better so endowed for the purposes of defence (the last two of which may be related to primitive pimping, a correlate of primitive prostitution). So perhaps there was not only the fairly well understood female prostitution analogue of exchange of food for sexual access (an evolutionarily ancient behavior), but also a more tacit and less obvious pimping correlate (with evolutionarily ancient origins, but more recently evolved 'refinements') …

But probably these aspects did not apply to earlier hominins because they were physiologically and technologically limited to power scavenging in larger groups, rather than hunting alone or in pairs or small groups, and probably much more importantly, because they were physiologically and technologically limited to defending themselves from large carnivores in larger groups, being unable to do so consistently alone or in pairs or small groups.

It seems that cooked starchy foods have a significantly worse effect in generating dental caries than raw starchy foods, and also that cooked starchy foods have a higher Glycemic Index than raw starchy foods, these two reports being consistent regarding the increased early breakdown of complex carbohydrate into simple sugars.
Also, there is the possibility that the cooking modifies the food fiber such as to reduce its cleaning brushing action.]

7.2 Presentism: over-extrapolation of evidence towards conformity with our own lives and lives of our times (Walker 2007)

Walker A. Chapter 1 Early hominin diets. In, Ungar PS. (Ed) (2007) Evolution of the human diet The known, the unknown, and the unknowable. Oxford University Press.

"Both paleontologists and historians know of the danger of the pull of the present (presentism), whereby the past can be distorted by our knowledge and immersion in present-day circumstances. It is extremely important not to extend modern human behaviors far into the past without good evidence. Several ideas and cases in this book suggest that the pull of the present is particularly acute. The use of modern hunter-gatherers [/ gatherer-hunters] as analogs is one such case, where these people are now mostly found in extreme environments. They mostly have not been able to extend into more favorable ecosystems occupied by other people, and they cope with ecological harshness by sophisticated cultural adaptations."

7.3 Survey of 104 gatherer-hunter groups; dubious dietary data; meat nutritionally secondary; plant food nutritionally primary; extreme habitat exceptions; seasonal exceptions; insects; infanticide; cannibalism; group size; division of labour (Hayden 1981); hunting as an escape from drudgery by males; hunting as a proxy display of martial power, and meat as its symbol; maintenance of territory; meat sharing in bonding; cuckoldry, prostitution, and pimping; far too little mention of defense needs

Hayden B. Chapter 10 Subsistence and ecological adaptations of modern hunter/gatherers. In, Harding RSO, Teleki G (Eds) (1981) Omnivorous primates Gathering and hunting in human evolution. Columbia University Press. New York

"This survey of hunter/gatherer subsistence habits is not exhaustive, for such an undertaking would be monumental; I hope, however that what follows is generally representative of hunter/gatherers in their major types of habitats at various levels of technological development. ......
Because of the exploratory nature of this review, groups were not selected by rigidly systematic or random sampling. Comprehensive coverage of the world's hunter/gatherers was impossible, so the sample cuts across several major cultural areas, geographical zones, and environmental regions (fig. 10.1).
[There are 34 groups from within a few hundred km of the western coast of North America, the southern area of that slightly preponderating; 6 groups from inland Canada; 9 groups from South America, no groups from higher Andean regions, 3 groups from a fully tropical climate, 3 groups from inland temperate climates, and 3 groups near Patagonia (the southernmost extremity); 14 groups from Australia, 7 groups from the wet tropics, 4 groups from the dry center, 2 groups from the far southeastern coast and Tasmania (temperate with substantial rainfall), and 1 group from a dry steppe area; 5 groups from Africa, 3 groups from wetter warmer central areas, and 2 groups from dryer, cooler southern areas; 4 groups from wetter, warmer southeast Asia; 1 group from temperate drier southeast Asia; the Ainu from Japan; and 3 groups from coastal eastern Russia]

A major problem was that data in the various ethnographies were seldom directly comparable. ......

Still more serious was the lack of comparable descriptions of data-gathering techniques. Some ethnographies were compiled from the front porch on mission or cattle stations by talking with natives about traditional activities. Other reports were written by field-workers who lived at contemporary campsites but obtained only verbal information about subsistence forays. Very few ethnographers actually accompanied natives on such forays. (It is remarkable how often the statement appears that natives ate "undetermined" amounts of food as snacks while foraging.) Some authors give no indication as to how their information was obtained. Given this state of affairs, a considerable amount of intuitive extrapolation was required to make the data at all useful for comparative purposes ......

For instance, while there is a positive correlation between precipitation and resource availability in Australia from dry to moderately wet areas (Birdsell, 1953, 1975), this relationship begins to reverse itself with high precipitation values, to the point that in rain forests the terrestrial fauna available to hunter/gatherer groups is negligible. ......

In support of this classification, we find that the rain forest of Tasmania has been virtually uninhabited (Hiatt, 1976; Jones, 1968:207-10), and there is no evidence that the

densest rain forests of the Congo basin, the Amazon basin, and Borneo were occupied before the advent of fishing and horticultural technology.
[But in rainforests, such evidence does not remain, it decomposes very quickly.] ......

A third point meriting attention is that all hunter/gatherers for which we have reasonable data are often viewed as unrepresentative because they live in "marginal" or "peripheral" habitats (e.g. Tanaka, 1976:114). This position is clearly untenable. Observations indicate that some coastal Australian and West Coast American groups lived as what can only be described as ideal hunting/gathering conditions into the late nineteenth or mid-twentieth century, and information on these groups is often useful for studies such as this.
[Hayden fails to justify this claim by detailing what diets were available there...]
Moreover, areas where dry farming is entirely marginal, because of low precipitation or difficulty tilling, often have the highest animal biomass and high plant productivity (e.g. savannahs) and can thus support thriving hunter/gatherer populations.
[But are the plants there edible, suitable in whole-diet context, and palatable for humans?] These are in many respects optimal environments for hunter/gatherers, even though agriculturalists tend to live there only when all other options are exhausted (Butzer, 1971a).
[Note the tautological (circular reasoning) nature of the argument here stemming from the assumption that hunted animal flesh is of great value relative to alternative plant foods, such as those chosen for propagation by agriculturalists...]

...... Reports that give relative percentages or actual weights of edibles consumed are exceedingly rare (e.g. lee, 1969; Rogers, 1972), and even in these cases no consideration is given to the often substantial "snacks" eaten while people forage. ......

Despite shortcomings in the quality of the data base, several basic principles of subsistence have been proposed by lee and others. These can be summarized as follows, with the reservation that major modifications may be necessary when nutritional values are eventually recalculated on the basis of caloric intake.

1. ...... Gathered vegetable foods provide the most reliable resource base (Woodburn, 1968a; Tanaka, 1976), followed by shellfish and fish, and then by mammals, which are unpredictable because of their high mobility. [Fails to address insects] Although meat may be highly desired, Lee (1968) maintains that it is obtained opportunistically and is not counted upon for subsistence needs where other resources are available.
[Very often meat is valued less for its nutritional value as for its cultural value as evidence of a hunter's martial powers in mating displays and territorial boundary establishment and maintenance, as an outlet for aggression and dominance drives, and in requiring the hunting activity as an escape from less exciting domestic chores.]

2. In the ethnographic sample covered in table 10.4 [20 groups from around the world, selected on an unknown basis...], hunting is of least importance (by weight) to most hunter/gatherer groups.

3. At high latitudes, where plant foods are exceedingly scarce, fishing and hunting must, by default, contribute more to the total diet ......

4. On the basis of relative weights (or other unspecified criteria), hunting among generalized hunter/gatherers of the temperate and tropical zones provides anywhere from 10% to 60% of the total diet, with modal, mean, and median values for all the sampled societies converging at about 35%.

Finally, in addition to the usual categories of foods consumed by humans, there are two habitually neglected categories: insects and earth. Tacit assumptions, perhaps [very probably…] stemming from dietary ethnocentrism, seem to relegate these items to insignificant roles in descriptions of hunter/gatherer subsistence. Among a number of groups, however, insects and insect products are major seasonal staples. In the Southeastern Highlands of Australia, the seasonal emphasis on the fatty, nutritious bodies of the Bogong moth has resulted in labeling of their human consumers as "moth hunters" (Flood, 1973). Intensive exploitation of insects or insect products appears to be largely confined to tropical and subtropical regions and is perhaps most pronounced in the dry areas of these regions. ……
Also surprising is that some groups apparently consume large amounts of clay which has been carefully ground, dried, and made into "cakes" and baked. ……

Anthropophagy [cannibalism] is also unexpectedly widespread among hunter/gatherers, at least in Australia and probably other areas as well (e.g. Schultz, 1891:234; Forrest, 1875:44; Basedow, 1925; Spencer and Gillen, 1899:474-5; Davidson, 1935:153; Steward, 1938:20; Osgood, 1936:31).
This behavior seems to have been associated in most cases with the ritual killing and consumption of close relatives or enemies, but sometimes it occurred at times of extreme starvation. The extent to which such activities contributed to subsistence remains unknown.
[Cannibalism has been a lot more prevalent than most of us probably thought, see Constantine (2006), and it might best be viewed as only the most extreme end of a range of cases, habits and customs of humans eating unnatural/ suboptimal food due to hunger caused by population growth…]

Group size. ……

Briefly, the most important parameters are: resource density, resource nucleation, resource reliability, the importance of hunting, techniques of hunting, the nature of game hunted, and food-sharing practices. ……

Important reasons lead me to expect that the most efficient subsistence economy occurs in groups of *minimum* size (Hayden, 1978).

Because substantial energy is used in travel and transportation of food (see the section …..), large aggregations of people are likely to rapidly exhaust food resources within the foraging range of campsites, thereby forcing the entire group to move frequently.
[This assumes that weaponization is sufficient as for the people to be not very vulnerable to predation, an assumption we may take for granted with present-day gatherer-hunters, and historically and pre-historically recent humans and hominids, and that I posit is also correct as a necessary precondition for the earliest hominids to survive anywhere without climbable trees very readily to hand (see section 1.18, chapter 1 above).
Unfortunately, here as generally elsewhere in the literature, the defense needs of hominins (hominids early and late, and humans) are not properly dealt with.] ……

In terms of the principle of least effort (Zipf, 1949), as well as the greater risk of disease, the increased competition for resources, and the geometrically increasing potential for conflict and dispute as groups grow larger (Alexander, 1974:328; Harris, 1971:224,227; Rappaport, 1968:8), we are logically led to expect a grouping tendency that hovers around the smallest self-sufficient reproductive unit – i.e., the nuclear family.
That this is perfectly feasible even in poor areas is demonstrated by Silberbauer's observations that the G/Wi disperse into nuclear-family foraging units both seasonally and in times of stress in order to make more economical use of scarce resources

(1972:297). [The need for defense against large carnivores needs to be considered here – in later times of better weaponization carnivores probably learned to avoid humans more than they would have hominids in earlier times, substantial (but less so) weaponization being probable then also notwithstanding] ......

Since living in nuclear family groups is feasible, and since many groups spend at least part of their annual cycle in this state, it comes as a surprise that groups of 25-50 are characteristic for hunter/gatherers (table 10.7). What factors can account for such unexpectedly large group sizes? ...... while there is no apparent ecological reason why the gathering of plants and other staple foods should result in significant grouping behavior. ......
[And no mention of defense needs at all in the following 6 pages…] ......

Based on these observations, I submit that communal hunting occurs predominantly in areas where meat is an absolute survival necessity, because this tactic is more reliable than individual hunting and probably yields more meat per day per group. Most likely, communal hunting is not practiced otherwise because of its high cost in time and effort [and reduced opportunities for cuckoldry?...] ......
If the reports of greater and more reliable yields in communal hunts are accurate and if communal hunting had no disadvantages, every hunting group in the world presumably would practice communal hunting on a regular basis. Since only groups that inhabit resource-poor areas do so, there must be some disadvantage to the technique.
[reduced opportunities for cuckoldry?...] ......
If other resources such as vegetable foods are plentiful, it is relatively unimportant that game be procured at every outing.
[There is also the issue of what varying/variable needs there are for maximizing defensive presence (including re-affirmative proxy displays of martial power through hunting, as well as negotiation and combat engagements) at the territory periphery.] ......

Other examples indicate, however, that additional factors may stimulate communal hunting tactics.
Game drives occurred in the Australian Western Desert only after rains, when prey were both numerous and clustered (Gould, 1966c:44; 1969:262).
Boas (1964:101) noted that the central Eskimo resorted to individual stalking only when deer were scattered; and when game is highly dispersed, as in the Kalahari area, it usually cannot be hunted by communal drives (Silberbauer, 1972:290-1). ......

Indeed, there is no good indication in the literature that hunting and gathering in small groups occurs for any other reasons than companionship and safety.
[At last, the merest mention of defense needs….] ......

The arguments just discussed beg the question of why peoples who hunt most efficiently as individuals nonetheless live in groups. ...... there is another type of “cooperation” in which each individual hunts alone (or with a companion) and pools his yield with other group members on his return. ...... It makes sense to maintain groups that can field enough hunters to make at least one kill per day highly probable.
[And what of the need for men/ a man to be with the women for defense? Maybe men/ a man who had brought more than their ‘fair share’ of hunted meat back already, such performance constituting a proxy display (i.e. mating, as well as intra-group competitive/dominance) of martial power, and who would then have greater opportunities for cuckoldry?... The less successful hunters then having, as well as a duty to fulfill to balance the food contributions, a very easily rationalized displacement activity to engage in to avoid cuckoldry-related combat?] ......

Moreover, I have not addressed the question of why high resource density should lead to large group sizes. Such densities certainly permit large group sizes, but they hardly compel them. It is therefore necessary to assume that other factors are also important. The problem is too complex to deal with here.
[Could consider the opportunity to promote intergroup peace – though this may later be maintained at the expense of an increase in infanticide – and this arrangement would minimize chance or otherwise encounters of men from one group with women from another, which encounters if including violence would be more difficult to attribute to any particular group due to so many of them being locating within easy striking (the travel part of it) range.] ……

tendency of hunter/gatherers throughout the world to maintain group sizes in the "magical range" of 25-50 members (Lee and De Vore, 1968b)
…… When small groups do occur, it appears that generally all able adult males hunt every day – a situation not characteristic of richer environments.
[And territories being very large make opportunistic or chance attack by other humans on women and children less likely – the converse being the case in the richer environments.]
…… It is also important to note that other proximate factors – such as defense or avoidance of conflict, sanitary conditions, and deaths – can influence group movements.
[This is the second of the only merest mentions of defense needs here… badly in need of elaboration…] ……

Settlement stability may be more closely related to plant food availability (see, e.g., Lee, 1969). ……

The vast majority of plant foods used by modern hunter/gatherers are simply not toxic in either their raw or roasted forms.
[Are some legumes a significant exception to this? Maybe not given their general substantial role as a staple through primate evolution…]

Coursey (1973) argues, for instance, that the use of a simple digging stick had much more profound effects on the ability of humans to make use of plant resources than did the use of fire. ……

It is interesting how frequently hunters in band societies are prevented by strict codes of conduct from eating their own kills, or end up with nominal portions. This can best be interpreted as a method of rigidly reinforcing the pooling of individual subsistence efforts. [This is not an "interpretation", it is merely a re-statement – the question of interpretation, including why the practice exists, remains unanswered. Tacit payment for cuckoldry, to females (prostitution) or males (pimping) and the promotion of fully cooperative and committed behavior of group members in potentially fatal combat against other groups, need to be considered….] ……

Based on observations that meat often is not shared with women,
[The exact circumstances where it is and isn't are probably important to know…]
Bowdler (1976) and Mackay (1976) infer that dietary regimes among male and female members of hunter/gatherer societies differ markedly, and then speculate about the potential effects of these dietary differences on social bonding among males and on other intragroup phenomena. ……

Given the ethnographic interest in emic [relating to data analysis using functional elements of behavior or categories of people] values associated with meat, it is curious that equally abundant and explicit observations on preferences for fats have gone virtually unnoticed. …… Could it be that the meat was actually valued little, whereas the fat contained by the prey was the real prize? This might explain the anomalous cases where "valuable" carcasses are abandoned after the kill because of inadequate fat content (Tindale, 1972:248), and other cases where normally shunned animals (e.g., cougars and eagles) are eaten when sufficiently fat (Drucker, 1951). ……

In general, one might logically expect that opportunities for selection are rather limited in areas that are low in resource diversity and density, while greater selectivity can presumably be exercised in environments that offer abundance and diversity.
At high latitudes and in desert regions, groups probably had to exploit a higher proportion of available edible items in order to survive and probably had to exploit virtually all edible resources during regularly recurring periods of instability.
In rain-forest habitats, where seasonal variation is less pronounced, the picture is much less clear. ……

This avoidance [of raptors as food] may be due to their "foul" taste [ammonia, sulfides?], often mentioned in ethnographies; ……

The traditional view of labor division is that women gather plant foods whereas men hunt animal foods. In reality, this dichotomy is far too simplistic. Women gather small animals as well as plants, and in some instances even participate in capturing various prey; men, despite devoting most of their attention to hunting, may also gather plant foods to be eaten on the spot on hunting trips. ……
First, some division of labor may be needed to exploit any highly diverse resource base. …… Second, men may be innately more predisposed to hunting, while women may be behaviorally more predisposed to gathering. The contention here is that psychological and physiological differences orient females to gathering and males toward hunting (Watanabe, 1968). ……
Third, as I have suggested elsewhere (Hayden, 1972), since women produce children and maintain infanticide practices, sexual division of labor may function primarily as a population control mechanism, in which the differential allocation of tasks ensures that, on the average, women cannot cope with more than one dependant offspring at a time no matter what the degree of sedentism or the richness of the environment.
[Is infanticide the preferred option to fatal/lethal combat? Is one or the other (or cannibalism otherwise arranged) necessary in the absence of losses to predation, disease or natural disaster?]
This would partly account for the apparent life of ease that males sometimes enjoy in comparison to females. …… These hypotheses are not mutually exclusive; on the contrary, all three may be mutually reinforcing. ……

The common claim that generalized hunter/gatherers spend, on the average, about 2-5 hours per day obtaining food seems to be an accurate statement. No allowance is made in these estimates, however, for other factors that might influence energy expenditure, such as food preparation, storage, and equipment maintenance. ……

Many contemporary hunter/gatherers, such as the !Kung, do not occupy marginal environments and may, in fact, live in conditions close to the "ideal" hunting and gathering environment; they are probably not atypical of prehistoric groups.
[Note the circularity/tautology here regarding the optimality of the diet composition…]
……
Eating while foraging is much more prevalent among generalized hunter/gatherers than has usually been assumed.
This probably has had an important skewing effect on most assessments of nutritional adequacy and food preferences in the literature."
[It is important to note *both* nutritional adequacy and food preferences.]

7.4 Californian and Australian gatherer-hunters; diet mostly vegetarian; egalitarian meat sharing (Gould 1981)

Gould RA. Chapter 11 Comparative ecology of food-sharing in Australia and Northwest California. In, Harding RSO, Teleki G (Eds) (1981) Omnivorous primates Gathering and hunting in human evolution. Columbia University Press. New York

"As suggested earlier, a simple use of analogy from current to past behavior is not adequate, especially when dealing with the behavior of the earliest hominids. ……

Schaller argues that bringing a carcass to the lair would endanger the young by attracting other carnivores and scavengers. From an ecological and evolutionary point of view, such behavior would be maladaptive because it would threaten the survival of the offspring and thus the reproductive survival of the species (Schaller, 1973:275).
These same ecological constraints would have applied to early hominids.

Before the use of fire, they could not have effectively defended their home base from predation, especially at night, when considerable predation occurs. The use of fire changed the ground rules for survival, enabling early humans for the first time to bring their kill to camp and share it there without undue risk from predation.

Current hunter/gatherers all use fire and are relatively free of this constraint.
[Better differentiation of the *use* and the *making* of fire would be an improvement here]
……
The diet is primarily vegetarian. …… Men hunt but with generally poor success. Both men and women collect small game, which provides the only protein available most of the time.
[This latter part of the statement is quite incorrect – it would be impossible for a substantial (probably sufficient) amount of protein not to come from the plant foods.]

…… the Western Desert Aborigines could be described as gatherer/hunters, since the bulk of their diet on most occasions consists of plant foods. Yet the amount of time and energy expended makes hunting a major subsistence activity. ……

Grinding stones are the most common class of appliance in the Western Desert, where they are used to process all edible grass seeds. ……

When a group of men hunts a kangaroo or other large animal, the man who kills the animal is the last to share, and he sometimes receives only the innards. ……
Certain classes of the hunter's kin, such as his father-in-law and brothers-in-law have first choice of the portions, followed by other classes of kin, such as elder and younger brothers, and, last of all, the hunter himself.
These shares are transported back to camp and, in turn, divided by each sharer among his own parents, wives, and children. ……

Although relatively poor in acorn-bearing oak trees compared to other parts of the state, northwestern California contains three species that bear well and provided a staple food [acorns] for the aboriginal Tolowa: valley oak, Quercus lobata; canyon oak, Q. chrysolepus; and tanbark oak, Lithocarpus densiflora. …… the groves on the flats were heavily exploited.

…… Aggrandize behavior increases within a society in direct proportion to the opportunities for optimal harvesting with minimal risk by individual family or household groups.

The extent of sharing above the level of the individual family is important in a society in direct proportion to the risk involved in obtaining basic resources within a given ecosystem.
[And to needs for defense from other human groups…]

7.5 Kalahari G/Wi gatherer-hunters; diet mostly plant foods; seasons; water from melons; seasonal plant food plenty allows male freedom from gathering; nitrogen metabolism endproducts diuretic; power scavenging; hide acquisition and use; hunting ages and activities; male prestige of hunting; high symbolic value of meat; substantially/ sufficiently egalitarian sharing of meat (Silberbauer 1981)

Silberbauer G. Chapter 12 Hunter/gatherers of the Central Kalahari. In, Harding RSO, Teleki G (Eds) (1981) Omnivorous primates. Gathering and hunting in human evolution. Columbia University Press. New York

"When the aboveground plants are blighted by frost, the number of edible species is reduced by 75%, and the band must rely on those which have usable bulbs, tubers, or roots. ……

The G/Wi suffered heavy losses from smallpox and poliomyelitis in the early 1950s. ……

Gathered plants constitute the main part of the subsistence base (Table 12.2). In terms of mass, these provide 75-100% of the diet, depending on the time of year and the fortunes of hunters.

Plant foods are the main (and sometimes only) source of fluids during the greater part of the year, when no rainwater is to be had. ……

In the seasons of scarcity, however, edible plants are so sparsely scattered that only the combined efforts of husband and wife are sufficient to meet the household's needs, and in these times men do as much gathering as their wives and older unmarried daughters.

The mass of plant food gathered each day is fairly constant throughout the year, at 3.5-5kg per person. Tanaka (1976:112), however, estimated average intake to be 800g. The discrepancy may be due to Tanaka's having subtracted the amount of plant food eaten to meet fluid requirements. ……
Melons are an exception [regards susceptibility to spoilage], and when tsamas (Citrullus lanatus) are plentiful, gathering need be done every second day and the intake is often considerably above 5kg. Good tsama seasons, however, are not common (Leistner (1967:106) estimated that only one year in five brings abundance), and gathering is normally a chore taking up 4-8 hours of each woman's day.
[I recall a Radio National program in which Kalahari Bushmen women recounted having obtained their moisture only from melons and other food, not from drinking water, when living in that gatherer-hunter style.]

Food plants are ranked by the G/Wi in terms of their thirst- and hunger-allaying properties, the ease with which they can be utilized and, least important, their flavor. ……

Springhares are hunted by means of long, barbed probes thrust into the warrens to hook and hold the prey fast until it can be dug out. Running-down, spearing, and clubbing are less commonly used techniques.

Other predators, notably lions, are occasionally robbed of their kills, … …… Some two or three times a year, men encounter lions with fresh kills and drive them off by rushing up and furiously waving [weapons, we assume?...] and shouting.
[What of lionesses, which normally do the hunting and killing?...]
The trick apparently lies in correctly judging the moment; if challenged too early in his meal, a lion will defend (and perhaps extend) his dinner, and if left until sated and lazy he is likely to get angry.
[The former seems correct, the latter seems incorrect – rather, the more sated, the more prepared to leave?...]
Scavenging in this manner is a great game.
Vultures dropping down tell the tale of a fresh kill, and if the distance is not too great, all able-bodied men grab their spears and clubs and go tearing off through the bush, whooping and yelling in high excitement.
The wives' responses vary from a ho-hum indifference to their men's silliness to obvious fears of imminent widowhood.
On most occasions the men return empty-handed and rather sheepishly explain that the kill was too old or too small. ……

The hides of many of the antelopes which are killed are made into leather for clothing and a variety of other articles. ……

A man is at the peak of his hunting career in his late 20's. With good sight, hearing, and wind and with his strength undiminished, he has the confidence and experience to handle every situation he is likely to encounter. At 35 he is past his best and begins to restrict his hunting, and within about 5 years will be forced to rely on springhares and trapping and be able to succeed in only the easiest bow-and-arrow hunting.
[Is this really the case, or rather, he prefers to leave warfare and combat and territory marking and defense to younger fellows who are keen to prove themselves therein?
Compare 40 year-olds in athletic events: Sebastian Coe, still competitive in running the mile; Dave Scott, coming-back at the Hawaii Ironman Triathlon (the world championship of the 3.8km swim, 180km ride, 42km run distance) to a second place overall at 40y.
That bow-and-arrow hunting should be as physically demanding as implied is due to the requirement for following game intoxicated by the poison arrows that are used – presumably the larger the game, the less intoxicated due to dose:size ratio, and the further the distance that pursuit is required for] ……

… my point is that hunting does not take up the greater part of their [(G/Wi men)] time, and they have no expectation that it should. …… It is evident that hunting is neither a full-time, nor even a consistently followed, occupation among G/Wi men but is subject to seasonal restrictions, competing demands on available time, and an absence of inclination to seize every opportunity presented. ……

Even in January, when meat was most plentiful, it contributed a maximum of 10% by weight of the food eaten in that month. In September, by contrast, it made up only 0.75% of the diet.

If meat were the only source of protein, individual monthly intake would have to be 3.27 kg to meet the calculated adult requirement of protein (Thomas and Corden, 1970:54). Less than this amount is eaten in the 3 months of early summer (September-November) when available plant food is restricted almost entirely to roots and tubers of low nutritive value
[This is probably incorrect – macronutrient density of any sort (and particularly of protein) is far from necessarily able to be logically equated with overall nutritive value,

though this is very commonly done by those less competent in thinking about such matters, and doubtless other matters also, in a multifactorial way].

Depressed levels of protein intake could be advantageous in this very dry season, as the end-products of nitrogen metabolism are diuretic.
[at the very least, they require to be held and excreted in solution, which uses up water].
A carbohydrate diet, on the other hand limits ketosis and dieresis, which is a desirable circumstance when body fluid must be conserved. ……

The G/Wi do not have to hunt in order to eat; the predominance of plant foods in their diet weakens the subsistence tie that so closely interrelates the state of other predator populations with that of their prey. ……

The G/Wi do not make heroes of their hunters, but hunting is nevertheless a prestigious activity. …... It [(meat)] has a symbolic value far greater than its nutritive value or the pleasure it imparts to the palate. …… And yet a tasty melon or a bowl of morama nuts is consumed with about the same enthusiasm and satisfaction as a generous portion of venison.
[cooked, not raw, venison, note; and raw, not cooked melon and nuts(?) – also it is probably psychosocially and culturally significant that usually men make the fire that cooks the meat, which obviously enough is some sort of display of skill and power over nature, and probably also the making of fire has a quality of 'magic' about it…]
[Silberbauer lists the portions distributed in the sharing out of a killed gemsbok, and although the sharing is not fully egalitarian, not only regarding the visitors, guessing offspring numbers, it seems egalitarian enough within those it was distributed amongst, and, very importantly, that a multiplicity of such sharings within significantly overlapping subgroups within the group would result in a network of sharing equal enough in overall distributive effect to cause the whole community to be bonded together well enough to fight potentially and actually fatal/lethal battles in a fully cooperative and committed manner.] ……
The men's feast is given by the hunter, acting alone or in concert with his partner, who announces it and issues invitations. …… The quality of the relationship is not expressed by the size or value of the gift, but by the frequency of the exchange. ……The value of gift-giving as a means of creating and maintaining social solidarity is amply documented in anthropological literature. As discussed earlier, gift exchange, with its interpersonal specificity, operates as a differentiator within the G/Wi kinship system.
By enabling people to discriminate among those grouped together in the broader kin categories and to express more precisely the quality of their relationship, exchange serves as a supplementary source of information to overcome redundancy and to more finely order the society.
[I think Silberbauer misses the very important, and probably much more important, point I've noted in my comment immediately above, and he does not realize it at all anywhere in his summing up the place of hunting in the G/Wi life.]

7.6 Mbuti pygmies, Zaire; interaction with agricultural civilization; diet seasonality; hunting; meat-sharing protocol includes all group members; prestige of elephant killing (Harako 1981); big-game hunting as a proxy display of martial power; no mention of infanticide, warfare, cannibalism, or disease

Harako R. Chapter 13 The cultural ecology of hunting behavior among Mbuti pygmies in the Ituri Forest, Zaire. In, Harding RSO, Teleki G (Eds) (1981) Omnivorous primates Gathering and hunting in human evolution. Columbia University Press. New York

"It hardly needs saying that further anthropological research on hunter/gatherers is urgently needed at this time, because all such populations are declining rapidly and undergoing considerable change in behavior and culture. ……

Today villages are seldom isolated in the forest. Instead they are situated along roads, with slash-and-burn cultivation extending 2-3 km in all directions, encircling the village(s). ……
The Ituri Forest is more of an evergreen forest with no grasses growing on its floor.

…… In the major rainy season, most of the Mbuti bands return to their usual campsites near villages. These "base camps" are located on the outskirts of the secondary forest, and the surrounding primary forest constitutes the hunting range of the band.
While living at the base camp, the Mbuti depend largely on the food produced by the agriculturalists.
[No detail given of that food…]
The nomadism of Mbuti bands seems to have an annual cycle: during the major rainy season they stay near the village(s) at the border of their home range, while during the dry season and the minor rainy season they move into the interior of their hunting range.

…… seasonal changes are slight ……

Each band regularly associates with and depends on one village, and it is usually the tribal name of this village that the band adopts. ……

Dogs are the only domesticated animals used by the Mbuti. ……

The Mbuti commonly hunt by one of three methods: net hunting, bow-and-arrow hunting, and spear hunting. ……

When no game is encountered on a hunting expedition, the Mbuti collect other foods – such as plants, nuts, honey, termites, insect larvae – and check what resources will soon be available at various locations. ……

The average duration of spear-hunting expeditions is 11 hours. Upon their return to camp, spear hunters usually show great fatigue and sink exhausted around the fire – a situation that rarely happens after net or bow hunting expeditions. When spear hunters have been unsuccessful despite much work and effort, they tend to be unusually quiet in the evening, neither singing nor dancing. This high energy expenditure results in part from the need to travel long distances to good hunting areas, and in part from the necessity to move about constantly during the hunt. The distance travelled to the hunting site is often as much as 10km as the crow flies, and probably several times that much in reality. These are the reasons, in addition to the danger, that old persons and children (and myself) were not allowed to go spear hunting. ……

Spearing a wounded animal requires great bravery, according to the Mbuti, and so the owner of the second spear has high priority when the meat is distributed. ......
In most instances those who own and use nets, spears and arrows are also those who wound or kill the game, so it is clear that the net hunter, the first spear hunter, or the first arrow hunter owns the game.
Net borrowing is common among the Mbuti, however, and the rule is that the borrower gets one leg and the net owner the rest. ......
Spear or arrow borrowing is rare, but when it happens the owner of the equipment gets only forelegs and the owner retains the rest.
The difference here is probably related to degrees of prowess and risk involved, with individual variability greatest in archery and spear hunting.

Another convention is that the person who makes the morning hunting fire before a net hunt gets all the heads of animals captured that day.

The archer whose arrow first hits the prey becomes the owner if no one else wounds or kills it. But the hunter who makes the second arrow attack receives the loins and haunches.

When a dog tracks and brings down the prey after it was wounded by the first and second arrows, the dog's owner keeps the loins, the owner of the second arrow receives a section of the hips, and the owner of the first arrow gets the rest. ......

Similar conventions apply to spear hunting. The hunter who thrusts or throws the first spear to wound the prey has primary ownership, and the hunter using the second spear receives the loins. In the case of an elephant kill, the hunter using the second spear also gets the head and neck, although the primary owner can retain the trunk if he wants it. ......
After transportation to the base camp, the primary owners of carcasses or carcass sections redistribute portions until all band members have received a share. There are no prescribed rules for redistribution, and the Mbuti frequently remark that this depends on a person's generosity.
In fact, however, the primary owner does not physically participate in the further sectioning of carcasses. He sits instead at the public fire and feigns indifference.
Such behavior is standard among hunters who have just returned from a successful expedition. The person most responsible for the kill speaks little or not at all, and the other hunters must describe the details of the expedition.
When hunters return from a successful day without bringing the game to camp, women sometimes learn what happened only by the appearance of the weapons.
Thus, even though primary ownership of the game is formally established, the impression is given that the catch actually belongs to all band members, and everyone expresses pleasure in the successful hunt.
Alternatively one could say that the personal rights established by the rules of initial distribution function mainly as a formal way of socially honoring the successful hunter.

...... Villagers brought plantains, cassava, loincloths, pans, and leaf tobacco to the Mbuti camp and in return received cuts of duiker meat (excluding the heads, internal organs, abdominal walls, toes and fur).

Each carcass owner is the principal in the exchange, receiving village goods on behalf of all band members. He retains membership only of the loincloth, but others frequently borrow even this item. ......

Of all the hunting methods, spear hunting appears to have the most festive atmosphere, and such expeditions are greatly anticipated by band members. …… After a kill the first spear hunter takes the leading part in singing and dancing, with others harmonizing to his melodies. Most of the songs are extemporaneous, the objective being to recite the roles played in the hunting episode: finding the prey and how it reacted, the stealthy approach and spear thrust, the animal's strength and fighting ability, the prowess needed to track and kill it, and so on. During this time the hunter is regarded as a hero, but only in singing and dancing scenes; at all other times he resumes his normal role.
[Ostensibly or purportedly, but maybe not quite…]

…… The fame of a hunter who can kill elephants spreads far and wide, and he is honored both within and beyond his own band. All the bands and villages in an area can identify the best spear hunters. In fact, only through skillful spear hunting can a male Mbuti achieve such distinction and be called by the honorary name of *tuma*.
In contrast, an adept spear hunter who has not yet killed an elephant has a reputation only within his own band, where all members keep track of each person's hunting record. The reputation of a *tuma* spreads much farther, perhaps because other bands as well as villagers are routinely invited to elephant feasts.
[And the function of hunting prowess being a proxy display of martial power can be seen to very probably be happening here at various levels both within and between groups…]
…… This situation changes dramatically, however, when interactions between Mbuti bands and villages occur. The leadership of the *tuma* becomes more overt, in part because he acts as negotiator in economic and social contexts. ……

In the absence of archaeological evidence from the Ituri region, there is no alternative to ethnographic reconstruction of Mbuti prehistory, but I stress that the following sketch is highly speculative. ……
Prehistoric bands containing 30-60 individuals, or about 5-15 families, probably subsisted largely on plant foods gathered by women and men and on small and medium-sized game hunted by men equipped with bows and arrows.
[Obviously enough the spear is a much more basic technology and it is very very likely to have predated the bow and arrow.] ……
Hunting was probably emphasized during the dry season, with exploitation of plant resources predominating in the major rainy season, at which time social interaction within and between bands was perhaps most frequent.
[Unfortunately, no mention made here by Harako of infanticide, warfare, cannibalism, or disease, and their variation over various times…]

7.7 Hadza of Africa; seasonal food types; tubers, wild vs. cultivated (Schoeninger 2007)

Schoeninger MJ. Chapter 9 Reconstructing early hominin diets Evaluating tooth chemistry and macronutrient composition. In, Ungar PS. (Ed) (2007) Evolution of the human diet The known, the unknown, and the unknowable. Oxford University Press.

In other words, the modern day Hadza live in a "mosaic" habitat of open grassy plains, woodland/savanna, and brushy/wooded riverbeds.

In general, their dry-season diet consists of large animals, tubers, baobab fruit pulp and seeds, some berries, and honey, which is unavailable some years but plentiful in others (A.N. Crittenden, personal communication, 2004).

Their wet-season diets consist of small animals (hornbills, hyraxes and dikdik), tubers, many fruits, baobab seeds and honey. …… …, early hominins probably ate liquid honey and bee larvae at all stages of larval development (Skinner, 1991). ……

Throughout Africa today, the tubers of many types of domestic and wild plants provide important energy sources for human foragers (Vincent, 1985b) as well as for small-scale agriculturalists (Newman, 1975).

Recently tubers have appeared prominently in several different scenarios about human evolutionary history, including the Grandmother Hypothesis (O'Connell, Hawkes and Blurton-Jones, 1999), hypotheses involving the use of cooking (Wrangham et al., 1999), possible non-C_3 foods eaten by South African australopithecines (Peters and Vogel, 2005; Sponheimer et al 2005), and as fallback foods (Laden and Wrangham, 2005). ……

Most of the wild tubers belong to the Leguminosae [C_3 or C_4 ???] (Vincent, 1985a, 1985b) whereas most of the cultivated tubers belong to the Convolvulaceae [C_3 or C_4 ???]. Further, the wild tubers grow much more deeply in the soil (up to 3m) whereas the cultivated tubers grow superficially.
In terms of appearance, agricultural tubers vary somewhat across species, although all have fairly thin outer coverings.
In contrast, the wild tubers vary dramatically in superficial appearance and have extremely thick, almost barklike external coverings. Each must be peeled before consumption …
[Which is very probably easily enough done, even if just by smashing the tuber up with rocks and using the sharp edges of naturally shaped rocks.]

7.8 Australian Aborigines; diet and season; tubers; native millet, a summer staple (O'Dea 1991) fixation on animal hunting and failure to address well-documented plant food use as proportion of diet; failure to address fallback foods; hunting as a proxy display of martial power; territory acquisition and defense

O'Dea K (1991) Traditional diet and food preferences of Australian Aboriginal hunter-gatherers. Phil Trans R Soc London B 334: 233-241.

"Until European settlement of Australia 200 years ago, Aborigines lived as nomadic hunter-gatherers all over the continent in widely varying geographic and climatic conditions. ……
[Note: The richest lands of which they had taken from them the earliest in history, generally, the most marginal lands of which they had left to them, and most of Australia is considered marginal land relative to many other countries.]
Little reliable information is available about the Aborigines of southeastern Australia as their land was the first to be taken over by European settlers for farming …..

Aborigines from all over Australia were omnivorous, deriving their diet from a wide range of uncultivated plant foods and wild animals.

The composition and diversity of the food supply, and the relative proportions of plant and animal foods, were greatly influenced by both the season of the year and the geographic location.

….. Larger tribal gatherings for traditional ceremonies were only possible if there was sufficient food available to support the larger group.
Notable instances where this occurred regularly were in the alpine region of southeastern Australia during the summer months, when the Bogong moth provided a plentiful food supply (Flood 1980), in southern Queensland when nuts from the Bunya pine were ripe, and historically along the Murray river in southeastern Australia where hundreds of Aborigines feasted on freshwater crayfish or fish at different times of the year (Sahlins 1972).

…… Aborigines in different parts of Australia had practices that could be described as a semi-agricultural approach to several important plant foods (Flood 1983, Kirk 1981). In Northern Australia, the long yam (Dioscorea transversa) is a dry-season staple, and there are numerous reports of the top of the tuber being left attached to the tendril of the vine when the rest of the tuber was dug out, to ensure that the yam would grow again in the following year. Yams were also planted on off-shore islands, presumably as a 'reserve' food supply. Fruit seeds spat out at campsites, especially if they landed on the 'compost' of other food debris, may well have been the source of native fruit trees frequently found at regularly used camps.

The seed-collecting practices of Aborigines in the Darling River basin of western New South Wales have been thoroughly documented (Allen 1974).
Although they relied on the river to supply their diet for much of the year (fish, shellfish, birds, aquatic plants), in the summer the staple of their diet was the seeds of the native millet.
They developed an ingenious method of harvesting, and would gather the green grass before the seed was ripe, leave it in stacks for the seeds to ripen and dry, and then thresh it and collect the seeds.
Archeological evidence shows that seed grinding in the Darling River basin goes back 15,000 years (Flood 1983).

….. They needed to know when mammals, reptiles, birds and fish were most likely to be 'fat'…..

All animals were potential food sources: mammals, birds, reptiles, insects and marine species (mammals, reptiles, fish, crustacean and shellfish).

Everything edible on an animal carcass was eaten, including muscle, fat depots and internal organs (although usually not intestinal contents) ….

There are numerous reports of Aborigines eating 2-3kg of meat at one long sitting ….. on those irregular occasions when it was available ….. converted into fat and deposited as adipose tissue, thereby providing an energy reserve to tide an individual over periodic food shortages. ….
The most highly prized components of the Aboriginal hunter-gatherer diet were the relatively few energy-dense foods: depot fat, organ meats, fatty insects and honey ……
Although many animals were actively hunted at those times of the year when their fat depots were largest …..
[Unfortunately O'Dea makes no mention of fatty nuts here, in particular the Macadamia nut and Bunya nut…]
[Nor does O'Dea address the issue of protein as being more than 50% of calories in humans being metabolically toxic, and experienced as illness, as the body is unable to dispose of the nitrogenous waste product (ammonia) left after the carbon, oxygen and hydrogen part of the amino acids has been removed for provision of energy. This in the context of others having reported the actual diet of some Australian Aborigines as being quite predominantly made up of plant foods – despite the popular (and with some academics, apparently) fixation on kangaroo/wallaby, crocodile and goanna hunting]

Discussion

K. O'Dea. The available data since European contact indicate that severe food shortages did occur in association with prolonged droughts in arid areas of Australia (primarily inland). …..
[Which would cause the consumption of less preferred fallback foods to increase as a proportion of diet – and if the Aborigines were observed at these times a faulty unrepresentative impression of their diet might result – added to the fact that the arid areas are of course the marginal and less preferred areas and not representative to start with regards (all of) the diet(s) of Australian Aborigines.]

A. Whiten (Scottish Primate Research Group, University of St Andrews, U.K.). We call the mode of foraging 'gathering and hunting', ……
[Note: Implying that more food is obtained by the gathering than by the hunting, hence 'gathering and hunting' rather than the more commonly and usually mistakenly used 'hunting and gathering.']

[The successful or necessary use of big game hunting in the provision of regular daily protein-sufficient meals to the youngest weaned children in a hominid group depend on (at least) the following factors, and they themselves depend to varying extent on each other:

1. Whether the protein could be more easily, palatably or nutritiously (i.e. nutrients other than protein/ amino acids) be acquired elsewhere.
2. Whether the whole group or only some part of it is involved in the hunting of individual animals, that is men, women and children working together albeit in differing roles; all the men working together; small groups of men working together, individual men working alone.

3. The tools used: clubs, thrown stones, spears (very probably a quite early invention), arrows (a very late invention in human evolution), poison arrows (a very very late invention), traps/snares.
4. The geography: plains, valleys, cul-de-sacs, cliffs.
5. The type of ground cover: grass, shrubs, trees, forest, jungle.
6. Temperature: cold enough, or not, to enable meat to be kept for later consumption.
7. Humidity: dry enough, or not, to enable meat to be dried and kept for later consumption.
8. Defensibility of the carcass: from dangerous predators and scavengers.
9. The size and strength of the hominins in conjunction with their weapons and cooperative defensive fighting ability.
10. Whether the big-game hunting served some other non-nutritional (i.e. psychological, social or cultural) purpose, such as being a proxy display of martial power useful in establishing and maintaining a group's territory, mating displays to females, and displays to other males in the group.]

[The thinking about the role of hunting in hominid and human life is usually quite defective in its consideration of all of the above factors together. For example, the following work by Hawkes, O'Connell and Blurton Jones:]

7.9 Hadza of Africa; large prey as common goods with extensive sharing and continuity of supply (Hawkes, O'Connell, Blurton Jones 1991); failure to consider big-game hunting as a proxy display of martial power with value within the group as a mating and power display and as an outlet for competitive and aggressive urges, and between groups in the acquisition and defense of territory

Hawkes K, O'Connell JF, Blurton Jones NG (1991) Hunting income patterns among the Hadza: big game, common goods, foraging goals and the evolution of the human diet. Phil Trans R Soc London B 334: 243-251.

"Many accounts of the early stages of human evolution assume that the practice of hunting and scavenging large animals arose and persisted because of nutritional advantages hominid hunters earned for themselves and their families (Isaac 1978).

[They studied the Hadza who, unlike other low-latitude foragers operating without firearms or dogs, take big game to the exclusion of small-bodied prey, and reported that:]

Sharing large kills reduces the risk of meatless days for big game hunters, and obviates the problem of storing large amounts of meat.
It may be unavoidable if large carcasses cannot be defended economically against the demands of other consumers. If so, then large prey are common goods.
A hunter may gain no consumption advantage from his own big game acquisition efforts. We use Hadza data to model this 'collective action' problem, and find that an exclusive focus on large game with extensive sharing is not the optimal strategy for hunters concerned with maximizing their own chances of eating meat. Other explanations for the emergence and persistence of this practice must be considered.

[They failed to consider that if the group needed to defend itself successfully against other groups of humans, or against predators, then the health and strength of all of the group would be best maximized, and also, that the cooperativity and cohesion of the group would be best maximized – and that the sharing of the large game, by individuals, who had procured it alone with the use of poison arrows, would maximize these (particularly as poison arrow hunting is done by stealth, which is facilitated by *solitary* operations). These considerations seem fairly obvious ones to make......

Also seemingly fairly obvious is the consideration that if the protein and energy needs of the growing children (the energy needs being more likely to be the limiting factor, the protein needs being low relative to energy, and the P/E need ratio low relative to that of most other animals) were being met sufficiently, mostly or entirely (and any of these is likely enough...) by the foods gathered by the women and children, then the inevitable random periodic unavailability of big-game meat would not be a problem regards child nutrition.

Less obvious, but not greatly so to anyone with a reasonably substantial knowledge base in the behavioural sciences, is the possibility that, defense from predatory animals having been better achieved, and the hominid/human population having therefore grown, and competition for preferred foodstuffs having therefore increased, and the necessity for defense from rival groups of hominids/humans having therefore increased, that the taking of big game could very plausibly serve as a proxy for intra-specific combat prowess / martial power in the minds of not only one's own group members, but also members of other neighboring groups, while also serving the purposes of an outlet for male intra-specific competition and/or aggression (general to most mammals) that would not do the counterproductive greater-than-natural damage to one's own group members feasible with weaponry, and an effective practice of the use of weapons and aggression that would then be directed at rival groups of hominids/humans.]

7.10 Amazonian Araya, Parakana, Arawete (neighboring), and Mayoruna natives; different carbohydrate-rich staples and tooth decay with roasted corn; arbitrary prey type selection and avoidance and cultural distancing; fat- and protein-rich palm larvae (Milton 1991); high dietary proportion of animal food (including human-like monkeys) associated with high degree of inter-tribal warfare and killing; tenability of arbitrary prey type selection indicates energy- not protein-limitation of available foods

Milton K (1991) Comparative aspects of diet in Amazonian forest dwellers. Phil Trans R Soc London B 334: 253-263.

"In non-human primates, strong intraspecific aggression between groups is the norm; as members of the same species, they occupy the same dietary niche, and intense competition is expected to prevail. .... human groups in the Amazon Basin show a strong tendency to repel or discourage other human groups from utilizing their supplying area."

[Milton (1991) provided data on three contemporary primitive Amazonian rainforest Indian groups living in neighboring areas with the central villages being roughly 60, 100 and 140km from each other in a triangle, and inland some 200 miles and next to rivers, and a fourth group (Mayoruna) that lived much further away (1,000+ miles) inland and next to a river, listed last below:

Arara: Carbohydrate staple sweet potato;   primary prey monkeys;   avoided prey deer and tapir,

Parakana: Carbohydrate staple bitter manioc;   primary prey tortoises;   avoided prey deer, monkeys, cracid gamebirds and macaws and,

Arawete: Carbohydrate staple (roasted) corn; primary prey large birds; avoided prey deer and tapir,

Mayoruna: Carbohydrate staple sweet manioc; primary prey peccaries; avoided prey snakes, felids, anteaters, coatis, various monkeys and birds, and squirrels

Milton concluded that:]

"Primary crops, as well as animal types most utilized as prey, were found to differ markedly between groups.
Although some differences can be accounted for by general environmental factors, no compelling single environmental factor can explain why any one group could not behave dietarily in ways more similar to another.
Many of these intergroup dietary differences appear to represent a type of cultural character displacement that aids in distinguishing the members of one group from another. As all human groups, through the medium of culture, are actual or potential occupants of the same dietary niche, each group may distance itself from potential dietary rivals through cultural conventions.
This behaviour may be justified, as the lack of overlap between forest-living groups in combination with generally intense intergroup hostility suggests that the biomass and distribution patterns of critical dietary resources in this environment may set limits to viable population size for particular areas."
[Note: To make this a little clearer, the dietary differences help to establish other groups as different and therefore more readily subjected to the high level of intergroup violence that the Amazon Rainforest primitive aborigines are now fairly well known for, relative to other primitive hunter-gatherers and (maybe not insignificantly if killing in one context (killing animals, maybe particularly animals that look like humans i.e., monkeys), makes it more likely in other contexts, i.e., killing humans) gatherer-hunters.]

[That these prey-selective practices would *still* be indulged in, and could *still* be tenable, fairly clearly indicates that energy not protein is *still* (Milton's discussion indicates she does not see this other than for the period before farming of carbohydrate staples) the limiting factor in this environment (in the absence of the deliberate farming of the carbohydrate staples that (now) occurs).
This is notwithstanding that in the three closely adjacent areas the populations of the hunted animals in one area would be replenished to some extent from the two neighboring areas where they weren't hunted, perhaps more so than from neighboring areas not of these three groups' territories.]

"The Parakana also eat insects, particularly palm larvae which are a rich source of fat. Analyses show fat contents (dry mass) from 35% to 69% for various types of palm larvae, which also happen to be high in protein (range 22-39% dry mass (K. Milton, unpublished data)). ….. Clastres (1972) noted that the non-horticultural Guayaki actively cultivated palm larvae sites throughout the forest, such that this energy-rich dietary resource was available to them throughout the year. ……

An informant who lived with one Guaja [Brazilian Indian nomadic gatherer-hunters] group for 14 months after first contact stated that their travel patterns revolved around stands of babacu palms, the mesocarps and kernels of which provided the bulk of their intake with respect to calories."
[In nearly all cases, if mesocarp and kernels provide the bulk of calories, then human protein requirements will be met or exceeded incidentally…]

[Milton stated that the Arawete, with their dietary staple of preliminarily roasted and stored corn, had by far the worst dental condition, with rampant dental caries, possibly due to abrasion, or carbohydrate deposition at the gum-tooth margin, on which bacteria would act to produce acid. It has been reported (see also chapter 1, section 1.14) that cooked starchy foods have a significantly worse effect in generating dental caries than raw starchy foods, and also that cooked starchy foods have a higher Glycemic Index than raw starchy foods, these two reports being consistent regarding the increased early breakdown of complex carbohydrate into simple sugars. Also, there is the possibility that the cooking modifies the food fiber such as to reduce its cleaning brushing action.]

References for chapter 7

Constantine N (2006) A history of cannibalism. Chartwell Books Inc., Book Sales Inc., Edison, New Jersey, USA.

Gould RA. Chapter 11 Comparative ecology of food-sharing in Australia and Northwest California. In, Harding RSO, Teleki G (Eds) (1981) Omnivorous primates Gathering and hunting in human evolution. Columbia University Press. New York

Harako R. Chapter 13 The cultural ecology of hunting behavior among Mbuti pygmies in the Ituri Forest, Zaire. In, Harding RSO, Teleki G (Eds) (1981) Omnivorous primates Gathering and hunting in human evolution. Columbia University Press. New York

Hawkes K, O'Connell JF, Blurton Jones NG (1991) Hunting income patterns among the Hadza: big game, common goods, foraging goals and the evolution of the human diet. Phil Trans R Soc London B 334: 243-251.

Hayden B. Chapter 10 Subsistence and ecological adaptations of modern hunter/gatherers. In, Harding RSO, Teleki G (Eds) (1981) Omnivorous primates Gathering and hunting in human evolution. Columbia University Press. New York

O'Dea K (1991) Traditional diet and food preferences of Australian Aboriginal hunter-gatherers. Phil Trans R Soc London B 334: 233-241.

Milton K (1991) Comparative aspects of diet in Amazonian forest dwellers. Phil Trans R Soc London B 334: 253-263.

Schoeninger MJ. Chapter 9 Reconstructing early hominin diets Evaluating tooth chemistry and macronutrient composition. In, Ungar PS. (Ed) (2007) Evolution of the human diet The known, the unknown, and the unknowable. Oxford University Press.

Silberbauer G. Chapter 12 Hunter/gatherers of the Central Kalahari. In, Harding RSO, Teleki G (Eds) (1981) Omnivorous primates. Gathering and hunting in human evolution. Columbia University Press. New York

Walker A. Chapter 1 Early hominin diets. In, Ungar PS. (Ed) (2007) Evolution of the human diet The known, the unknown, and the unknowable. Oxford University Press.

Chapter 8. The disappearance of large animals in recent prehistory and The Overkill Hypothesis

8.1 Summary of chapter 8 p375

8.2 Overkill Hypothesis; debate; disappearance of large animals; timing; climate change; human migration and population increase; faulty ethnographic analogy; archaeological bias (Webster 1981) big-game hunting as a proxy display of martial power; territory acquisition and defense p376

References for chapter 8 p383

8.1 Summary of chapter 8

[I interpret the meaning of the disappearance of large animals in recent prehistory to be that human predation probably was involved in their disappearance, but probably as a lesser factor.
Human predation may well have driven certain animal populations already under severe natural stress over the brink of extinction, but that predation probably only either merely accelerated a process that was already destined to culminate in extinction at a not much later point in time, or just tipped a precarious balance over to the side of progression to extinction.
However, my opinion on this is not a strong one, as I do not have an expert understanding of the data, which is incomplete and complex, such that the experts are themselves in disagreement – and as the matter is not crucial to my overall analysis of the information addressed here in this book I have chosen to forego acquiring an expert knowledge of what data there is.
It could be that that hunting may have been a bigger contributory factor in the disappearance of large animals in recent prehistory than Webster (1981) posits as below here and as I tend to agree with, because hunting of large animals is not only a nutritional activity – it also serves as a proxy display of martial power to all three of: members of other groups of humans (with whom territory is contestable and very probably was contested, in the context of an expanding population); male members of one's own group (in establishing prestige, and very probably also some sort of hierarchical position, the reported and ostensibly credible egalitarian nature of gatherer-hunter and hunter-gatherer groups notwithstanding); and females (as a mating display leading to increased and varied sexual activity).
It is therefore possible that many more large animals were killed than were required merely to meet (/complete) nutritional requirements – possibly many times over that amount.

However, as detailed and explored in other chapters of my work here, it does not appear that the extent of meat eating that this contributed to our genetic evolution was great enough for long enough to contribute much change in our nutrition-related genes. Though, there are likely to be some small or very small number of people in some populations who have DNA-*degenerative* genetic mutations that reduce their ability to absorb, transport, store or use vitamin B12, due to there being an amount of it in the diet (for some thousands of years), of some gene-lines, that was probably far in excess of the amount of vitamin B12 in the diet of our evolutionary ancestors of the last 6 million years. Notwithstanding my own demonstrated ability to have survived a 20+ year experimental exclusively raw exclusively plant food (vegan) diet with no supplemental vitamin B12, apparently without damage to my nervous system (see also chapter 10), there are very probably some small or very small number of present-day humans who would experience

possibly disastrous damage to their nervous system if they were to attempt to replicate my experiment.
Such people are much less likely to be found in populations that have lived for some dozens (or maybe less – in more extreme circumstances only one or two generations might be needed) or hundreds of generations on low-meat diets in historical times (times after the extinction of the large mammals occurred, when agriculture became prominent) because selection of the fittest under various selective stresses would have weeded those DNA-degenerative changes reducing the ability to absorb, transport, store or use vitamin B12, back out of the gene pool, preventing generational transfer of such genes along into present-day populations.

In any case, in the absence of a clear family history of some number of generations on both sides of an infant's family having *all* lived and thrived mentally and physically on a very low animal food diet, infants should not be subjected to a purely vegan diet or breast milk – their diet should have either some substantial animal food content or a vitamin B12 supplement. Infants with nervous system damage from a purely vegan diet have been clearly established in the medical journal literature via reported case histories.]

8.2 Overkill Hypothesis; debate; disappearance of large animals; timing; climate change; human migration and population increase; faulty ethnographic analogy; archaeological bias (Webster 1981) big-game hunting as a proxy display of martial power; territory acquisition and defense

Webster D. Chapter 14 Late Pleistocene extinction and human predation: a critical overview. In, Harding RSO, Teleki G (Eds) (1981) Omnivorous primates Gathering and hunting in human evolution. Columbia University Press. New York

"The Overkill Hypothesis. ……

"Overkill" has two possible implications that are not mutually exclusive:
human predator populations were so large in proportion to their various prey species that constant predation caused extinction;
human predators developed such effective hunting techniques that they were able to kill wastefully – i.e. to destroy many more animals than they required.
In either case, predation levels exceeded the reproductive and survival potentials of the prey populations. ……

The central and undeniable fact on which the overkill hypothesis rests is the disappearance of a wide variety of fauna, most conspicuously large herbivores and associated carnivores, during the late Pleistocene and early Holocene times (Martin, 1967). ……

Although extinction is a common evolutionary pattern, Martin (1958, 1967) identifies what he thinks are unusual characteristics of the late Pleistocene-early Holocene extinction pattern:
1. Particularly heavy loss of large animal genera, especially large herbivores with adult weights of over 50kg.
2. The narrow chronological span in which extinctions occurred.
3. The phenomenon of removal without replacement, resulting in "empty" ecological niches.
4. The lack of clear-cut evidence for associated drastic environmental changes that could explain disappearance of so many genera, especially since many forms had successfully survived previous multiple glaciations.

Confronted with these apparently unique circumstances, Martin seeks a unique cause, and invokes man the hunter, maintaining that all of the extinction episodes coincided either with the first appearance of human populations (e.g., in Australia, the New World, and several island environments) or with a climax period of hunting efficiency (e.g., in Africa)

Table 14.1 The timing of major faunal extinctions according to Martin

Old World continental extinctions	
Africa	40,000-60,000 years before present
Northern Eurasia/Europe	11,000-13,000 ybp
Southwest and Southeast Asia	?
Australia	ca. 13,000 ybp
Insular extinctions	
West Indies	"mid-postglacial" [Approx. 6,000 ybp ?]
New Zealand	ca. 400-900 ybp
Madagascar	800 ybp
New World continental extinctions	
North America	11,000-12,000 ybp
South America	10,000-11,000 ybp

Data are from Martin (1967) and Mosimann and Martin (1975). ……

In regard to the overkill hypothesis, most anthropologists familiar with the literature on hunter/gatherers would certainly not subscribe to the myth of primitive man as always a natural conservationist who carefully husbands his resources (and especially his faunal resources); the literature contains too many instances to the contrary. ……

It seems undeniable that hunting, and especially the hunting of large animals, frequently results in some level of overkill in the sense that more prey individuals may be taken than are energetically necessary.
Various factors that can lead to wastage include: ……

4. Killing strategies that destroy more animals than can possibly be utilized (e.g., mass kills.
[As a subset of which I add, killing big-game animals as a proxy display of martial power, which could potentially be very important – consider the possible deterrent effect of large numbers of large, strong victims left littering the territory as a warning to other groups that attack or intrusion would be perilous, along with excesses due to intragroup proxy displays of martial power]. ……

6. Selective utilization only of preferred edible portions of acquired prey.

7. The killing of some animals for by-products, such as hide, bone, or horn, rather than for food. ……

Several distinct positions have emerged with regard to the causes of late Pleistocene-early Holocene extinctions:

1. Humans, through their hunting activities, were *directly* responsible for the destruction of late Pleistocene fauna – that is, without their presence the animal populations would not have become extinct when they did.

2. Human predation had no significant effect on the prey species; extinction was caused by some natural factors(s) such as environmental change or disease.

3. Ecological stress, probably related to environmental change, reduced animal populations to the point where human predation pushed them over the brink to extinction.

4. *Indirect* effects of human presence, such as the introduction of non-human predator or competitor species (e.g., dogs, rats), or modification of habitats (e.g., through burning or agricultural activities) caused extinction in the absence of natural ecological pressures.

5. Latent effects of selective human predation caused new patterns of intense, destructive competition *among the prey species themselves*, driving some species to extinction (see Krantz (1970) for an elaboration of this ingenious argument). ......

Most [(archaeological)] sites are special-purpose kill or butchering sites. Our sample is thus biased toward specific functional types of sites which are relatively easy to find, and in which remains of megafauna are probably overrepresented, besides being more obvious and more durable than the remains of small fauna or plant resources. ......

There is no question that traditional use of ethnographic analogy has often been uncritical, unconvincing and unproductive. ......

Complicating the picture is evidence for continent-wide climatic changes near the beginning of the last glacial period (ca. 60,000 BP) which undoubtedly affected, in ways still poorly understood, both floral and faunal populations (Clark, 1970:105-7) ......

[Regards Europe] Guilday (1967:136) points out that, of the large mammal genera that were lost, fully 50% died out in the Gunz glaciations and 25% in the Riss-Wurm; and of the 32 genera which still existed in Europe during the late Pleistocene, only three, or 9% (some would count four or five genera) became extinct.
Yet it was precisely during this latter period that, according to excellent archaeological evidence, the advanced hunting cultures flourished, some of them highly specialized in terms of prey species, with remarkably dense populations.

Butzer (1964:393) speculates that there may have been a 10-fold increase in population between middle and late Wurm times in Western Europe. Such dramatic demographic growth must have been related to increasingly efficient techniques of mass hunting, which are archaeologically well attested.
[This might be simplistic - could the increasing capacity for language possibly have increased defense from animals, and decreased warfare and infanticide, instead of or in addition to hunting techniques, be factors?]

Kowalski has remarked on the peculiar pattern of European extinctions: Reindeer, red deer, aurochs, wild horses – all extensively hunted – were able to survive, whereas some mammals, which were probably more difficult for primitive man to kill (mammoth, wooly rhinoceros, cave bear) or were of no interest to him (lion, hyena), became extinct. (1967:349-50)
[But, these species were probably preferred targets for proxy displays of martial power, precisely because they *were* more difficult to kill...] ......

Nevertheless, Jones remains open-minded with regard to the causes of extinction, and does not rule out the possibility that human populations caused extensive environmental disturbances, particularly through use of fire. ......

New Zealand was inhabited by many species of large flightless birds (moas) that apparently proved ideal prey for early hunters, beginning at or slightly before 1,000 AD.

Their remains are conspicuous in midden deposits for about five centuries, at which point the last survivors were killed (Martin, 1967:104).

A similar late penetration of Madagascar exposed a vulnerable fauna, of which seven genera of lemurs, two species of tortoises, one species of pygmy hippopotamus, and two species of flightless birds disappeared (Martin, 1967:111).

In the West Indies the casualties included a tortoise and several small mammals, birds and lizards (Martin, 1967:109). ……

Extinction characterizes the entire North American Pleistocene as well as the late Pliocene. "As now known, the fossil record indicates that 18% of the genera of large mammals recorded from North America did not survive the late Pliocene; approximately 23% made their last appearance in late Blancan time (early Kansan); 18% in Rancho La Brean time; and 30% in postglacial time" (Axelrod, 1967:2).

Martin (1967:77-78) points out what he feels are a number of distinctive features concerning the rash of extinctions (at ca. 12,000-11,000 BP) at the end of the North American Pleistocene, in addition to its surprising magnitude.

First, differential extinction occurred: all the mammals known to be involved were large herbivores (except for two carnivores – both sabertooths – and one omnivore – a bear).

Second, the extinctions involved loss without replacement; that is, vacant ecological niches were not filled by other or new species, as would normally be the case.

Third, the time range postulated for extinctions is narrow (see Table 14.1). Finally, most of the animals involved had already successfully survived multiple climatic changes associated with previous glaciations.

The only event that can satisfactorily resolve these puzzling features is, according to Martin, the appearance of human hunting cultures in a wave of explosive colonization at the very end of the Wisconsin glacial period (Martin, 1973; Mosimann and Martin, 1975).
……
Another consideration in this chronological realignment [by Martin, of mastodon extinction date and (possible shortening of) duration of extinction process] was, I would guess, the uncomfortable proximity of the 8,000 BP date to the beginning of the Altithermal, a period of about 7,500-4,500 BP when worldwide temperatures were somewhat higher than they are today (Antevs, 1955; Baumhoff and Heizer, 1965; Bryan and Gruhn, 1964). ……

Guilday (1967) has emphasized, however, that factors of preservation, deposition, and research orientation have produced a chronological sample that is not representative of the actual extinction processes.
Dated fossil finds usually come from archaeological sites or from late- or post-Wisconsin bog or cave deposits providing unusually good environments for preservation. "As a result we have a relatively rich fauna from a limited horizon, with two sharply artificial ["artificially sharp" is closer to what was meant, I think] cutoff points associated with the appearance and disappearance of the Paleoindian hunting tradition" (Guilday, 1967:135).

…… Wilkinson (1973) has pointed out that although we have a rough idea of the terminal dates for widespread extinction, we know almost nothing about the chronological onset of massive, late Pleistocene extinction, which is essential for any understanding of it.

His own evaluation of the available data suggests that extinctions began well before 20,000 BP indicating long-term patterns of extinction incompatible with the overkill scenario
[But that early date is well inside the last glacial period, so it's not particularly consistent with theories of purely or mostly climatic cause either?] ......

A cornerstone of the overkill hypothesis is the apparent extinction of herbivore megafauna while smaller mammalian and nonmammalian fauna were relatively unaffected in the New World. ......
After a recent study of Pleistocene bird extinction, Grayson (1977) concluded that although extinctions increased throughout the Pleistocene, fully 45% of the missing avian genera disappeared between about 13,000 and 10,000 BP.
Of the progressive mammalian extinctions that occurred throughout the Pleistocene, about that same percentage falls at the very end of the epoch.
This evidence indicates stresses other than human predation. ......

Axelrod (1967), among others, proposed that small mammals may possess adaptive buffers against rapid environmental changes, and that differential extinction could thus occur in the absence of human predation.
[More detail is required here, for it to be evidence against the overkill theory]......

These positions can be labeled the *late entry hypothesis* and the *early entry hypothesis*.

The former is the more traditional, and its proponents hold that rapid population movements colonized the New World at the very end of the last (Wisconsin) glacial period. ......
Alternatively, the early entry hypothesis postulates a much earlier initial colonization (while not ruling out later, additional ones). ......

Cultural (and probably physical) diversity was well developed over the entire New World by about 9,000 BP (MacNeish, 1976), and the emergence of regional cultures with highly specialized adaptations is most explicable given a long, rather than a short, time perspective. [but still sufficiently feasible in a shorter time to be plausible enough?] ......
..., I feel that stratigraphic, technological, and radiocarbon evidence from his [(MacNeish's)] sites clearly indicates human presence [(in Peru)] as early as 15,000 – 10,000 BP. ......
If early human populations were already exploiting South American environments by ca. 13,000 – 10,000 BC[(= 15,000 – 12,000 BP)], how much earlier was the initial colonization of North America? ......
The point I want to stress here is that either interpretation [(initial colonization of North America at 50,000 - 40,000 BP, or, by 20,000 BP)] - is virtually lethal to the overkill argument as presented by Martin because, even if we use his own extinction chronology, human predator populations were present long before the proposed extinction "event" at ca. 12,000 – 11,000 BP.
[Webster's assertion here is quite incorrect, population *growth* and *levels* are in question, and they are very much part of the question of whether and to what extent the overkill hypothesis is valid.] ......

One obvious weakness in this concept [a continuous wave of a rapidly expanding population, as part of Martin's overkill hypothesis] is its blatant artificiality and its oversimplification of the patterns of population expansion into a vast expanse of vacant habitats. Even if rates of human population growth and density increase were sufficient to form such a front, humans would obviously not distribute their movements continuously over the varied landscapes of a continent. These movements would instead be unevenly

channeled by optimal environmental characteristics such as resource and shelter availability, and ease of movement.
Many areas marginal to human survival would be neglected or avoided.
Such selective patterns of movement would not generate the enormous predation pressure along a hypothetical front, and extinction of fauna would no longer be plausible.
[This is not necessarily correct - if other food resources were the population-limiting factor , and large animal hunting was largely a proxy display of martial power that would induce competing groups to move apart rather than fight, and also help to resolve intragroup power dynamics by being a proxy outlet for aggression, competition and dominance urges, and a source of group bonding through meat-sharing that would help to bond the group into a fully cooperative and committed fighting force, then it could still cause, or be a contributing cause to, extinction, and not necessarily at the population front.] ……

The notion that hunter/gatherers emphasize hunting over all other subsistence activities has recently been called into question (Bicchieri, 1972; Lee and De Vore, 1968a).
Primarily as a result of modern ethnographic studies, many have adopted the counterposition that, for hunter/gatherer economics in general, energy derived from large fauna is subordinate to other energy sources, particularly plant foods.
[Needs to mention that this applies not only to energy, but also to protein.] ……

For example Silberbauer (1972) found that the Central Kalahari G/Wi, who apparently have more large prey at their disposal than the !Kung, also rely heavily on plant foods although they lack a plant resource as productive as the mongongo nut.
According to Silberbauer, the dry-season plant intake (6-8 lb per capita per day) functions in large part to supply required liquids in an environment singularly lacking in water resources.
Thus the G/Wi and !Kung diets do resemble one another in that both are heavily dependent (at least during some seasons of the year) on plant food. ……

Despite problems of sampling and interpretation with recently acquired ethnographic data, many anthropologists have adopted radically different perspectives on hunter/gatherer subsistence from those held a generation ago, particularly with regard to the importance of hunting. A consensus has developed that Lee (1968) is correct in his assessment, based upon cross-cultural surveys, that caloric intake from hunting is almost universally much less important than that derived from gathering of plant food. ……

Since we have so few studies of hunter/gatherer subsistence, and since these are often from regions which in terms of available megafauna are marginal compared to the Pleistocene environments featured in the overkill hypothesis, other interpretations are possible. ……

More sober considerations of human subsistence costs, in terms of time and energy, make it clear that in fact hunter/gatherers like the Bushmen are much *less* efficient than many primitive agriculturalists.
[But this might only apply if purely nutritional bioenergetic aspects are considered – if social aspects such as defense of territory and intragroup dynamics are included big-game hunting might be found to be more efficient than otherwise apparent.] ……

Although the Plains Indians provide us with perhaps the most archetypal models for "big-game" hunting societies, it is clear that they were by no means fully dependent on megafaunal predation, or even on hunting. I know of no measurements of energy intake for any Plains society, but plant foods seem to have been an important component of the diet, especially when processed for use over the long winters when mobility was limited and bison hunting difficult (Hoeble, 1960; Spencer and Jennings, 1965; Wedel, 1961). Not only were wild plants used, but maize was acquired from sedentary farming populations during visits or through raiding or extortion. ......

Apart from the traditional and questionable myth that humans are predators par excellence, I suggest that for Mosimann and Martin the reasoning is essentially circular. Massive extinction occurred and (in their view) the only unique factor was a human predator population, which must therefore have been responsible. ......

As Birdsell himself noted: "To assume that a native population would not begin to colonize adjacent favorable and unspoiled hunting territories prior to attaining a maximum carrying capacity is untenable" (1957:54). ......

My impression of the North American simulation model [(Mosimann and Martin's)] is that the values attached to its various components are virtually geared to produce an extinction outcome and that, in general, little attention was devoted to critically evaluating their accuracy or consistency with known data. Although I cannot offer an alternative set of simulations here, it is apparent that slight modifications in basic values will yield the result of no extinction.
[Firstly, Mosimann and Martin should have done those 'sensitivity analyses', secondly, Webster should have presented here the "slight modifications" mentioned.] ......

Axelrod's main argument is that there are severe stresses on reproduction, particularly among large mammals, if conditions of high climatic equability quickly shift toward greater seasonality (i.e., marked and abrupt fluctuation in temperature).
Since Darwin's time, botanists and biologists have been disturbed by the problem that many floral and faunal assemblages [groupings] include species which would today be segregated into tropical and temperate environments (Axelrod, 1967:12-21).
One explanation offered by Axelrod (1967:5-8) is that these ancient assemblages existed during periods of high equability: "Under conditions of high equability, temperatures are neither so high as to discourage temperate forms, nor so low as to prevent the success of tropical ones." He concludes that "a change in the factor that explains the association of animals and plants of tropical and temperate requirements in Quaternary and Tertiary communities may be responsible for the widespread selective extinctions of the Quaternary." ......
Wilkinson (1973) maintains that high continentality was characteristic of the final (Wisconsin) glacial episode in North America, and that extinction probably began much earlier than has been thought. ......

Conclusions.

My own conclusion is that the overkill controversy is premature, in the sense that our present level of knowledge is inadequate to allow formulation of productive questions about the causes of extinction processes, much less to provide convincing answers. ......
The overkill hypothesis, especially as elaborated by Martin (1967) and Mosimann and Martin (1975), is seriously flawed, both in terms of its basic assumptions and in the nature of its use of the supportive data. ......

My preference, after a critical review of the evidence, is the conservative position that if human predation was involved in the terminal Pleistocene extinctions of megafauna, it was in the nature of an indirect cause.
Human predation may well have driven certain animal populations – already under severe natural stress – over the brink of extinction, but its basic effect was merely to accelerate a process already under way."

[I tend to agree with this, but that hunting, particularly as an oft-used proxy for display of martial power, rather than for food, was a bigger contributory factor than Webster posits.] [This controversy seems little closer to resolution today in 2010, from looking at various websites that a Google search using the terms "hunting overkill theory" and "Pleistocene extinction sequence" returned. For example, Prof. DK Grayson of Washington U at:
http://www.anthro.utah.edu/PDFs/grayson--jar07.pdf
elaborates on the unsatisfactory and insufficient nature of the data available on just what species went extinct at what times and at what rates of population decrease.]

References for chapter 8

Grayson DK (Prof. of Washington U) (2010) at:
http://www.anthro.utah.edu/PDFs/grayson--jar07.pdf

Webster D. Chapter 14 Late Pleistocene extinction and human predation: a critical overview. In, Harding RSO, Teleki G (Eds) (1981) Omnivorous primates Gathering and hunting in human evolution. Columbia University Press. New York

Chapter 9. Examples of human performance: Dave Scott, vegan Hawaii Ironman Triathlon (3.8 km swim, 180 km cycle, 42 km run) 6 times world champion; Carl Lewis, vegan 100 m & 200 m sprint and longjump world champion

Dave Scott won the Hawaii Ironman Triathlon (The accepted world championship of Ironman and ironman-distance Triathlons), a 3.8 km swim followed by a 180 km bicycle ride followed by a 42 km marathon run, six times, as a pure vegan, that is, not eating any animal flesh, eggs or dairy products. (Endurance performance)
He is reportedly (I have been unable to contact him directly) no longer a vegan, I surmise due to further considerations of the vitamin B12 issue, but this does not negate the fact that he dominated that extreme event for so long as a vegan. He is quoted as saying something to the effect that the claim that animal protein was necessary for human health was utter nonsense – and it would seem somehow unreasonable to argue with that, given the evidence of his performance?...
Carl Lewis reached the peak of his performance as a 100m & 200 m sprint and longjump world champion(s) and Olympic gold(s) medalist as a vegan. (Explosive performance)
Also, web-search "vegetarian athlete" and find Ed Moses, Martina Navratilova, etc, etc…

Chapter 10. An example of human performance after 20 years of experimental exclusively raw vegan diet with absolutely no supplemental vitamin B12: The author, in the top 15% of the 2007 Ironman Korea Triathlon (Swim 3.8 km, bicycle 180 km, run 42 km), a lifetime best at age 45 carrying hamstring and intervertebral disc injuries, and nervous system apparently still intact. The details.

(The following is repeated in chapter 1 section 1.14 above)

[I have myself carried out a personal *experiment* (*not* a belief system, please note, other than maybe a little, very early on…) for more than 20 years (from 1986 to 2008) of eating only a raw vegan (exclusively plant food) diet, eating only 2 meals per day, late brunch and dinner, without any supplementation of vitamin B12 (and this, the lack of any vitamin B12, and very probably this alone is the risky part of the experiment, and I do not advise being without some source of vitamin B12 as an ongoing thing, and I make explicit here that it would be unethical to subject a child to that).

To start taking vitamin B12 supplementation in 2008 was virtually the only change I have made to this, and this to reduce risk, because I had achieved informatively valuable endpoints already by that stage of the experiment.

I stake all the credibility that my qualifications in science can muster on my claim that this is the absolute truth, with virtually no deviation or exception whatsoever, other than maybe 10 times or less a tiny scrap of cheese (to total a combined weight of certainly less than 10g, and very probably only a fraction of that) that was missed by my very thorough scrutiny of a salad made by someone else may have found its way into me.

Nor was there any oral sex other than less than half a dozen instances, to provide whatever small amount of vitamin B12 might come from that (And no, not due to lack of sex drive (The presence of females at university and elsewhere in fact inflamed my sex drive to an annoying and distracting extent (frequent erections and carnal thoughts)), rather due to focus on large workloads, restraint of the urge, and lack of fixed abode such as would better facilitate entering into an intimate life partnership).

In general vinegar flies were prevented from laying their eggs in my food supplies, the very large majority of which were not organically grown for budgetary reasons, and although some very very small amount of vitamin B12 probably was supplied by vinegar fly or other insect eggs and larvae, it would certainly have been orders of magnitude below that amount that apes get incidentally in nature from only their plant food, without even considering what they get from adult insects.

However, I have throughout eaten a large green salad with virtually every meal (only two per day, late brunch and dinner), and green leaves are the main source of cobalt atoms that are not already incorporated in vitamin B12 – therefore it is possible, indeed probable, that my large intestine bacteria did manufacture some vitamin B12 from that cobalt, which found its way down a concentration gradient in the moist raw-food-derived large intestinal contents, back up through the ileocecal valve and into the small intestine just on the other side of that valve, where vitamin B12 is absorbed by the human body.

My vitamin B12 levels for most of that 20-plus year experimental period were 40-100 picomolar (normal range 150-450 picomolar) (around 50 picomolar for at least the last 10 years of that period) and my homocysteine levels were 80-100uM (normal range less than 12uM).

Regards body muscle and nervous system health:

At the end of that period of 20-plus years of raw vegan diet unsupplemented by any vitamin B12, at the age of 45 years, I finished the 2007 Ironman Korea Triathlon (a 3.8 km swim followed by a 180 km bicycle ride followed by a 42 km marathon run, the same distance as at the famous Hawaii Ironman Triathlon) at the 15th percentile, that is, with 15% of the field in front of me and 85% of the field behind me, using a cheaper bicycle. That is my whole of life best objectively measured athletic result, notwithstanding that it was the one most trained for (for three months of set regime coming off a substantial mix of add-lib running and cycling for a base) and tried hard for. I have never manifested athletic champion genetic ability, and have carried since the age of 18 hamstring and hip joint region damage with adhesions and a partly rehabilitated prolapsed lumbar intervertebral disk. I was in employment, though on less than the normal 30 (about 24) teaching hrs/week, as an ESL English teacher throughout the time of the Ironman training.

Working on farms harvesting fruit in between contiguous years of university study, I have consistently averaged 6 full workdays per week, for the 3 month periods, and my

reputation as a hard and good worker on those many farms and amongst those I've worked alongside is good, better than average, though I have no proof of this to offer here (farmers, Mareeba, Queensland: Howe, Nastase, Rogato, Morgante, Alba, Wever, Bosnic, Di Stefano, Hartley, Del Ben, Mete, Iacutoni, Ravanello, Christensen, Cazzola, Nucifora). I always had nothing to eat and only water to drink till lunchtime at midday while working on these farms – lunchtime was usually one hour, but sometimes only half an hour – when lunchtime was only half an hour I cut my leafy green consumption to one mouthful at each of the start and finish of the meal, and spent the rest of the short time eating peanuts and rolled oats (chewed dry and properly), and fresh raw mangoes (eaten more quickly than I like to (which is slowly), but not bolted) – my work performance in the afternoons was, like that in the mornings, above average.
Sometimes (say twice a week) I ran for 10-odd km after work, with a 1-hr-+ bicycle ride thrown in maybe once a week after work. Yes, the evening meal included a good gutful of mangoes, but the oily nut and seed intake was no more than 100-odd grams – and, somewhat annoyingly, it was usual for me to have more body fat (though not that much) at the end of the harvest season than at the start of it – mangoes are not as concentrated in energy as bananas, but they are at the high end of the range for fresh raw fruits…

And the old saying "Hunger is the best sauce" was well borne out at those farm work-day lunch hours, the raw peanuts and dry rolled oats tasting *incredibly* good, to an extent that the average westerner with their sense of taste deranged by food preparation and flavouring techniques, and their lack of proper physiological hunger due to chronic overeating relative to requirements, would be far from able to comprehend – and this is/was not due to me talking myself into it – as a psychology graduate I am fairly well armed against that – it was very real, and so marked as to be fairly surprising.

Since the start of the experimental raw vegan diet 20-plus years ago my resting heart rate has almost never been above 48 beats per minute, with the minimum ever a couple of 39 bpm amongst the 42bpm of my fitter early focal triathlon period in 1986 to 1991 – by most standards that indicates, in the absence of some pathological process, a pretty good level of fitness.
Only now at age 48 and a half years does it appear that I must concede some barely noticeable ageing, with my failure to go under 40 minutes for the 10km 2010 Bridge to Brisbane fun run (40:55) on training that I think would have sufficed in 2008 (when I did 39:49), notwithstanding a little extra upper body muscle and fat (1 kg ?), this compared to my personal best of 38:10 in 2001 and comparable half marathon personal best of 1:27 in 2001 and wind(gale)-adjusted equivalent in 2006 of 1:28 (this would plausibly adjust back to 1:27 or less (for the howling gale), the personal best time I had trained fairly hard for), though at these times I had been doing more running and there was less (say 1-2kg) of upper body muscle and fat. Half marathons in 2010 of 1:34 for both of the only two.

There has throughout the experiment virtually always been mostly daily enthusiastic physical exercising, running and martial arts mostly, with a variable good serving of bicycling (road and MTB) thrown in when convenient.

Regards nervous system i.e., brain health:

In 1996, 10 years into the experimental raw vegan diet, I achieved approximately (the results distribution graph was very small and not lined so as to facilitate reading it very accurately) the 15th percentile of the GAMSAT test that Australian university students must sit to gain entry to postgraduate medical schools to study medicine to be medical doctors – approximately 25% of those who sit the test are, after an interview assessment, admitted to the courses. Had I completed senior high school maths and physics before sitting that exam, which I hadn't, due to being an almost singularly unapplied and rudderless student in high school (as a junk-food eating omnivore, notwithstanding the lack of almost any steering committee…), I would almost certainly have achieved a score higher again than the 15th percentile.

(High school physics and Pure Maths were completed in 2000, 14 years into the experimental raw vegan diet, at the Queensland TAFE's Open Learning Centre, with the highest possible grade category of the three passing categories being achieved, with a bare minimum of tutor assistance).

In my most recently completed university degree, Master of Public Health, at one of Australia's most highly ranked universities, the University of Queensland, in 2008, just after 22 years of the experimental diet, and some months into vitamin B12 supplementation (which did not result in any detectable difference in subjective feeling, notwithstanding me being of course non-blinded to my conditions), my grade point average was 6, the possible maximum being 7 (though I strongly objected to the marking of work in one subject (graded 5) for which I had probably better qualifications for the assessable items than did the course tutor and coordinator, and in which it seemed my rightfully destructive criticism of the poor work we were required to write about was poorly received due its antipathy to the networking practices often employed by the incompetent in academia in securing each other's little niches in the avoidance of the dreaded manual labour…) though that was done on only a part-time basis while I carried out another research project, a monograph on the biomedical significance of homocysteine (which is raised in vitamin B12 or folate deficiency, among other conditions), which will soon be finished and uploaded to the Multifactor Health website, sections of it being there already.

And the state of my brain function may be further assessed by examination of this writing here.

I have done most of my university degrees, commenced in 1991 after 5-odd years of raw vegan diet, at an overload of subjects (none repeated subjects), for example 50% overload in the third year of the BSc, 66% overload of PGDip PH&TM and second year psychology (this did have some, but not a full, content overlap in the statistics area), 75% overload (all subjects passed, at a D or 6/7 gpa, though I did have some prior knowledge) in one semester of Master of Medical Statistics and Master of Nutrition & Dietetics (which degree I was prevented from finishing, during the very *un*objectively assessed practical placements, the final semester of the course, by a most silly little coterie of liars and mental midgets (such as professor(!!...) _oger _ughes, very probably the stupidest PhD candidate that Prof Kerin O'Dea and Dr Shawn Sommerset have ever nursed through that degree (I challenge them to deny it…), and the stupidest blockhead I've ever had as a Head of Department or lecturer anywhere in my university studies, who would not have relished my adversarial and much better qualified and competent presence in the field of public health nutrition that he has elected to network his easiest possible to arrange of paychecks in…) such as are not in short supply in the dietetic profession, but that's another story and I'm sure the intelligent reader could surmise readily enough the general theme of my idiot-inflicted problems, due to my vegetarian"ism", there, for example by fellow student _llison _piegel (previous 'leadership' course attendee somewhere…) who achieved (we corrected each other's exam paper) 6 out of 24 for a multiple choice clinical dietetics exam of 24 questions with 4 answers to choose from per question (the result you'd be most likely to get if you knew nothing *at all* about the examined material and selected answers totally randomly – my result was 22/24, the highest result in the class on that occasion…)).

Of course these academic challenges were far from easy for me – I submit that very few people would have found them easy… And the regular daily physical exercise was continued alongside that.

Regards the eating/meal schedules:

Since 1986 or so I've virtually only ever eaten two meals a day, late brunch and dinner, with nothing other than water in between (and this water usually mostly in summer, the meals having their own generally nicely sufficient water content), excepting a couple of liters of fruit juice in the middle of the 6-odd hour Ironman training rides and the Ironman race (No-one else does these things without taking in some food along the way either, generally…) and a couple of marathon (42 km) runs and one 9 hour running traverse end-to-end of Korea's Baekunsan Provincial Park's mountain ridge system, and a 6 hour run followed by 7 hour fastish walk to traverse the even more formidable Jirisan National Park end-to-end in one day (Yes, as a pure run, which was the intention, Jirisan beat me, but the one day full traverse just in itself is nevertheless quite a challenge – Hwaomsa Temple up to Nogodan along to Cheonwangbong peak (1,950m) down to Daewonsa Temple, if anyone's interested, and I'd advise not calculating your estimated time of completion solely on the basis of the map distance…).

I do spend some substantial time eating my food, say a total of 3 hours a day (but almost nil time in preparation and washing up…), but I am a slow eater, by choice, because I enjoy what I am eating, and do not like to rush my meals, and I have seen many others eat the foods I linger over in a mere fraction of the time I spend.

Even so, far from always feeling hungry, often enough I choose lower-calorie foods to prevent my tendency to overeat (which I have had since a child, though it has never lead to obesity as formally defined, only to, say, 5 kg of excess fat, mostly for some periods of childhood and adolescence) from adding unwanted and physiologically unnecessary fat to my carcass – for example, although I like bananas well enough, I usually reserve them for dinners before a race or long bicycle ride (more than two hours) or run (more than one hour).

I have only very, very infrequently eaten dried fruit in the 20-plus years of the experimental diet (not at all for most of the first 15 years – I did fall for the dried persimmons in Korea in 2000, and again in 2006-7 and 2009-10, but not heavily), mostly lately in an effort to be a bit more relaxed about life, but I have stopped this again now (for the present, I will have some more dried persimmons, hopefully, and Chinese dates, in the future…) as only a small amount of dried food at the end of a meal was adding unwanted body fat to me, and was reducing a little my appreciation of raw fresh fruit, which I usually find delicious, not merely tasty.

Furthermore I generally do not eat more than 100-200 grams of nuts and oily seeds plus the dry weight of the soaked legumes with grains per day, as to do so would be a gross excess of protein intake, and also bring my energy intake to an excess – and I can crack brazil nuts, almonds, macadamia nuts etc, with naturally shaped stones, faster than I can eat them – so obviously Professor Wrangham's (of Harvard University…) statements above in chapter 1 section 1.14 and in his book Catching Fire How Cooking Made us Human (2009), that humans cannot meet their caloric requirements from raw plant foods, are so far wrong as to be laughable, if the possible reinforcement and spreading of this gross ignorance were not a very serious matter regarding the public health.

In short, If I was to spend more than approximately one hour per day eating the more concentrated raw natural plant foods available to me, at the speed that I've seen many people eat them, I would become more and more overweight by the week… if I was to spend two hours per day doing that, I would become morbidly obese, despite my regular exercise, soon enough!...

To be properly rigorous, I must note that while I was contemplating possible DNA hypomethylation from low vitamin B12 and cancer risk and how much longer I should continue the absolute non-intake of vitamin B12, I suffered nonseminomatous testicular cancer in 2008, and took B12 injections mainly with a view to reducing the risk of thrombosis from high homocysteine in the operation to remove the testicle, and I now use some supplemental B12.
Testicular cancer is not known to be especially related to vitamin B12, and my case had many other known and plausible factors contributing to its probable cause (genetics, father and his brothers have had cancer, also my mother and her sister and father; slightly small and malformed left testicle; lifetime bicycle riding with long triathlon training rides in the bent-over 'aero' position shortly before the cancer diagnosis; repeatedly kicked groin in martial arts training; long hours seated on a cushion in sweat-producing summer heat working at a computer shortly before the diagnosis; eating many washed but unpeeled Korean apples; refraining from masturbation in the context of having no sexual relationship; having a negative view of my genitals; overwork; , ,).
As detailed in another document on the Multifactor Health website, I/ my body did (without chemotherapy or surgery) beat an immediate relapse of the cancer in my lymph nodes, and my white blood cell counts were unaltered by the B12 intake, arguing against B12 being the reason for the victory. Bear in mind that some cancers are associated with generalized DNA *hypo*methylation, but specific DNA *hyper*methylation, but other cancers, the converse, …]

Chapter 11. Breast Milk: nearly always the right concentration in nearly all climatic conditions

Most experts agree that breast milk is the best food for infants because it's naturally suited for infants, being naturally evolved to have the right amount of protein, fat, sugar, vitamins, minerals, water and antibodies for infants (and is low in contaminants, generally).

Most experts agree that exclusively breast fed infants should not be given water to drink, even in very hot and dry conditions because the greater amount of milk that the infant drinks in very hot and dry conditions will be appropriately diluted with the mother's own body water, if she is not severely dehydrated (as shown by studies of infant urine osmolality in very hot and dry conditions).

Chapter 12. Sense of taste: natural taste; habit-formed taste; proper hunger; Sense of smell and visual aesthetics

Regarding the natural taste of humans, the first thing to note is how very flexible our sense of taste is to the forming of habits.

Consider that one's first taste of a cigarette (its smoke, in inhalation) is almost universally revolting. Note that cigarette smoke is very obviously not what we are naturally evolved to breathe, and is unarguably toxic. However, after that first experience, psychological and social factors, in conjunction with the drug action of nicotine, after the forming of the smoking habit, cause the taste of cigarettes to become pleasurable.
The drinking of most alcoholic drinks is very similar in this regard.

Now, if one can develop, through a habit which is not started until many years of life have gone by, a sense of pleasure from the taste of things which fairly obviously naturally revolt us, then how much easier is it to develop a sense of pleasure from the taste of things that are not outright and obvious poisons, but only second- or third-rate as foods, when the habit has been developed from very early childhood? *Much* easier, fairly obviously.

I'll begin by considering human taste for the high-protein foods meat, and nuts and seeds (roughly 20% each in protein on average, the nuts and seeds being more variable in this than the meats) (And the considerations apply to legumes also, but for convenience and on the basis of more similarly low carbohydrate contents, I will deal with meat and nuts and seeds.
Now, note that for many of us raised in the west on a diet based on cooked and usually garnished (salt, pepper, chili, garlic, curry, herbs etc) meat, raw nuts and seeds, even encountered later in life and much less frequently than the cooked meat, are much more palatable than raw meat.
We must note that the fair comparison of taste even just in any person's present state of habituation, and not adjusting for that habituation, is raw nuts and seeds compared with raw meat, cooked nuts and seeds compared with cooked meat, cooked and salted nuts and seeds compared with cooked and salted meat, and cooked, salted and garnished nuts and seeds compared with cooked, salted and garnished meat.
Obviously enough, a comparison of raw nuts and seeds with cooked, salted and garnished meat would not be fair, or rather, not be sensible.

If we are naturally predisposed to eat more meat than nuts and seeds, then why doesn't raw meat taste as good as raw nuts and seeds, to people who have been brought up eating far more cooked meat than raw nuts and seeds (And note that this applies also to those people who have been brought up eating far more cooked plant foods than raw plant foods)? Is it just *totally* habit or what they are used to (raw nuts and seeds, and cooked meat), or does this taste preference suggest that our natural predisposition is towards eating much more nuts and seeds than meat?

An alternate suggestion that must be considered is that nuts and seeds might be rarer in nature than meats and more valuable than meat but only as a lesser fraction of the protein food eaten, the greater fraction better being meats – this would be analogous with apes and most monkeys having a preference for fruits over leaves, but the natural situation being that fruits are always in undersupply regards desire, while the leaves are always available, the net result being that for most species of apes and monkeys, more green leaves are eaten than fruit, this being possibly the better diet for them than having an unlimited supply of fruit available. Countering this suggestion is the fact that our anatomy, our teeth, lack of claws or talons, and long digestive tract are similar to those of our very predominantly plant food eating nearest relatives, the apes and monkeys, and not to the sharp pointy teeth and claws or talons, and short digestive tracts of the more naturally predominantly carnivorous animals.

An interesting and relevant phenomenon that I have experienced is that, as a long term (now more than 20 years, but the phenomenon came to be experienced much earlier, as dietary and taste habits were reformed) eater of raw un-garnished foods, when starting meals with some green leafy vegetables (not enough to meet my protein requirements) and following this with high-protein (average 20% protein) raw nuts and seeds, if I do not keep introducing different nuts and seeds for flavor variety, my taste tells me when I have eaten enough protein to meet my requirements – more when exercise has been such that recovery and adaptation would need more protein, and less otherwise – the amount indicated by my taste is in very good agreement with the scientific facts of protein

requirement (20grams of protein per day if it has the perfect amino acid complement and is perfectly digested and absorbed, 40grams of protein per day recommended as the minimum, for a grown adult in a normal, stable physical exercise situation, as those perfections do not usually apply).
My natural taste for protein reaches extinction if and when I have slowly and properly chewed and therefore fully tasted all of the nuts or seeds eaten – I know from my taste that I have eaten enough protein to meet my requirements, and that it is time to move to the foods that will complete my energy requirements, mostly carbohydrate bearing fruits (nuts and seeds having a substantial amount of fat in them will have made a substantial initial contribution to the energy requirement also, as well as some very variable carbohydrate content).
If I keep turning to new nuts and seeds (and there are many tasty nuts and seeds to turn to…) through gluttony, seeking more pleasure than is sensible, through the new flavors that are alongside the taste of protein, then I can easily go on to consume two or three times the amount of protein I really require.
Almost invariably I feel most healthy when I limit my consumption of protein foods to that which meets but not much exceeds my physiological protein requirement, even if consuming several tasty varieties which tempt me to overeat for (immediate) pleasure. The phenomenon is not as clearly demarcated when I consume raw soaked grains (dry protein content 10%, the balance being much more carbohydrate than fat).

To those who say that this may be due to individual idiosyncrasy of my biochemistry or physiology, I can only say, do the scientific thing – give it fair trial – get properly lean, rehabilitate your taste buds from the effects of garnishing and flavoring (salt, pepper, garlic, chili, herbs), be physiologically hungry at meals, not just eating for pleasure, overcome any psychological barriers and biases, and see for yourself.
And note that among old sayings, one of the very truest is “Hunger is the best sauce” – without that proper physiological hunger at mealtimes, you’ll just be chasing your tail with excessive variety and garnishings and flavorings, seeking pleasure that is not reasonably or rightfully yours to experience…

Relating this to the question of what is the natural food of humans – carnivores almost certainly do not experience protein taste satiation when they have met their protein requirements – they go on to eat until they will meet their caloric (energy) requirements also – 60 kg cats and dogs do not start looking around for carbohydrate sources after eating 50 grams of meat at each of two daily meals. They will meet their energy requirements, as well as from fat (saturated fat, and with cholesterol), from breaking down the protein (amino acids) that is not required for rebuilding proteins, into an energy source molecule and a nitrogen-containing waste molecule (ammonia) that will eventually (as a different molecule (urea)) have to be excreted by the kidneys. Some animals (for example sharks) are naturally benefited by high blood nitrogen levels of this source, and others are not. I suggest we are not, and that is why I feel better when I limit my protein intake to what I need for body protein maintenance or building, and why my sense of taste tells me fairly unmistakably when this has occurred in a meal that is eaten in a way that allows my taste buds to properly experience the protein that is being eaten.

The sweetness preference in the human taste, even though some people do not like extremely sweet foods, is still prevalent enough to clearly indicate that we are evolved to seek fruit as a large part of our diet. However, in nature, there is not always the abundance of sweet ripe fruit readily to hand that there is now in modernized societies, and the triplication of the salivary amylase enzyme for the breakdown of starch into glucose and smaller glucose polymers (small starch molecules) is a strong suggestion that we are also evolved to eat starchy foods like starchy root vegetables and cereal grains, where the salivary amylase releases glucose in the mouth to provide a sweet taste. The high resistance to extinction during a meal of the sweetness preference in taste (that most of us can eat quite a lot of sweet food before losing pleasure from doing so) suggests that we are evolved to be suited to a diet with a large proportion of fruit, but it should be kept in mind that the high resistance to extinction of the sweetness taste preference may well be evolutionarily due to sweet fruit not being always fully available in the natural environment, such that other things (beneficial by way of contributing to a healthier complement of food in the diet) such as green leaf, flower or shoot vegetables, seeds, nuts, grains and starchy root vegetables, with or without some deliberately eaten animal matter as well as that eaten incidentally along with the plant matter, must be eaten.

The fact that we also, like the chimpanzees, like the taste of fat, in contradistinction to gorillas, who do not like the taste of fat, suggests that we are also evolved to be suited to a diet with a substantial proportion of fat. Similarly to sweetness preference in taste, if maybe subordinate to it, the high resistance to extinction during a meal of the fat taste pleasure (that most of us can eat a fair bit of fatty food before losing pleasure from doing so) suggests that we are evolved to be suited to a diet with some fat at some times, one having some proportion of oily nuts and seeds, but it should be kept in mind that the high resistance to extinction of the fat taste pleasure may well be evolutionarily due to oily nuts and seeds not being always fully available, at least without some effort, in the natural environment.

In nature a meal may not always contain significant fat or much carbohydrate, but the optimal meal may well contain a certain proportion of protein:fat:carbohydrate.
It seems to me that the best proportion is roughly: only enough protein to meet our protein requirements; the amount of fat that comes with the meeting of those requirements from protein sources that include a good proportion of oily nuts and seeds alongside whatever legumes and cereal grains may also contribute to the protein source mix; and the balance of our caloric (energy) requirements from fruits (which would therefore meet also roughly all our water requirements).

Such a mixture provides a slower uptake of dietary carbohydrate from the digestive tract, and some accompanying fat, and one effect of this is a feeling of well-being due to that steady drawn-out delivery of fuel to the body. Also as detailed below in chapter 19 on weight loss diets, because fat metabolism is activated, when the carbohydrate supplied by the meal is no longer available for ready fuel use, fuel use can switch to depot fat (in our fat stores) more smoothly and with less sense of lethargy and hunger or craving for food than would occur if the meal had had no fat.

It is important to understand that my statements regarding this feeling of wellbeing and its proposed reasons are not negated by reports of a feeling of wellbeing reported by people on the Atkins high protein diet.
The feeling of wellbeing reported by people on the Atkins high protein diet is due to their blood sugar curves previous to the Atkins diet having been wildly fluctuating through (refined or high Glycemic Index) carbohydrate being ingested without being accompanied by any fat to slow down its absorption into the bloodstream, or to prime fat metabolism to make access of depot fat for use as energy easier for the body.
Refined or high Glycemic Index carbohydrate being ingested without being accompanied by any fat to slow down its absorption into the bloodstream, in the context of a low-fat diet, causes initially rearing blood glucose levels, causing rearing insulin levels, which cause subsequent plunging blood glucose levels, in the presence of low blood fat levels, leaving regular periods of time when both blood glucose and blood fat and fat metabolism are low, causing fatigue and hunger, the latter causing overeating in an attempt to regain some short-term blood energy fuel.

Also, the Atkins diet gives a slow and relatively constant feed of fat into the bloodstream that blunts the hunger that occurs when (refined or high Glycemic Index) carbohydrate is ingested without being accompanied by any fat to slow down its absorption into the bloodstream.
One big problem with the Atkins Diet is that in the long run, atherosclerosis is very likely to occur from the saturated fat, cholesterol, and oxidatively damaged unsaturated fat in which it is high.

Condiments/seasoning such as sugar, salt, pepper, garlic, chili and herbs make food taste good/better and rather than their effect being mostly to provide added nutrients and to be good for health, they pervert the natural sense of taste that should be enough for a properly hungry person to enjoy natural human foods, and often have undesirable and/or unknown pharmacological (drug) activity.
Salt raises blood pressure, which contributes to atherosclerosis, heart disease and stroke. Garlic, chili, curry and pepper irritate the gastrointestinal tract, exacerbating GIT disease, and increasing the chance of cancer there, and of undesirable GIT permeability.

People whose sense of taste has been perverted by habituation to the use of these unnatural, abnormal or undesirable flavourants will find food in its natural state not tasty enough, and will have to endure a period of reduced pleasure from eating before their sense of taste has been rehabilitated and they find food in its natural state tasty (again, or maybe for the first time in their life…).

Our sense of smell is physiologically very closely related to our sense of taste, and therefore very much subject to the force of habit in what we perceive as pleasant in the smell of food – in other words, just as the sense of taste may be very deranged from what it is naturally (by evolution) constituted to be, so may the sense of smell. However, there is one interesting commonality that seems to have prevailed – that as perfumes humans favor those scents that are very closely related to fruit scents, and hardly at all related to animal food scents. Bear in mind that scent preference, like almost all other features of us and any other animal, is selected for in evolution for a reason related to our survival as a species, not for no reason at all or mere artistic flair or caprice – the fairly obvious and very plausible reason for the evolution of that scent preference over other scents seems to be that we are by nature evolved to eat fruit as a large proportion of the diet.

Consider also the reason for the almost universal visual pleasure that humans get from gemstones, with their colours and regular geometric shapes. Again bear in mind that visual aesthetic preferences, like almost all other features of us and any other animal, is selected for in evolution for a reason related to our survival as a species, not for no reason at all or mere artistic flair or caprice – and again a very plausible reason for the evolution of that visual aesthetic preference seems to be that we are by nature evolved to eat fruit, with their colours and regular geometric shapes, as a large proportion of our diet.

Regarding the natural taste of humans for green leaf, flower or shoot vegetables, my own experience is that since my sense of taste has been rehabilitated from the deranged state an early life diet of cooked, garnished and junk foods had rendered it into, I find that if some logistic reason prevents me from eating something like 200 g of green leaf, flower or shoot vegetables (perhaps in some brunchtimes of occupationally restricted length), then at the next meal my sense of taste clearly urges me to eat an amount of green leaf, flower or shoot vegetables such as to make up for the lack at the previous meal.
However, the green leaf, flower or shoot taste pleasure is subordinate to the sweetness taste preference and the fat taste pleasure, very probably because evolutionarily these were more readily available than sweet or fatty foods, and therefore needed less drive to

ensure their being eaten, as they would be eaten somewhat by default in the natural absence of enough sweet and fatty foods to produce complete satiation.
This aspect of natural taste pleasure for natural foods was the last to be properly rehabilitated in the course of my excluding cooked and junk foods – replacing sweets/candy and other sweet junk foods with fruits was the first, and experienced as more straightforward and easy to accomplish, and finding the right dietary place for fatty foods was a little more subtle and involved than for the sweet foods. Rehabilitating the taste for and acceptance of green leaf, flower or shoot vegetables requires us to somewhat mimic in our eating habits the natural environment where green leaf, flower or shoot vegetables are immediately available, but the sweet fruits and fatty/oily foods are less available – this can be done as noted elsewhere here, by making the effort to eat some green leaf, flower or shoot vegetables at the start of the meal (which is good for dental health as these foods have no acidity which many fruits have, nor significant carbohydrate for oral bacteria to turn into acid, which most fruits have, and as they have tooth minerals (calcium and magnesium) to provide less of a concentration gradient out of the teeth than some other foods do), and some second in the meal with the smaller amount of more protein-rich foods, and some last in the meal to knock off the teeth some of the remains of the other foods. After a while this practice will no longer require effort, it will have become a beneficial habit that one will reap the benefit of.

Chapter 13. Dental health, optimization of digestion, and food group eating order

13.1 The causes of dental disease p393

13.2 Individual dental health management requirements p394

13.3 Nutrition/ diet in dental health management: food type eating sequence p394

13.4 Physical exercise in dental health management p396

13.5 Pharmacy in dental health management p397

13.1 The causes of dental disease:

Genetic: People are born with different genetic tendencies towards dental disease, including those involving enamel strength, tooth spacing, saliva acidity, saliva volume, mouth versus nose breathing, and taste preferences for carbohydrate versus protein or fat.

Developmental/Environmental: Due to differences in their environment, from conception onwards, people develop different tendencies towards dental diseases, including those involving tooth enamel strength, tooth spacing, saliva acidity, saliva volume, mouth versus nose breathing, and taste preferences for carbohydrate versus protein or fat.

Imperfections in dental prosthetics (false teeth plates, caps, bridges, veneers, implants) both in the manufacture and fitting (by dental personnel), and in the use and maintenance (by the person having them) can contribute to further dental disease. Also, inherent limitations in the suitability of a prosthetic for the particular patient can contribute to further dental disease, even if the particular prosthetic was logically the best choice, and more so if it wasn't logically the best choice for the particular patient.

13.2 Individual dental health management requirements.

Any one person's appropriate dental health management, with a view to providing acceptable dental health for the remainder of their life, depends on their genetic and developmental factors (which combine to produce their present state of dental health), and what level of dental health they deem acceptable for the remainder of their life. Also on how much money they are able or desire to spend on remedial measures, some of which can be very expensive.

Those experiencing serious dental decay/disease in the first half of their life expectancy, before the age of forty years, need to make serious changes in what they are doing in order to have a generally acceptable level of dental health for the rest of their life.

Also, those experiencing more moderate dental decay/disease in the first quarter of their life expectancy, before the age of twenty years, need to make serious changes in what they are doing in order to have a generally acceptable level of dental health for the rest of their life, because dental disease usually accelerates, not only in the physical fact(s) of it, but very much so in the degree to which the person notices the unpleasant effects of it.

Those experiencing more moderate dental decay/disease in the first half of their life expectancy, before the age of forty years, need to make at least some changes in what they are doing in order to have a generally acceptable level of dental health for the rest of their life.

13.3 Nutrition/ diet in dental health management: food type eating sequence

The details of what nutrients in different foods contribute to tooth decay / caries is included in the following description of the best order in which to eat the foods at a meal. To the extent to which one deviates from the general sequence of four general food types or groups, one is reducing one's resistance to tooth decay. However, some deviations from the sequence are (much) more important than others (see note near end this section). To the extent to which one has dental decay problems, if one's order of eating foods differs from the following, one may change towards the following to reduce their dental decay problems.

A good sequence/order of eating the foods in a meal is to:

First (perhaps most importantly, for dental health), eat some green leaf/flower vegetables (lettuce, cabbage, broccoli, cauliflower) (alone and without dressings, many of which contain vinegar which is acidic and therefore damage tooth enamel, all of which contain other food substances best not packed into tooth spaces at the start of a meal)
because,

this gives a pre-meal brushing-off of bacteria, & when this material packs into any spaces between or in the teeth, it is not acidic to start with, and nor does it have significant carbohydrate for oral bacteria to turn into acid, so that is the best start for the teeth,
and,
leafy greens have substantial calcium and magnesium, and a variable amount of fluoride, so that in general these minerals will be leached more slowly from the tooth enamel, down a more shallow concentration gradient (Such flows are only very small at any one time, but over years they are significant), than it would for other foods on average,
and,
the lack of easily digestible carbohydrate will prevent premature triggering of release of the gastric contents into the duodenum (small intestine),
and,
leafy greens do not have a lot of water and so will not dilute the gastric digestive juices very much, so that protein eaten immediately thereafter will still be well digested by concentrated gastric juices that will be allowed time to act properly in breaking down dietary protein.

Second, eat the protein and fat source(s) (legumes (beans, peas, lentils, chick peas etc), nuts, seeds, meats, dairy products (milk, cheese, butter etc), eggs, and even grains (wheat, oats, rye, barley, rice etc) and grain products such as breads because although they have more carbohydrate than fat they have roughly 10% protein) (some more leafy greens can be eaten with these; raw root vegetables, slowly and thoroughly chewed with these, also) because,
the protein needs concentrated gastric juices, that will be allowed time to act properly in breaking it down, and fat mixed with carbohydrate before it enters the duodenum (small intestine) slows down the absorption of sugars into the bloodstream, creating a much more stable blood sugar curve, that will have a meal-increased higher blood fat curve alongside it. Also, protein lodging in any remaining spaces between or in the teeth has amino acid parts that buffer acidity, and bacteria take longer to cause a problem by acting on this material than by acting on carbohydrates.

Third, eat the carbohydrate (and/or acid) sources (fruits, and starchy root vegetables such as potatoes and sweet potatoes, carrots having only one half the starch/energy of these), because,
they are easily digested and trigger release of the gastric contents into the duodenum (small intestine), and dilute the gastric juices (as fruit and root vegetables have a lot of water), decreasing their ability to break up dietary protein,
and,
this way the carbohydrate is released into the small intestine more slowly than it would be if it was eaten on its own, and it will also be absorbed from the small intestine more gradually, leading to more stable blood sugar levels, that will be accompanied by beneficial blood fat levels (see also chapter 19 here on weight loss).

After eating the carbohydrate source, particularly if it has any substantial acidity of its own (tastes even slightly sour/tart)(citrus (limes, lemons, grapefruit, oranges, mandarins, tangerines, cumquats, lemonades, pomeloes), pineapples, most stonefruit (peaches, nectarines, plums), some grapes and apples, etc) rinse the mouth with water 1-3 times.

Last, finish with more green leaf/flower vegetables (lettuce, cabbage, broccoli, cauliflower),
because,
they are nature's toothbrush; fruits are not as they have substantial carbohydrate and, more often than not, some acidity.

<u>Most important</u> are:
To begin each meal with green leaf vegetables alone for a few well-chewed mouthfuls; thereafter some green leaf vegetables and some root vegetables raw and therefore slowly ingested due to the time taken to chew them properly may be added to the concentrated protein sources (likewise well chewed to maximize digestion).
To not add fresh fruit (and even dried fruit, because the sugars are a signal to the stomach that hastens the release of the stomach contents into the small intestine, possibly before the proteins have been digested as well as possible in the stomach) to the foregoing, rather to eat it after the foregoing, in proper slowness as noted above in order not to cause premature dilution of stomach digestive enzymes and premature passing of the stomach contents into the small intestine.
To end each meal with green leaf vegetables alone for a few well-chewed mouthfuls.

Immediately after each meal, rinse the mouth, brush the teeth, use a proxy brush (preferably) and/or, as suits, dental floss between all teeth, and rinse again to finish.

Use of a high-pressure 'water-pick' will further enhance cleaning between the teeth.

Of course, do not eat between meals, as this disrupts the effects outlined above.

The fiber in natural fruit and vegetables can prevent really thick deposits of almost pure carbohydrates (and of protein and fats/oils) on and in between teeth, by physically brushing some of it off, and by being in amongst the carbohydrate, in effect diluting it.

13.4 Physical exercise in dental health management

Physical exercise contributes to making a healthy physiological hunger, which helps natural foods taste as good as they can taste, and ought to taste. This will help one to enjoy eating natural foods, which on average, and in any case if eaten in the right sequence, are less damaging to tooth enamel than most unnatural foods.

A healthy physiological hunger probably enhances salivation (saliva production and release), which is good for tooth enamel health.

It is preferable in general for tooth enamel health to breathe through the nose, rather than through the mouth, because this keeps the mouth and especially exposed tooth surfaces moist, – this is less important for short intervals, such as a few minutes of mouth breathing, regularly, or for the occasional longer race or some-such, but in general try to breathe through the nose where possible, and to moisten the mouth with saliva in between mouth breathing.

Avoid becoming dehydrated during physical exercise, and otherwise, as the dry mouth experienced in association with this doubtless reduces saliva flow onto the tooth enamel surfaces, which reduces tooth enamel health and resistance to decay/ caries.

Chewing of natural foods including a substantial amount of sufficiently fibrous raw fruit and vegetables (grated if need be) can provide the right stimulus to keep/make the teeth solidly placed in their sockets – lack of sufficient chewing can lead to or contribute to the teeth becoming looser in their sockets.

Chewing enhances salivation (saliva production and release), which is good for tooth enamel health.

13.5 Pharmacy in dental health management.

Fluoride is a necessary mineral, in the right proportion of replacement of hydroxyl in tooth enamel hydroxyapatite, for tooth enamel hardness and resistance to decay. If your local water and soil (and therefore foods) and your imported foods together do not provide sufficient fluoride to augment your genetic and developmentally resultant tooth enamel strength, then for the best resistance to decay you will require some supplementary fluoride (not too much, because this causes the tooth enamel to become too hard, that is, to become brittle, and liable to breakage. And other bones too.).

Fluoride painting from your dentist, fluoride mouthwashes (not to be swallowed, as they are somewhat, if cumulatively, poisonous in that sort of dose) and fluoride tablets, are all possible ways to do this.

It is difficult to give exact recommendations for dose, due to individual genetic, developmental and environmental differences, as well as to lack of specialist knowledge by myself, but finding out your own water supply content is a good place to start (However, if your diet has a healthy amount of fresh fruit and vegetables, you won't actually be drinking much tap water unless (maybe) you are extremely physically active, so then you are back to trying to get some idea of your food/soil fluoride content).

In any case, the extremes of over-intake of fluoride become visible in mottling of tooth enamel, but trying to titrate fluoridation against tooth mottling is probably not advisable.

Perhaps finding out the recommended range of dentist applied fluoride painting frequency, and choosing the low or high end (or middle) of that range after consideration of your own state of tooth enamel, and finding out some idea of your water and food fluoride content (i.e. through asking your dentist, local council, university etc).

If possible, I would prefer to apply not only fluoride, but also magnesium and calcium, and perhaps other elements as well, via a paint or mouthwash, but I do not know if these are available, or chemically possible in the one mix at the desirable concentrations, because they might precipitate out in solution as the same sort of fairly insoluble mineral as is in our teeth (the protein part of teeth notwithstanding) – perhaps separate applications would be necessary - this is worth asking about, and you may eventually get a more satisfactory response on this than what I have provided here.

Whether the teeth have been fully (there is some small(?) amount of exchange of tooth enamel with surrounding fluids and tissues ongoing throughout adult life) formed and grown into the mouth obviously changes the considerations here, as children with teeth yet to erupt require internal application/ ingestion of fluoride and related minerals, not (or as well as, for those that have erupted) external application.

There are various mixtures based on cow milk protein (casein) on the market for application to the external tooth surfaces for periods of time much longer than for mouthwashes – their comparative merit is unclear.

Chapter 14. Ecology, ethics and animal welfare/ rights

By now in 2010 it seems pretty well agreed that it is better ecologically to consume plant protein sources directly ourselves rather than feed them to animals first and then eat the animals, because only a fraction of the farmland area is necessary that way, and less fertilizer and farm chemical waste pollutes the environment.

However, there is the question of the transport of food from one place to another, such that foods produced from different soils can be beneficially combined in the diet to reduce the likelihood of any mineral or associated vitamin deficiency. Until the stage is reached where all soils are analyzed and their mineral or related deficiencies are corrected by addition of minerals or other (organic, in the C, H, O, N, S sense) soil components, some amount of exchange of foods between geological regions so that diets may be drawn from them seems desirable.

Also, there is the question of the need of developing or poorer countries to sell their products to the wealthier countries in order to develop. Also, the question of what natural resources particular countries have available to them to market to other countries. Until we can get to the stage where we divide up the wealth/resources more or less equally according to need among all people on earth, it seems unreasonable not to buy food from developing countries whose export economy is centered on food.

Although in many developing countries animal foods are seen as a desirable luxury that was previously unavailable due to poverty (and in those countries, where life is harder, life, both human and of other animals, is often cheaper…), in some 'developed' countries there is an increasing consideration of the ethics of the unnecessary slaughter of animals for food.
Vegetarianism has different motivating factors for different people, but a strong component overall is that vegetarians usually don't eat meat because they don't like people to kill or be cruel to animals, and they don't think eating meat is necessary for good health, or even beneficial to health.

As I think I demonstrate pretty conclusively in this book here, this latter is almost certainly true, other than the need for some small amount of vitamin B12, which can be obtained fairly ethically from free-range eggs or some dairy products, or brainless (and therefore painless) animals like oysters and most other mollusks, or nearly brainless animals like insects and crustaceans (which are very probably only barely aware that they are even alive, and almost certainly not cognizant of the difference between life and death, and can be (not always are, possibly) collected and quickly dispatched without them experiencing emotional stress. (And vitamin B12 supplements and injections are readily available.)

In particular, the cage-farming of animals (particular examples being chickens farmed for eggs, and pigs), seems to many, myself (well professionally qualified to roughly quantify the distribution of suffering on the Earth) included, to be one of the cruelest, or maybe the most cruel practice of humankind at the moment and for some decades now.
I am still amazed, or at the very least greatly dismayed if hopefully not rendered utterly hopeless of humans as a species ever learning to overcome their greatest defects, that people can be so mentally deficient as to attempt to justify keeping any animal with a brain and obvious capacity for emotion and need for locomotion, in a small cage for effectively all its life.
In this case, and also pretty much for most humans asked about life in general, the quality of life before death matters much more than whether death comes earlier or later.

It is very probably not psychologically healthy for humans to knowingly participate in the cage-farming of animals, whether directly, or indirectly through the paying of others to do this dirtiest of all works by buying the products (cage-farmed eggs, pork etc).
Surely psychological health requires us to identify the severest and most easily addressed of these practices and put a stop to our support of them, and then make improvements to our practice over time by ceasing support of the next severest and most easily addressed (it seems to me that cage-farmed eggs and pork are the best places to start – for everybody's sake, people should make a very good start by stopping buying them).

Also, it seems fairly obvious to me that unethical treatment of animals induces the unethical treatment of humans – if it's ok to treat animals in the way we do as societies, then it is entirely consistent with that treatment to treat other humans as badly as we can get away with. Those who would argue otherwise should read some very basic psychology about psychological rationalization, the almost universally prevalent human characteristic of employing faulty reasoning in trying to justify our unjustifiable actions or thoughts to ourselves.

Apart from cow milk being an inferior food for humans, as I think I demonstrate well enough elsewhere in this book, there is also the matter of whether the immediate separation and slaughter of newborn calves, and the problems involved in the genetically oversized udders, are ethically acceptable.

Another issue of considerable importance is that school tuckshops and hospitals (their cafeterias, other stores, and vending machines) should not sell unhealthy junk food, because it's unethical/immoral; because unhealthy eating habits acquired during childhood tend to carry on through adult life, and contribute greatly to disease and premature death; and because such sale tends to reinforce the stupidifying and infantilizing effect of junk food advertising on the intellect, as schools and hospitals are an information authority and seen as such, and school is a pervasive influence for much of one's formative years, and hospitals are often visited for very significant health-related reasons.

Chapter 15. The psychology of diet optimisation

No pain, no gain.
However, we each have different amounts of pain/discomfort in our lives from different sources. So it can be difficult to take on some more discomfort in the short term, even if we know there will be less discomfort/ more pleasure in the long-term.
Think about how to best handle the best amount of short-term pain/discomfort in order to best reduce the amount of long-term pain/discomfort in total (including dental and associated pain/discomfort).
Suitable (not negative/silly) rewards in the short-term may need to be used.

The habits of a lifetime are hard to break, but once broken, they are gone, and you reap the benefits/pleasure.

There is pleasure and satisfaction from breaking bad habits in itself – the feeling of successfully controlling oneself for the better, instead of being a slave to bad habits unfortunately acquired.

Don't punish yourself mentally for your bad habits – perhaps the most fundamental (not fundamentalistic…..) law of science is **"Every action causes an equal and opposite reaction, and every reaction requires an equal and opposite set of preceding causes."** Logically, that is to say that at least nearly everything about us is causally predetermined according to the law of cause and effect – we can't help the way we are. However this does not negate the fact that the subjective feeling of being in control and breaking a bad habit will make us feel more satisfied with that, nor does it negate that being very much a good thing for us, and for others, particularly if we can help ourselves and others to focus on the remedy for bad things rather than looking to place blame for them. Better to say **"I've/you've been a fool in this regard up till now, but what have been the causes of this, and how do I/you improve in this regard from now?"**, than to say **"I've/you've been a fool in this regard, and it's all my/your fault, and I/you deserve to feel bad about it and to be punished for it." Forgive yourself, and forgive others**, it's logical, even if giving unpleasant feedback to ourselves or others can be beneficial in the (very) short term and is natural to some extent.

Two steps forward, one step back, two steps forward, better than none forward at all. **The people who sell junk food** are not really interested in **improving your quality of life**, they're interested in **getting your money** and improving their bank accounts, while actually **reducing the quality of your life**. You can **derive pleasure from not being made a fool of** in this way.

The people who sell junk food use **'clever' psychology** to get you to buy their products, by making you **think and behave more foolishly** than you would otherwise. You can **derive pleasure from not being made a fool of** in this way.
The whole barrage of this sort of advertising has the effect of making the whole population think and behave more foolishly than they would otherwise – given the problems we collectively face on the planet, this is not a good thing, because we need all the brainpower possible.

"Hunger is the best sauce" is an old saying that is very true – when we are physiologically hungry, not just **eating out of boredom or seeking more pleasure**, the taste of food is greatly enhanced. Eat only when you are hungry. If you aren't hungry, don't eat, or don't eat much, or eat mostly salad vegetables.

Seek to identify those areas in your life where you are not gaining the pleasure/ satisfaction you could/should, and to identify those parts of your use of food which are negative counterproductive attempts to compensate for that lack of pleasure/ satisfaction. Then act to remedy those circumstances, or to find a more positive productive way of compensating.

There are many easily accessed ways of using time that are more pleasurable and satisfying than eating unhealthy junk food, or eating excessive amounts of otherwise fairly healthy food. These include quality reading, quality TV/ movie viewing, a course of study (TAFE short/long/lifestyle/interest courses/ university), sketching, painting, sculpture, pottery, photography, learning/playing a musical instrument, walking, bicycling, humanitarian activity, etc, etc, etc……. **Don't be using your time so poorly that eating something becomes a way of filling in unoccupied time.**

If you're really more tired than hungry, your body probably really wants rest, and not food - try having the extra rest and not the food, and see if that makes you feel better (It almost certainly will….).

When one eats food that has been **treated to make it much more tasty than natural food**, one's **taste bud receptor molecules become down-regulated** (reduced in number) from the excessive stimulation in the same way that a heroin addict's natural endorphin receptor molecules become down-regulated by the excessive stimulation that heroin causes in making its pleasurable effect. And in the same way, that due to that reduction in number of receptor molecules, natural life (without heroin) is unnaturally unpleasant to the heroin addict, and life with heroin eventually not much more pleasant than natural life is to non-addicts, **natural foods become unnaturally unpleasant to those whose taste buds are accustomed to food that has been treated to make it much more tasty than natural food, and the unnaturally treated food eventually tastes more ordinary**.

Given enough time not eating unnaturally tasty food, taste receptor molecules will become up-regulated (increase in number) again to their natural level, and **natural foods will come to be experienced as very nearly as tasty (that is, very tasty)** as the unnaturally tasty foods were when the taste receptors had been down-regulated.

The same (analogous) thing happens with regard to our psychological perception of the different pleasure associated with eating different foods.

Learn how best to say no to well-meaning friends – explain your decision, try to carefully, gently and gradually **raise them up in their lives, rather than letting them bring you down in your life**.

Chapter 16 Natural and synthetic food contaminants; organic food; parasites

Some important food (constituents / substances / molecules) other than the major nutrient groups are antioxidants, antinutrients, allergens, toxins, carcinogens, irritant biochemicals, pharmacologically active biochemicals.

Some important unhealthy contaminants of foods are: pesticides (herbicides, fungicides, bactericides, antibiotics, vermicides), farm hormones, funguses[i]/moulds, bacteria, parasites, parasite eggs, viruses, toxins, preservatives, artificial colors, artificial flavors, irritant condiments, pharmacologically active condiments, allergens, carcinogens, leached food container metals and dyes, motor vehicle exhaust substances, industrial pollutants, pollutant metals, and plastics molecules….

If cooked food is insisted on instead of raw food, food is generally better for health stewed, boiled, poached or steamed, because roasted, baked, broiled/grilled, fried and sauteed foods have unhealthy molecules in them, including: carcinogens, such as Polycyclic Aromatic Hydrocarbons (PAH), mutagenic heterocycles, benzene derivatives, and nitrites); causes of atherosclerosis such as oxidatively damaged unsaturated fat; Maillard Products (, which incorporate amino acids, notably lysine, that are thereby not absorbable by the gastrointestinal tract ('GIT')); acrylamides, of as yet unknown activity.

It should be borne in mind that raw flesh often carries parasites that can bore into various parts of the body, such as the brain and liver.

Also, raw eggs may carry pathogens, and eating a lot of them can cause a biotin deficiency.

Plant foods that are acquired from near the ground (such as root vegetables and many green leaf, flower or shoot vegetables), such as to be near to animal feces should be washed properly, or peeled, if eaten raw, to remove parasites that can bore into various parts of the body, such as the brain and liver.

Organically grown food is probably better on average than food grown with chemicals, but soil that is deficient in nutrients should have those nutrients enriched. Atoms are atoms and molecules are molecules, regardless of how they are sourced (i.e., metals) or produced (molecules) – it is the surrounding context of other atoms and molecules that may be important, in comparing the different foods that contain them.

Apples and pears are said to be among the worst hit with pesticides and should be peeled.

There is a lot more that could be beneficially written here on this subject, but that will have to wait for a later edition.

Chapter 17 Balanced meal and balanced diet; vegetarian diet and health; caloric restriction and longevity

It seems to me that, in general, on average:

A balanced meal includes a (preferably a mixture of) concentrated protein source(s) such as flesh, eggs, dairy products, nuts, seeds, legumes or grains; a fat source (which is often also a protein source); at least 200 g of green leaf or flower vegetables for an average-sized adult; and root vegetables and fruit (, in the right proportions (, that is, increasing down this list by weight, and such that there is enough root vegetables and fruit (roughly 1-1.5 kg) that the meal roughly meets water requirements (clear urine often enough))).

A balanced diet includes a variety of protein sources, fat sources, green leaf or flower vegetables, root vegetables, fruit, and a source of vitamin B12.
In more detail:
A balanced diet includes a variety of protein sources (such as flesh, eggs, dairy products, nuts, seeds, legumes or grains); a fat source (, which is often also a protein source); an average of at least 500 g / day of green leaf or flower vegetables; root vegetables and fruits; in the right proportions (, that is, increasing down this list by weight, and such that there is enough root vegetables and fruit (roughly 2-3 kg/ day) that the diet roughly meets the water requirements); and in the right schedule (, that is, in balanced meals). Also, and a source of vitamin B12 (e.g. flesh/insects, eggs, dairy products, vitamin supplement)

Is a vegetarian diet healthier than an omnivorous diet?
A vegetarian diet is not necessarily healthier than an omnivorous diet.
For personal health it is more important that the diet have protein, fat, carbohydrate, vitamins, minerals and antioxidants in the right proportions and combinations, and be not much damaged by any cooking process (roasted, baked, broiled/grilled, fried and sautéed food is oxidatively damaged food is oxidatively damaged food is oxidatively damaged), than whether or not the diet is vegetarian or omnivorous.
That is to say that it is more important that the diet, and the meals, be composed of a smallish amount of protein and fat foods (such as legumes, nuts, seeds, grains, eggs, dairy products and meats), a largish amount of green leaf vegetables, and the largest amount of fresh fruits and root vegetables (with or without grains, by caloric content not by weight) as the major carbohydrate source.

Maybe the single most important dietary factor for long life is whether one eats and absorbs too many calories worth of food relative to the energy one requires.

The criteria that any member of the general public needs to understand and apply to their assessment of themselves is actually very basic and low-tech - If one cannot see some muscle and ribs under the external body fat, then one is too fat, unless one is expecting a period of unavoidable food shortage sometime soon, or in pregnancy, when a *little* extra body fat may be desirable.

Just very roughly, it seems to me that the difference to lifespan is likely to be at least 10 years in lifespan between good leanness and having 10 kg of excess fat, and more for having more excess fat.

Excessive leanness can be assessed on the basis of physical weakness, especially in walking or running, which are very basic natural things that all humans have evolved to do, and need to do, irrespective of our preferences for other sorts of physical exercise.
And by hormonal derangements – in particular growth retardation, and retardation of menarche or cessation of menstrual period.

Chapter 18 Individual health vs. population health - deleterious mutations and the effect of medical treatments and nutritional supplements on their retention and prevalence rate in the gene pool; Individuality within species, populations

If a disease is not always fatal, then it is multifactorial in nature. In infectious disease, for example, not only the inoculation load (the number of infectious disease viruses, bacteria or parasites that make up the incoming load of infectious organisms) but other factors also will contribute to whether the infected person lives or dies.
Some of these other factors will be environmental, for example, stress, nutrition, and chemicals.
Genes are other factors that contribute to the outcome of life or death for the person with the disease. And *many* genes will contribute to the outcome, some more strongly than others.
Deleterious mutations are continuously coming into the human gene pool in a fairly random fashion – all animals/organisms have a particular mutation rate related to the characteristics of their DNA and DNA repair mechanisms.
When disease kills a person before they have reproduced, on average, the death removes a genetic code that overall has more deleterious mutations than the average genetic code among people making up the population. It removes more of the genes more strongly involved in the disease course, but it also removes some of the genes involved less strongly in the course of the disease.
So the net amount of deleteriously mutated DNA in the population is determined by the balance between the rate that random mutations come into the gene pool, and the rate at which death (i.e., from disease) before reproduction removes deleterious mutations from the gene pool.
When any intervention, be it medical or nutritional, results in preventing disease from killing people before they are able to reproduce, at a population level the result is that the gene pool comes to include more deleterious genetic mutations.
The more such interventions are successful, and the longer they go on for, the more deleterious mutations come to be included in the gene pool.
If we go on generally intervening to prevent disease killing people before they are able to reproduce, without genetic engineering of humans, eventually the human genetic code will become so riddled with deleterious mutations that it will become a major and eventually uniformly fatal disease(s) in itself.
My personal preference, and I doubt cessation of intervention to prevent disease killing people before they are able to reproduce will become the popular preference, is for ongoing intervention plus genetic engineering of humans.

Now, different people have different geological and dietetic evolutionary backgrounds. We should not make too much of this, but it probably applies more to humans than it does to other animals.
As dealt with elsewhere in this book, some lines of humans had a high vitamin B12 intake in those (evolutionarily relatively short) periods of our history when we ate a lot of animal food and less plant food (including the colder portions of the most recent ice age). So, even though later on animal food consumption was significantly decreased back down from those evolutionarily uncommon highs, some lines of the human race will have retained through natural selection some deleterious mutations decreasing the ability of some few individuals to absorb, transport or/and store vitamin B12 (In the same way that we totally lost the ability to synthesize vitamin C, which most other animals still have, because evolutionarily our diet contained such a high amount of vitamin C).
So, have the Swiss, or people of Swiss ethnic background, from living in the iodine-deficient mountain regions, on average, less iodine absorption-, transport- and metabolism-related deleterious mutations in their gene pool, because they were more severely selected against in natural selection than in populations elsewhere? Have they,

from eating the dairy foods that genetically prolonged their lactase function into adulthood, more calcium absorption-, transport- and metabolism-related deleterious mutations in their gene pool, because they were less severely selected against in natural selection than in populations elsewhere?
What about the coastal peoples eating marine mollusks and crustaceans with their high load of nutrient minerals – would nutrient metal absorption-, transport- and metabolism-related 'deleterious' mutations in their gene pool, actually be protective from the excessive metal nutrient levels found in these foods?
Would people with an evolutionary background in the selenium deficient Keshan area of China (intake 3-22 ug Se / day) fare extra poorly if living in the selenium-richest Enshi area of China (intake 3,200-6,700 ug Se /day) or at the coast and eating a lot of marine mollusks and crustaceans, from selenium toxicity?

It seems that many of these questions may soon be answerable through genetic epidemiology, and answerable at the individual level with genetic testing, such that people will be able to know what changes in their nutrient intake they might benefit from.

In the meantime, on the basis that deficiency is more likely to be a problem than toxicity from excess, I think that on average a vitamin and nutrient supplement of very roughly 1/2 of the RDAs might be a good idea for all people to consume, once a day or two, with a meal, to account for unknown gene differences and differences in produce nutrients due to geographic regions of origin, and that individuals should take their genetic evolutionary history into account where feasible in adjusting their intakes of particular nutrients.
However, the vitamin and nutrient supplement should not be seen as a replacement for healthy whole foods, because many studies using supplements of isolated nutrients have found that the benefits of the nutrients in that context have been nowhere near as great as the benefits associated with the whole food sources of the same amount of the nutrient.

Chapter 19. Eating for weight (surplus body fat) loss, blood sugar and fat levels, fat metabolism and body fat mobilization for fuel

In nature a meal may not always contain significant fat or much carbohydrate, but the optimal meal may well contain a certain proportion of protein:fat:carbohydrate.
It seems to me that the best proportion is roughly: only enough protein to meet our little protein requirements of 20 g/day of pure amino acids in the right proportion for an average adult (minimizing toxic nitrogenous waste from metabolizing excess protein); the amount of fat that comes with the meeting of those requirements from protein sources that include a good proportion of oily nuts and seeds alongside whatever legumes and cereal grains may also contribute to the protein source mix; and the balance of our caloric (energy) requirements from (raw) starchy root vegetables and fruits (which would therefore meet also roughly all our water requirements) (note that the fructose content of fruits is partly what makes them generally low-GI (Glycemic Index, measure of the speed of absorbtion of sugars into the bloodstream from the gastrointestinal tract) foods, which makes them, as for many other good reasons as elaborated on elsewhere here, and overall, a good choice as a major portion of the diet, contrary to what the now fairly popular if not quite so infallable Mercola website states (August 2011) in failing to differentiate fruit from fructose in commercial corn-syrups and their products, which is not a trivial failure.

Such a mixture provides a slower uptake of dietary carbohydrate from the digestive tract, and some accompanying fat, and one effect of this is a feeling of well-being due to that steady drawn-out delivery of fuel to the body. Also, because fat metabolism is activated, when the carbohydrate supplied by the meal is no longer available for ready fuel use, fuel use can switch to depot fat (in our fat stores) more smoothly and with less sense of lethargy and hunger or craving for food than would occur if the meal had had no fat.

It is important to understand that my statements regarding this feeling of wellbeing and its proposed reasons are not negated by reports of a feeling of wellbeing reported by people on the Atkins high protein diet.
The feeling of wellbeing reported by people on the Atkins high protein diet is due to their blood sugar curves previous to the Atkins diet having been wildly fluctuating through (refined or high Glycemic Index) carbohydrate being ingested without being accompanied by any fat to slow down its absorption into the bloodstream, or to prime fat metabolism to make access of depot fat for use as energy easier for the body.
Refined or high Glycemic Index carbohydrate being ingested without being accompanied by any fat to slow down its absorption into the bloodstream, in the context of a low-fat diet, causes initially rearing blood glucose levels, causing rearing insulin levels, which cause subsequent plunging blood glucose levels, in the presence of low blood fat levels, leaving regular periods of time when both blood glucose and blood fat and fat metabolism are low, causing fatigue and hunger, the latter causing overeating in an attempt to regain some short-term blood energy fuel.
Also, the Atkins diet gives a slow and relatively constant feed of fat into the bloodstream that blunts the hunger that occurs when (refined or high Glycemic Index) carbohydrate is ingested without being accompanied by any fat to slow down its absorption into the bloodstream.
One big problem with the Atkins Diet is that in the long run, atherosclerosis is very likely to occur from the saturated fat, cholesterol, and oxidatively damaged unsaturated fat in which it is high.

The one valuable piece of information that can be derived from the Atkins Diet experience is that having a *totally* fat free diet is *not* the best way to shed excess body fat, because fat metabolism is reduced with every new load of dietary carbohydrate, and when the carbohydrate is used up the blood will be low in not only sugar but also fat, such that before depot fat is accessed for energy, enough time goes by with the person in a state of both low blood sugar and low blood fat/ fat metabolism that they feel substantially weak, irritable and hungry.

So my suggestion for diet modification for the loss of excess body fat is to reduce the dietary carbohydrate, and to keep the protein and fat content the same as I recommend for the general diet (per day): 100-200 grams combined weight of nuts, seeds, cereal grains and legumes (with or without animal foods: flesh, eggs, dairy products), at least 500 g of green leaf, flower or shoot vegetables, and enough fruit and starchy root vegetables to not quite make the total energy/calorie intake equal to the total energy/calorie use.
In general, the food should be raw wholefoods.
And it should be properly chewed.

The foods in a meal should be eaten generally in the order described in chapter 13 here above, to optimize the delivery of sugar and fat from the digestive system into the bloodstream, to optimize dental health, and to optimize digestion of proteins:
Some green leaf, flower or shoot vegetables first for dental health, high protein and fat foods next (allowably interspersed with more green leaf, flower or shoot vegetables) for optimal digestion by concentrated digestive juices over a sufficient time, then starchy root vegetables and fruit, and a little more green leaf, flower or shoot vegetables last to knock the residue from the other foods off the teeth for dental health (followed by rinsing, brushing and picking the teeth after every meal, of course).

In general there should be only two meals per day, brunch as close to noon as feasible and tolerable, and dinner in the evening after some session of physical exercise.
Three meals otherwise.
Nothing but water (if necessary) in between meals.

Chapter 20 Homocysteine, vitamin B12 and folate and associated foods, atherosclerosis

Homocysteine ("plasma total") at levels of 200-500 uM in a few severe genetic metabolic diseases is associated with premature atherosclerosis, and likely to be a major part of the cause of it in patients with these diseases, due to interference with disulfide bonding in (particularly the formation of) structural proteins in tissues, themselves already (in most cases, the main disease type) compromised by the reduced cysteine levels caused by the block in metabolism of homocysteine to cysteine, one of three available metabolic pathways for converting ostensibly useless homocysteine back into something useful (cysteine or methionine).

Various authorities have posited that with a population normal range given as 8-12 uM, levels of 12-20 uM, which are associated with atherosclerosis, are a contributory cause of that atherosclerosis. Other authorities, myself included, think it more likely that the homocysteine at this level is only a marker for other things that are the real causes of this atherosclerosis – lack of folate and fruit & vegetables – and disease process components.

Folate, found in higher levels in plant foods, and vitamin B12, found virtually only in animal foods, are equally involved in reducing homocysteine levels through a particular metabolic pathway. Vitamin B6 is involved in another pathway, but it is plentifully present in both plant and animal foods.
Exclusive vegans with no vitamin B12 supplementation may have homocysteine levels as high as 100uM.
There have been no reports in the scientific journal literature of vegans and vegetarians having more atherosclerosis, usually, in fact, the opposite, of them having less atherosclerosis.
Studies to date have incompetently failed to assess whether a case of raised homocysteine level was due to low vitamin B12, from eating less animal food, or due to low folate, from eating less plant food. Given the large amount of reference there has been in the mass media to animal fat and fruit and vegetables regards atherosclerosis, I will not insult the reader's intelligence by spelling out why it would be crucial to determine whether raised homocysteine levels in atherosclerosis cases were due to low vitamin B12 or due to low folate. I will however mention that animal foods are themselves a marker for greater oxidative cooking damage to foods (i.e., their unsaturated fat) and plant foods are themselves a marker for less oxidative cooking damage to foods.

My advice to vegetarians is that if your homocysteine levels (say 12 – 20 or 30 uM) are due to low vitamin B12 levels, and your diet is composed mostly of raw or not very heat-damaged plant foods, that level of homocysteine is very unlikely to contribute anything other than maybe a tiny, tiny, utterly insignificant part of any atherosclerosis that you may develop (or have had already). I recommend not worrying about it at all.
If such levels were due to low consumption of plant foods and therefore low folate, then worry about atherosclerosis is entirely appropriate and dietary change (*not* merely folate supplementation) is very much called for.
Exclusive vegans with higher levels of homocysteine, up to 100 uM, are very probably much more at risk of damage to their nervous system, and maybe of getting some cancers related to global DNA hypomethylation, than they are of serious atherosclerosis, but given some other bad factors (genetic or environmental) atherosclerosis could become problematic despite the high levels of protective factors in diets with a lot of raw fruit and vegetables. Unless a person wants to participate in a somewhat risky personal experiment and contribute some more data, I recommend that they should have enough vitamin B12 supplementation to bring blood B12 levels into the low 'normal' range, and to bring their homocysteine levels roughly under 20 uM.

Chapter 21 Afterthoughts: The China Study; Ornish's and Esselstyn's heart disease reversal diets; Inflammation Index and other potential refinements of diet

The China Study (Campbell & Campbell, 2006, Benbella Books) warrants a much more detailed treatment than that I provide here, which will be forthcoming. In very brief, a major epidemiological study comparing many different regions of China (constituting a 'natural experiment' due to the little immigration into and emigration out of the different regions, with their different diets) found that even at lower levels of animal-food consumption, lower consumption of animal-food was associated with lower heart disease and cancer. There is doubtless much of great value in this study. However, it does not consider the oxidative damage to unsaturated fat that various methods of cooking cause. (The result is that the C=C double bonds open up to include oxygen atoms in epoxides and peroxides, which are the same type of chemicals that macrophages cause when they lyse things with the 'oxidative burst' of hydrogen peroxide that is the mainstay of their activity, these chemicals then attracting other macrophages to the site, which is a main part of the process of the generation of atherosclerosis (arterial hardening and plaqueing)) Nor does it consider that to some extent animal products are actually a marker for more oxidatively damaging cooking methods (such as frying, roasting, grilling, sautéing).

Both Ornish's (1990, Dr Dean Ornish's Program for Reversing Heart Disease, Ballantine Books) and Esselstyn's (2007, Prevent and Reverse Heart Disease, Penguin Books) warrant a more detailed treatment than that I provide here, which will be forthcoming. While they offer much of value in improving on the traditional western diet, unfortunately they too do not consider the oxidative damage to unsaturated fat that various methods of cooking cause, and that to some extent animal products are actually a marker for more oxidatively damaging cooking methods. Consequently they have un unreasonably severe and inadequately differentiated negative view of *all* dietary fats.

I only make a bare mention of an interesting development/ possibility currently part of the SelfNutritionData website: the Inflammation Index, derived from an algorithm I have not acquired the details of, which incorporates various food-related factors to produce an overall number quantifying the potential contribution of the food to the body's general state/level of inflammation. My general recommendations on diet composition would give a very favorable overall Inflammation Index. Something to find out more about?...

Marine versus land polyunsaturated fats are another issue, and, other than having read some studies that did not convince me due to methodological shortfalls/ publication bias (studies finding some difference in their results get published more often than studies finding no difference in their results, at least for smaller, investigatory studies), and I will say no more for now other than that we are a land animal, not an aquatic or marine animal, other than perhaps to a small extent those people coming from a long line of coastal or island ancestors eating a high proportion of seafood in their diet – and seafood itself differs from the coast to the deep sea, and by latitude... And that the question of the suitability of a food for humans must be resolved through accounting for multiple factors in the particular food, in the meal of which it is to be a part, and in the diet of which it is to be a part – which is far and away most often not at all done... More study required.

Would eating only the minimum necessary amount of protein, and ensuring it was as fully digested as possible by chewing it fully (and mechanically grinding/pasting it before eating if necessary), and by eating it near the beginning of a meal, and eating the meal slowly, result in less polypeptides able to affect the function of the immune system getting through the always somewhat permeable intestinal walls into the bloodstream, and might this be to the benefit of health conditions involving overactivation of the immune system? (Some inflammatory diseases respond quite well to fasting...)

Right of reply offers to Professors/Doctors whose work has been critiqued in my work here, and invitations to comment, and all such replies and comments

In general, the right of reply to criticisms, and an invitation to comment on my work here, is extended to authors contributing to the book Evolution of the Human Diet The Known, the Unknown and the Unknowable (2007), edited by Peter Ungar. This is mostly because I take it to be the overall best collection of authoritative opinions on the subject up to that time (and search of the Journal of Human Evolution from 2006 to the present (2010) revealed little to add to it).
I undertake to contact at least one author of each of the chapters (nearly all) of the book that I have reviewed and critiqued here, well before the end of November 2010, and hope that they will pass on the communication to their coauthors.
I may contact other authors whose work I have dealt with here, but many will have retired and not be easily contactable by email (those whose email addresses I could not acquire readily from the internet have a "(?)" after their names below – if anyone is able to pass this on to them, please do so) – all of them are similarly accorded right of reply and extended the invitation to comment – I hope that those I contact will contact those that they know should have been contacted but that I have not explicitly noted here.
There will be several others extended an invitation to comment, and if any of those contacted know of anyone who is eminently suitable to do so, that I have not named, please pass the communication on to them.
I think this is all that could be reasonably asked of me by way of subjecting my work here to authoritative scrutiny – and I also think it surpasses what often constitutes peer (i.e., network…) review.

So, such authors will be provided with a copy of this work in early November 2010, and any response they provide at any time thereafter, including an indication of intent to respond at some stated or unstated time in the future, will be added here, verbatim as provided, no matter how negative, immediately below within 2 weeks of my receiving it.

Any response from me, here, to such response from any of them will likewise be communicated to them within 2 weeks of my making it. And any response from them to that response, likewise. And so on until all novel points, rebuttals, arguments or theories have been included accordingly here.

To start with I will place a very tentative limit of 100 pages in total for all of us together, expecting this to be more, and probably much more, than ample, and ask respondents, regardless of how savagely stung by my critique of their work, to please keep responses concise, and to a maximum of around your own fraction of the 100 pages, at least to start with, giving all the others the same opportunity for timely responses, to avoid me having to post them elsewhere with a web address for them only here.

Of course you may all decline to comment at all – that is your choice, but I have made what offer I could reasonably be expected to…

Those I will contact, if I can locate their email addresses from the web, are, starting with those from the Contributors list in Ungar (2007): RJ Blumenschine of Rutgers U, HT Bunn of U Wisconsin, L Cordain of Colorado State U, DJ Daegling of U Florida, SB Eaton of Emory U(? other email), FE Grine of Stony Brook U, AG Henry of George Washington U(? now sent), JE Lambert of U Wisconsin, J Lee-Thorp of U Bradford(? now sent), WR Leonard of Northwestern U, PW Lucas of George Washington U, CR Peters of U Georgia(?), BL Pobiner of Rutgers U &

Smithsonian I, AL Rector of Arizona State U, KE Reed of Arizona State U, ML Robertson of Northwestern U(?), D de Ruiter of Texas A & M U, MJ Schoeninger of UCSD(? now sent), J Sept of Indiana U, JJ Shea of Stony Brook U, JJ Snodgrass of U Oregon, M Sponheimer of UCB, MF Teaford at Johns Hopkins USchMed, PS Ungar of U Arkansas, A Walker of Pennsylvania State U(?), B Wood of George Washington U(? now sent), R Wrangham of Harvard U;
D Chivers of U of Cambridge, C Stanford of U SoCalLA;
G Teleki ? (retired?), RSO Harding ? (retired?)
K O'Dea of U South Australia, R Hughes of U Sunshine Coast, S Sommerset of Griffith U, B Desbrow of Griffith U, G Marks of U Queensland;
L Walsh of U Queensland, D Roberts prosthodontist,
ME Shils of Cornell U, M Shike of Cornell U, AC Ross of Pennsylvania State U, B Caballero at Johns Hopkins U, RJ Cousins at U Florida;
P Insel of Stanford U(?), RE Turner of U of Florida, D Ross of Cal. Inst. Hum. Nutr.;
E Whitney of Florida State U (formerly?), SR Rolfes of Florida State U (formerly?).

As at Nov 22 2010 (and now at Feb 18 2011), responses have been received from 4 of the above people, in the order I've presented the dialogues below:

Professor David Chivers of U of Cambridge,

Professor Kerin O'Dea of U of South Australia,

Professor Laurence Walsh of U of Queensland,

Professor Richard Wrangham of Harvard University.

The following was a response (18 November 2010) from Professor David Chivers of the University of Cambridge to a communication from myself offering right of reply and inviting comment, on 11 November 2010:

David

Many thanks for your mail and very impressive book. I have skimmed through
but no obvious comments arise, but I am currently overwhelmed by teaching
and other duties

best regards

David

The following was a response (14 November 2010) from Professor Kerin O'Dea of the University of South Australia to a communication from me offering right of reply and inviting comment, on 11 November 2010:

Thank you David. I won't comment – I think we have different perspectives on this.

But thank you for giving me the opportunity.

Regards

Kerin O'Dea

I replied on the 14th November as follows:

Hello Kerin,
That's your choice, but the point is very much to attempt to sort out the collective meaning of various evidence considered from various perspectives, to enable a clearer view of which opinions and theories are more likely to be correct. Disagreements of opinion (and there are many in this field) are unlikely to be satisfactorily resolved merely through hand-holding and exchange of warm fuzzy sentiment, network-building and grant receipt-facilitating though these common research methodologies may be...
In particular, if you like to detail, as little or as much as you like, in what way(s) you think my analyses and assessments might be incorrect, well, perhaps that's what science is best about really, if the aim is to reduce the suffering of those worst afflicted by getting as close to the truth of the relevant questions as possible, and the offer will remain open to you more or less indefinitely - again, full verbatim reproduction of your reply, no matter how negative, at the end of my book, within two weeks of you supplying it.
Regards,
David Vance.

And there is no response to this as yet.

The following was a response (14 November 2010) to a communication received from Professor Laurence Walsh of the University of Queensland, Australia, on about 9 November 2010, which it seems, assuming Hotmail is not at fault, I clumsily deleted by mistake (but which Prof Walsh would still have in his Sent email list, though there is fairly obviously nothing of importance that does not appear in the reproductions of other communications that do appear below):

Hello Laurence,

Thanks for your response, although I sent you the book prematurely in an unfinished form mistakenly in a multiple-address email - nevertheless, the substance you replied to has not changed significantly.
Sorry to have delayed my response to thank you for your response and reply to it.
Your original response, has been deleted by something (I hope not myself), yet is not available from the Hotmail Deleted file as the other things I know I've deleted recently are...

Anyway, you noted that cheese eaten at the end of a meal can be beneficial because it has buffering of acidity and calcium provision functions (correct me if I'm wrong).
My response to that is that cheese is probably an inferior food in any case, due to excessive (e.g. four times the human (milk, added after the fact on 19 Nov)) protein:energy ratio, and saturated fat and cholesterol content, also, that the teeth should be properly cleaned after eating anything (things should not be eaten in between meals), so buffering and calcium content is much more useful at the start of a meal than at the end of it. Also, that for best digestion proteinaceous foods are best eaten near the start of a meal, so that they can be in the stomach long enough and in a less diluted gastric juice (acid and enzymes) medium, and so that they can slow the release of carbohydrate (eaten later in the meal, but still going into the mixture of stomach contents coming from the whole meal) into the small intestine, giving rise to best blood combined carbohydrate and fat levels.

Also you stated that the salivation response to sour foods is greater than it is to chewing. Thanks for that, and just as well too, with the acidity of the sourness attempting to eat into the enamel...

I'm happy to take any further responses from you, as noted.

Regards,
David Vance.

Professor Walsh's response to that was:

Thanks David

like all things there is a best sequence hence the proteins then carbohydrates then the cheese – at least in terms of oral health!

good luck with the book

kind regards

Laurence

My response to that was:

Thanks Laurence,

Beyond noting that my aim is to integrate all aspects, oral and other, of health, I've nothing further to add.
Anything further you add will be included, the offer being open more or less indefinitely.

Regards,

David Vance.

The following was received from Professor Richard Wrangham of Harvard University, on 12 November 2010, other than my words in response in square brackets (DV) [Richard, abc … xyz.] being interspersed among his words now here in quotation marks (RW) "abc … … xyz" and the resultant below being sent back to Prof Wrangham. Corrections of my own words made after the facts of the communications are identifiable as "(correction: abc … xyz.)" – it can readily be seen that none of these would alter the substance of the discussion much, although the omission (in haste, not due to lack of knowledge, as my book itself would make clear) of legumes from the food types is something I should have been corrected on. RW's refs. to my page numbers are now +2:

(RW) "Dear David Vance,
Thank you for your courtesy in inviting a reply to your in-depth analysis. You have certainly read a lot of material!
I note that on pages 45, 172 and 387 you refer to my "statements here and in his book Catching Fire How Cooking Made us Human (2009), that humans cannot meet their caloric requirements from raw plant foods", and you propose that I am wrong. This raises two points.
(1) I don't remember saying that "humans cannot meet their caloric requirements from raw plant foods", but maybe you can find somewhere I did so. If so, I apologize for my mistake. Actually I know perfectly well that people can survive and thrive on a raw diet, provided that it is a modern urban context using the global food resource, domesticated foods, etc. What I remember saying is that (a) people who eat 100% raw foods often do not meet their caloric requirements; (b) no humans are known to have lived successfully off 100% raw foods in the wild; and (c) based on the available evidence so far, humans would not be able to live successfully off 100% raw foods in the wild. I would be most interested if you have evidence against these points."

(DV) [Richard, in chapter 16 of Ungar (Ed, 2007) you say the following:
"(p315 lines 1-2) It is also known that a diet of raw plant food creates substantial energetic problems for humans under even the best conditions. ; (lines 25-26) Second, even under the most favorable conditions people who attempt to restrict their diets to raw

food do not thrive. ; (41-3) In sum, the sparse current evidence suggests that raw-food diets produce inadequate energy for humans even under excellent conditions, at least when the diet is dominated by plant items. (p316 lines 1-3) Third, even the most committed raw-foodists find it difficult to keep to their regime because they are consistently hungry even when they eat as much as possible." All of these exact quotations of you appear on p168 of my book. "(p316 line 15) Humans are not known to be able to survive on raw food, which suggests that …" which exact quotation of you appears on p173 of my book.
Furthermore, chapter 1 (Quest for Raw-Foodists) of your (2009) book, Catching Fire How Cooking Made us Human (which book I will deal with shortly in a separate review), is devoted to attempting to establish just this fallacy, which very much constitutes one of the crucial bases for your theory.
Furthermore in noting failure to meet caloric requirements it is necessary to different (correction, differentiate) cases of anorexia nervosa from the whole group, or the analysis will be fatally flawed.]

(RW) "You very sensibly say that a diet based around oily nuts could provide adequate calories. But you doubtless realize that no wild habitat has been found where such a food source is available for sufficient duration to sustain people. It would be lovely to find one, and the question would then be whether such a habitat would ever have been sufficiently widespread to have evolutionary significance."

(DV) [Richard, I did not say that a diet "based" on or "around" oily nuts could provide adequate calories – I have made it abundantly clear in my book that a diet including oily nuts and oily (and other) seeds and cereal grains and legumes and starchy root vegetables and fruit, even allowing for a good serve of green leaf vegetables alongside, would provide adequate calories – and that, obviously enough, the proportions of these in meals and in the diet would vary, including by season – and as for other animals, seasonal surfeits and deficits apply.
As to what habitats have been found providing a sufficiency of these, you fail to note that the point is not the lack of habitats found, the point is the lack of any attempts to trial a (sensibly-proportioned, unlike some of the very defective studies you've cited in chapter 1 of your (2009) book…) raw-food diet of wild foods in any habitat. Lack of evidence is not the same thing as evidence of a lack. Gatherer-hunters use fire for many reasons, not the least of which is for technological/skill/magic social status, among flavor and temperature reasons.
Also, the knowledge of how to utilize the wild foods available is obviously not to be assumed in setting up any study, as selecting appropriately from among a complex variety of foods is obviously the reason the orangutan has such a large brain, as it obviously doesn't have such a large brain for social interaction – and the wild human/hominin situation would be likely to have even more dietary complexity, including the use of tools from the earliest of times.
Furthermore, you fail to note that the large majority if not all of the richer habitats have been taken over by agriculture, leaving only more marginal areas in a wild state.]

(RW) "(2) The repetition on pp 45, 172 and 387 suggests that your book is still in rather an early stage. Would it be more sensible to invite comments when you are nearer completion?"

(DV) [Richard, the book is in exactly the form that I want it. Repetition of crucial points in appropriate contexts is not at all a bad thing (and at times unavoidable, as something may be relevant to a plurality of topics), and where you find it, it is quite deliberate.]

(RW) "Good luck with your project."

(DV) [Richard, thank you, and good luck with the evolution of your own knowledge in the area of human diet, and I hope I may be, by way of my book, of some assistance with that.]

The following response to the above communication was received from Professor Richard Wrangham of Harvard University, on 16 November 2010, other than my words in response in square brackets (DV) [Richard, abc … xyz.] being interspersed among his words now here in quotation marks (RW) "abc … … xyz" and the resultant below being sent back to Prof Wrangham:

(DV) [Richard, in response to your communication below, as before, my responses are interspersed in your communication in square brackets, as this sentence is.]

(RW) "Comments re David Vance

Email, Nov 15 2011

"Richard, in chapter 16 of Ungar (Ed, 2007) you say the following:
"(p315 lines 1-2) It is also known that a diet of raw plant food creates substantial energetic problems for humans under even the best conditions. ; (lines 25-26) Second, even under the most favorable conditions people who attempt to restrict their diets to raw food do not thrive. ; (41-3) In sum, the sparse current evidence suggests that raw-food diets produce inadequate energy for humans even under excellent conditions, at least when the diet is dominated by plant items. (p316 lines 1-3) Third, even the most committed raw-foodists find it difficult to keep to their regime because they are consistently hungry even when they eat as much as possible." All of these exact quotations of you appear on p168 of my book. "(p316 line 15) Humans are not known to be able to survive on raw food, which suggests that …" which exact quotation of you appears on p173 of my book."

Thanks for finding these. In case there is any misunderstanding, I hope you appreciate that my statements (which as you will know are based primarily on the Giessen raw-food study, albeit supplemented by every other study I found) refer to populations. You sound to be a very physically fit individual who does well on a raw diet, and you write almost as if you are personally affronted by the idea that people do not do well on raw food diets in general!"

(DV) [Richard, if there is a tone of affront it is less personal and more due to:
The affront felt by a scientist committed to the fullest rigor in science when the appropriate degree of rigor has obviously not been applied to producing a work of science, particularly when that work is by someone who holds an academic position the prestige of which is not backed up by the quality of the work, and particularly when the author manifests assumptions of having a superior grasp of a topic on which it seems likely to me that my own understanding is in fact superior, my lack of a prestigious (or any, as yet…) academic position notwithstanding;
The affront felt by someone who sees publication of poorer work than their own, achieved not by scientific merit, but rather by network connections, when their own work goes unpublished due to not being part of such networking, particularly if it points out important flaws in the networkers' work;
The affront felt by a worker in health science working towards the reduction of the suffering of those beings (of whatever species) on earth who are suffering the most, to the waxing lyrical about dietary habits (born far more of cultural evolution than physiological evolution), the pandering to which via the cage-farming of animals is one of the very greatest and most easily preventable sources of suffering on the planet, couched in pseudoscience born of culture not of reason;]

(RW) "At any rate, please make clear in your book, as I hope is clear from a full reading of the passages that you cite, that I was summarizing the data on average."

(DV) [Richard, no that isn't clear at all – all of your statements there, in the full context of all of what surrounds them, are clearly made in the absolute – there are no references to exceptions or averages – and clearly you are very wrong, and as a scientist should admit that you are wrong, and very wrong, because it is a matter crucial to your theory…]

(RW) "(To avoid the kind of problem that you have drawn attention to, perhaps I should have written "It is also known that *on average* a diet of raw plant food creates substantial energetic problems for humans under even the best conditions." And so on for the other lines.)"

(DV) [Richard, no, that would not do either – if one is talking about the best conditions, the only way that could be sensibly interpreted is that the diet is suitably composed with respect to proportions made up by the different food groups: nuts, oily seeds, cereal grains, other seeds, starchy root vegetables, (calorific) fruit, and green leaf vegetables – any other meaning would be plainly silly and irrelevant – you were wrong, and should admit you were wrong…]

(RW) "Your general tone is that my conclusions were not merely wrong but somehow biased and/or corrupt."

(DV) [Richard, it does indeed seem to me that, like many other scientists who apparently have not a good grasp of psychology, such as might be applied to understanding their own mental processes regarding their cherished enculturated habits better, you are indeed biased by your cherished psychosocially and culturally programmed habits of carnivory and eating cooked food. I note you haven't even got as far, in all your years of work as a primatologist, and now venturing into human nutritional science (a competency background of mine…), as to simply try *yourself* various *sensible* formulations of raw food diet…]

(RW) "But do you know of any studies that contradict the findings of the Giessen study? I did not you citing any such research. Sorry if I missed it – please send if possible."

(DV) [Yes, my own personal 20-year experiment, as detailed in chapter 10 of my book. Sorry if you missed it… Do you think it would achieve publication in the 'peer-reviewed' scientific literature without me being part of some little network?....]

(RW) " "Furthermore in noting failure to meet caloric requirements it is necessary to differentiate cases of anorexia nervosa from the whole group, or the analysis will be fatally flawed."
If a significant proportion of the subjects in the Giessen study were anorexic, yes, that would be a problem, I agree. The care that the Giessen researchers took, and the consistency of their results (and other studies) in finding caloric shortages among raw-foodists would make a hidden role for anorexia surprising, but it is theoretically conceivable."
(DV) [Richard, having looked at the journal articles (Koebnick et al 2005 and the abstract for 1999) and a website by the authors, it is obvious that your ability to make authoritative comments on the "care that the Giessen researchers took", doubtless due to your lack of epidemiological and population research methodology background and competence, is none too good – there is no mention of the Giessen researchers screening out anorexia nervosa patients, and in fact in the population there is a continuum from milder (under-)eating disorder cases through to more severe ones, and the Giessen methodology for acquiring their subjects would have not at all removed any other than the most severe staggering cases of (under-)eating disorder from the subject group – furthermore, the association with underweight is further confounded by the fact that roughly 50% of the subjects took up the raw food diet as a way of dealing with a pre-existing disease.
Furthermore, regards the Evo Diet Experiment, as mentioned in your (2009) book, it is the most basic of knowledge that salt retention (the need for retention in the body of more water to hold excess table salt (NaCl, from the usual western diet) in solution before it can be excreted, actually) would account for maybe most of the loss of body weight when it was no longer necessary on the Evo Diet – failure to address this aspect indicates a serious lack of competence in the science of physiology.
Furthermore, consistency of results means nothing if all are making the same mistakes, which is not a very rare event, actually…]

(RW) "On the broader topic of the reliability of the information on the effects of a raw diet I take your general point that this area is not very studied yet. I completely agree with you that much remains to be learned. In my attempt to review this area from as many directions as possible I found what I considered some remarkable evidence all pointing in the same direction, but we still need a prospective study in which individuals start on a cooked diet and are then medically monitored as they eat increasing amounts of raw food in their diet. In my lab we are thinking about trying to design such a study in my lab, but it is a major undertaking – and my main work is as a primatologist. Hopefully the ideas presented in *Catching Fire* may lead others to undertake such studies even if my own group cannot find the resources."

(DV) [Richard, it's actually a very *easy* and *in*expensive undertaking – just do it yourselves! Gain the basic competence in nutrition necessary to formulate the appropriate composition by food groups of the diet, and eat the diet yourselves! If you are unable to gain that basic competence in nutrition, I have it and am available to help you with that, for a modest fee and the accordance of authorship in the doubtlessly resultant journal article…]

(RW) " "the point is the lack of any attempts to trial a … raw-food diet of wild foods in any habitat."
Yes! I greatly look forward to such a study (and have tried to encourage some efforts in that direction).

"selecting appropriately from among a complex variety of foods is obviously the reason the orangutan has such a large brain"
You would find a lot of dispute on that point among the people who study primate brain evolution. It is only one of numerous possibilities; nothing is "obvious" yet about that topic."

(DV) [Richard, yes, it is possible that the orangutan's large brain is a holdover developed for e.g. social cognition in an ancestral social primate, but note that the gorilla's social complexity is not very high either – and in any case failure to address these aspects, as is common enough in the literature, is indeed a major failure – in the literature what I find mainly is the assumption that the large primate brain is due to higher social function, without any reference to or explanation for the nonconformity to this of both the orangutan and the gorilla…]

(RW) " "Furthermore, you fail to note that the large majority if not all of the richer habitats have been taken over by agriculture, leaving only more marginal areas in a wild state."
This is an interesting topic. I agree with you about most hunter-gatherer habitats being marginal. (But not all: e.g. those in the Okavango in Botswana, Asmat area of New Guinea, and Murray-Darling basin of southeast Australia were high-quality in many ways.) But the problem that I come up against is that great habitats like the Nile Valley or the Ganges Basin offer plant resources that mostly need to be cooked. What is your idea for an optimal pre-agricultural habitat where humans could have lived raw?"

(DV) [Richard, somewhere with relative plenty of some suitable combination of nuts, oily seeds, cereal grains, other seeds, starchy root vegetables, fruits and green leaf vegetables – when I say relative, I mean in the context of our prehistory – where such a criterion might be met today I do not know, as I am far from expert in botany, ecology and geology.]

(RW) "P 221-222 Hunting by chimpanzees a response to food shortage? Actually the evidence mostly points in the opposite direction. See Gilby & Wrangham 2007, attached."

(DV) [Richard, it is quite consistent among the literature I've reviewed in my book that chimpanzees hunt mostly when preferred food is in short supply.
In any case, in the article of Gilby and yours that you've cited just here, after controlling for the effect of the number of adult males in the group being observed, the odds ratio for hunting rather than not hunting on encounter of red colobus monkeys, although statistically significant (not very, p = .013) due to the number of encounters being large at 790, was only 1.01, that is, a 1% increase! Big deal!..
Furthermore, the actual success rate of the hunting, as a measure of the actual commitment of the hunt, rather than it maybe just being activity for excitement, with time and energy on one's hands (like young male elephants hassling rhinos for no apparently good reason other than practicing hassling, or humans ditto…) needs to be a focus of analysi(e)s. Also colobus characteristics such as tree of occupation characteristics, area of occupation relative to area of chimpanzee occupation, age (regarding dependant young) distribution, pregnancy, vocalization, and nutritional state.
Furthermore, as someone perhaps better qualified than yourself in such matters, I state that when you write an abstract for such a journal article, you need to include the quantification of the effect (association) size (in this case the measly 1%, but the 95% confidence interval for the odds ratio (which needs to be provided, but you didn't…) would very likely have included a lower value of less than 1.005, that is, an increase of a tiny 0.5%...), not only its statistical significance – with large enough numbers even the most *virtually in*significant effects (associations) are *statistically* significant.
Now, in that work of Gilby and yours you stated a U-shaped relationship between nutritional abundance and hunting activity was indicated – why do you not state so here – all you speak of here is one end of that curve – which in any case, once the tiny effect size is known, would not be a U shape, but the only (correction: 'only the') first half of the U, finishing in a tiny rise so utterly trivial that the naked eye would not even be able to see it…. And accordingly Fig 1 in your work is a gross misrepresentation of the actual results – by eye, one would not be able to see *any* rise at the right hand side of the probability line, because what rise there is, is so tiny – and by the way, a probability of 1 means that the event *always* happens – your representation of an odds ratio of 1.01, a 1% increase in probability, as a 100% certainty (probability = 1) is wrong to the point of being bizarre – which I would expect anyone who had passed an introductory course in statistics at university to pick up in very short order…
These certainly look like misrepresentations for the sake of refusing to admit your mistakes, both here in this dialogue, and also in your theorizing on the subject of the larger discussion as addressed in your and my books.]

(RW) "NB In various sites, such as Kibale, chimps eat almost no insects – we have excellent data on that point after 20+ years of observation."

(DV) [Richard, for a start, what eats wild figs, eats insects, incidentally. And ditto for other wild plant matter, particularly fruit. I'm astounded that you would attempt to argue this. You may have a lot of data, but its excellence in this respect is obviously far from established…]

(RW) "p. 222 "suggestive of predation not being naturally a strong primary drive." That is a difficult concept. Everyone who studies chimpanzees is impressed by their interest in meat."

(DV) [Richard, from your own (2009) book Catching Fire (p142): "Consider chimpanzees, who hunt little and whose raw-food diet can safely be assumed to be similar to the diet of australopithecines [This latter claim is very defective in its failure to consider differences in grass seed and starchy root vegetable use, which are far from trivial matters…]. At Ngogo, Uganda, chimpanzees hunt intensely compared to other chimpanzee populations, yet males still average less than three minutes per day hunting."]

(RW) "True, the Gombe studies can be criticized for the banana period. But we now have lots of data from provisioned and unprovisioned sites. Not only do chimps like to hunt (when they have the energy to do so) but also, once meat has been captured, the intensity of interest is huge."

(DV) [Richard, interest is intense because the lean seasons, in which most hunting does take place, are so lean that growing juveniles actually cease to grow – interest is intense because hunger is intense…]

(RW) "Read Goodall's (1986) book, which captures the interest in meat for any population."

(DV) [Richard, ditto.]

(RW) "Remember, females and young will work hard to just get a few drops of blood as they fall to the leaves on the forest floor!"

(DV) [Richard, ditto.]

(WR) "It is interesting that chimps don't prey on some animals that they might be expected to, such as bats in weaver nests and maybe monitor lizards. But with regard to fish, remember that chimps definitely are averse to eating animals that they find dead; and catching fish live would be very difficult. Chimps do eat lots of other things – the list of species is somewhere between 50 and 100 by now if I remember right."

(DV) [Richard, as noted in my book p223 citing van Lawick-Goodall (1968), whose work you must be aware of, that chimpanzees have been well tested to show very little interest in fresh carcasses of usual prey animals placed trailside, while they are very interested in fruits offered similarly, surely must suggest that meat is not a preferred food?...]

(RW) "p. 222 "Wrangham's more detailed picture (1975) based on 82 of the most reliable predation records in the Gombe Stream Research Centre files, covering four different periods [we need to know the total time to consider the correctness of the calculation!...We also need to be able consider any trend across the four different periods!...]
I am not sure what you are referring to here. I did calculate predation rates with respect to time, and everyone else does so too, as you can see in Stanford, Mitani and Watts etc."

(DV) [Richard, what I am referring to here, which would be obvious enough, I would hope, to anyone with sufficient a background in research methodology and data analysis to carry out an appropriate such work, is that one needs to know the representativeness of the periods from which the data that was analyzed came – so one needs to know which months of which years the data came from – and also, one needs to be able to compare the four different periods from which data came with each other – otherwise very important data detail is missing/ hidden/ obscured…]

(RW) "p. 165
"It should likewise cause a substantial drop in the amount of time that individuals spend chewing, with large consequences for the species activity budgets (Wrangham and Conklin-Brittain, 2003).
[But cooking also requires time; and the smoke may attract unwanted attention that then has to be guarded against; and the food being cooked is less immediately portable than it was before should exiting the area become necessary or strategic; and food can be processed for young children using naturally shaped stone and wood tools…]"
The only thing that I disagree with here is your use of the word 'But'. You could equally well have said 'And as Wrangham expressly notes, cooking also requires time (etc)…' Something we agree on!"

(DV) [Richard, no, I think the correct word is "but", because your implication, from the context of the surrounding text, is not one of merely a rearrangement/reallocation of the energy budget, which I would roughly agree with, but a saving in the energy budget, which is far from obvious or reasonably assumable…]

(RW) "p. 165 "[Firstly it needs to be noted that the most valuable, and valued (sought after) foods for chimpanzees are not "concentrated" foods, they are fruits."
Not really. When chimpanzees hear evidence of a predation (or see red colobus in a vulnerable position) >95% of the time they leave their fruit-eating to go and try to get meat. On the occasions when they don't, there are rather obvious reasons for it (e.g. the chimp is too timid.)"

(DV) [Richard, again, from your own (2009) book Catching Fire (p142): "Consider chimpanzees, who hunt little and whose raw-food diet can safely be assumed to be similar to the diet of australopithecines [This latter claim is very defective in its failure to consider differences in grass seed and starchy root vegetable use, which are far from trivial matters…]. At Ngogo, Uganda, chimpanzees hunt intensely compared to other chimpanzee populations, yet males still average less than three minutes per day hunting." Now, are you trying to say that chimpanzees know where to look for fruits but are totally clueless about where they might find colobus by either looking or listening?...]

(RW) "p. 165 "Fourthly, many individual members of a chimpanzee group will actually eat almost no meat, and the large majority eat only a very small amount (less than 5% of calories) of meat (due to their inability (i.e. in males) to catch it, or their disinclination (i.e. in females with offspring, due to the danger) to catch it, or to their inability to successfully beg for it from those who have caught it), whereas other foods are eaten in far more equal proportions among members of a group."
That's all roughly correct. But it is not an argument against meat being valuable."

(DV) [Richard, for the Galapagos Islands tortoises (normally virtually totally herbivorous) that ate their dead fellows and so were saved the same fate of starvation because all the grass had been eaten, that meat was valuable – that is not the same thing as it being either preferred or optimal… and the same applies to apes, and to humans, if you could but consider it aside from the cherished dietary habits that you have been habituated, socialized and enculturated into (please excuse the element of technical redundancy here, but these are very important concepts in understanding why we do what we do…)]

(RW) "p. 175 "In Chapters 2 and 3 Wrangham again fails to so much as even mention the alternative to cooking that is well known to all scientist/academics in the knowledge area – that food processing by the use of naturally shaped stone and wood tools, which wild chimpanzees and even some wild monkeys are well documented to do, provides the same enhanced breakdown and selective consumption of components of otherwise difficult foods."
(1) True, I say little about this topic in Chs 2-3, but I do mention it (p. 49 in Ch 2).
(2) Surely it is important to distinguish between tools used to get access to food (which is what chimps and capuchins do) and tools used to physically process food (which I think is unknown in wild primates)."

(DV) [Richard, surely the difference between cracking open a nut and belting something to crush it (which latter I seem to recall some ape populations and some monkeys have been recorded as doing also) is largely a matter of semantics in "accessing" vs. "processing"? For the purposes of the larger questions, it's all much of a continuum – surely it's a very small evolutionary step for an australopithecine to go from cracking nuts to grinding seeds and mashing tubers, and an obviously very plausible one?...]

(RW) "(3) You refer to "the same enhanced breakdown and selective consumption of components of otherwise difficult foods". But this is a guess: very little work has been conducted to assess the relative amount of 'breakdown' of food due to non-thermal vs. thermal processing, so you cannot legitimately claim "the same enhanced breakdown". Rachel Carmody has been doing some new research on this topic at Harvard, and will shortly publish results that you should be interested in."

(DV) [Richard, on p79 of your 2009 book Catching Fire, with regards to pythons eating meat you state that cooking and grinding had equal (12.3%, 12.7%) effects on increasing digestibility, the effects being additive – how this relates to humans and our different foods is in question, but it is very relevant that humans are very obviously designed to chew, whereas pythons absolutely do not chew at all… In any case, I personally know the difference between defecating lumps of improperly/un-chewed starchy root vegetables and passing properly-chewed stuff that certainly looks like more digestion has taken place, and I have experienced the amount of energy available to me from properly chewed potatoes and sweet potatoes – much the point is modern people being habituated to not chewing food properly…]

(RW) "p. 174 "Wrangham (2009) in his book Catching Fire How Cooking Made us Human, is just as fatally flawed in his arguments that we are evolutionarily genetically physiologically adapted to a diet of cooked food high in animal food" This puzzles me. Where did I say that we are "genetically physiologically adapted to a diet … high in animal food"? Are you mixing me up with Stanford?"

(DV) [Richard, I wrote the meaning, not a quotation – if it sounds a little clumsy due to redundancy, bear in mind that as noted I wrote my book with the average person in mind, so some things will be redundant for some readers, while helping others to understand. If that's not what you were referring to, your writings certainly give me the impression that you think we are by evolution physiologically adapted to a diet high in animal food – have I formed or received the wrong impression?...]

(RW) "p. 239 "In any case it is quite evident that the proportion of chimp diet, at this location, constituted by vertebrates has not been at all properly quantified – and what quantification that can be readily done from the data, as here, suggests that the proportion of chimp diet (in this location) constituted by vertebrates of any size taken together is very low, and vary variable according to environmental conditions."
That's right. In Kanyawara I did not feel ready after 3 years of data (with chimpanzees that were still in the process of being habituated) to make a calculation of the amounts eaten, since there could have been many sources of error. But now that we watch our chimps for around 3000 hours per year I can be confident that meat-eating in Kanyawara is much less than in Ngogo, Gombe or Tai. The reason is probably partly because our community has fewer males, and partly because they have a forest that is less productive of fruit."

(DV) [Richard, the last sentence here contradicts your earlier assertion that a plenitude of fruit gives rise to increased hunting…]

(RW) "The role of particular individuals could be important too. See Gilby et al 2008."

(DV) [Richard, agreed, and I guess you'll agree that the Pumphouse Gang baboon study on this aspect will remain a classic of primatology forever – notwithstanding other work I've not yet read…]

(RW) "p. 117 "Instead Wrangham and colleagues claim that a dietary change to increased tuber
consumption
[It seems to me much more likely that tuber consumption had been increasing and/or substantial for a long time before the evolution of H. erectus – the early hominin can see food acquisition of tubers by other animals, and it can readily evolve the use of the digging stick to facilitate their acquisition for itself….]
enabled the evolution to H. erectus."
That is not what we intended. Our idea was that USOs were already a part of the diet, but that cooking – of both USOs and meat – enabled the evolution of *Homo erectus*. As various colleagues and I have argued elsewhere, USOs were likely critical in the diets of pre-humans."

(DV) [Richard, so Bunn got that (among many other things…)(conveniently, for his theorizings?) wrong, misrepresenting you – and we have the correction from you now. And this is an example of just what I've avoided in my book by using direct quotations of what people actually did say, rather than rephrasing it to something else…]

(RW) "p. 224
"Excluding the third method, which by definition is always successful, Gombe chimpanzees share meat at an average rate of once every 8.6 minutes.
Requests, which involve active solicitation with postures, gestures and vocalizations, occur every 6.5 minutes and are successful every 22.5 minutes (Teleki, 1975)."
Much has been discovered since Teleki's brief study. For example Gilby, I. C. 2006. Meat sharing among the Gombe chimpanzees: harassment and reciprocal exchange. Animal Behaviour, 71,953-963."

(DV) [Richard, I'll have to read that later – the harassment I could hardly doubt, particularly given the hunger of the lean season when most hunting occurs…]

(RW) "p. 166
[Wrangham's reference to Gould's (2002) statement that speciation can occur in less than 25,000 years, is meaningless without reference to the generation time of the species referred to. Also the average time of speciation, or the distribution of speciation times, were they known, might be more relevant. Also differentiation of the time required for DNA-degenerative change, time required for removal of that, time required for selection-from-DNA-variety change, and time required for neogenesis of new function, is relevant. (see also section 3.2, p267 below)]
Generation time does matter, fair enough. But the fact remains that evolution can be very fast. I give a couple of examples in *Catching Fire*."

(DV) [Richard, fine, but those differing DNA mechanisms are crucial to understand, in their differences – as noted, I'll be providing a detailed critical review of your book Catching Fire in the near future.]

(RW) "p. 200
"But except in the most extreme environments, edible plants are abundant and available in sufficient variety to allow a wide range of choices. From fruits, nuts, beans, roots, tubers, stalks, bulbs, berries, melons, and gums, leaves, fungi and flowers a highly diversified, nutritionally rich, and often extremely tasty diet can be obtained.
[And I agree, whereas Wrangham (2007, see section 1.14 here above) does not…]"
That's right: I clearly disagree with the notion of "nutritionally rich", unless that means diverse with respect to micronutrients. My experience tells me that the notion of a ready supply of food is derived from the industrialized and developed world. It does not apply to wild primates (and other animals), for which we have abundant evidence of extra food causing a rise in population growth rates and population size, and numerous others kinds of evidence showing that they live on the energetic edge. It does not apply to hunter-gatherers, for whom similar kinds of data abound. Nor does it apply to much of the third world, where people routinely go to bed hungry every night."

(DV) [Richard, I come back a little from my stance on this, but only insofar as I think that there would have been an average sufficiency such as human population numbers/density were probably able to be very roughly what they are/were for other apes – yes, I agree that all wild primates, and many other animals, reproduce such as to increase in number up to the carrying capacity, interacting with the level of predation by cats etc., and disease (don't forget these), where they live, roughly, as you put it, "on the energetic edge."]

(RW) "p. 20
"Humans, and therefore all of our evolutionary ancestors since the common ancestor of humans and chimpanzees, still climb trees quite well, just not as well as the apes do - certainly modern humans and all of our hominin ancestors climbed trees well enough to be harvesting and eating fruit from them as a daily staple."
If you mean that fit young males could climb up and throw fruits or branches down, I would agree. But if you suggest that humans fed like apes, i.e. each for themselves, which would mean that babies, old men, middle-aged mothers etc were required to climb for themselves, your proposition seems unreasonable. You are doubtless a very fit and able male, but think of others! Even nowadays, the major source of mortality for some hunter-gatherer men is falling out of trees, and they do not perform the kind of climbing that chimpanzees do. Among the Efe pygmies that I worked with, they stay close to the trunk, often use vine ropes, etc. And old men get young men to go up on their behalf."

(DV) [Richard, I'm a genetically physically average 49 year old male – old enough to be a grandfather, old enough to be more or less redundant for the survival of my progeny, in the prehistorical and evolutionarily relevant contexts, and very few wild animals (and I assume this applied to our evolutionary ancestors also) die of pure old age – also I carry permanent lumbar disk and hamstring tear & adhesion injuries – I have known many genetically physically average enough 60 and 70 year olds who could climb trees well enough to harvest fruit - and the use of a forked stick for pulling in fruit/branches would have been available and almost certainly used by even the earliest hominins – and in those days, hominins were either fit, or maimed; no-one got unfit simply through lack of physical exercise, as is often the case these days… And yes, I don't doubt there were various divisions of labour such as to reduce risk, and this would include the riskier climbing, which is far from all climbing for harvest… If the Efe pygmies couldn't climb better than me, there must have been something the matter with them, or with the observation of them, because I've seen film footage of people whose climbing ability makes mine look pretty basic, although mine is much better than the average westerner, simply because I've actually done some…]

(RW) "p. 175
"In chapter 4 Wrangham fails to understand that humans still climb trees well enough to make nests in them as apes do"
Same kind of comment. I have on occasion reached a chimpanzee nest and lain in it. But I cannot imagine being able to construct one. The balancing alone would be very difficult, and the strength required to bend and break branches would be beyond me, and I think would have been beyond me even when I was at my fittest."
(DV) [Richard, one needs to know what you were fit for – if it was for walking all day, that is not relevant to branch-breaking – if it was for climbing trees for hours of every day, then your upper body strength would have been very much greater – it might take longer but I'm pretty sure I could make a nest, in the right type of tree, and it would be strength, not balance, that would limit me – notably, it seems you've not yet tried to make a nest yourself, even after all these years of primatology…]

(RW) "You list an additional series of "flaws" here. To pick up on a couple… (1) I agree that it is fun to think about being able to sleep on the ground without fire. But remember that many nights are moon-less. Are you really prepared to try relying on sentries and access to climbable trees on a pitch-black night with lions roaring close by? And again, this is not just for fit males. The old, the young and the weak have to be able to be safe too!"

(DV) [Richard, for survival of the species, in wild nature, the weak do not have to be safe – in fact they are usually eliminated by one circumstance or another, and this very often includes the aged.
If the lions are roaring, I (hominid) won't need the sentries to wake me to climb the tree. In a big enough group (and we (hominids) are in larger groups, because we are smaller and less well armed, than recent gatherer-hunters), which is much the point in considering the survival methods of our earliest hominid ancestors, yes, we rely on multiple sentries, to wake us appropriately, and then we defend as a group (with our weapons of throwing stones, clubs and pointed sticks/ primitive spears) such as to allow the mothers and children to get up the tree(s) first. Or we are sleeping on a barely-accessible ledge, as the gelada baboons do for safety.

Maybe on pitch-black nights we slept in the trees, or spelled sentries more often so they could be sharper in their vigilance, and posted more sentries, or had more people awake more often.]

(RW) "(2) I pointed out that fire does not have to be made when first used (see CF p. 191: "The steps to managing fire need not have involved the difficult process of deliberately making it." (3) The question of the proportions of archaeological sites that do and do not have evidence of fires is really difficult to answer. One person's evidence does not satisfy another; some sites are preserved in such a way that it is unreasonable to expect evidence; and the number of sites in any one time-depth is very few by the time you reach the Lower Paleolithic."

(DV) [Richard, nevertheless the quantification, including with 'sensitivity analyses' that address levels of evidence, needs to be made in order to have a meaningful discussion about the matter.]

(RW) "Of course you are right that I am challenging received wisdom about the antiquity of fire, and many such as yourself think I am wrong."

(DV) [Richard, surely by now you must know that I am less a follower of received wisdom, and more an analyzer of evidence…]

(RW) "But at least the biological evidence that I have assembled is causing some archaeologists to look again at the evidence."

(DV) [Richard, the biological evidence that you have "assembled" is a human digestive tract suited for a relatively smaller volume of food than the digestive tracts of the apes – as I think I've substantially demonstrated in my book, this is more likely to have been facilitated by a diet composed of suitable and readily enough available proportions of nuts, oily seeds, cereal grains, other seeds, (correction: and legumes) starchy root vegetables and fruit, along with some amount of leafy green vegetables, with the use of primitive tools for digging, cracking, grinding and mashing, and drying for storage and rehydration via soaking as required, than to have been facilitated by the use of fire for cooking.]

(RW) "It is not an open-and-shut case, as most archaeologists agree. The exciting scientific approach is to stay open-minded and see what the next decade of evidence brings. If a boundary period can be found before which the evidence for control of fire is satisfyingly absent, and after which it is strongly present, our knowledge of the past will be improved. But during the time since I began using biology to attempt to reconstruct the past, the strong evidence for the use of fire has gone from 400,000 ya to almost 800,000 ya. Who knows what the future might bring?"

(DV) [Richard, firstly, a very small number of sites indicates, much less proves, little about the general state of fire use in the periods of interest/ debate. Quantification, and identification of possibly relevant site discovery and preservation biases are necessary…

It must be borne in mind that fire that is merely controlled (not made) is at risk of being frequently extinguishing by rain, and in some wet seasons in some locations would only have been able to be maintained by storing dry timber/fuel in a particular place(s), near to another particular place where the fire could be maintained – regardless of where these two types of sites were relative to the actual place of the use of the fire for living, it seems that some degree of sedentariness or localization would be necessary to be able to ensure the maintenance of the fire in this way – in which such cases fires used for living by hominids should have been recurrent enough at locations for evidence to potentially be findable. Otherwise, after the rain, another lightning strike or volcano eruption (volcanoes might have been relatively permanently available in some areas) must be waited for to start the next fire to be captured and maintained.]

(RW) "That is my general approach. This is a fascinating area, and I am open to correction. Maybe I am wrong that cooking strongly tends to increase the amount of energy that we obtain from our food. But that is the way that the evidence points right now. If experiments can be done to show I am wrong, fair enough! That's science. But until then, the energy theory of cooking seems to me a really interesting way to re-thinking one of the most important things we do."

(DV) [Richard, I think it the evidence is that cooking does increase the amount of energy that we modern humans obtain from a significant amount of our food.]

(RW) "Likewise, maybe I am wrong that humans are biologically adapted to eating food cooked. But again, that's where the evidence points."

(DV) [Richard, no it doesn't – at best the evidence does not conclusively contradict your pet theory – which is not the same thing as the evidence pointing to your pet theory being correct…]

(RW) "If you know of evidence that contradicts that position, let's hear it!"

(DV) [Richard, haven't I done that in my book? Surely you don't want me to repeat it all again here!... Death by filibuster?!?... Crucially, raw foods, in various sensible proportions by food group, with the use of primitive tools for digging, cracking, grinding and mashing, and drying for storage, and rehydration via soaking, as required, can meet the requirements! There is no need for cooking to meet energetic requirements; there is much better evidence for early tool use than there is for early fire use…]

(RW) "There will be intriguing puzzles to come, such as how humans ate with their small teeth and guts in the Lower Paleolithic if they were not using fire."

(DV) [Richard, the teeth and guts aren't very small, only a little so – rather than fire they more likely used primitive tools for digging, cracking, grinding and mashing, and drying for storage, and rehydration via soaking, as required, to eat suitable proportions of (raw) nuts, oily seeds, cereal grains, other seeds, (correction: and legumes) starchy root vegetables and fruit, along with some amount of leafy green vegetables…]

(DV) [Richard as before, any response you make to my responses here will be included verbatim and in total, no matter how negative, at the end of my book.]

Professor Wrangham's response on 18 November 2010 to the above was:

> Dear David,
> Thanks for this. It is interesting to see your perspectives. We could
> doubtless continue the discussion for a long time.
> Can I mention one point. You seem to think that I am committed in some
> personal way to carnivory (and to cooking!). I do not feel in the
> least committed to either. Although I am not a full vegetarian (I eat
> fish and eggs) I do not like to eat animals that I would not be
> personally prepared to kill, and I have not eaten mammals since 1977.
> I will move on at this stage, and wish you luck in writing your book.
> Richard

And my immediate response to that (other than additions made after the fact as "(abc … xyz)" was:

Hello Richard,

I am pleased to hear your ethics regards eating animals are better than the impression you might forgive me for forming/ receiving as a result of the frequent sensationalization and exaggeration of chimpanzee carnivory - yes it was newsworthy on discovery, and worthy of study, but you will probably understand my annoyance at every last drop of publication potential being improperly milked out of it, with some very very negative effects on animal welfare resulting.
I thank you for, and acknowledge those ethics.

If you do care to respond to my communication you should discard the previous one and use the one attached here (this is what appears above here), as I have corrected a mistake made in haste (the omission of legumes from the list of food types) and improved something else (the deletion of a mistake in reasoning of mine, and its replacement with the list of colubus characteristics that should have been addressed in the study of chimpanzee hunting which I specifically criticize here above).

Yours in science,

David Vance.

Made in the USA
Monee, IL
05 November 2026

67378971R00240